ALGEBRA 2 INTERACTIVE STUDENT GUIDE

FOR THE COMMON CORE

mheonline.com

Send all inquiries to:
McGraw-Hill Education
8787 Orion Place
Columbus, OH 43240

ISBN: 978-0-02-143917-1
MHID: 978-0-02-143917-6

Printed in the United States of America.

2 3 4 5 6 7 8 9 QTN 19 18 17 16

Contents

Chapter 4 Quadratic Functions

Chapter 5 Polynomial Functions

Chapter 6 Radical Functions

Chapter 10 Trigonometric Functions

Each chapter of the *Common Core Interactive Student Guide* has features to help you succeed.

Chapter Overview

Each **Chapter Focus** section gives you an overview of what you will learn. The key standard from each lesson is listed along with preview questions. As you work through each chapter, revisit this section to complete each question.

CCSS SMP 1 For the given right triangle, write a function, $f(x)$, for finding the hypotenuse.

Lesson Overview

Each **lesson** gives you ample opportunity to explore concepts and deepen understanding.

A strong emphasis on modeling helps you connect mathematics to the real world. Throughout the program you will develop, test, and refine models to more accurately represent a real-world situation.

b. **USE A MODEL** Write a quadratic function that describes the relationship between the dimensions of the enclosure and its area, $A(x)$. State the domain of the function. CCSS F.IF.5, SMP 4

Assessment Practice

Performance Tasks at the end of each chapter allow you to practice abstract reasoning, perseverance, and problem solving on tasks similar to those found on Common Core assessments.

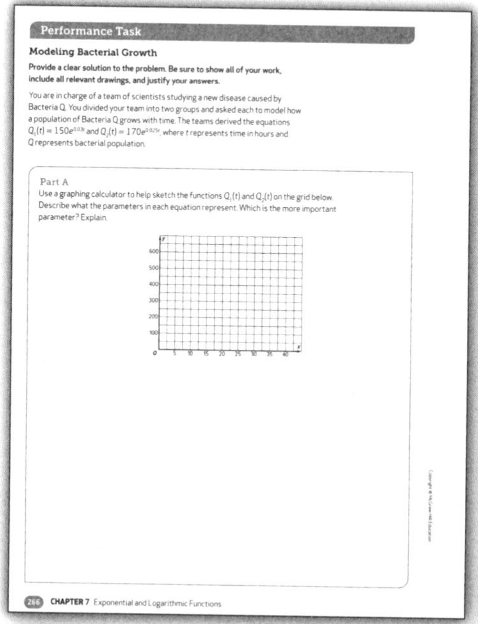

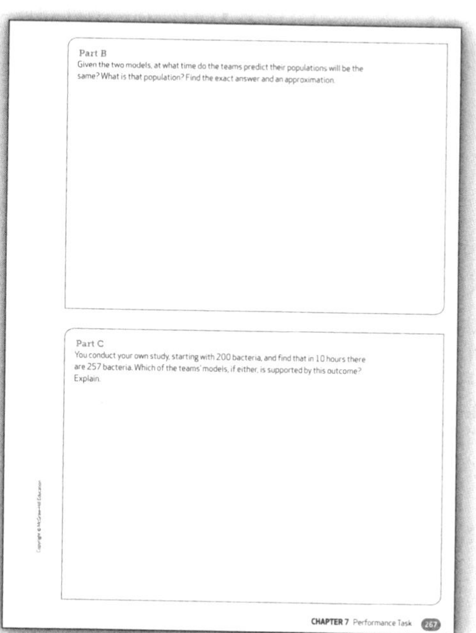

At the end of each chapter, there is a **Test Practice** section that is formatted like the Common Core assessments. Use the test practice to familiarize yourself with new kinds of assessment questions.

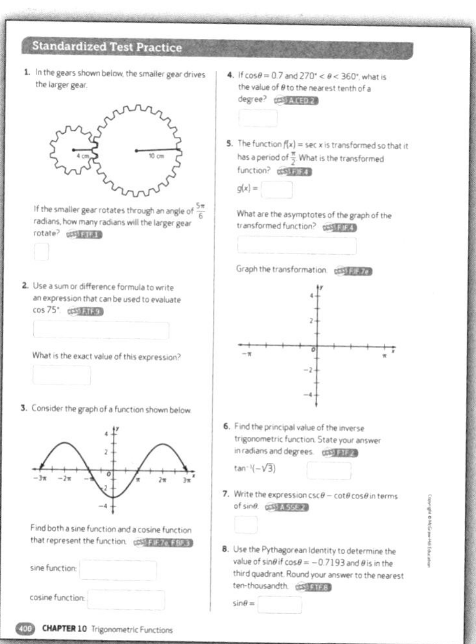

Guide to Developing the Standards for Mathematical Practice
This guide provides a standard-by-standard analysis of the approach taken to each Mathematical Practice, including its meaning and the types of questions students can use to enhance mathematical development.

The Common Core State Standards are made up of:

- the **Standards for Mathematical Content**, which detail what students should learn.
- the **Standards for Mathematical Practice**, which describe how students should approach mathematics.

The goal of the practice standards is to instill in ALL students the abilities to be mathematically literate and to create a positive disposition for the importance of using math effectively.

What are the Standards for Mathematical Practice?

1. Make sense of problems and persevere in solving them.
2. Reason abstractly and quantitatively.
3. Construct viable arguments and critique the reasoning of others.
4. Model with mathematics.
5. Use appropriate tools strategically.
6. Attend to precision.
7. Look for and make use of structure.
8. Look for and express regularity in repeated reasoning.

Why are the Standards for Mathematical Practice important?
The Standards for Mathematical Practice set expectations for using mathematical language and representations to reason, solve problems, and model in preparation for careers and a wide range of college majors. High school mathematics builds new and more sophisticated fluencies on top of the earlier fluencies from grades K–8 that centered on numerical calculation.

1 Make sense of problems and persevere in solving them.

Mathematically proficient students start by explaining to themselves the meaning of a problem and looking for entry points to its solution. They analyze givens, constraints, relationships, and goals. They make conjectures about the form and meaning of the solution and plan a solution pathway rather than simply jumping into a solution attempt. They consider analogous problems, and try special cases and simpler forms of the original problem in order to gain insight into its solution. They monitor and evaluate their progress and change course if necessary. Older students might, depending on the context of the problem, transform algebraic expressions or change the viewing window on their graphing calculator to get the information they need. Mathematically proficient students can explain correspondences between equations, verbal descriptions, tables, and graphs or draw diagrams of important features and relationships, graph data, and search for regularity or trends. Younger students might rely on using concrete objects or pictures to help conceptualize and solve a problem. Mathematically proficient students check their answers to problems using a different method, and they continually ask themselves, "Does this make sense?" They can understand the approaches of others to solving complex problems and identify correspondences between different approaches.

What does it mean?	What questions do I ask?
Solving a mathematical problem takes time. Use a logical process to make sense of problems, understand that there may be more than one way to solve a problem, and alter the process if needed.	• What am I being asked to do or find? What do I know? • How does the given information relate to each other? Does a graph or diagram help? • Is this problem similar to any others I have solved? • What is my plan for solving the problem? • What should I do if I get "stuck"? • Does the answer make sense? • Is there another way to solve the problem? • Now that I've solved the problem, what did I do well? How would I approach a similar problem next time?

2 Reason abstractly and quantitatively.

Mathematically proficient students make sense of quantities and their relationships in problem situations. They bring two complementary abilities to bear on problems involving quantitative relationships: the ability to decontextualize—to abstract a given situation and represent it symbolically and manipulate the representing symbols as if they have a life of their own, without necessarily attending to their referents—and the ability to contextualize, to pause as needed during the manipulation process in order to probe into the referents for the symbols involved. Quantitative reasoning entails habits of creating a coherent representation of the problem at hand; considering the units involved; attending to the meaning of quantities, not just how to compute them; and knowing and flexibly using different properties of operations and objects.

What does it mean?	What questions do I ask?
You can start with a concrete or real-world context and then represent it with abstract numbers or symbols (decontextualize), find a solution, then refer back to the context to check that the solution makes sense (contextualize).	• What do the numbers represent? What are the variables, and how are they related to each other and to the numbers? • How can the relationships be represented mathematically? Is there more than one way? • How did I choose my method? • Does my answer make sense in this problem? • Does my answer fit the facts given in the problem? If not, why not?

3 Construct viable arguments and critique the reasoning of others.

Mathematically proficient students understand and use stated assumptions, definitions, and previously established results in constructing arguments. They make conjectures and build a logical progression of statements to explore the truth of their conjectures. They are able to analyze situations by breaking them into cases, and can recognize and use counterexamples. They justify their conclusions, communicate them to others, and respond to the arguments of others. They reason inductively about data, making plausible arguments that take into account the context from which the data arose. Mathematically proficient students are also able to compare the effectiveness of two plausible arguments, distinguish correct logic or reasoning from that which is flawed, and—if there is a flaw in an argument—explain what it is. Elementary students can construct arguments using concrete referents such as objects, drawings, diagrams, and actions. Such arguments can make sense and be correct, even though they are not generalized or made formal until later grades. Later, students learn to determine domains to which an argument applies. Students at all grades can listen or read the arguments of others, decide whether they make sense, and ask useful questions to clarify or improve the arguments.

What does it mean?	What questions do I ask?
Sound mathematical arguments require a logical progression of statements and reasons. You can clearly communicate their thoughts and defend them.	• How did I get that answer? • Is that always true? • Why does that work? What mathematical evidence supports my answer? • Can I use objects in the classroom to show that my answer is correct? • Can I give a "nonexample" or a counterexample? • What conclusion can I draw? What conjecture can I make? • Is there anything wrong with that argument?

4 Model with mathematics.

Mathematically proficient students can apply the mathematics they know to solve problems arising in everyday life, society, and the workplace. In early grades, this might be as simple as writing an addition equation to describe a situation. In middle grades, a student might apply proportional reasoning to plan a school event or analyze a problem in the community. By high school, a student might use geometry to solve a design problem or use a function to describe how one quantity of interest depends on another. Mathematically proficient students who can apply what they know are comfortable making assumptions and approximations to simplify a complicated situation, realizing that these may need revision later. They are able to identify important quantities in a practical situation and map their relationships using such tools as diagrams, two-way tables, graphs, flowcharts and formulas. They can analyze those relationships mathematically to draw conclusions. They routinely interpret their mathematical results in the context of the situation and reflect on whether the results make sense, possibly improving the model if it has not served its purpose.

What does it mean?	What questions do I ask?
Modeling links classroom mathematics and statistics to everyday life, work, and decision-making. High school students at this level are expected to apply key takeaways from earlier grades to high-school level problems.	• How might I represent the situation mathematically? • How does my equation or diagram model the situation? • What assumptions can I make? Should I make them? • What is the best way to organize the information? What other information is needed? • Can I make a good estimate of the answer? • Does my answer make sense?

5 Use appropriate tools strategically.

Mathematically proficient students consider the available tools when solving a mathematical problem. These tools might include pencil and paper, concrete models, a ruler, a protractor, a calculator, a spreadsheet, a computer algebra system, a statistical package, or dynamic geometry software. Proficient students are sufficiently familiar with tools appropriate for their grade or course to make sound decisions about when each of these tools might be helpful, recognizing both the insight to be gained and their limitations. For example, mathematically proficient high school students analyze graphs of functions and solutions generated using a graphing calculator. They detect possible errors by strategically using estimation and other mathematical knowledge. When making mathematical models, they know that technology can enable them to visualize the results of varying assumptions, explore consequences, and compare predictions with data. Mathematically proficient students at various grade levels are able to identify relevant external mathematical resources, such as digital content located on a website, and use them to pose or solve problems. They are able to use technological tools to explore and deepen their understanding of concepts.

What does it mean?	What questions do I ask?
Certain tools, including estimation and virtual tools, are more appropriate than others. You should understand the benefits and limitations of each tool.	• What tools would help to visualize the situation? • What are the limitations of using this tool? • Is an exact answer needed? • How can I use estimation as a tool? • Can I find additional information on the Internet? • Can I solve this problem using another tool?

6 Attend to precision.

Mathematically proficient students try to communicate precisely to others. They try to use clear definitions in discussion with others and in their own reasoning. They state the meaning of the symbols they choose, including using the equal sign consistently and appropriately. They are careful about specifying units of measure, and labeling axes to clarify the correspondence with quantities in a problem. They calculate accurately and efficiently, express numerical answers with a degree of precision appropriate for the problem context. In the elementary grades, students give carefully formulated explanations to each other. By the time they reach high school they have learned to examine claims and make explicit use of definitions.

What does it mean?	What questions do I ask?
Precision in mathematics is more than accurate calculations. It is also the ability to communicate with the language of mathematics. In high school mathematics, precise language makes for effective communication and serves as a tool for understanding and solving problems.	• How can the everyday meaning of a math term help me remember the math meaning? • Can I give some examples and nonexamples of that term? • Is this term similar to something I already know? • What does the math symbol mean? How do I know? • How do the terms in the problem help to solve it? • What does the variable represent, and in what units? • Does the question require a precise answer or is an estimate sufficient? If the answer needs to be precise, how precise? • Have I checked my answer for the correct labels?

7 Look for and make use of structure.

Mathematically proficient students look closely to discern a pattern or structure. Young students, for example, might notice that three and seven more is the same amount as seven and three more, or they may sort a collection of shapes according to how many sides the shapes have. Later, students will see 7×8 equals the well remembered $7 \times 5 + 7 \times 3$, in preparation for learning about the distributive property. In the expression $x^2 + 9x + 14$, older students can see the 14 as 2×7 and the 9 as $2 + 7$. They recognize the significance of an existing line in a geometric figure and can use the strategy of drawing an auxiliary line for solving problems. They also can step back for an overview and shift perspective. They can see complicated things, such as some algebraic expressions, as single objects or as being composed of several objects. For example, they can see $5 - 3(x - y)^2$ as 5 minus a positive number times a square and use that to realize that its value cannot be more than 5 for any real numbers x and y.

What does it mean?	What questions do I ask?
Mathematics is based on a well-defined structure. Mathematically proficient students look for that structure to find easier ways to solve problems.	• Can I think of an easier way to find the solution? • How can using what I already know help solve this problem? • How are numerical expressions and algebraic expressions the same? How are they different? • Can the terms of this expression be grouped in a way that would allow it to be simplified or give us more information? • How can what I know about integers help with polynomials? • What shapes do I see in the figure? Could a line or segment be added to the figure that would give us more information?

8 Look for and express regularity in repeated reasoning.

Mathematically proficient students notice if calculations are repeated and look for general methods and shortcuts. Upper elementary students might notice when dividing 25 by 11 that they are repeating the same calculations over and over, and conclude they have a repeating decimal. By paying attention to the calculation of slope as they repeatedly check whether points are on the line through (1, 2) with slope 3, middle school students might abstract the equation $\frac{y-2}{x-1} = 3$. Noticing the regularity in the way terms cancel when expanding $(x - 1)(x + 1)$, $(x - 1)(x^2 + x + 1)$, and $(x - 1)(x^3 + x^2 + x + 1)$ might lead them to the general formula for the sum of a geometric series. As they work to solve a problem, mathematically proficient students maintain oversight of the process while attending to the details. They continually evaluate the reasonableness of their intermediate results.

What does it mean?	What questions do I ask?
Mathematics has been described as the study of patterns. Recognizing a pattern can lead to results more quickly and efficiently.	• Is there a pattern? • Is this pattern like one I've seen before? How is it different? • What does this problem remind me of? • Is this problem similar to something already known? • What would happen if I…? • How would I prove that? • How would this work with other numbers? Does it work all the time? How do I know? • Would technology help model this situation? How?

Number and Quantity
The Complex Number System N-CN Perform arithmetic operations with complex numbers. 1. Know there is a complex number i such that $i^2 = -1$, and every complex number has the form $a + bi$ with a and b real.
2. Use the relation $i^2 = -1$ and the commutative, associative, and distributive properties to add, subtract, and multiply complex numbers.
Use complex numbers in polynomial identities and equations. 7. Solve quadratic equations with real coefficients that have complex solutions.
8. (+) Extend polynomial identities to the complex numbers.
9. (+) Know the Fundamental Theorem of Algebra; show that it is true for quadratic polynomials.
Algebra
Seeing Structure in Expressions A-SSE Interpret the structure of expressions. 1. Interpret expressions that represent a quantity in terms of its context.* a. Interpret parts of an expression, such as terms, factors, and coefficients.
b. Interpret complicated expressions by viewing one or more of their parts as a single entity.
2. Use the structure of an expression to identify ways to rewrite it.
Write expressions in equivalent forms to solve problems. 4. Derive the formula for the sum of a finite geometric series (when the common ratio is not 1), and use the formula to solve problems.*
Arithmetic with Polynomials and Rational Expressions A-APR Perform arithmetic operations on polynomials. 1. Understand that polynomials form a system analogous to the integers, namely, they are closed under the operations of addition, subtraction, and multiplication; add, subtract, and multiply polynomials.
Understand the relationship between zeros and factors of polynomials. 2. Know and apply the Remainder Theorem: For a polynomial $p(x)$ and a number a, the remainder on division by $x - a$ is $p(a)$, so $p(a) = 0$ if and only if $(x - a)$ is a factor of $p(x)$.
3. Identify zeros of polynomials when suitable factorizations are available, and use the zeros to construct a rough graph of the function defined by the polynomial.
Use polynomial identities to solve problems. 4. Prove polynomial identities and use them to describe numerical relationships.
5. (+) Know and apply the Binomial Theorem for the expansion of $(x + y)^n$ in powers of x and y for a positive integer n, where x and y are any numbers, with coefficients determined for example by Pascal's Triangle.
Rewrite rational expressions. 6. Rewrite simple rational expressions in different forms; write $\frac{a(x)}{b(x)}$ in the form $q(x) + \frac{r(x)}{b(x)}$, where $a(x)$, $b(x)$, $q(x)$, and $r(x)$ are polynomials with the degree of $r(x)$ less than the degree of $b(x)$, using inspection, long division, or, for the more complicated examples, a computer algebra system.

*Mathematical Modeling Standards

7. (+) Understand that rational expressions form a system analogous to the rational numbers, closed under addition, subtraction, multiplication, and division by a nonzero rational expression; add, subtract, multiply, and divide rational expressions.
Creating Equations* A-CED Create equations that describe numbers or relationships. 1. Create equations and inequalities in one variable and use them to solve problems.
2. Create equations in two or more variables to represent relationships between quantities; graph equations on coordinate axes with labels and scales.
3. Represent constraints by equations or inequalities, and by systems of equations and/or inequalities, and interpret solutions as viable or non-viable options in a modeling context.
4. Rearrange formulas to highlight a quantity of interest, using the same reasoning as in solving equations.
Reasoning with Equations and Inequalities A-REI Understand solving equations as a process of reasoning and explain the reasoning. 2. Solve simple rational and radical equations in one variable, and give examples showing how extraneous solutions may arise.
Represent and solve equations and inequalities graphically. 11. Explain why the x-coordinates of the points where the graphs of the equations $y = f(x)$ and $y = g(x)$ intersect are the solutions of the equation $f(x) = g(x)$; find the solutions approximately, e.g., using technology to graph the functions, make tables of values, or find successive approximations. Include cases where $f(x)$ and/or $g(x)$ are linear, polynomial, rational, absolute value, exponential, and logarithmic functions.*
Functions
Interpreting Functions F-IF Interpret functions that arise in applications in terms of the context. 4. For a function that models a relationship between two quantities, interpret key features of graphs and tables in terms of the quantities, and sketch graphs showing key features given a verbal description of the relationship.*
5. Relate the domain of a function to its graph and, where applicable, to the quantitative relationship it describes.
6. Calculate and interpret the average rate of change of a function (presented symbolically or as a table) over a specified interval. Estimate the rate of change from a graph.*
Analyze functions using different representations. 7. Graph functions expressed symbolically and show key features of the graph, by hand in simple cases and using technology for more complicated cases.* b. Graph square root, cube root, and piecewise-defined functions, including step functions and absolute value functions.
c. Graph polynomial functions, identifying zeros when suitable factorizations are available, and showing end behavior.
e. Graph exponential and logarithmic functions, showing intercepts and end behavior, and trigonometric functions, showing period, midline, and amplitude.
8. Write a function defined by an expression in different but equivalent forms to reveal and explain different properties of the function. a. Use the process of factoring and completing the square in a quadratic function to show zeros, extreme values, and symmetry of the graph, and interpret these in terms of a context.
b. Use the properties of exponents to interpret expressions for exponential functions.

9. Compare properties of two functions each represented in a different way (algebraically, graphically, numerically in tables, or by verbal descriptions).

Building Functions F-BF

1. Build a function that models a relationship between two quantities.
   b. Combine standard function types using arithmetic operations.

Build new functions from existing functions.

3. Identify the effect on the graph of replacing $f(x)$ by $f(x) + k$, $k\,f(x)$, $f(kx)$, and $f(x + k)$ for specific values of k (both positive and negative); find the value of k given the graphs. Experiment with cases and illustrate an explanation of the effects on the graph using technology.

4. Find inverse functions.
   a. Solve an equation of the form $f(x) = c$ for a simple function f that has an inverse and write an expression for the inverse.

Linear, Quadratic, and Exponential Models F-LE

Construct and compare linear and exponential models and solve problems.

4. For exponential models, express as a logarithm the solution to $ab^{ct} = d$ where a, c, and d are numbers and the base b is 2, 10, or e; evaluate the logarithm using technology.

Trigonometric Functions F-TF

Extend the domain of trigonometric functions using the unit circle.

1. Understand radian measure of an angle as the length of the arc on the unit circle subtended by the angle.

2. Explain how the unit circle in the coordinate plane enables the extension of trigonometric functions to all real numbers, interpreted as radian measures of angles traversed counterclockwise around the unit circle.

Model periodic phenomena with trigonometric functions.

5. Choose trigonometric functions to model periodic phenomena with specified amplitude, frequency, and midline.*

Prove and apply trigonometric identities.

8. Prove the Pythagorean identity $\sin^2(\theta) + \cos^2(\theta) = 1$ and use it to calculate trigonometric ratios.

Statistics and Probability

Interpreting Categorical and Quantitative Data S-ID

Summarize, represent, and interpret data on a single count or measurement variable.

4. Use the mean and standard deviation of a data set to fit it to a normal distribution and to estimate population percentages. Recognize that there are data sets for which such a procedure is not appropriate. Use calculators, spreadsheets, and tables to estimate areas under the normal curve.

Making Inferences and Justifying Conclusions S-IC

Understand and evaluate random processes underlying statistical experiments

1. Understand statistics as a process for making inferences about population parameters based on a random sample from that population.

2. Decide if a specified model is consistent with results from a given data-generating process, e.g., using simulation.

Make inferences and justify conclusions from sample surveys, experiments, and observational studies

3. Recognize the purposes of and differences among sample surveys, experiments, and observational studies; explain how randomization relates to each.

4.	Use data from a sample survey to estimate a population mean or proportion; develop a margin of error through the use of simulation models for random sampling.
5.	Use data from a randomized experiment to compare two treatments; use simulations to decide if differences between parameters are significant.
6.	Evaluate reports based on data.
Using Probability to Make Decisions S-MD	
6.	(+) Use probabilities to make fair decisions (e.g., drawing by lots, using a random number generator).
7.	(+) Analyze decisions and strategies using probability concepts (e.g., product testing, medical testing, pulling a hockey goalie at the end of a game).

Equations and Inequalities

CHAPTER FOCUS Learn about some of the Common Core State Standards that you will explore in this chapter. Answer the preview questions. As you complete each lesson, return to these pages to check your work.

What You Will Learn	Preview Question
Lesson 1.1: Real and Complex Numbers	
CCSS N.CN.1 Know there is a complex number i such that $i^2 = -1$, and every complex number has the form $a + bi$ with a and b real. CCSS N.CN.2 Use the relation $i^2 = -1$ and the commutative, associative, and distributive properties to add, subtract, and multiply complex numbers.	CCSS SMP 7 Is the square root of a negative number a real number? Explain. CCSS SMP 2 Can a real number a also be considered a complex number?
Lesson 1.2: Solving Equations	
CCSS A.CED.1 Create equations and inequalities in one variable and use them to solve problems.	CCSS SMP 1 Solving an equation involves isolating a variable, but how would you solve an equation with a variable on both sides? $2x + 3 = x - 5$ CCSS SMP 1 If a number is multiplied by 3 and then decreased by 5, the result is 8. Write an equation to represent this relationship.
Lesson 1.3: Solving Absolute Value Equations	
CCSS A.CED.1 Create equations and inequalities in one variable and use them to solve problems. CCSS A.CED.3 Represent constraints by equations or inequalities, and by systems of equations and/or inequalities, and interpret solutions as viable or nonviable options in a modeling context.	CCSS SMP 6 If the absolute value of a number is its distance from zero on the number line, why does $\lvert x \rvert = 3$ have two solutions? CCSS SMP 2 The absolute value of the difference between a number and 2 is 6. Write an absolute value equation to represent the scenario.

What You Will Learn	Preview Question
Lesson 1.4: Solving Inequalities	
CCSS A.CED.1 Create equations and inequalities in one variable and use them to solve problems. CCSS A.CED.3 Represent constraints by equations or inequalities, and by systems of equations and/or inequalities, and interpret solutions as viable or nonviable options in a modeling context.	CCSS SMP 3 Aditya says the solution to the inequality $3 - x < -2$ is $x < 5$. Do you agree? If not, what is the solution? Explain. CCSS SMP 2 A phone company charges \$20 in addition to \$0.10 per minute for a month of service. How many minutes can be used without exceeding \$50 of charges? CCSS SMP 1 A fabric company has 1000 yards of fabric in stock. How many 3-yard orders can they fulfill?
Lesson 1.5: Solving Compound and Absolute Value Inequalities	
CCSS A.CED.1 Create equations and inequalities in one variable and use them to solve problems. CCSS A.CED.3 Represent constraints by equations or inequalities, and by systems of equations and/or inequalities, and interpret solutions as viable or nonviable options in a modeling context.	CCSS SMP 6 An inequality can have an infinite number of solutions. How would you describe the solutions to the inequality below in words? $-5 < x < 5$ CCSS SMP 2 Write and solve an absolute value inequality that represents all numbers that are less than 6 units from 1. Graph your solution on a number line. −6 −5 −4 −3 −2 −1 0 1 2 3 4 5 6 7 8

1.1 Real and Complex Numbers

Objectives

- Understand the concepts of imaginary and complex numbers
- Use arithmetic properties to perform operations with complex numbers

CCSS STANDARDS

Content: N.CN.1, N.CN.2
Practices: 2, 3, 4, 5, 6, 7
Use with Lesson 4-4

The search for solutions to equations such as $x^2 + 1 = 0$ led mathematicians to the discovery of imaginary numbers. The **imaginary unit *i*** is defined by the rule $i^2 = -1$. The number i is the principal square root of -1; that is, $i = \sqrt{-1}$.

Pure imaginary numbers are square roots of negative real numbers. For any positive real number b, $\sqrt{-b^2} = \sqrt{b^2} \cdot \sqrt{-1}$ or bi.

EXAMPLE 1 Products of Pure Imaginary Numbers (CCSS N.CN.1, SMP 3)

a. **CRITIQUE REASONING** One of the properties of square roots is $\sqrt{a} \cdot \sqrt{b} = \sqrt{ab}$, provided $a > 0$ and $b > 0$. Verify the property using $a = 4$ and $b = 9$. Repeat the process with $a = -4$ and $b = -9$, then once more with $a = -9$ and $b = 4$. Does the property hold true when a and b are both negative numbers or when one is negative and the other positive?

b. **MAKE A CONJECTURE** Consider the results of **part a**, and make a conjecture about the value of $\sqrt{a} \cdot \sqrt{b}$ for real numbers a and b when $a < 0$ and $b < 0$.

A real number and a pure imaginary number are not like terms and cannot be combined into a single term. Instead, an expression that has both a real and an imaginary term can be written as a complex number.

KEY CONCEPT Complex Numbers

A **complex number** is any number that can be written in the form $a + bi$, where a and b are real numbers and i is the ____________. We refer to a as the real part of the complex number, and refer to b as the imaginary part.

The Commutative and Associative Properties of Multiplication hold true for complex numbers, i.e., $a + b\boldsymbol{i} = b\boldsymbol{i} + a$. To add or subtract complex numbers, add or subtract the real parts, and add or subtract the imaginary parts, $(a + b\boldsymbol{i}) + (c + d\boldsymbol{i}) = (a + c) + (b + d)\boldsymbol{i}$.

Two complex numbers are equal if and only if their real parts are equal and their imaginary parts are equal. That is, $a + b\boldsymbol{i} = c + d\boldsymbol{i}$ if and only if $a = c$ and $b = d$.

EXAMPLE 2 Add and Subtract Complex Numbers CCSS N.CN.1, N.CN.2

a. **MAKE A CONJECTURE** Simplify $(5 + 2i\sqrt{3}) + (-5 - 2i\sqrt{3})$. For real numbers, the sum of opposites a and $-a$ is equal to zero. Does a similar relationship appear to exist for complex numbers? What is the opposite of $a + bi$? CCSS SMP 3

b. **REASON QUANTITATIVELY** Subtraction of complex numbers is performed in the same way as subtraction of real numbers. To subtract one complex number from another, add the opposite of the complex number being subtracted. Simplify $(4 + 2i) - (3 - 4i)$. What is the opposite of $(3 - 4i)$? Show your work. CCSS SMP 2

c. **REASON ABSTRACTLY** What is the sum of $a + bi$ and $a - bi$? What about the sum of $a + bi$ and $-a + bi$? Are these complex numbers opposites of each other? If not, is there anything unique about their sum? CCSS SMP 2

The Distributive Property of Multiplication holds true for complex numbers. To multiply complex numbers, apply the Distributive Property, and then replace any occurrences of $\boldsymbol{i}^2$ with -1.

EXAMPLE 3 Multiply Complex Numbers CCSS N.CN.2

a. **REASON QUANTITATIVELY** Complete to show the product and justify each step. CCSS SMP 2

$(1 - 2.3i)(-3 + 2i) =$

$1(-3 + 2i) -$ ______ $=$ ______

$-3 + 2i +$ ______ $=$ Multiply.

$-3 + 2i + 6.9i + 4.6 =$ ______.

______ Add real and imaginary parts.

b. USE STRUCTURE The numbers $a + bi$ and $a - bi$ are called **complex conjugates** of each other. What can you say about the product of a pair of complex conjugates? State the result as a mathematical formula. Can you think of a good name for this formula? Recall that $(a + b)(a - b) = a^2 - b^2$ is the difference of squares formula. CCSS SMP 7

PRACTICE

CALCULATE ACCURATELY **Perform each arithmetic operation. Show your work.** CCSS N.CN.1, N.CN.2, SMP 6

1. $(1.7 + 3.3i) + (-2.6 - 3.8i)$

2. $(1.7 + 3.3i) - (-2.6 - 3.8i)$

3. $(7 + 3i)(-2 - i)$

4. $(7 + 3i)(7 - 3i)$

CALCULATE ACCURATELY **Find real numbers a and b so the equation is true.** CCSS N.CN.1, N.CN.2, SMP 6

5. $(4 - i\sqrt{3})(3 + bi) = 6 - 11i\sqrt{3}$

6. $\left(\frac{5}{2} + \frac{3}{2}i\right) - 3(a + bi) = 1 + 3i$

7. $(9 + i)(a + bi) = 82$

8. $(a - i\sqrt{2})(4 - i\sqrt{2}) = 10 + bi$

9. Cameron says, "A number can be both real and complex." Li Jing claims that this is not possible. CCSS N.CN.1

   a. CRITIQUE REASONING Who is correct? Explain your reasoning. CCSS SMP 3

   b. Change the wording in Cameron's statement so that Li Jing's conclusion is correct.

10. You can perform many calculations with complex numbers using technology, such as a graphing calculator. CCSS N.CN.2

a. **USE TOOLS** Use technology to find the following values. Write your answers in the form $a + bi$. CCSS SMP 5

$(1 + i)^2$ ______ $(1 + i)^4$ ______

$(1 + i)^6$ ______ $(1 + i)^8$ ______

$(1 + i)^{10}$ ______ $(1 + i)^{12}$ ______

b. **MAKE A CONJECTURE** Based on your results, make a conjecture. 

c. **CALCULATE ACCURATELY** Find the product $(1 + i)^2 = (1 + i)(1 + i)$ by hand. Show your work. CCSS SMP 6

d. **REASON ABSTRACTLY** Explain how you can use your result from **part c** to determine the values of $(1 + i)^4, (1 + i)^6, \ldots$. CCSS SMP 2

11. **CRITIQUE REASONING** Thomas says that the expression $8i(3 - 2i) - 24i$ does not represent a real number. Is he correct? Justify your answer. CCSS N.CN.1, N.CN.2, SMP 3

12. **USE A MODEL** In an electrical circuit with alternating current, the voltage is equal to the product of the current and impedance, and these quantities can be represented by complex numbers. Note that engineers use j as the imaginary unit, with $j^2 = -1$. If an AC circuit has a voltage of $9 - 2j$ and an impedance of $3 + 5j$ ohms, what is the current? Explain your reasoning or show your work. (Hint: represent the current as $x + yj$ amps, and find the values of x and y.) CCSS N.CN.1, N.CN.2, SMP 4

13. **USE STRUCTURE** Every complex number can be written in the form $a + bi$ where a and b are real numbers. Find real numbers a and b to write $(4 - 2i)(2 + i) + 3i$ in the form $a + bi$. CCSS N.CN.2, SMP 7

1.2 Solving Equations

Objectives

- Create equations in one variable and use equations to solve problems.
- Solve equations using the Properties of Equality.

Content: A.CED.1
Practices: 1, 4, 6, 7, 8
Use with Lesson 1-3

An **algebraic expression** consists of one or more numbers and variables along with one or more operational symbols. An **equation** is a mathematical sentence stating that two algebraic expressions are equal. A **solution** of an equation is a value from the replacement set that makes the equation true. You can solve an equation by using the **properties of equality** and creating a series of equivalent equations.

EXAMPLE 1 Verbal to Algebraic Equation CCSS A.CED.1

EXPLORE The difference of 3 times a number and 2 is divided by the cube of the number, and the quotient equals 15.

a. **INTERPRET PROBLEMS** Write an equation to represent the verbal sentence, using x as a variable. CCSS SMP 1

b. **COMMUNICATE PRECISELY** Explain why $3x - \frac{2}{x^3} = 15$ is not the correct equation for the verbal sentence in **part a**. CCSS SMP 6

c. **INTERPRET PROBLEMS** Write an expression for each of two different ways to interpret the equation "3 times a number minus 2 equals 15." CCSS SMP 1

d. **COMMUNICATE PRECISELY** For each equation you wrote in **part c**, write a verbal sentence that would be represented by the equation and would not be subject to multiple interpretations. CCSS SMP 6

e. **COMMUNICATE PRECISELY** Why is it important that verbal sentences be written as clearly as possible and not in such a way that multiple interpretations are possible? Use values of x for the equations you wrote in **part c** to justify your reasoning. CCSS SMP 6

EXAMPLE 2 Use Properties of Equality to Solve an Equation CCSS A.CED.1

a. **PLAN A SOLUTION** What Properties of Equality can you use to solve the equation $5x + 7 = 38$? What are the operations that you must perform? CCSS SMP 1

b. **USE STRUCTURE** Solve for x in $5x + 7 = 38$ by writing a series of equivalent equations. Justify each step. CCSS SMP 7

c. **DESCRIBE A METHOD** How can you check that your answer in **part b** is correct? CCSS SMP 8

EXAMPLE 3 Solve a Formula for a Variable CCSS A.CED.1

A regular polygon has congruent sides and congruent angles. The formula for finding the area of a regular polygon is $A = \frac{1}{2}aP$, where a is the apothem and P is the perimeter of the polygon. The apothem is the perpendicular bisector from the center of the polygon to any side.

a. **USE STRUCTURE** Solve the equation $A = \frac{1}{2}aP$ for a. What does this new equation allow you to do? CCSS SMP 7

b. **CALCULATE ACCURATELY** This polygon is a regular hexagon. The length of each side s is 3 cm, and its area A is 23.4 cm^2. Use your result from **part a** to find the length of *the apothem*. CCSS SMP 6

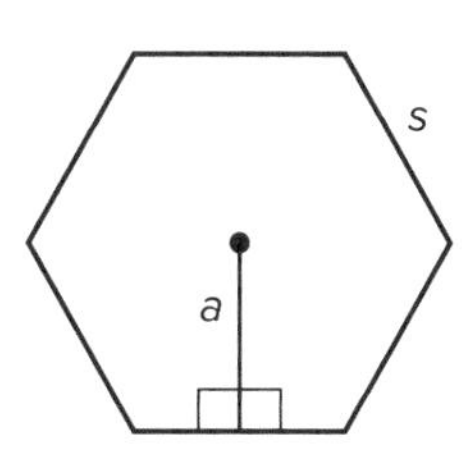

c. **INTERPRET PROBLEMS** Suppose this hexagon is inscribed in a circle, as shown on the right. What is the area of the circle to the nearest tenth? Justify your answer. CCSS SMP 1

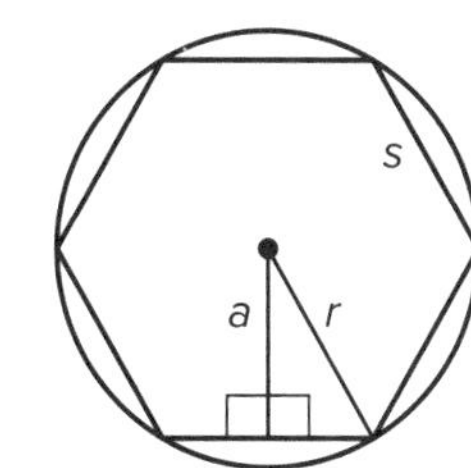

PRACTICE

1. If 10 times a number is increased by 5, the result is 12 less than 9 times the number.

   a. **INTERPRET PROBLEMS** Write an equation that represents this relationship. CCSS SMP 1

   b. **USE STRUCTURE** Solve the equation for n and identify the properties of equalities used to write each equivalent equation. CCSS SMP 7

2. Use a Property of Equality to complete the following statement. Use both equations to create an equivalent equation.

   a. **USE STRUCTURE** If $2x + \frac{5}{2} = a$ and $a = -2(5x + 1)$, then … CCSS SMP 7

   b. **USE STRUCTURE** Which Property of Equality justifies your response to **part a**? CCSS SMP 7

   c. **CALCULATE ACCURATELY** What is the solution to the equation in **part a**? Justify your answer. CCSS SMP 6

3. The formula for the temperature in degrees Celsius in terms of the temperature in degrees Fahrenheit is $\frac{5}{9}(F - 32) = C$.

   a. **USE STRUCTURE** Solve the equation for F. What operations do you have to perform? CCSS SMP 7

   b. **CALCULATE ACCURATELY** Use your result from **part a** to determine the temperature in degrees Fahrenheit when the Celsius temperature is 30. CCSS SMP 6

   c. **USE STRUCTURE** If a temperature has the same measure in degrees Fahrenheit and degrees Celsius, write and solve an equation for the temperature in degrees Fahrenheit. CCSS SMP 7

4. The figure on the right shows a field consisting of a rectangle with two congruent semi-circles at each end. The formula for the perimeter of this field is $P = 2(\ell + \pi r)$, where ℓ is the length of the rectangle, and r is the radius of the circle.

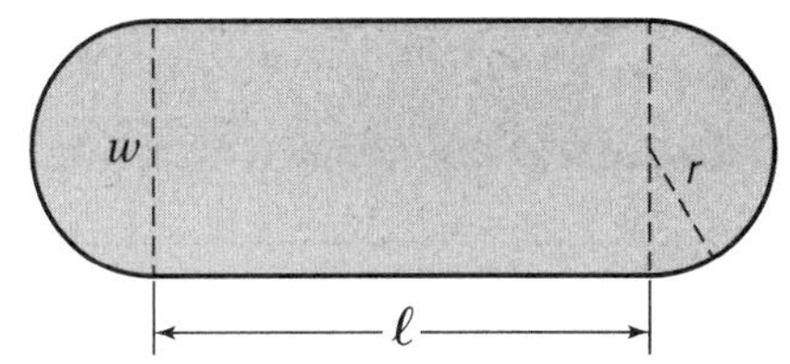

a. **CALCULATE ACCURATELY** If $P = 261.3$ m and $\ell = 60.0$ m, what is the value of r in meters to the nearest tenth? CCSS SMP 6

b. **EVALUATE REASONABLENESS** How can you check your answer for **part a**? Does your answer check? If not, explain the difference. CCSS SMP 8

5. Allyson and Bryan are each driving to the state capital. The trip takes 2.5 hours for Allyson and 4.0 hours for Bryan, since he lives 80 miles farther away from the capital. Bryan drives 5 mph faster than Allyson.

	D (distance)	r (rate)	t (time in hours)
Allyson	D	r	2.5
Bryan	$D + 80$	$r + 5$	4

a. **USE A MODEL** The equation $D = rt$ can be used to determine each distance. Model the relationship between Allyson's distance and Bryan's distance by writing an equation using the variable r to represent Allyson's rate. CCSS SMP 4

b. **CALCULATE ACCURATELY** Solve the equation and determine Allyson's average speed and her distance from the state capital. Justify your answers. CCSS SMP 6

c. **INTERPRET PROBLEMS** Based on your answer to **part b**, what information can you determine about Bryan's drive to the capital? Justify your answer. CCSS SMP 1

1.3 Solving Absolute Value Equations

Objectives

- Represent constraints of absolute value equations.
- Write and solve absolute value linear equations.

CCSS STANDARDS

Content: A.CED.1, A.CED.3
Practices: 1, 2, 3, 4, 6, 7
Use with Lesson 1-4

The **absolute value** of a number is its distance from 0 on a number line. If the absolute value of a number is a, then the number is either a or $-a$.

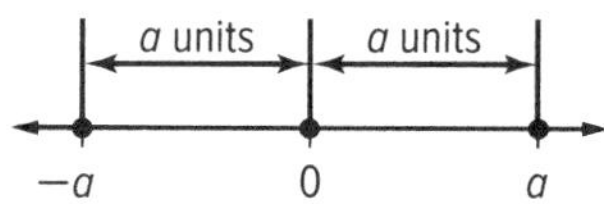

When solving absolute value equations, you must consider two possibilities. The quantity inside the absolute value symbols could be negative or positive, so solving $|x - c| = d$ means solving $x - c = d$ and $-(x - c) = d$ to determine the possible values of x. In some cases, solutions to the absolute value equation do not satisfy the constraints of the equation. These solutions are called extraneous solutions, and occur when the solution results in an absolute value equation that is less than zero.

Properties of Absolute Value

- $|a| \geq 0$
- $|a| = a$ if $a \geq 0$
- $|a| = -a$ if $a < 0$

EXAMPLE 1 Writing Absolute Value Equations CCSS A.CED.3, SMP 6

COMMUNICATE PRECISELY **Write an absolute value equation to represent each situation. Then solve the equation and discuss the reasonableness of your solution given the constraints of the absolute value equation.**

a. The absolute value of the difference between 3 times a number and 2 is 7.

b. The absolute value of the sum of 4 times a number and 7 is the sum of 2 times a number and 3.

c. The sum of 7 and the absolute value of the difference of a number and 8 is negative 2 times a number plus 4.

EXAMPLE 2 Solve an Absolute Value Equation CCSS A.CED.1

A cereal manufacturer sells boxes of cereal that are labeled as weighing 14.7 oz. The packaging equipment is accurate to within 0.35 oz.

a. **USE A MODEL** Write an equation to represent the situation. Explain your reasoning.  CCSS SMP 4

b. **CONSTRUCT ARGUMENTS** Solve the equation and justify each step. State the maximum and minimum weight for the boxes of cereal. CCSS SMP 3

EXAMPLE 3 Effect of Constraints on Absolute Value Equations

USE STRUCTURE **Find the number of solution(s) for each equation.** CCSS A.CED.3, SMP 7

a. $5 = |x + 2| + 7$

b. $|x + 3| + 2 = 2$

c. $|x - 5| - 7 = x - 12$

d. **MAKE A CONJECTURE** Compare your results with those of other students. Then make a conjecture about the number of elements in the solution set of an absolute value linear equation. CCSS SMP 3

PRACTICE

REASON QUANTITATIVELY **What is the solution set for each equation? Check each solution.** CCSS A.CED.3, SMP 2

1. $|2x| - 6 = 0$
2. $|2x + 1| = 9$
3. $|3w + 9| - 4 = 5$
4. $|x - 1| + 3 = 5$
5. $8|7x - 3| - 7 = 73$
6. $|5 - x| = x + 3$
7. $|3 - x| + 1 = 1$
8. $8 - |5 - x| = 2|5 - x| - 10$
9. $6 + |5x + 2| = 0$
10. $|x + 2| = -2 - x$

11. **COMMUNICATE PRECISELY** A toy store sells bags of play sand that are labeled as weighing 35 lbs. The equipment used to package the sand produces bags with a weight that is within 8 oz of the label weight. Write and solve an absolute value equation to determine the maximum and minimum weight for the bags of play sand. Justify each step in the solution. CCSS A.CED.1, A.CED.3, SMP 6

12. **REASON ABSTRACTLY** If three points *a*, *b*, and *c* lie on the same line, then *b* is between *a* and *c* if and only if the distance from *a* to *c* is equal to the sum of the distances from *a* to *b* and from *b* to *c*. Write an absolute value equation to represent the definition of betweenness. CCSS A.CED.1, SMP 2

13. **USE A MODEL** A quality control inspector at a bolt factory examines random bolts that come off the assembly line. All bolts being made must be to a tolerance of 0.04 mm. The inspector is examining bolts that are to have a diameter of 6.5 mm. Write and solve an absolute value equation to find the maximum and minimum diameters of bolts that will pass his inspection. CCSS A.CED.1, A.CED.3, SMP 4

14. **INTERPRET PROBLEMS** Ben makes a rectangle that is 12 inches by 30 inches. Sue wants to make a rectangle having a length 8 inches longer than its width. If she wants the perimeter of her rectangle to be within 20 inches of the perimeter of Ben's rectangle, what are the minimum and maximum values for the width of her rectangle? What are the corresponding length and perimeter? CCSS A.CED.1, A.CED.3, SMP 1

15. **CRITIQUE REASONING** Megan and Yuki have been asked to solve the equation $|x - 9| = |5x + 6|$. Megan says that there are 4 cases to consider because there are two possible values for each absolute value expression. Yuki says that they only need to consider 2 cases. With which person do you agree? Will both girls get the same solution(s)? CCSS A.CED.3, SMP 3

16. **COMMUNICATE PRECISELY** Write an absolute value equation to represent each situation. Then solve the equation and discuss the reasonableness of your solution given the constraints of the absolute value equation. CCSS A.CED.3, SMP 6

a. The absolute value of the difference between a number and twice the number plus 2 is 4.

b. The absolute value of the sum of twice a number and 2 is the number minus 3.

c. Negative three times the absolute value of the difference between a number and 2 is the sum of 1 and five times the number.

1.4 Solving Inequalities

Objectives

- Create and solve problems using inequalities in one variable.
- Represent constraints by equations or inequalities.

Content: A.CED.1, A.CED.3
Practices: 1, 2, 3, 4, 6, 7, 8
Use with Lesson 1-5

Inequalities can be solved using properties similar to those used for solving equations. For example, adding the same amount to or subtracting the same amount from each side of an inequality preserves the inequality. However, multiplying or dividing both sides of an inequality by a non-zero number only preserves the inequality if the number is positive. If the number is negative, the inequality symbol must be reversed.

EXAMPLE 1 Evaluating the Reasonableness of Solutions CCSS A.CED.3

Each shelf of a shelving unit can hold a maximum of 80 pounds. Ryan wants to determine the number of 5-pound bags of flour he can put on one shelf that already has 33 pounds of canned fruit.

a. USE A MODEL Write an inequality to determine the number of bags Ryan can place on the shelf. CCSS SMP 4

b. USE STRUCTURE Solve the inequality you wrote in **part a**. State any constraints that are needed. CCSS SMP 7

c. USE A MODEL On another shelf Ryan would like to put 14.5 oz boxes of cereal. Write and solve an inequality to determine how many boxes Ryan can place on the shelf. State any constraints that are needed. CCSS SMP 4

d. COMMUNICATE PRECISELY Is the answer to **part c** reasonable? Explain how constraints on the situation other than those provided above affect the solution. CCSS SMP 6

e. COMMUNICATE PRECISELY Is the answer to **part b** reasonable in the context of the situation? Explain your reasoning. CCSS SMP 6

EXAMPLE 2 Model with Inequalities CCSS A.CED.1

An ice rink offers open skating several times a week. The table shows the cost of a session for members and non-members. An annual membership costs $60.

Open Ice Skating Sessions	
members	$6
non-members	$10

a. **USE A MODEL** Maria plans to spend no more than $90 on skating this year and attend 3 of the monthly ladies day half-price special sessions. Define a variable then write and solve an inequality to find the number of sessions she can attend without buying a membership. CCSS SMP 4

b. **USE A MODEL** Female members get in free for the ladies day special sessions. Write and solve an inequality to find the number of sessions Maria can attend if she becomes a member of the skate club and spends no more than $90. CCSS SMP 4

c. **CONSTRUCT ARGUMENTS** Would you recommend that Maria becomes a member of the skate club? Explain. CCSS SMP 3

d. **REASON QUANTITATIVELY** Without any specials, what number of skating sessions is a better deal for members? Explain how you know. CCSS SMP 2

EXAMPLE 3 Interpreting Viable Solutions CCSS A.CED.3

Ahmed will bicycle to his friend's house, leave the bike, and then walk home. The walk will be along a path that is half the distance he bicycles. He will average 15 miles per hour on the bicycle, and 3 miles per hour walking.

a. **USE A MODEL** Ahmed wants the entire trip to take less than 45 minutes. Let d represent the distance that Ahmed walks. Write an inequality that represents this situation. Explain your reasoning. CCSS SMP 4

b. **REASON QUANTITATIVELY** Solve the inequality and show your work. Explain what the answer means in terms of the situation. CCSS SMP 2

c. **USE STRUCTURE** Suppose Ahmed gets a flat tire halfway along his bike ride, and he must walk half the distance he would have bicycled. If the entire trip can still be completed in less than 45 minutes, write and solve a new inequality to find the change in the length of the path. CCSS SMP 7

PRACTICE

REASON QUANTITATIVELY Solve each inequality. CCSS A.CED.1, SMP 2

1. $6x + 15 > 27$

2. $-3y - 5 \geq 10$

3. $-7z - 1 < -8$

4. $-\frac{1}{3}m + 1 < 5$

5. $-6x + 5 \geq 3x - 13$

6. $-3y + 15 \geq -6y$

7. $\frac{1}{2}x - 2 \leq x + 1$

8. $-2z + 2 > -z + \frac{1}{2}$

9. **REASON QUANTITATIVELY** Cheyenne pays \$8 to rent a wetsuit each time she goes surfing. She can buy the same wetsuit for \$120. Write and solve an inequality that can be solved to determine when the renting the wetsuit is more expensive than buying a new suit. Interpret your solution. CCSS A.CED.1, SMP 2

10. Kevin's scores on his first four vocabulary quizzes are 84, 93, 80, and 92. The highest possible score on any quiz is 100. His teacher offered to reward any student with an average score of 90 or higher after five quizzes with a free lunch. CCSS A.CED.3

    a. **USE A MODEL** Write and solve an inequality to determine the score Kevin needs on his fifth quiz in order to qualify for the reward. CCSS SMP 4

    b. **EVALUATE REASONABLENESS** Will Kevin qualify for the reward? Explain. CCSS SMP 8

    c. **REASON QUANTITATIVELY** Kevin's teacher decides to extend the reward offer to the first six quizzes. Write and solve an inequality to determine the total combined score Kevin needs on the fifth and sixth quizzes to qualify for the reward. Is it possible for Kevin to qualify for the reward? CCSS SMP 2

11. **CRITIQUE REASONING** Vivek's teacher made the statement, "Four times a number is less than three times a number." Vivek quickly responded that the answer is *no solution*. CCSS A.CED.3, SMP 3

a. Do you agree with Vivek? Write and solve an inequality to justify your answer.

b. What mistake did Vivek make?

12. **USE A MODEL** The table shows two ways the manager of a car dealership can pay the members of her sales force each month. Write an inequality that can be used to determine for what amount of sales s Plan A produces the greater pay. Then solve the inequality and interpret the solution. CCSS A.CED.1, SMP 4

Plan A	\$1800 plus 2% of sales
Plan B	\$300 plus 4% of sales

13. **EVALUATE REASONABLENESS** Students who correctly answer at least 90% of the 63 questions on a placement test will be allowed to take Honors Spanish next year. Write, solve, and interpret an inequality describing what numbers of correct answers will qualify students to take Honors Spanish. CCSS A.CED.1, SMP 8

14. **PLAN A SOLUTION** A food company is making a new nutritional breakfast bar. The bar is to have at most 20% of its calories from fat, and also less than 4 grams of total fat. If a gram of fat is about 9 calories, how many calories could the whole bar have? Justify your answer. CCSS A.CED.1, SMP 1

15. **INTERPRET PROBLEMS** A pizza party is being planned for the last day of class and the teacher wants to know if ordering eight pizzas will be enough. CCSS A.CED.1, SMP 1

a. If each student in the class will eat at most 3 pieces of pizza and the teacher will eat 3, write an inequality for the number of pieces of pizza p that will need to be ordered based on the number of students s.

b. If each pizza has 8 pieces and the class has 21 students, do you think ordering 8 pizzas be enough for the party? Explain your reasoning.

1.5 Solving Compound and Absolute Value Inequalities

Objectives

- Represent constraints using compound and absolute value inequalities.
- Use compound and absolute value inequalities to solve problems.

CCSS STANDARDS

Content: A.CED.1, A.CED.3
Practices: 1, 2, 3, 6
Use with Lesson 1-6

A **compound inequality** consists of two inequalities joined by the word "*and*" or the word "*or*". To solve a compound inequality, you must solve each part of the inequality. The graph of a compound inequality containing "*and*" is the **intersection** of the solution sets of the two inequalities, while the graph of a compound inequality containing "*or*" is the **union** of the solution sets of the two inequalities.

KEY CONCEPT Compound Inequalities

- A compound inequality containing the word *and* is true if and only if both of the inequalities are true.
- A compound inequality containing the word *or* is true if and only if one or more of the inequalities are true.

EXAMPLE 1 Solving Compound Inequalities

EXPLORE Solve and graph the compound inequalities. CCSS A.CED.3

a. **INTERPRET PROBLEMS** Find the solution of $2x + 5 \leq 3(4 - 0.5x)$ and $6(3x + 1) \geq 16 - 2x$, and graph your answer. CCSS SMP 1

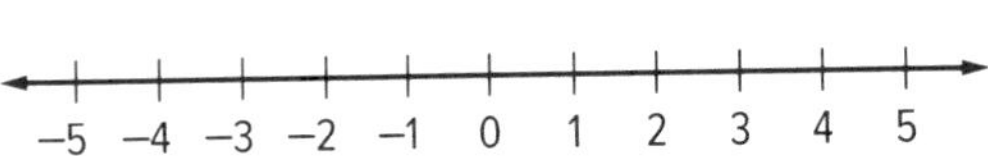

b. **INTERPRET PROBLEMS** Find the solution of $-0.5(5 - 4x) > 4x + 1$ or $-12(1 - x) > 2(6.5 - 4x)$, and graph your answer. CCSS SMP 1

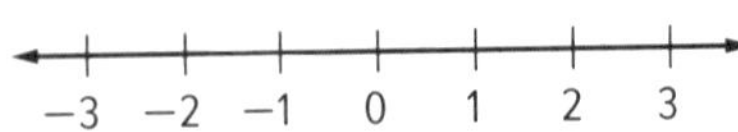

c. **CONSTRUCT ARGUMENTS** Compare the solution of $x < 5$ and $x < 3$, to the solution of $x < 5$ or $x < 3$. Explain the differences. CCSS SMP 3

An absolute value inequality can be solved by rewriting it as a type of compound inequality depending on the inequality symbol used.

KEY CONCEPT Absolute Value Inequalities

The absolute value inequality $|ax + b| > c$ is the same as the compound inequality $ax + b > c$ or $-(ax + b) > c$, which is equivalent to $ax + b > c$ or $ax + b < -c$.

The absolute value inequality $|ax + b| < c$ is the same as the compound inequality $ax + b < c$ and $-(ax + b) < c$, which is equivalent to $-c < ax + b$ and $ax + b < c$.

EXAMPLE 2 Solving Absolute Value Inequalities CCSS A.CED.3

a. PLAN A SOLUTION Solve the inequality $|4 - 5x| < 13$, and graph the answer. CCSS SMP 1

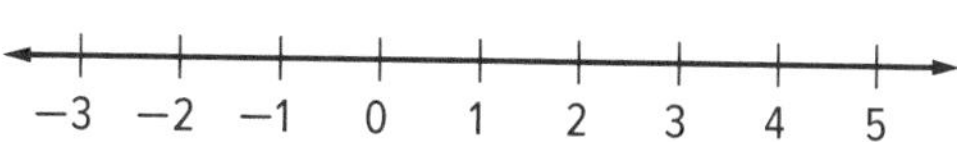

b. INTERPRET PROBLEMS The numbers -1.8 and 3.4 in **part a** divide the number line into three intervals. Choose a number from each interval to verify which interval(s) are the solution for the original inequality. CCSS SMP 1

EXAMPLE 3 Working with Absolute Value Inequalities

a. REASON ABSTRACTLY An object is launched into the air and then falls to the ground. Its velocity is modeled by the equation $v = 200 - 32t$, where the velocity is measured in feet per second and t is measured in seconds. The object's speed is the absolute value of its velocity. During what time intervals will the speed of the object be between 40 and 88 feet per second? CCSS A.CED.3, SMP 2

b. CRITIQUE REASONING Roberto claims that the solution to $|3c - 4| > -4.5$ is the same as the solution to $|3c - 4| \geq 0$, since an absolute value is always greater than or equal to zero. Is he correct? Explain your reasoning. CCSS A.CED.3, SMP 3

c. CONSTRUCT ARGUMENTS Discuss the number of solutions to the inequality $|ax + b| < c$ when $c > 0$ and when $c \leq 0$. CCSS A.CED.3, SMP 3

d. **CONSTRUCT ARGUMENTS** A student claims that for any non-zero constant b the solution to $|x| \leq |b|$ is the inequality $-b \leq x \leq b$. Do you agree? If not, give the correct solution. Explain your reasoning. CCSS A.CED.3, SMP 3

PRACTICE

1. **REASON ABSTRACTLY** Alamo Photo offers a high quality photo printer rental for $30 per month plus $0.05 per print. Bluebird Photo offers the printer for $20 per month plus $0.10 per print. Cactus Photo offers the printer for $5 per month plus $0.15 per print. For what number of prints per month is Bluebird Photo the most expensive plan? CCSS A.CED.1, SMP 2

2. **CRITIQUE REASONING** Miao is solving the compound inequality $8.4 \leq 2x + 3 \leq 12.9$. She simplifies the inequality and decides the solution set is $\{x \mid x \leq 4.95 \text{ or } x \geq 2.7\}$. Do you agree? Explain your reasoning. CCSS A.CED.3, SMP 3

3. **PLAN A SOLUTION** Solve $|2x - 1| < 5 + 0.5x$ and graph the solution set on a number line. Show your work. CCSS A.CED.1, SMP 1

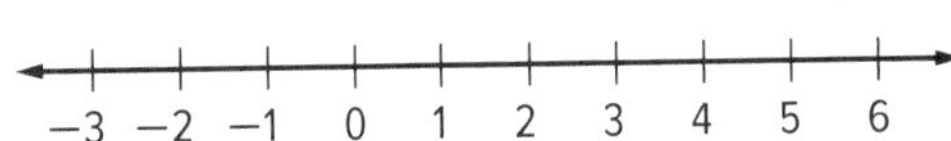

4. **CRITIQUE REASONING** Leonardo solves an inequality and determines that the solution is $x > 6.2$ and $x < 4.7$. How should he express his answer? Explain your reasoning. CCSS A.CED.3, SMP 3

5. **REASON ABSTRACTLY** Renata is going to invest $50,000 in two different options. Option 1 is a money market account that guarantees to pay 4% simple interest. Option 2 is a risky stock that could return as much as 24% on the investment. What amount does she need to invest in the stock so that the total value of both investments will be at least $56,000 and no more than $58,000 after one year if the stock returns its maximum interest? Explain. CCSS A.CED.1, SMP 2

6. INTERPRET PROBLEMS Solve the inequality $|3x + 1| > 2$ and graph the answer. Verify your solution by evaluating the inequality at a number from each interval. CCSS A.CED.3, SMP 1

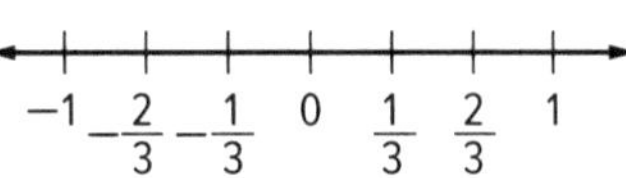

7. COMMUNICATE PRECISELY Solve the compound inequality $4 - 2x < |x| < 2x - 6$. Explain your solution process. CCSS A.CED.1, SMP 6

8. COMMUNICATE PRECISELY Solve the following compound inequality and graph the solution set on a number line. Show your work. CCSS A.CED.3, SMP 6

$$5 - 3x < 17 \quad \text{or} \quad -3 \leq 2x + 1 \leq 5$$

−10 −8 −6 −4 −2 0 2 4 6 8 10 12 14

Performance Task

Absolute Amusement

Provide a clear solution to the problem. Be sure to show all of your work, include all relevant drawings, and justify your answers.

You are on the roller coaster maintenance team. You need to adjust bolts on 5 different coasters, all of which require different integer-value lengths of wrenches, based on location. You must also follow safety procedures relating to the strain on the bolts.

Part A

i Use the information in the table to write inequalities for the wrench lengths that are needed for each coaster. List the solutions for each inequality.

Coaster	Wrench Lengths Required
A	Any length 2 through 8
B	Any length 5 through 9
C	Length 2 or 3
D	Any length 7 through 11
E	Any length 9 through 15

ii What is the minimum number of bolt wrenches you must take with you to adjust all the bolts on all of the coasters? Explain using the inequalities, and give an example of the wrenches you could bring.

Part B

The coasters use three grades of bolts. Grade A bolts have a strength of 60 ksi (kilopound per square inch, or 1000 psi). Grade B bolts have a strength of 80 ksi. Grade C bolts have a strength of 70 ksi. A set of bolts is needed to support the 5000 ksi structure of a roller coaster.

i The guidelines for installing bolts to support this roller coaster are that the bolts provide support of at least 20% over the minimum requirement, the total number of bolts should be more than 77 but not exceed 95, and no less than half the bolts should be grade B. Write a set of inequalities to represent the guidelines using *A*, *B*, and *C* for the number of each corresponding bolt.

ii If for this particular structure 60 grade B bolts, 10 grade C bolts, and 8 grade A bolts are used, are the guidelines for installing bolts met? State which guidelines, if any, are not met.

Part C

Industrial bolts such as those used on roller coasters are often alternately threaded for extra stability. This means that to loosen some bolts, you must turn to the right, and to loosen others, you must turn to the left. Engineers use the *signum function* to designate each type of bolt. The function equals -1 for $x < 0$, 1 for $x > 0$, and is undefined for $x = 0$.

Define and write the signum function. Sketch its graph on the coordinate grid.

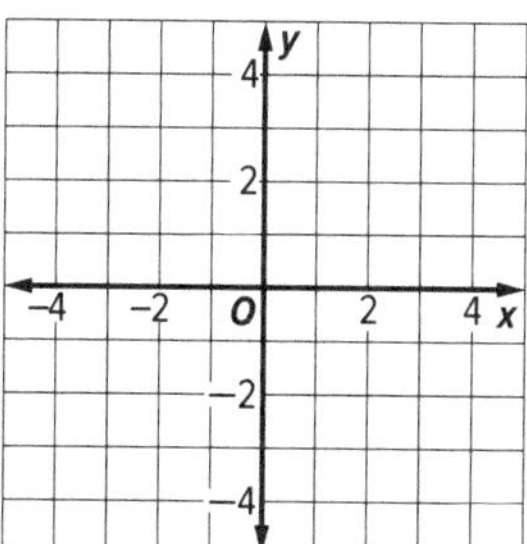

Performance Task

Piles of Peppers

Provide a clear solution to the problem. Be sure to show all of your work, include all relevant drawings, and justify your answers.

The pie chart shows hot pepper production each year on the Lyshi Hot Pepper Farm.

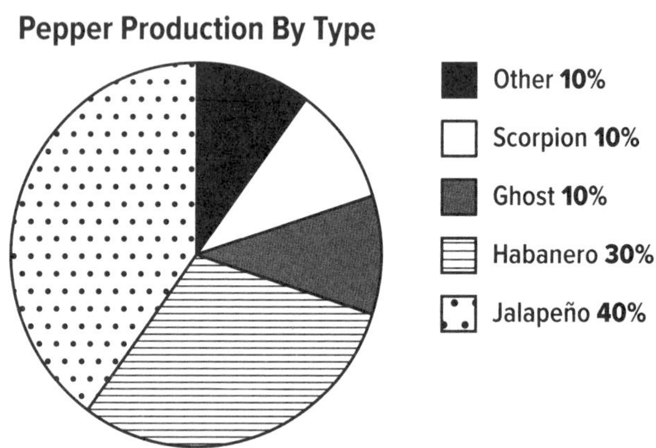

Part A

In 2013, the Lyshi Hot Pepper Farm produced 2.5 million jalapeño peppers. Determine the following:

i How many more habanero peppers were produced than ghost peppers?

ii How many specialty Anaheim banana chiles are produced every year if they make up 20% of the "Other" category?

Part B

The Lyshi Hot Pepper Farm makes a profit of \$0.06 for each jalapeño pepper produced, \$0.10 for each habanero pepper produced, \$0.20 for each ghost pepper, \$0.50 for each scorpion pepper, and a varying amount for each type of pepper in the "Other" category. If there were profits of \$1.65 million in 2013, what must have been the average profit per pepper in the "Other" category? Does this mean every pepper in the "Other" category results in that profit?

Part C

The Scoville scale measures the spicy heat level of a pepper. The units are in Scoville heat units, or SHU, which is related to the amount of a certain compound in the pepper. The inequality $|c - 40| \leq 10$ describes the possible heat levels of a cayenne pepper, where c is in 1000 SHU.

i The strongest habanero pepper is about 7 times hotter than the strongest cayenne pepper. Find the strongest heat level for a habanero pepper in SHU.

ii The range of the heat level for a habanero pepper is 12.5 times the range of a cayenne pepper. Write an absolute value inequality that describes the possible heat levels h of a habanero pepper, where h is in 1000 SHU.

Standardized Test Practice

1. Which of the following is a real number? CCSS N.CN.1, N.CN.2

$(1 + 2i)(1 - 2i)$ $17 - 17i$

$(3i - 7)(7 - 3i)$ $15 + 6i^4$

2. What real numbers a and b make the following equation true? CCSS N.CN.1, N.CN.2

$4(3 + ai)(2 + 5i) = 94 + bi$

$a =$ ☐ and $b =$ ☐

3. Write two equations that can be used to find the solution set of the equation $2x + 7 = |3x + 2|$. CCSS A.CED.3

☐

What is the solution set?

☐

4. The circumference of an official NBA basketball cannot be greater than 758.8 mm or less than 749.3 mm. Write an absolute value equation that represents the maximum and minimum circumferences. CCSS A.CED.1, A.CED.3

☐

5. 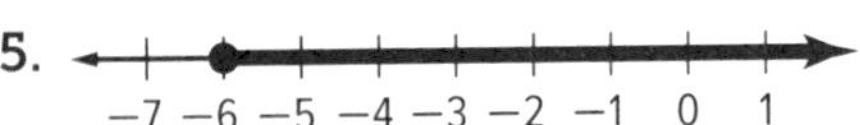

Which of the following compound inequalities could represent the solution set given by the graph? CCSS A.CED.1, A.CED.3

$x \geq -6$ or $x < 1$ $x < 0$ and $x \geq -6$

$x > -3$ or $x \geq -6$ $x \geq -6$ and $x > -2$

6. The formula for finding the area of a trapezoid is $A = \frac{a+b}{2}h$ where a and b are the bases and h is the height. The length of one base is 7, the height is 4, and the area is 34. What formula gives the length of the other base? CCSS A.CED.3

☐

Find the value of b: ☐

7. Declan and Carmen each ride their bikes for the same amount of time. Declan travels 4 miles farther, and 5 miles per hour faster, than Carmen. If Carmen travels 10 miles per hour, model the relationship between Declan's time and Carmen's time by writing an equation using the variable d for the distance Carmen traveled. CCSS A.CED.1, A.CED.3

☐

For how many minutes did each of them ride? ☐

8. Write an absolute value inequality for the graph below. CCSS A.CED.1, A.CED.3

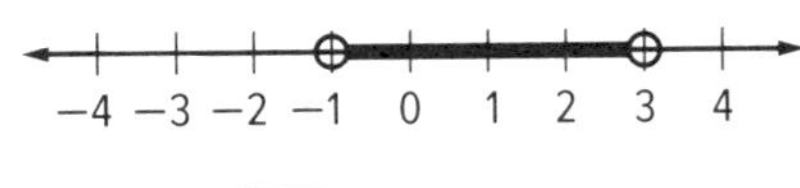

☐

9. Which of the following compound inequalities has the solution set represented by the graph? CCSS A.CED.3

$2x - 5 > 3$ or $5 - x > 8$

$5x - 1 > 9$ or $-x + 4 < 7$

$-6x - 12 > -24$ or $2x - 1 > -6$

10. Simplify the expression $4i(2 - i)^2 - (3 + 5i)$. CCSS N.CN.2

$4i(2 - i)^2 - (3 + 5i) =$ ☐

11. A vehicle begins driving west at 55 miles per hour at 2:00 p.m. If another vehicle leaves from the same place driving west at 70 miles per hour at 3:00 p.m., write an equation that can be used to find the number of hours t after 2:00 p.m. that it will take for the vehicles to meet. CCSS A.CED.1

At what time will the two vehicles meet?

☐ p.m.

12. Select the term that best describes each expression. CCSS N.CN.1, N.CN.2

Expression	Real	Imaginary	Complex
$(4 + i) - (2i + 5)$			
$\sqrt{(-3)(-3)}$			
$12i^{14}$			
$-9i^{15}$			
$(7i + \sqrt{3})(7i + \sqrt{3})$			
$2i(-6 + i) + 4(3i - 5)$			
$i^{13} + i^{17}$			

13. Does the statement have all real numbers as its solution? Choose Yes or No. CCSS A.CED.1

Statement	Yes	No
$\|x + 1\| > 0$		
$x < -3$ or $x > -10$		
$3x \geq 0$		
$\|x - 1\| \geq 0$		
$x < -3$ and $x > -10$		

14. The ideal temperature for roasting cocoa beans is within 14°C of 149°C.

a. Write an absolute value inequality to represent this situation. Then write an equivalent compound inequality and graph the solution. CCSS A.CED.1

b. A typical pod contains 20 to 50 beans. About 400 dried beans are required to make one pound of chocolate. Write an inequality to represent the number of pods p needed to make c pounds of chocolate. Explain your solution. CCSS A.CED

15. In the graph, point B is 1.5 times further from point A than 3 is. Label point A with its coordinate. Show your work. CCSS A.CED.1

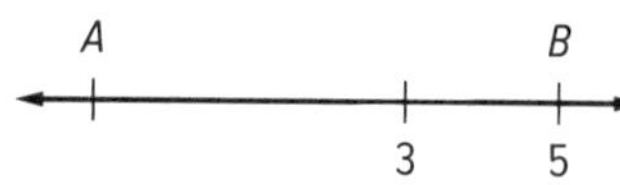

2 Linear Functions

CHAPTER FOCUS Learn about some of the Common Core State Standards that you will explore in this chapter. Answer the preview questions. As you complete each lesson, return to these pages to check your work.

What You Will Learn	Preview Question
Lesson 2.1: Linear Relations and Functions	
CCSS F.IF.4 For a function that models a relationship between two quantities, interpret key features of graphs and tables in terms of the quantities, and sketch graphs showing key features given a verbal description of the relationship. CCSS F.IF.9 Compare properties of two functions each represented in a different way (algebraically, graphically, numerically in tables, or by verbal descriptions).	CCSS SMP 7 These linear functions have different constant terms. How are these differences reflected in the graphs of these functions? $y = 2x + 3$ $y = 2x + 0$ $y = 2x - 3$
Lesson 2.2: Modeling: Linear Functions	
CCSS F.IF.4 For a function that models a relationship between two quantities, interpret key features of graphs and tables in terms of the quantities, and sketch graphs showing key features given a verbal description of the relationship. **Also addresses:** A.CED.4, A.SSE.1a, F.IF.5, F.IF.9	CCSS SMP 1 Barry uses the equation $d = 60t$ to represent the distance he traveled by car on a trip, based on the number of minutes driving. Solve the equation for t and interpret its meaning.
Lesson 2.3: Rate of Change and Slope	
CCSS F.IF.5 Relate the domain of a function to its graph and, where applicable, to the quantitative relationship it describes. CCSS F.IF.6 Calculate and interpret the average rate of change of a function (presented symbolically or as a table) over a specified interval. Estimate the rate of change from a graph. **Also addresses:** F.IF.4	CCSS SMP 2 What is the domain of this linear function? Suppose it is a model of the distance a car that travels 5 mi/h. How would you interpret the domain? $y = 5x$
Lesson 2.4: Writing Linear Equations	
CCSS A.CED.2 Create equations in two or more variables to represent relationships between quantities; graph equations on coordinate axes with labels and scales.	CCSS SMP 2 Which equation represents an increasing linear function? Which represents a decreasing linear function? How do you know? A. $y = -7x + 4$ B. $y = 7x - 4$

What You Will Learn	Preview Question
Lesson 2.5: Scatter Plots and Lines of Regression	
CCSS F.IF.4 For a function that models a relationship between two quantities, interpret key features of graphs and tables in terms of the quantities, and sketch graphs showing key features given a verbal description of the relationship.	CCSS SMP 6 Suppose you wanted to gather data on the number of hours studied versus the grade on a test. What do you think the graph of the data would look like? Describe the linear function that would best fit this data.
Lesson 2.6: Special Functions	
CCSS F.IF.7b Graph square root, cube root, and piecewise-defined functions, including step functions and absolute value functions. CCSS F.IF.5 Relate the domain of a function to its graph and, where applicable, to the quantitative relationship it describes. CCSS F.IF.4 For a function that models a relationship between two quantities, interpret key features of graphs and tables in terms of the quantities, and sketch graphs showing key features given a verbal description of the relationship.	CCSS SMP 6 What is the range of each function? If the range is not the same for each function, explain why. A. $y = x$ B. $y = \|x\|$
Lesson 2.7: Transforming Functions	
CCSS A.CED.2 Create equations in two or more variables to represent relationships between quantities; graph equations on coordinate axes with labels and scales. CCSS F.BF.3 Identify the effect on the graph of replacing $f(x)$ by $f(x) + k$, $k\,f(x)$, $f(kx)$, and $f(x + k)$ for specific values of k (both positive and negative); find the value of k given the graphs. Experiment with cases and illustrate an explanation of the effects on the graph using technology. Include recognizing even and odd functions from their graphs and algebraic expressions for them. **Also addresses:** F.IF.4, F.IF.9, A.CED.3	CCSS SMP 7 Describe how the graph of $g(x) = -(x - 1)^2 + 4$ is related to the graph of $f(x) = x^2$.
Lesson 2.8: Graphing Linear and Absolute Value Inequalities	
CCSS A.CED.3 Represent constraints by equations or inequalities, and by systems of equations and/or inequalities, and interpret solutions as viable or nonviable options in a modeling context.	CCSS SMP 4 A business uses the equation $R = 10 \cdot u$ to predict the amount of revenue earned R for selling any given units of a \$10 product. Write an inequality that can be used to determine the minimum number of units that must be sold in order to have earned revenue of at least \$10,000.

2.1 Linear Relations and Functions

Objectives

- Write equations of linear functions.
- Identify relationships that are linear functions and interpret their characteristics.
- Compare properties of two or more linear functions given in various forms.

CCSS STANDARDS

Content: F.IF.4, F.IF.9
Practices: 1, 2, 3, 4, 6, 7, 8
Use with Lesson 2-2

Two variables, *x* and *y*, can be related in various ways. Relations that have straight-line graphs are called linear relations and those that are not linear are called nonlinear relations.

EXAMPLE 1 Describing Temperature Using Relations CCSS F.IF.4, F.IF.9

EXPLORE **Viorel, Dante, and Malinda recently purchased space heaters to control the temperature in their sunrooms. They each initially set the temperature at 72° on Sunday at 9 am, and measured the temperature at 9 am each day for one week. They reported their data as follows:**

Viorel	Temperature reading was always 72 degrees.							
Dante	Temperature went up by 0.5 degrees each day.							
Malinda	**Day 0**	**Day 1**	**Day 2**	**Day 3**	**Day 4**	**Day 5**	**Day 6**	**Day 7**
	72	70	74	72	76	74	78	76

a. **COMMUNICATE PRECISELY** Sketch graphs that describe each of these three scenarios. Which appear to be linear? CCSS SMP 6

Viorel's Graph

Dante's Graph

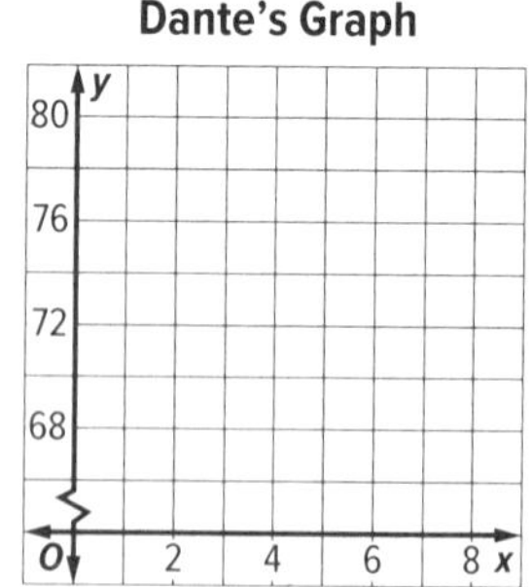

Malinda's Graph

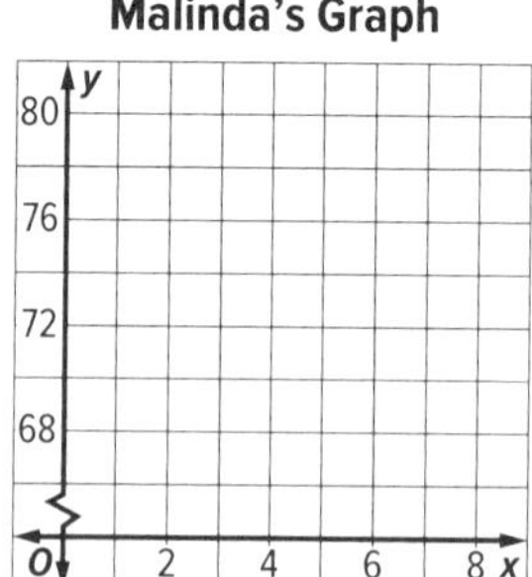

b. **COMMUNICATE PRECISELY** Describe how the graph of each person's data changes. CCSS SMP 6

c. **COMMUNICATE PRECISELY** What modification could you make to Dante's description that would result in a graph that is *decreasing* and linear? Explain. CCSS SMP 6

d. **REASON ABSTRACTLY** Where does each graph intersect the y-axis? What is the meaning of this point in the context of the problem? CCSS SMP 2

The graphs of the data sets for Viorel and Dante represent linear functions. A linear function is a function with ordered pairs that satisfy a linear equation that can be written as $f(x) = mx + b$, where m and b are real numbers. The equations of the functions represented by the graphs for Viorel and Dante are $f(x) = 72$ and $f(x) = 0.5x + 72$, respectively.

KEY CONCEPT Standard Form of a Linear Equation

The standard form of a linear equation is $Ax + By = C$, where A, B, and C are integers with a greatest common factor of 1, $A \geq 0$, and A and B are not both zero.

The y-coordinate of the point at which a graph crosses the y-axis is the y-intercept and the x-coordinate of a point at which a graph crosses the x-axis is called an x-intercept. Since two points determine a line, a linear function can be graphed by connecting the two intercepts of the line.

EXAMPLE 2 Describing Key Features of Linear Models CCSS F.IF.4

EXPLORE **Jodi and Alisha are reading novels for book reports. Jodi records the number of pages she has remaining to read after each day in the table below. Alisha records the number of pages she has remaining each day on the graph.**

Jodi:

Days	0	1	2	3	4	5
Pages	585	520	455	390	325	260

Alisha:

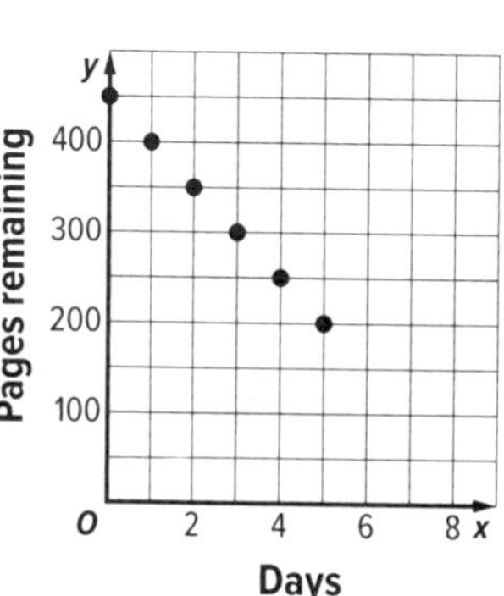

a. **INTERPRET PROBLEMS** Describe the function that models the number of pages remaining for each girl. CCSS SMP 1

b. **REASON ABSTRACTLY** What is the y-intercept for each function? Interpret its meaning in the context of the problem. CCSS SMP 2

c. **INTERPRET PROBLEMS** Write a linear equation in the form $f(x) = mx + b$ for the function that models the pages remaining for each girl. Then write each equation in standard form $Ax + By = C$. CCSS SMP 1

d. **REASON ABSTRACTLY** After how many days will each girl finish reading her book? What feature of the function represents this event? Explain your answer. CCSS SMP 2

e. **FIND A PATTERN** Which girl is reading faster and by how many pages per day? Support your answer. CCSS SMP 7

PRACTICE

1. **REASON QUANTITATIVELY** Name the x- and y-intercept for the linear equation given by $6x - 2y = 12$. Use the intercepts to graph the equation and describe the graph as *increasing, decreasing,* or *constant*. CCSS F.IF.4, SMP 2

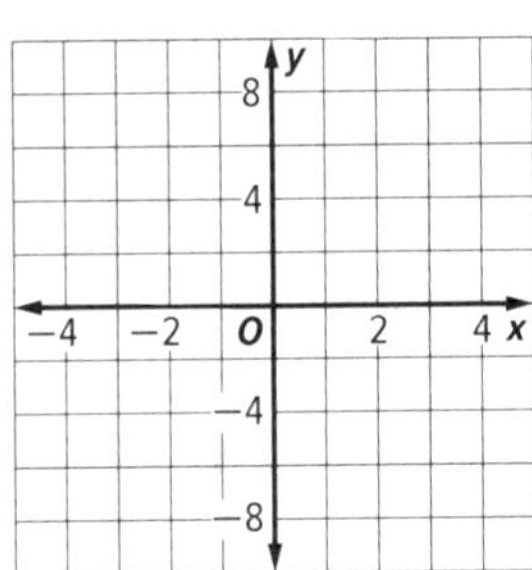

2. Kate claims that since $y = x + 1$ and $y = 3x + 2$ are both linear functions, the function $y = (x + 1)(3x + 2)$ must also be a linear function.

   a. **CRITIQUE REASONING** Disprove her statement. CCSS F.IF.4, SMP 3

   b. **CRITIQUE REASONING** Kate then claims that the difference of the two functions must be linear. Find the difference of the two linear functions, and determine if Kate is correct. CCSS F.IF.9, SMP 3

   c. **REASON ABSTRACTLY** Compare the features of the original two linear functions with the function found in **part b**. CCSS F.IF.9, SMP 2

3. Jacalyn is depositing money weekly into her savings account. The table shows the balance in her account after various weeks. CCSS F.IF.4

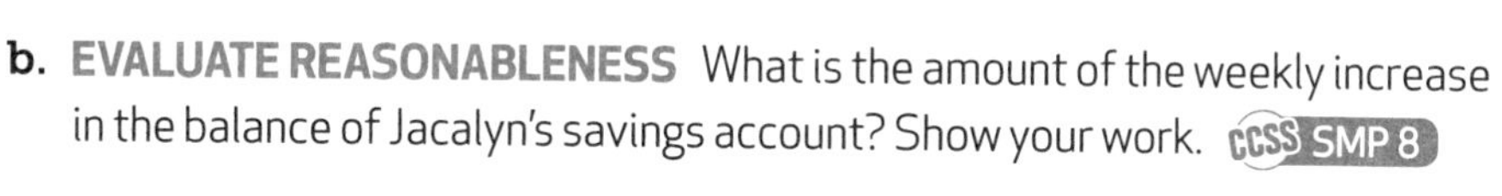

Week	0	3	6	9	12
Balance	\$350	\$440	\$530	\$620	\$710

a. COMMUNICATE PRECISELY Make a graph of the data. Describe a model that would represent the data. CCSS SMP 6

b. EVALUATE REASONABLENESS What is the amount of the weekly increase in the balance of Jacalyn's savings account? Show your work. CCSS SMP 8

c. USE A MODEL Write an equation for the function that models the balance of Jacalyn's savings account. CCSS SMP 4

4. REASON QUANTITATIVELY The equation and table below each describe a linear function. Which function has the greater y-intercept? Explain. CCSS F.IF.9, SMP 2

$3x + 4y = 8$

x	y
−2	15
−1	11
0	7
1	3

5. For a math club fund raiser, each of three club members sells 20 mugs. Their sales are described as follows. Reiko sold all of her mugs in 5 days and sold the same number each day. The number of mugs Cosmin had left after x days is given by the equation $y = -2x + 20$. The number of mugs that Sheena had left after each day are shown in the table. CCSS F.IF.9

Days	Mugs Left
0	20
1	15
2	10
3	5
4	0

a. INTERPRET PROBLEMS Write linear equations in the form $f(x) = mx + b$ for the functions that model the number of mugs remaining for Reiko and Sheena. CCSS SMP 1

b. REASON ABSTRACTLY What do the x-intercepts represent for each function? CCSS SMP 2

c. REASON ABSTRACTLY Which student took the longest to sell their mugs, and how long did it take? CCSS SMP 2

2.2 Modeling: Linear Functions

Objectives

- Interpret key features of the graphs of linear functions.
- Explain the relationship between the domain of a linear function and the quantitative relationship it describes.
- Compare two linear functions described in different ways.
- Solve a linear equation for one variable in terms of other variables.

CCSS STANDARDS

Content: A.SSE.1a, A.CED.4, F.IF.4, F.IF.5, F.IF.9
Practices: 1, 2, 4, 8
Use with Lesson 2-2

Linear functions are very useful in modeling a wide variety of scenarios. You will consider several such scenarios in this lesson.

EXAMPLE 1 Comparing Efficiency of Scooters CCSS F.IF.4, F.IF.9

EXPLORE Two new models of energy efficient scooters are now available. Model A has a 10-gallon gas tank and gets 22.4 miles per gallon. Model B has a 10-gallon gas tank. The gas usage for Model B is shown in the table.

Gas Used (gallons)	Miles Traveled
1	26.2
2	52.4
3	78.6

a. **REASON QUANTITATIVELY** For each model, write a linear equation in standard form that describes the relationship between gallons used and miles traveled. CCSS SMP 2

b. **INTERPRET PROBLEMS** Determine how many gallons of gas, to the nearest tenth of a gallon, it takes each model scooter to travel 100 miles. Explain your reasoning. CCSS SMP 1

c. **FIND A PATTERN** For each model, derive a general formula for determining the number of gallons necessary to travel m miles. CCSS SMP 8

d. **INTERPRET PROBLEMS** Graph each model on the same coordinate plane. Then compare the slopes of the two models. CCSS SMP 1

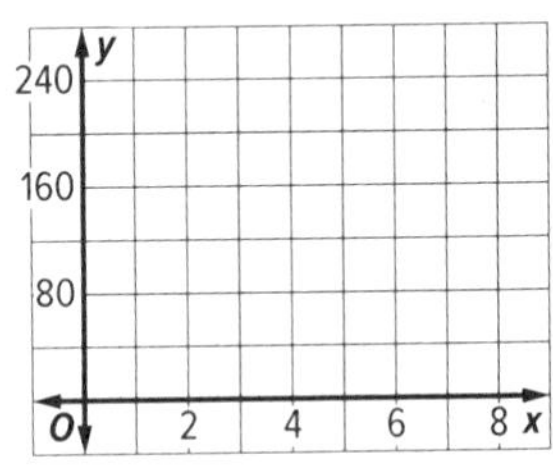

e. **USE A MODEL** Which of the two models is more cost effective to operate? Explain your reasoning. CCSS SMP 4

EXAMPLE 2 Distinguishing Among Linear Functions CCSS A.SSE.1a, F.IF.5

EXPLORE **Consider the following scenarios:**

I. "The temperatures dropped very quickly throughout the day."

II. "The free throw percentage gradually improved over the course of the 3-month season."

III. "The difficulty level of the game increased dramatically as the stages of the game increased from 1 to 100."

IV. "The price per pound of bananas has remained steady throughout the month of July."

a. **INTERPRET PROBLEMS** Graph each function on a separate coordinate plane. CCSS SMP 1

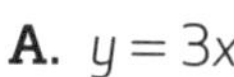

A. $y = 3x$ B. $y = 3$ C. $y = -3x$ D. $y = \frac{1}{3}x$

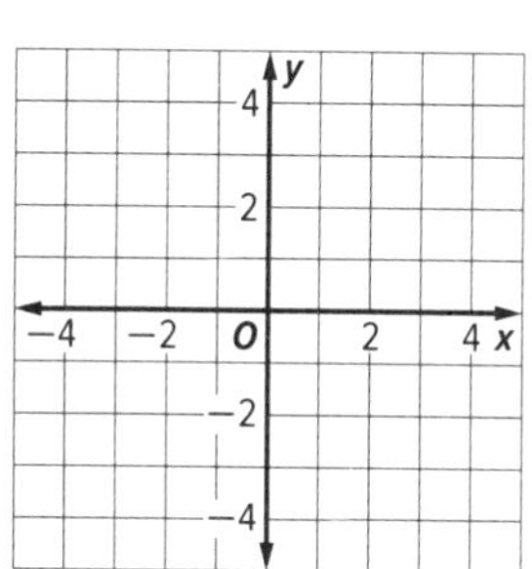

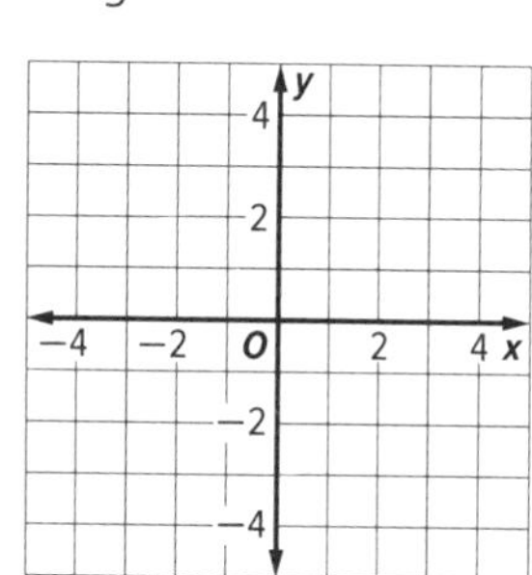

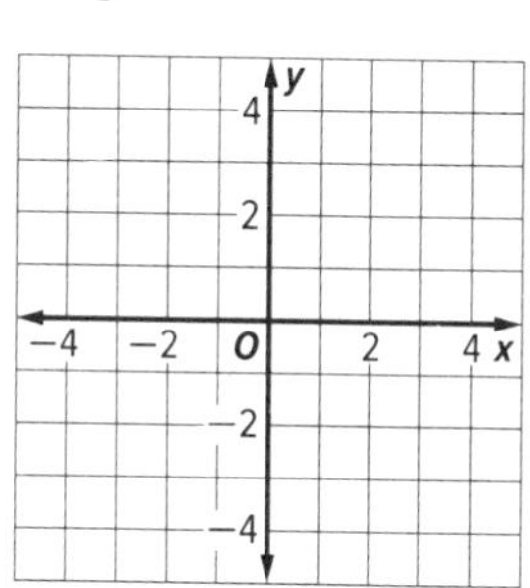

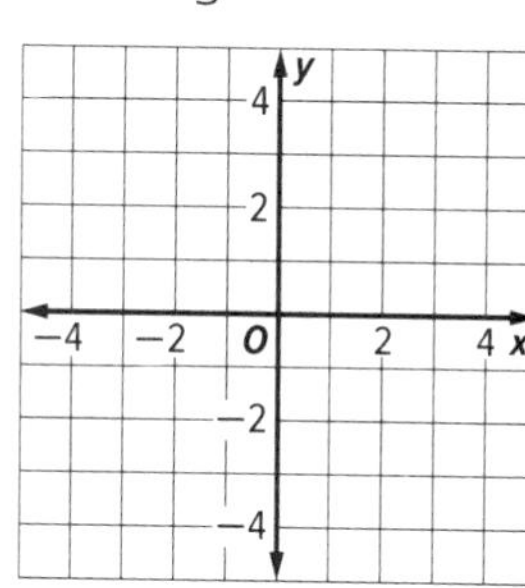

b. **PLAN A SOLUTION** Select the function from part a that best matches each scenario described. Include the appropriate units for x and y and the domain of each function. Explain your reasoning. CCSS SMP 1, SMP 2, SMP 4

c. **FIND A PATTERN** What does the coefficient of x appear to be doing to the graphs in **part a**? Using this interpretation, give an example of an equation that would describe the scenario, "The temperature dropped gradually throughout the day." CCSS SMP 8

EXAMPLE 3 Before and After Training CCSS A.CED.4

EXPLORE A new company claims that their innovative test-preparation strategy will dramatically increase one's score on a certain college entrance exam. The range of scores possible for the exam is 100 – 2000 points. To prove their claim, they administered the exam to a large group of people *before* they underwent the training. The participants then completed the training program and retook the exam. The equation relating their pre-training score, x, to the post-training score, y, is $100y - 94x = 12{,}000$.

a. **REASON QUANTITATIVELY** Solve the equation for y and interpret its meaning in context. Why is this useful? CCSS SMP 2

b. **REASON QUANTITATIVELY** Solve the equation for x and interpret its meaning in context. Why is this useful? CCSS SMP 2

c. **INTERPRET PROBLEMS** Determine the x and y intercepts and describe what they represent in the problem. Discuss whether or not each intercept makes sense in the context of the problem. CCSS SMP 1

d. **INTERPRET PROBLEMS** Determine the smallest and largest test scores one could expect if he completed the training program. Explain your reasoning. CCSS SMP 1

PRACTICE

1. **REASON QUANTITATIVELY** For each of the following linear functions, select the description that best matches each function. Then find the domain of each function. CCSS A.SSE.1a, F.IF.5, SMP 2

a. $y = -5x$

i. The total cost to manufacture a product is \$3 per unit plus a one-time fee of \$15.

b. $y = 20$

ii. The amount of debt a student has on a given day in March if he borrows \$5 a day for lunch

c. $y = 3x + 15$

iii. The length of a piece of fabric that originally measures 36 yards if it is sold in 2 yard increments

d. $y = -2x + 36$

iv. The temperature throughout the day of a freezer set at 20 degrees

2. Initially, Pool 1 is filled with 516 gallons of water and Pool 2 is filled with 720 gallons of water. CCSS A.CED.4, F.IF.4, F.IF.9

Pool 1

Time (minutes)	Water Remaining (gallons)
25	416
40	356
55	296
70	236

Pool 2

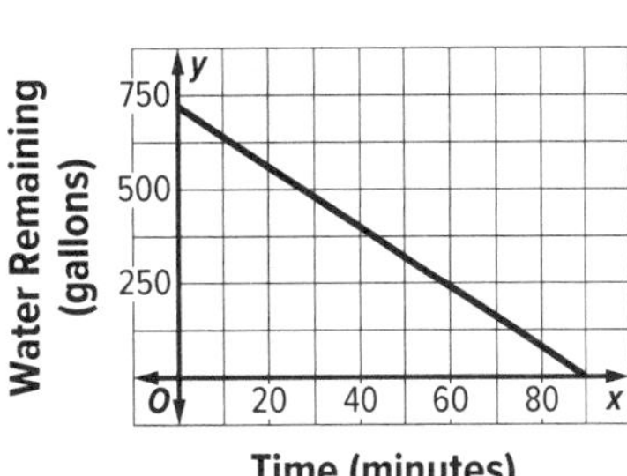

a. **REASON QUANTITATIVELY** For each pool, write a linear equation that describes the relationship between the number of minutes the pool has been draining and the number of gallons of water remaining in the pool. CCSS SMP 2

b. **USE A MODEL** Find the number of gallons of water remaining in each pool after 1 hour. Describe your solution process. CCSS SMP 4

c. **USE A MODEL** Which pool takes longer to drain? Explain your reasoning. CCSS SMP 4

2.3 Rate of Change and Slope

Objectives

- Compare the slope of a line to the rate of change of a linear function.
- Compare and interpret the slopes of two lines in context.
- Calculate and interpret average rates of change.

STANDARDS

Content: F.IF.4, F.IF.5, F.IF.6
Practices: 2, 4, 6, 7
Use with Lesson 2-3

A **rate of change** is a ratio used to compare the change in one quantity to the change in a second quantity. The denominator is the amount of change in an independent variable, $\triangle x$; the numerator is the amount of change in the dependent variable, $\triangle y$.

$$\text{rate of change} = \frac{\text{change in } y}{\text{change in } x}$$

The **slope** of a line is the ratio of the change in the y-coordinates to the change in the x-coordinate over a set interval. If a line passes through two points, (x_1, y_1) and (x_2, y_2), then its slope it given by:

$$\text{slope} = \frac{\text{change in } y}{\text{change in } x} \quad \text{or} \quad m = \frac{\square - \square}{\square - \square}$$

EXAMPLE 1 Interpreting Slopes CCSS F.IF.5, F.IF.6

EXPLORE Two plumbers each charge a fixed amount plus an hourly fee. The tables give some recent jobs each has done.

Paula's Plumbing and Pipes

Hours	2	3	5
Total Charged	\$80	\$105	\$155

Walter's Water Works

Hours	1	4	6
Total Charged	\$45	\$150	\$220

a. **USE STRUCTURE** Find the average rate of change in price for jobs completed by Paula's Plumbing and Pipes and Walter's Water Works. CCSS SMP 7

b. **USE STRUCTURE** Graph the data provided for each plumber and connect the data points. What do you notice? CCSS SMP 7

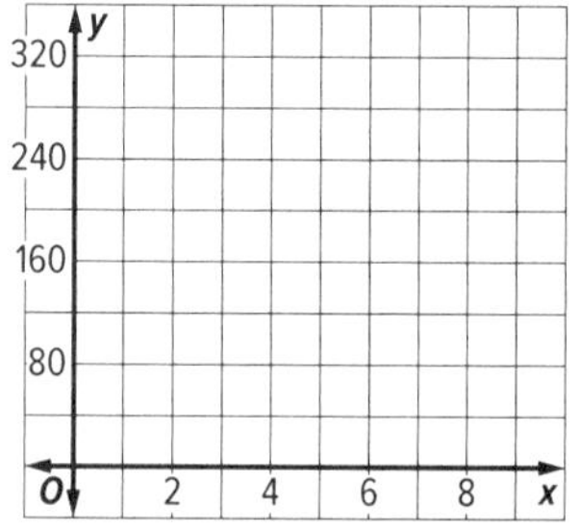

c. **USE STRUCTURE** Each set of points can be modeled with a linear equation. Discuss the similarities and differences between the equations of the lines that model the data for each company and the data presented in the table. Consider the slope, rate of change, domain, and intercepts in terms of their real-world applications. CCSS SMP 7

d. **USE A MODEL** Kaitlyn needs to fix a leaky faucet in her kitchen. She calls both companies to see how long it will take to complete the job. Who would you recommend she hire? Explain your reasoning. CCSS SMP 4

EXAMPLE 2 Slopes and Average Rates of Change CCSS F.IF.5, F.IF.6

Sam is preparing for a race. On Saturday she ran around the track using a timer to mark how long it took her to get to certain markers on the track. The graph below shows the results of her workout. The x-axis represents the time in minutes from when she started running. The y-axis represents the distance in meters she had run at that marker. Her total distance was 1600 meters.

a. **CALCULATE ACCURATELY** Find the slope of the line passing through the first two points. Find the slope of the line passing through the second two points. What do you notice? CCSS SMP 6

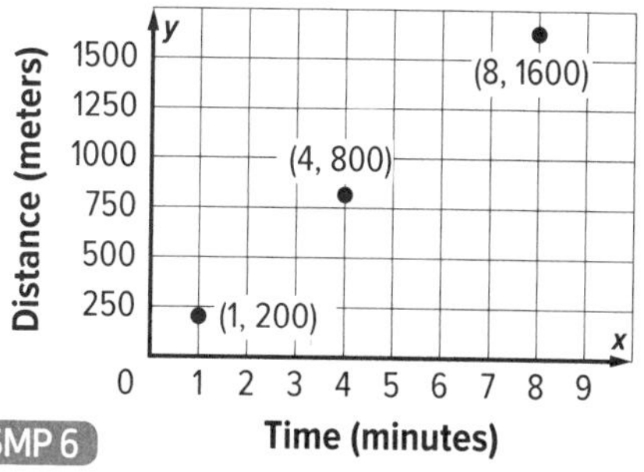

b. **COMMUNICATE PRECISELY** Interpret the slope as an average rate of change complete with proper units. Given the two slopes you calculated in **part a**, what does that tell you about the manner in which Sam ran? CCSS SMP 6

c. **USE A MODEL** Find the equation of the line that passes through the first and the third point. Does the line pass through the second point? Make sure to limit the domain of your line to fit the situation. CCSS SMP 4

EXAMPLE 3 Using Slope

Becky starts a babysitting service. She pays to take a certification course through a local hospital. Becky charges an hourly rate. The graph models the amount of profit, in dollars, Becky makes per hour spent babysitting. CCSS F.IF.4, F.IF.6

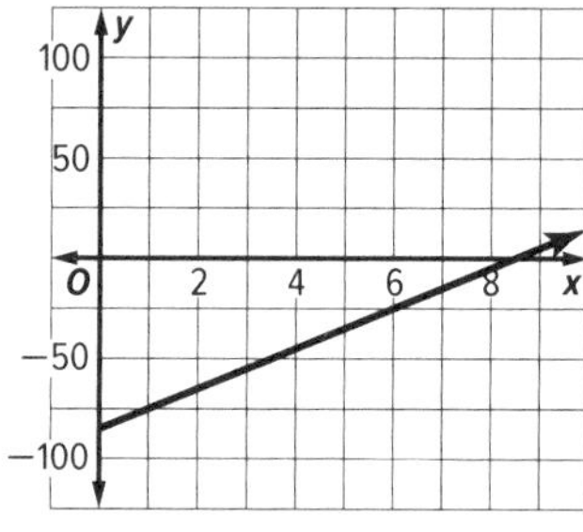

a. **USE A MODEL** Find the y-intercept and describe what it means in the context of the situation. Explain your reasoning. CCSS SMP 4

b. **USE A MODEL** Determine the average rate of change and describe what it means in the context of the situation. Explain your reasoning. CCSS SMP 4

c. **USE A MODEL** Determine the independent and dependent variables and then write a function relating Becky's profit for each hour she babysits. CCSS SMP 4

d. **REASON ABSTRACTLY** Determine when Becky's profit is negative, zero, and positive. CCSS SMP 2

PRACTICE

1. Javier has a bicycle-delivery business. He charges a flat fee to pick up each package. He also charges an additional rate per mile. The graph models the total amount, in dollars, Javier charges per mile to deliver a package. CCSS F.IF.4, F.IF.5, F.IF.6

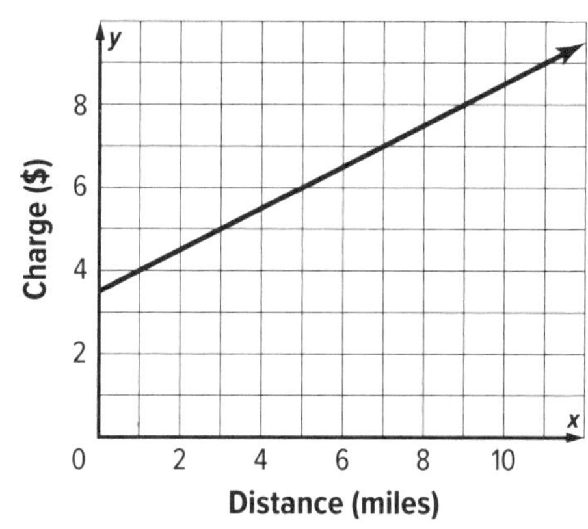

   a. **USE A MODEL** Find the y-intercept and describe what it means in the context of the situation. Explain your reasoning. CCSS SMP 4

   b. **USE A MODEL** Determine the average rate of change and describe what it means in the context of the situation. Explain your reasoning. CCSS SMP 4

   c. **USE A MODEL** Determine the independent and dependent variables and then write a function relating the total amount Javier charges to the distance traveled to deliver a package. What must be the domain of the function for it to model Javier's delivery business? CCSS SMP 4

2. A car initially traveling at 65 mph comes to a complete stop in 5 seconds. CCSS F.IF.4, F.IF.5, F.IF.6

a. **REASON QUANTITATIVELY** Find the average rate of change in the context of this situation. CCSS SMP 2

b. **USE A MODEL** Write and graph a function to represent this situation. Discuss restrictions to the domain so that is makes sense in the context of the problem. CCSS SMP 4

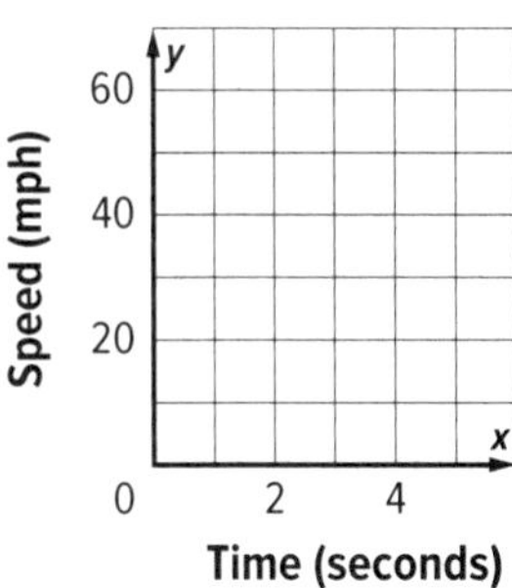

3. At midnight, Yasmina sets up a rain-collector—a tube that will measure any rain that falls. At 8 am the next morning, she sees that the water in the tube is up to the 51 millimeter mark. At 10 am, Yasmina sees that the water has reached the 85 millimeter mark. The rain stops falling at noon.

a. **CALCULATE ACCURATELY** Calculate the rate of change of the amount of water in the tube. CCSS F.IF.6, SMP 6

b. **USE A MODEL** Write an equation that models the situation. Explain. CCSS F.IF.6, SMP 4

c. **USE A MODEL** If the rain has been falling steadily since it started, then at what time did it begin raining? Explain. CCSS F.IF.4, SMP 4

d. **COMMUNICATE PRECISELY** Using the time the rain started and the time it ended, what would the domain of the function be? What would be the range? Discuss why the line cannot model rainfall once the rain stops. What would the graph look like after this point? CCSS F.IF.5, SMP 6

e. **COMMUNICATE PRECISELY** Explain why the equation of the line is only an approximation of the actual rainfall. CCSS F.IF.6, SMP 6

2.4 Writing Linear Equations

CCSS STANDARDS

Content: A.CED.2
Practices: 1, 2, 3, 5, 6, 8
Use with Lesson 2-4

Objectives

- Write linear equations in two variables to represent relationships between two quantities.
- Graph linear equations on the coordinate plane.

Given sufficient information about a line in the coordinate plane, you can write a linear equation to represent it. Stated below are important forms for an equation of a line.

The **slope-intercept** form of the equation of a line is $y = mx + b$ where m is the ______ and b is the ______.

The **point-slope** form of the equation of a line is $y - y_1 = m(x - x_1)$, where (x_1, y_1) are the ______ of a point on the line and m is the ______ of the line.

Any **vertical line** containing $P(x_1, y_1)$ has an equation of $x = x_1$.

Any **horizontal line** containing $P(x_1, y_1)$ has an equation of $y = y_1$.

EXAMPLE 1 Use slope and y-intercept to write the equations of the lines

CCSS A.CED.2, SMP 6

EXPLORE **Each line in the coordinate plane is represented by an equation.**

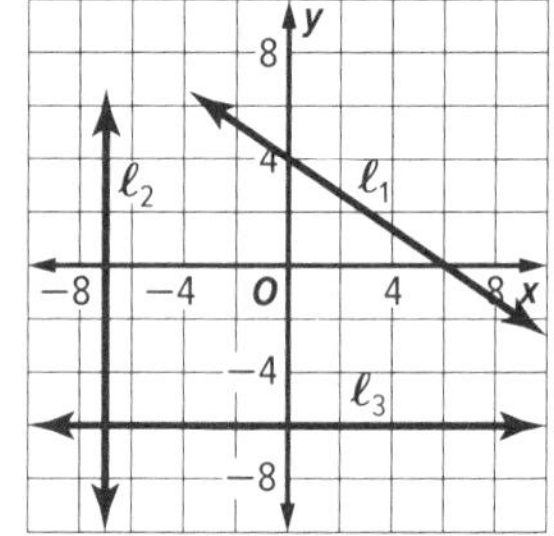

a. **CALCULATE ACCURATELY** Choose two points on ℓ_1, and write out the values needed to create the point-slope form of the equation of the line. Then, write out the point-slope form of the equation. Compare your answers with other students.

b. Substitute the values into the slope-intercept equation. Explain what the equation represents.

c. Find the slope of ℓ_2. What equation represents this line?

d. Find the slope of ℓ_3. What equation represents this line?

You can use point-slope form to find an equation of a line through two specific points. You must first find the slope. Then, substitute the coordinate point and slope into the point-slope form.

EXAMPLE 2 Writing an equation of a line given two points CCSS A.CED.2

a. **CALCULATE ACCURATELY** Write an equation for the line containing $L(-5, -6)$ and $M(3, 8)$. CCSS SMP 6

b. **FIND A PATTERN** In **part a** you chose one point in the point-slope form to find the equation. If you chose the other point, would the resulting equation be the same? Why? CCSS SMP 8

c. **DESCRIBE A METHOD** The equation for the line containing $A(-5, 2)$ and $B(4, 2)$ is $y = 2$. Without using the slope formula, how can you confirm that the equation of the line is $y = 2$? Give an example of two points on a vertical line and the equation containing those two points. CCSS SMP 8

d. **REASON ABSTRACTLY** Explain why either the point-slope form or the slope-intercept form can be used to find the equation of a line passing through the points (3, 1) and (0, 7). Use both methods to find the equation and verify that the two are equivalent. CCSS SMP 3

Parallel lines: Two distinct non-vertical lines are parallel if and only if they have the same ______. All vertical lines or horizontal lines are ______.

Perpendicular lines: Two non-vertical lines are perpendicular if and only if the product of the slopes is ______. Vertical lines and ______ are perpendicular.

EXAMPLE 3 Finding equations for parallel or perpendicular lines CCSS A.CED.2

Find an equation of the line specified. Graph the two lines in the coordinate plane.

a. **PLAN A SOLUTION** What is the slope of the line that is parallel to $y = \frac{1}{4}x - 2$? How can this information be used to find the line that is parallel and contains $P(2, 3)$? CCSS SMP 1

b. What is the slope of the line that is perpendicular to $y = \frac{3}{4}x - \frac{1}{4}$? How can this information be used to find the line that is perpendicular and contains $P\left(-\frac{9}{4}, \frac{1}{2}\right)$? CCSS SMP 1

c. **USE TOOLS** Graph the equations from **part a** in the coordinate plane. Why do the slopes of parallel lines have to be equal? CCSS SMP 5

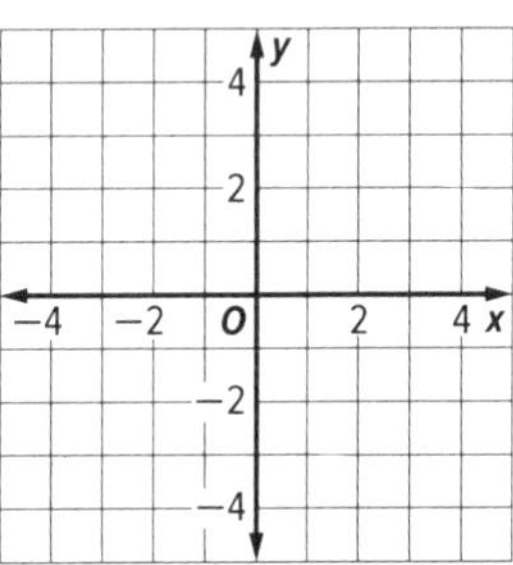

d. **USE TOOLS** Graph the equations from **part b** in the coordinate plane. What kind of angles are formed at the intersection between the two lines? CCSS SMP 5

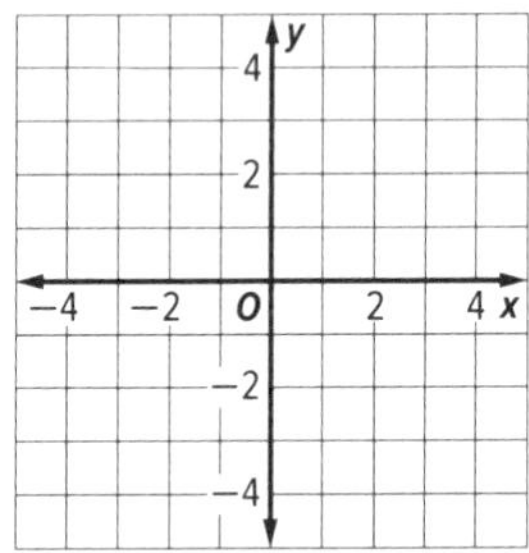

e. **CRITIQUE REASONING** Jude claims that two non-vertical lines are perpendicular if their slopes are "opposite reciprocals." In other words, if you take the reciprocal and change the sign of one slope, you get the other slope. Is Jude correct? Explain your reasoning. CCSS SMP 3

PRACTICE

1. **PLAN A SOLUTION** Write an equation for the line containing points $A\left(\frac{1}{2}, \frac{5}{6}\right)$ and $B\left(\frac{3}{4}, -\frac{1}{6}\right)$ CCSS A.CED.2, SMP 1

2. **PLAN A SOLUTION** Find the equation of a line parallel to $y = -\frac{3}{2}x + 3$ containing the $P(-2, 4)$ in slope-intercept form. CCSS A.CED.2, SMP 1

3. **REASON ABSTRACTLY** Find the value of k so that the graphs of $7y = kx - 10$ and $6x + 24y = 16$ are perpendicular. CCSS A.CED.2, SMP 2

4. **USE TOOLS** Sketch the graph of $y = 3x + 1$. Find the equations of 3 other lines that together with the line of $y = 3x + 1$ would form a rectangle. CCSS A.CED.2, SMP 5

5. **REASON ABSTRACTLY** In order to become a member at a local gym, there is a \$75 startup fee, then a \$15 fee every month. CCSS A.CED.2, SMP 2

   a. Write an equation in slope-intercept form modeling this situation. Graph the equation labeling the scales and axes.

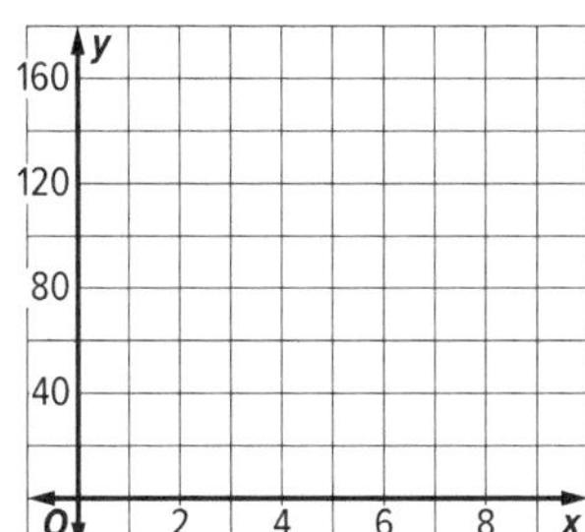

   b. A special promotion states that if you sign up today the startup fee will not be charged. How does the equation change?

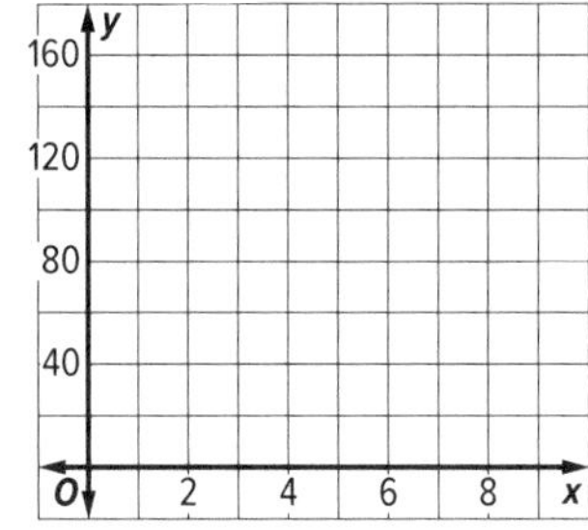

   c. Compare the graphs from **parts a** and **b**.

2.5 Scatter Plots and Lines of Regression

Objectives

- Represent the relationship between two quantities on a scatter plot and describe the key features.
- Write the equation for a line of best fit to represent the linear relationship between quantities on a scatter plot.

Content: F.IF.4
Practices: 2, 3, 4, 5, 7
Use with Lesson 2-5

A scatter plot is a tool for displaying data that has two coordinates. The data is plotted on a Cartesian plane as a collection of points. Scatter plots are useful for visually determining if there is a correlation between the two coordinates. The x and y coordinates have a positive correlation if y generally increases as x increases. The x and y coordinates have a negative correlation if y generally decreases as x increases. Correlations are considered strong or weak depending on how closely the coordinates relate to one another.

KEY CONCEPT

Label each scatter plot with the type of correlation it shows. Choose from strong positive, weak positive, strong negative, weak negative, or no correlation.

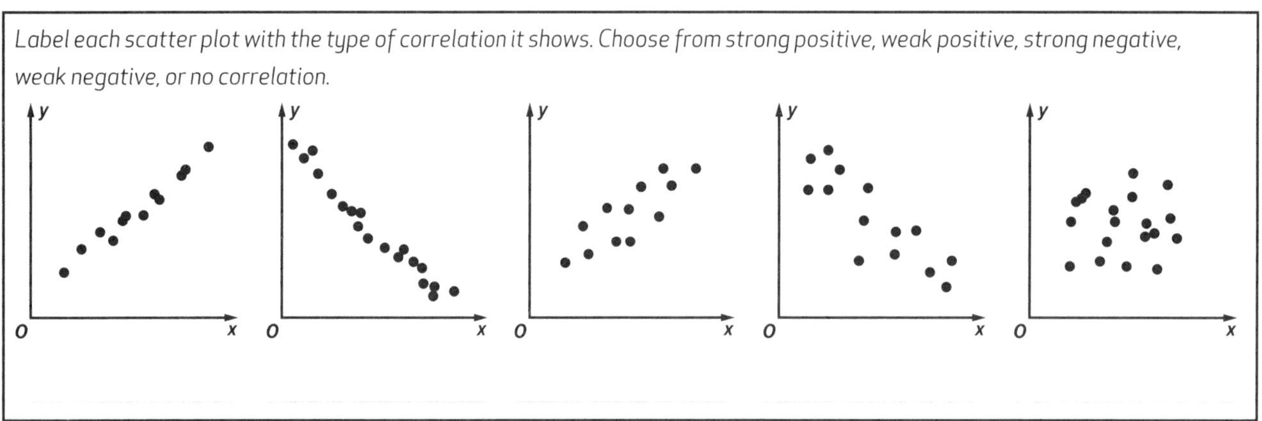

EXAMPLE 1 Use a Scatter Plot to Find Trends in Data CCSS F.IF.4

EXPLORE The table shows the enrollment of students in an SAT prep program over the past several years.

Years since 2007	SAT prep course enrollment
2	2892
3	3042
4	3087
5	3136
6	3296

a. **REASON QUANTITATIVELY** When making a scatter plot, which quantity would you graph along the x-axis and which would you graph along the y-axis? Why? CCSS SMP 2

b. USE A MODEL Plot the data on the graph at right and label your graph. Would you describe the data in your scatter plot as being a *strong positive, weak positive, strong negative, weak negative,* or *no correlation*? Explain your choice. CCSS SMP 4

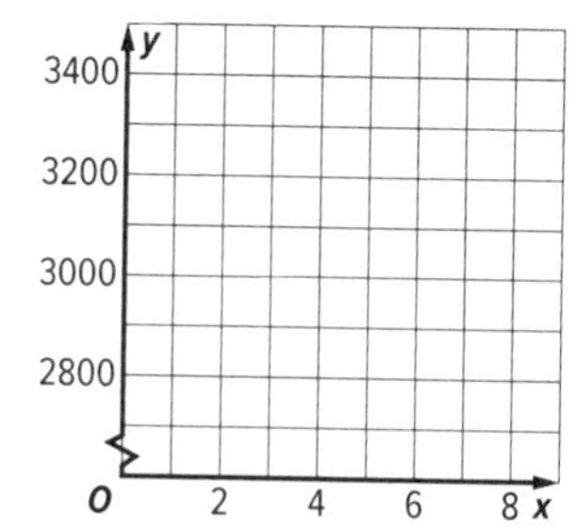

c. CONSTRUCT ARGUMENTS Scatter plots can help us notice trends in data because they provide a visual representation. Noticing these trends can help us predict data points that are not included in the original set. Based on the trend in the data, what would you say will happen with enrollment in another year? CCSS SMP 3

In **Example 1 part c**, we used a scatter plot to observe a trend in the data. Using that trend we were able to make an informal prediction for data that is not in the original set. We can make better predictions if we are more accurate about modeling the trend. To do this, we pick two points from the data set and use them to draw a line that seems to best model the data. This line is known as a **trend line**. The equation of the trend line can be used to find better predictions of data.

EXAMPLE 2 Model Home Prices CCSS F.IF.4

EXPLORE Home prices tend to be more expensive the closer they are to metropolitan areas. The table shows the relationship between the average price of a home and its distance from the city.

Miles from City, x	Price of Home, y
12	\$500,499
20	\$425,999
41	\$359,999
75	\$284,399
94	\$200,499
115	\$189,199

a. USE STRUCTURE Create a scatter plot using the data. Choose two points and use them to find an equation for a trend line. CCSS SMP 7

b. REASON QUANTITATIVELY Explain what your slope and y-intercept mean in the context of this scenario. CCSS SMP 2

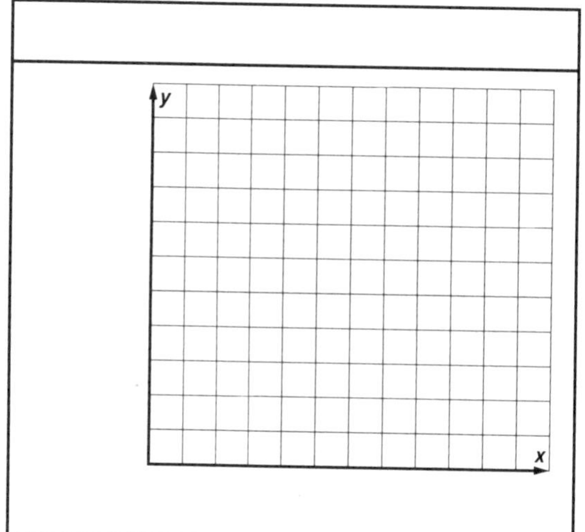

c. USE STRUCTURE What is a reasonable home price for an average home that is 50 miles from the city. How do you know? CCSS SMP 7

d. USE STRUCTURE Is the equation you found in **part a** reasonable for all values of x? Explain. CCSS SMP 7

EXAMPLE 3 Studying for a Final Examination CCSS F.IF.4

EXPLORE A teacher took a survey of some of his students in which he asked how many hours they spent studying for the final examination. He then created a table relating these values to the students' scores on the final. The data is given in the table below.

Hours Studying, x	3	2.5	3.25	4	4.5	5	4
Exam Score, y	67	55	70	81	88	96	79

a. USE TOOLS On your calculator, enter the hours studying (x) into a list. Enter the exam scores (y) into a second list. CCSS SMP 5

b. USE TOOLS Use your calculator to select these two lists and create a scatter plot using "Stat Plot." Sketch the scatter plot in the space below, including the axis labels and scale. CCSS SMP 5

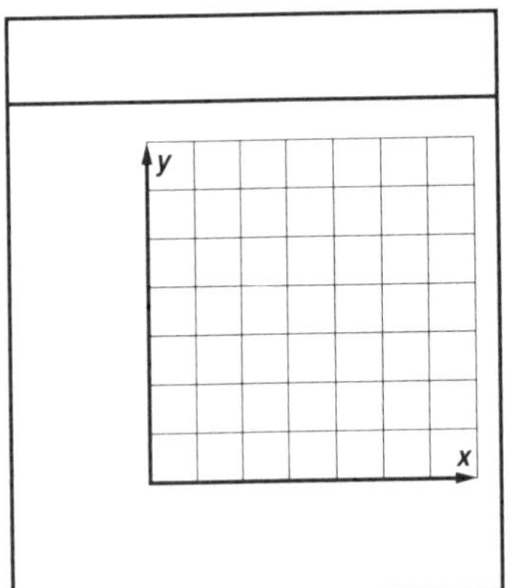

c. USE TOOLS Use the Stat tool on your calculator to find a "linear regression" or a line of best fit. What is the equation? Put this equation into the calculator so that it displays the scatterplot and the line of best fit on the same graph. Sketch the line of best fit on the scatter plot you created in **part b**. CCSS SMP 5

d. USE TOOLS The calculator should have given you a correlation coefficient along with the equation. What is the correlation coefficient, and what does it mean for the relationship between the data and the line of best fit? CCSS SMP 5

e. **REASON ABSTRACTLY** Use the line of best fit to predict the grade of a student who studies for 3.75 hours. How close would you expect this to be in reality? What shortcomings might this prediction suffer from? CCSS SMP 2

EXAMPLE 4 Predict College Enrollment CCSS F.IF.4

A college began tracking its enrollment in 2004. The table at right lists the enrollment from 2004–2010. Jayden and Leah are asked to determine the whole number of years, since 2004, that it will take for enrollment to exceed 25 million. Jayden inputs the data into his graphing calculator exactly as it appears.

College Enrollment	
Year, x	Number of Students, y
2004	17,272,000
2005	17,487,000
2006	17,759,000
2007	18,248,000
2008	19,103,000
2009	20,428,000
2010	21,016,000

Source: National Center for Education Statistics

a. **USE TOOLS** Use a graphing calculator to find the line of regression Jayden will use to solve the problem. Round values to the nearest hundredth. CCSS SMP 5

b. **USE STRUCTURE** Leah adjusts the values in the table before entering them into her calculator because she thinks it will simplify her work. Complete the table at right with the values Leah will enter into her calculator. CCSS SMP 7

College Enrollment	
Years since 2004, x	Number of Students, y, in thousands

c. **USE TOOLS** Use a graphing calculator to find the line of regression Leah will use to solve the problem. Round values to the nearest hundredth. CCSS SMP 5

d. **REASON QUANTITATIVELY** Explain how to use the equations for the line of regressions found in **parts a** and **c** to determine how many years since 2004 it will take for college enrollment to exceed 25 million students. CCSS SMP 2

PRACTICE

1. a. **CRITIQUE REASONING** Isaac made the scatter plot at right to show how his weight changed as he got older. Do you agree or disagree with the way he set up this scatter plot in order to represent this data? Explain your position. CCSS F.IF.4, SMP 3

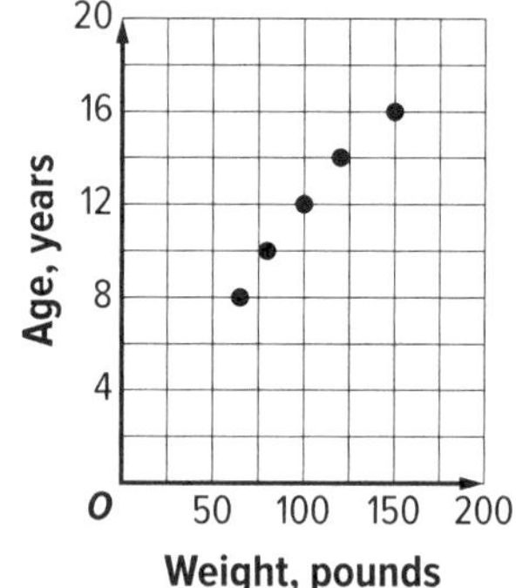

   b. **USE A MODEL** Would the equation for the line of regression for Isaac's weight data give a reasonable value for his weight at the age of 40? Explain. CCSS F.IF.4, SMP 4

2. **USE A MODEL** Consider the scatter plot relating the year to the tons of recycling in that year. CCSS F.IF.4, SMP 4

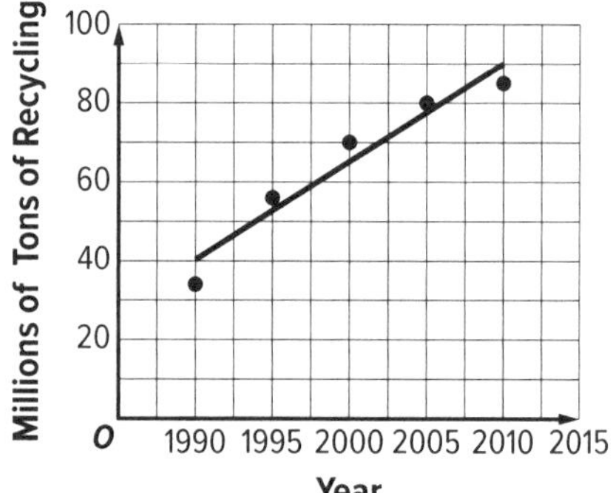

   a. Use algebra to find an equation of a line of fit on the scatter plot of recycling data.

   b. Use your equation to determine how much waste is expected to be recycled in 2015.

3. **USE TOOLS** The time it takes Jonathan to drive to school changes depending on when he leaves. The table below shows the data he collected during a week. CCSS F.IF.4, SMP 5

Number of minutes past 7:00 a.m. that Jonathan leaves, x	10	18	12	0	4
Drive time in minutes, y	21	25	21	16	18

   a. Use your calculator to create a scatter plot and a line of best fit on the same coordinate plane. Sketch the scatter plot in the space below, including the axis labels and scale. What is the equation of the line? What is the correlation coefficient?

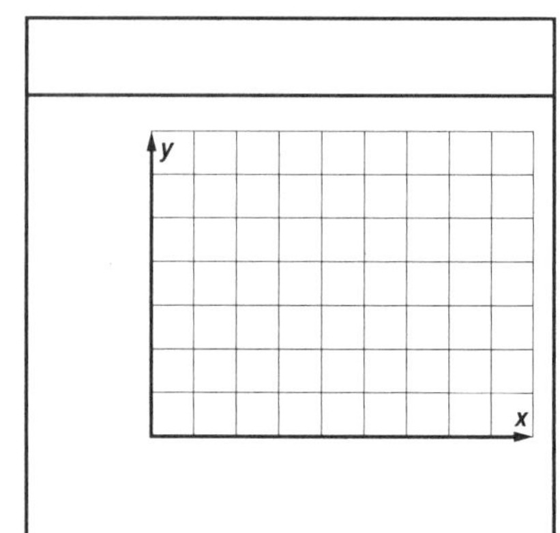

   b. Why does it make sense that the drive may take longer if Jonathan leaves later?

4. **USE A MODEL** The table shows the number of visitors to a ski park and the temperature it was outside on that day. CCSS F.IF.4

Ski Park Attendance	
Temperature, degrees F	Number of Visitors
20	600
25	620
30	650
35	585
40	502
45	565
50	380
55	362
60	403

a. **USE TOOLS** Use your graphing calculator to make a scatter plot of the data. Sketch the scatter plot in the space below, including the axis labels and scale. Describe the scatter plot as having *strong positive, weak positive, strong negative, weak negative,* or *no correlation* and explain your choice. CCSS SMP 5

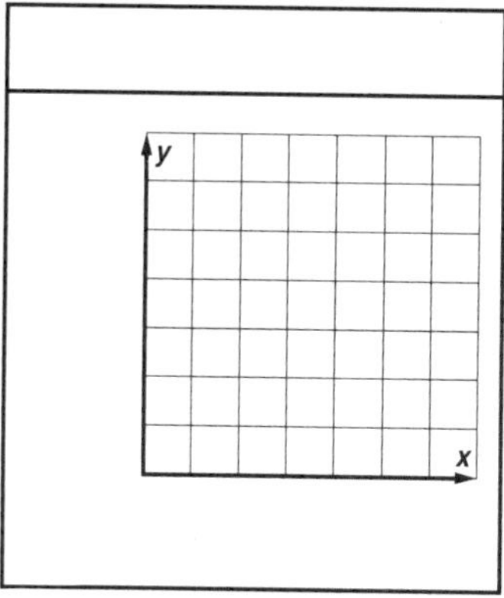

b. **USE TOOLS** Use your graphing calculator to find the line of regression for this data. Write the equation below. CCSS SMP 5

c. **REASON QUANTITATIVELY** What do the slope and y-intercept tell you about this data? CCSS SMP 2

d. **REASON QUANTITATIVELY** Does the equation for the line of regression return reasonable predictions for all values of x in the domain of the equation? Explain. CCSS SMP 2

e. **REASON QUANTITATIVELY** How many visitors should the park staff expect on 32-degree days? How do you know? CCSS SMP 2

2.6 Special Functions

Objectives

- Formulate equations of piecewise-defined functions from verbal descriptions and graphs provided within an applied context.
- Relate the domain of a piecewise-defined function to the quantitative relationship it describes.
- Graph piecewise-defined functions, showing their key features.

Content: F.IF.4, F.IF.5, F.IF.7b
Practices: 1, 2, 4, 5, 7
Use with Lesson 2-6

A function that is composed of two or more expressions is called a **piecewise-defined function**. On the graph of such a function, a dot indicates that a point is included in the graph, while a circle indicates that a point is not included in the graph. These functions are often defined using several linear functions.

A piecewise-defined function is written using a curly brace followed by the function definitions for each of the pieces and then the part of the domain on which that function definition applies. An example of a piecewise-defined function is shown below.

$$f(x) = \begin{cases} x & \text{if } x \geq 0 \\ -x & \text{if } x < 0 \end{cases}$$

EXAMPLE 1 Modeling Using Piecewise-Defined Functions CCSS F.IF.4

EXPLORE For a family reunion, the Cramers have reserved a banquet hall and intend to hire a caterer. The cost to reserve the hall is a flat fee of \$500. The catering cost per guest is \$17.50 for the first 40 guests and \$14.75 per guest beyond the first 40.

a. **REASON ABSTRACTLY** Write two expressions: one for the cost of the event if the number of guests is no more than 40, and one for the cost of the event if the number of guests exceeds 40. Explain your reasoning for both. Then, write a simplified piecewise-defined function describing the cost of the reunion. CCSS SMP 2

b. REASON QUANTITATIVELY What is a suitable domain for the function in this context? Explain. CCSS F.IF.5, SMP 2

c. USE A MODEL If the Cramers can spend up to $900 on the event, what is the largest number of guests that can attend? CCSS SMP 4

A step function is a common piecewise-defined function with a graph comprised of line segments. One specific step function is the **greatest integer function**, $f(x) = [\![x]\!]$. This symbol means that $f(x)$ is equal to the greatest integer less than or equal to x.

EXAMPLE 2 Modeling with Step Functions CCSS F.IF.4

EXPLORE An online gaming company charges players a monthly fee based on the amount of time per day they spend playing online games. The following graph shows the monthly charge based on the average number of hours spent online per day.

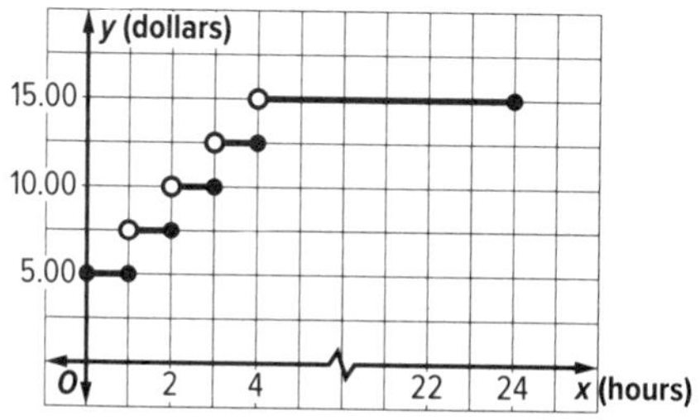

y (dollars)
15.00
10.00
5.00
O
2
4
22
24
x (hours)

a. REASON ABSTRACTLY The function graphed is a step function. Write a function to represent the graph. CCSS SMP 2

A competitor charges $1.40 per month for gamers who average less than 30 minutes of daily game play. Then they charge an additional $1.40 for any part of the next 30 minutes up to but less than 1 hour of total game play. Similarly, they continue to add $1.40 for each additional 30-minute interval up to 5 hours of daily game play, after which they charge a maximum fee of $15.40.

b. REASON ABSTRACTLY Formulate a greatest integer function describing the monthly charges. CCSS SMP 2

c. INTERPRET PROBLEMS Sketch a graph of this function and identify its domain and range. CCSS F.IF.7b, F.IF.5, SMP 1

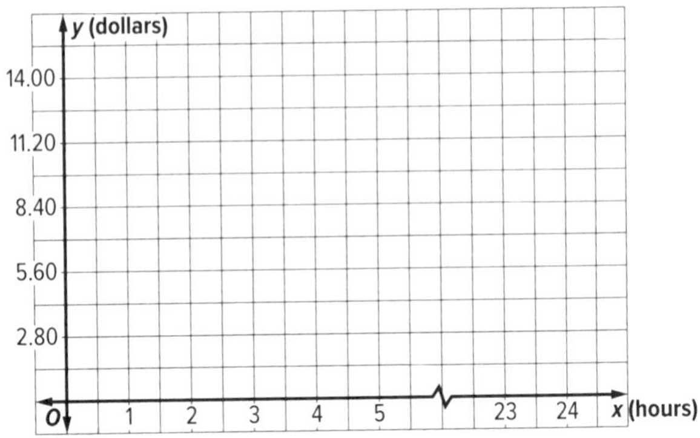

d. USE A MODEL Which company offers the better deal? Explain. CCSS SMP 4

Another common piecewise-linear function is the absolute value function, $f(x) = |x|$.

KEY CONCEPT Parent Function of Absolute Value Functions

Parent Function: $f(x) = \|x\| = \begin{cases} & \text{if } x < 0 \\ & \text{if } x = 0 \\ & \text{if } x > 0 \end{cases}$	Domain: Range: Intercepts: Not defined:

Note that $-x$ is not a negative number. In this first piece of the function, x itself is negative, so $-x$ is positive. This matches what we know about $|x|$ never being negative.

EXAMPLE 3 Modeling Using the Absolute Value Function 

EXPLORE A snow storm begins at 7 A.M. as light flurries, then gradually changes to blizzard-like conditions, and then tapers gradually as flurries until it is over. The rate at which it snows is described by the function $s(t) = -0.75|t - 3| + 2.5$, where t is the number of hours it has been snowing since 7 A.M. and $s(t)$ is measured in inches per hour.

a. **PLAN A SOLUTION** Graph the function. What is a suitable domain for the function in this context? Explain. CCSS F.IF.4, F.IF.5, SMP 1, SMP 2

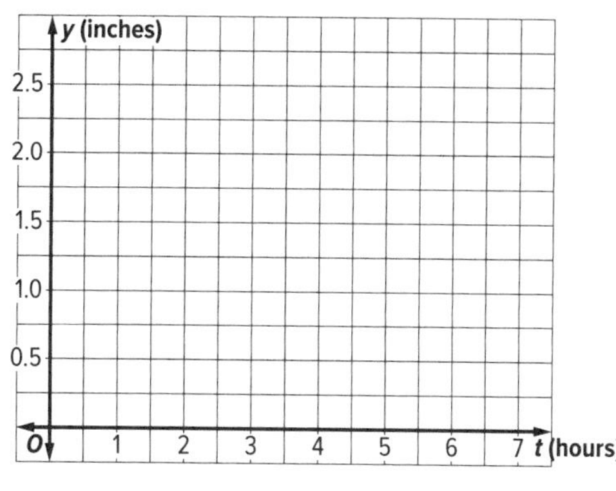

b. **USE A MODEL** At what time(s) is the snow falling the fastest? CCSS SMP 4

c. **USE A MODEL** At what time(s), if any, is the snow falling at a rate of 1 inch per hour? CCSS SMP 4

d. **USE A MODEL** At what times is the snow falling at a rate of at least 1.75 inches per hour? CCSS SMP 4

PRACTICE

1. The approval rating $R(t)$, measured as a percent, of a class officer during her 9-month term starting in September is described by the following graph, where t is the time in office.

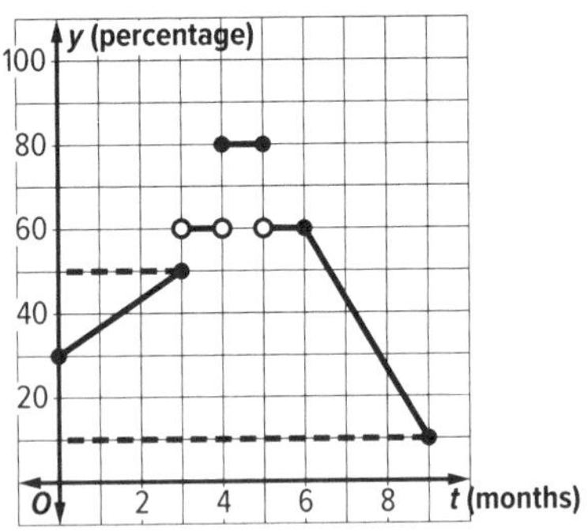

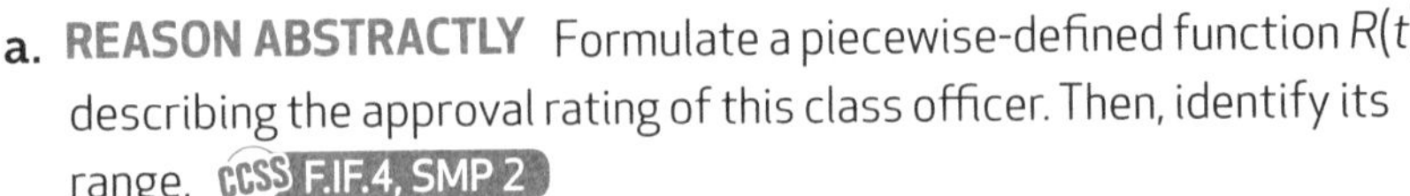

a. **REASON ABSTRACTLY** Formulate a piecewise-defined function $R(t)$ describing the approval rating of this class officer. Then, identify its range. CCSS F.IF.4, SMP 2

b. **REASON QUANTITATIVELY** During which months is the approval rating increasing? CCSS F.IF.4, SMP 2

2. Consider the functions $f(x) = 3[\![x]\!]$ and $g(x) = [\![3x]\!]$ for $0 \leq x \leq 2$.

a. **PLAN A SOLUTION** Graph each function. CCSS F.IF.7b, SMP 1

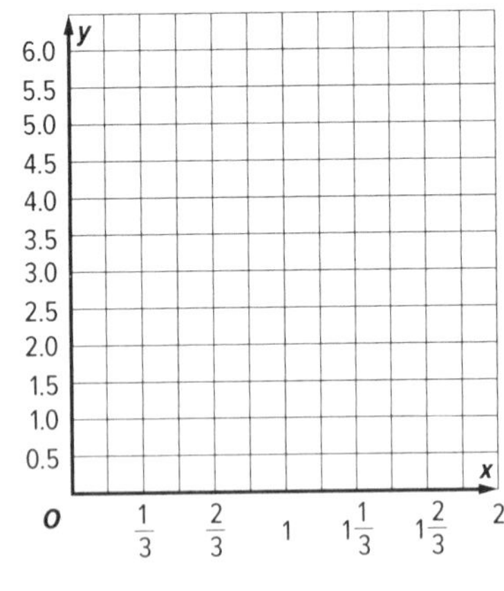

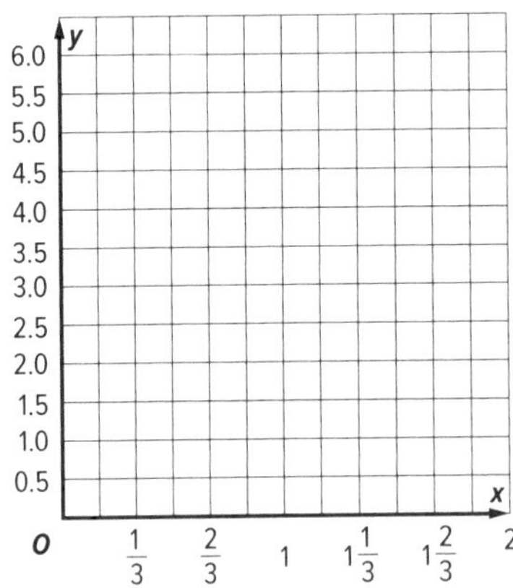

b. **REASON ABSTRACTLY** What effect does this 3 appear to have on the graphs? CCSS F.IF.4, SMP 2

c. PLAN A SOLUTION Consider the functions $f(x) = 4[\![x]\!]$ and $g(x) = [\![4x]\!]$ for $0 \leq x \leq 2$. Graph each function. CCSS F.IF.7b, SMP 1

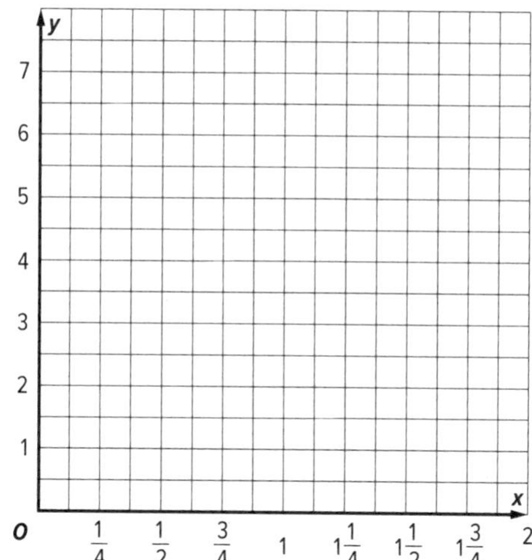

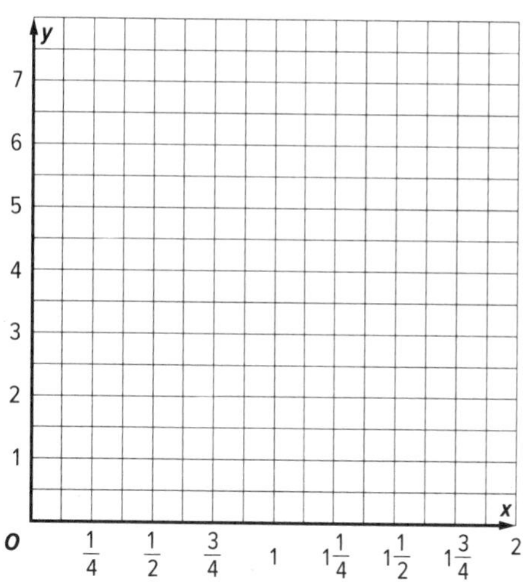

d. REASON ABSTRACTLY What effect does this 4 appear to have on the graphs? CCSS F.IF.4, SMP 2

e. FIND A PATTERN Generalize your findings from **parts a** through **d** to explain the differences between $f(x) = n[\![x]\!]$ and $g(x) = [\![nx]\!]$ for $0 \leq x \leq 2$, where n is any positive integer greater than or equal to 2. CCSS F.IF.4, SMP 8

3. USE TOOLS Use a graphing calculator to graph "the absolute value of the greatest integer of x", or $f(x) = |[\![x]\!]|$. Explain why the graph makes sense. CCSS F.IF.4, SMP 5

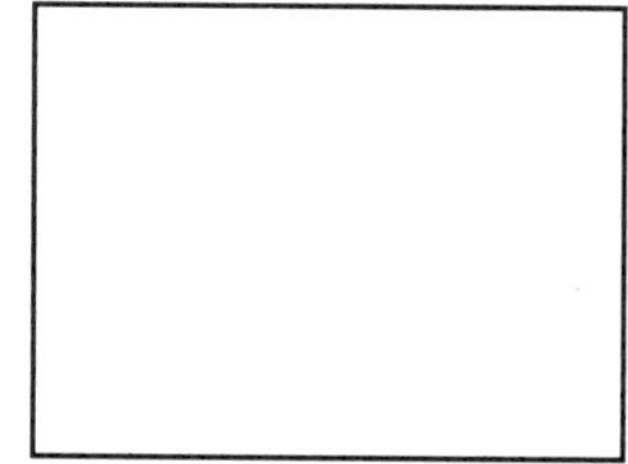

2.7 Transforming Functions

Objectives

- Identify key features of a family of graphs.
- Recognize and use parent functions.
- Describe the effects transformations have on functions.

CCSS STANDARDS

Content: F.BF.3, F.IF.4, F.IF.9
Practices: 1, 3, 4, 6, 7, 8
Use with Lesson 2-7

A **family of graphs** is a group of graphs that displays one or more similar characteristics. The **parent graph** is the simplest of graphs in a family and is obtained from the **parent function,** the most general function. This function can be transformed to create many other functions of similar characteristics.

KEY CONCEPT Parent Functions

Constant Function

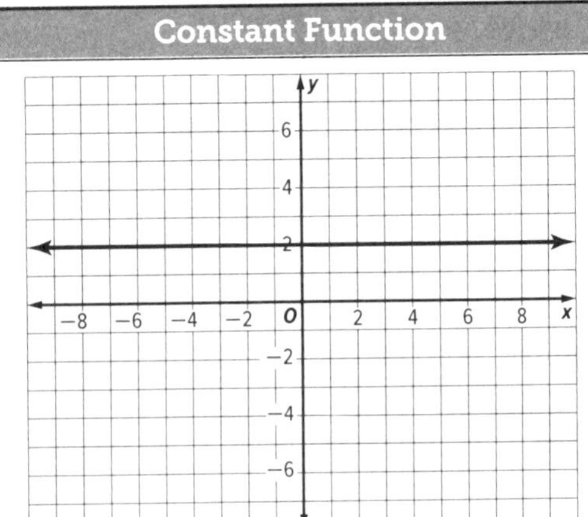

The parent function of the constant function is $f(x) = a$, where a is a real number. The domain is ____________, and the range is the number ______.

Identity Function

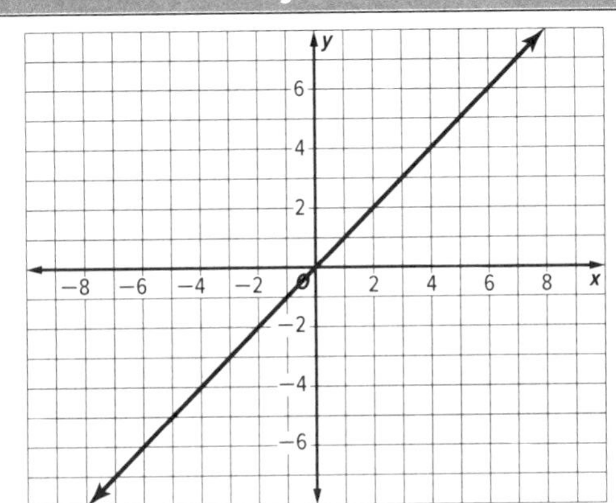

The identity function, $f(x) = x$, is the parent function of most linear functions. It passes through points with coordinates of the form (a, a). The ____________ of this function are all real numbers.

Absolute Value Function

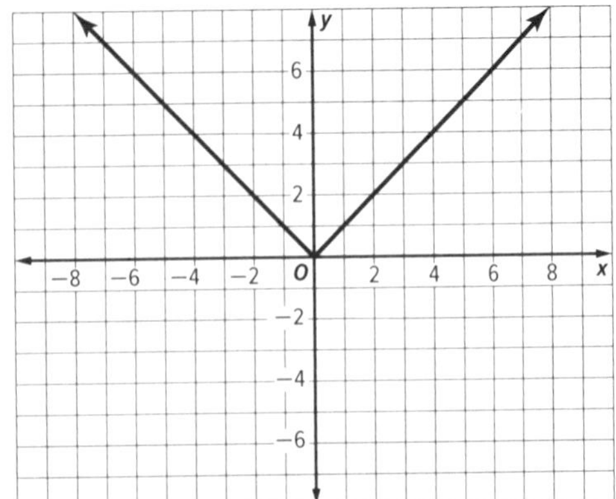

The parent function of absolute value functions is $f(x) = |x|$. It has a ________ of all real numbers and a ________ of all real numbers greater than or equal to 0.

Quadratic Function

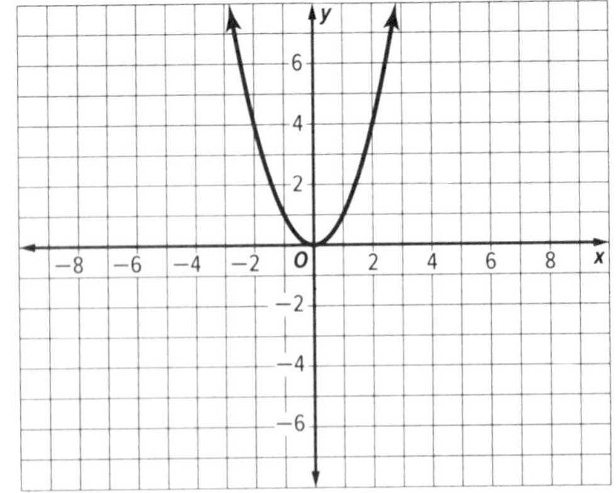

Quadratic functions are derived from the parent function $f(x) = x^2$. The domain is the set of ____________, and the range is the set of ____________ ____________.

EXAMPLE 1 Investigating Translations CCSS F.BF.3, F.IF.9

EXPLORE For parts a–d, refer to the tables and graphs below. (Note: *h* and *k* are real numbers.)

Table A: $f(x) + k$

x	$f(x) = x^2$	$f(x) = x^2 + 5$
−2	4	9
−1	1	6
0	0	5
1	1	6
2	4	9

Table B: $f(x) - k$

x	$f(x) = x^2$	$f(x) = x^2 - 2$
−2	4	2
−1	1	−1
0	0	−2
1	1	−1
2	4	2

Graph C: $f(x - h)$

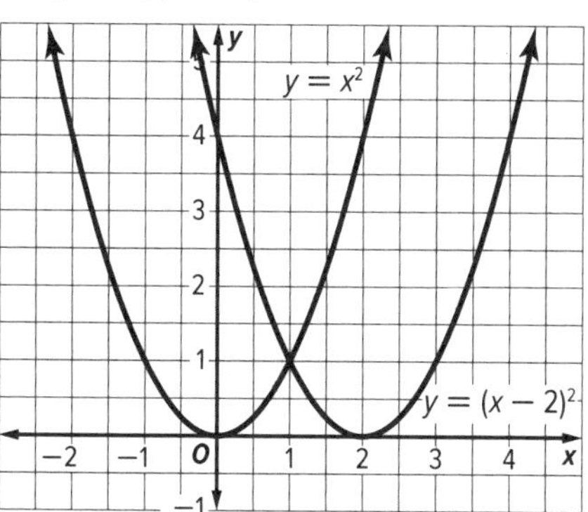

Graph D: $f(x + h)$

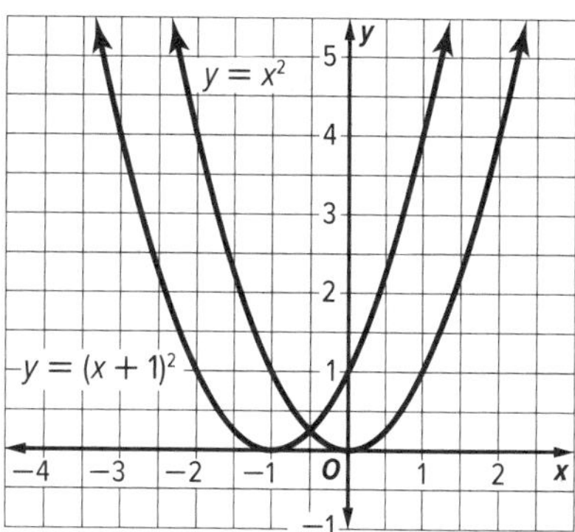

a. **USE STRUCTURE** Graph the parent function and transformation for the data in each table. Use arrows to show the direction in which the graphs move from the original position of the parent graph. CCSS SMP 7

Graph A:

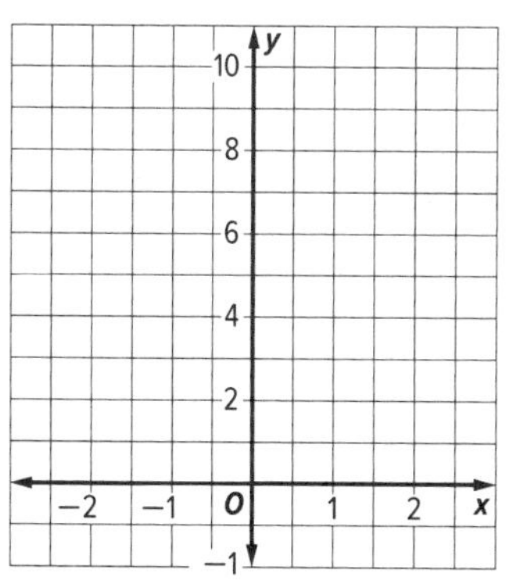

Graph B:

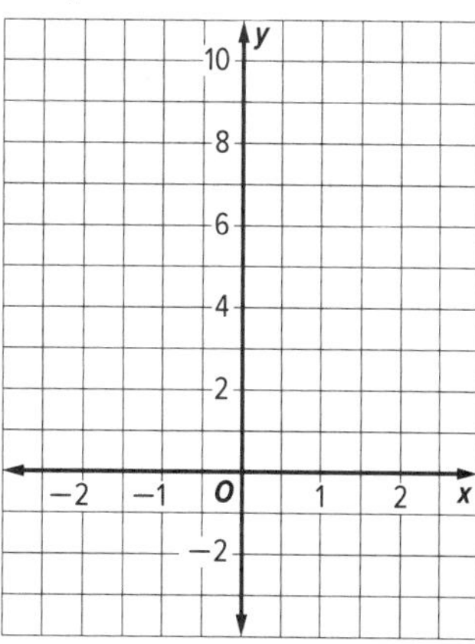

b. **EVALUATE REASONABLENESS** List two additional examples of a quadratic function that cause the parent function to move vertically up or down. Check using your graphing calculator. CCSS SMP 8

c. **INTERPRET PROBLEMS** What effect does adding or subtracting some value of k have on the graph of the parent function? CCSS SMP 1

d. **INTERPRET PROBLEMS** Examine graphs C and D. Compare the effects the transformations had on the parent function, citing specific examples of changes in coordinates. CCSS SMP 1

EXAMPLE 2 Dilations and Reflections CCSS F.BF.3

a. **CALCULATE ACCURATELY** Graph $f(x) = 4|x|$, $g(x) = 1.5|x|$, $v(x) = \frac{1}{4}|x|$, and $w(x) = -0.8|x|$ along with the parent graph of the absolute value function. CCSS SMP 6

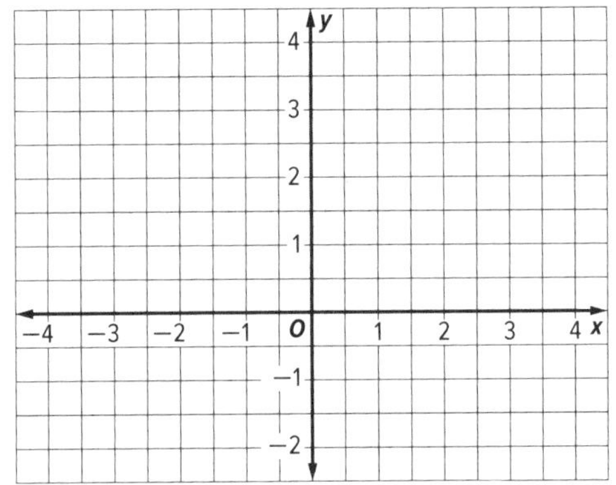

b. **MAKE A CONJECTURE** Based on the graph, what assumption can you make about the effect of a when the function has the form $af(x)$? CCSS SMP 3

c. **EVALUATE REASONABLENESS** Using your graphing calculator, graph the following functions and complete the table. CCSS SMP 8

Function	Effect
$f(x) = -\lvert x\rvert$	
$f(x) = \lvert 2x\rvert$	
$f(x) = \lvert \frac{1}{2}x\rvert$	
$f(x) = -x^2$	
$f(x) = (3x)^2$	
$f(x) = 0.75x^2$	
$f(x) = bg(x)$	
$f(x) = g(bx)$	

EXAMPLE 3 Modeling with Transformations CCSS F.IF.4, F.BF.3

Kassie is remodeling her kitchen with new square tiles and wants to know the maximum size of tile she can use to cover the floor.

a. **USE A MODEL** What type of function should she use to represent the size of her kitchen floor? Why? CCSS SMP 4

b. **USE STRUCTURE** Kassie uses the function, $f(x) = -1.25(x - 1)^2 + 18.75$ (x is the size of the tile in square feet), to represent the area of her kitchen floor. Describe the transformations she applied to the parent function in creating her function. CCSS SMP 7

PRACTICE

1. **COMMUNICATE PRECISELY** Graph the following function, $f(x) = -|3x + 5| - 1$. Describe the transformation of the parent function $g(x) = |x|$ that produces the graph of $f(x)$. What are the domain and range? CCSS F.BF.3, F.IF.4, F.IF.5, SMP 6

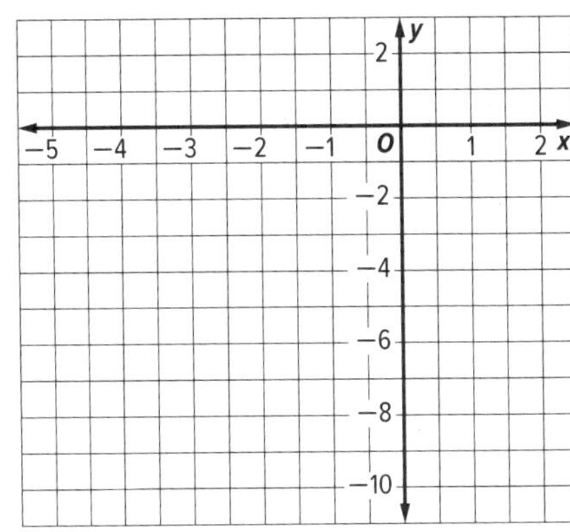

2. **CONSTRUCT ARGUMENTS** In transformations, there can also be a reflection in the y-axis, denoted by $f(-x)$. Apply this rule to the parent graphs of the absolute value and quadratic functions. What do you notice? Explain. CCSS F.BF.3, SMP 3

3. **COMMUNICATE PRECISELY** What determines whether a transformation will affect the graph vertically or horizontally? Use the family of quadratic functions as an example. CCSS F.BF.3, SMP 6

2.8 Graphing Linear and Absolute Value Inequalities

Objectives

- Graph linear and absolute value inequalities.
- Model real-world situations using linear and absolute value inequalities.
- Interpret solutions as viable or not.

CCSS STANDARDS
Content: A.CED.3
Practices: 1, 2, 4
Use with Lesson 2-8

A **linear inequality** is an inequality that describes a half-plane with a **boundary** that is a straight line. The boundary is defined by the inequality's related equation and separates the coordinate plane into two regions.

EXAMPLE 1 Graph a Linear Inequality CCSS A.CED.3

EXPLORE **Mary sells giant chocolate chip and peanut butter cookies for \$1.25 and \$1.00, respectively, to her classmates. She wants to make at least \$25 a day.**

a. **REASON ABSTRACTLY** Write a mathematical expression to represent the number of cookies Mary needs to sell each day. Describe how you would graph the inequality and then graph it. Identify any additional constraints on the situation. CCSS SMP 1, SMP 2

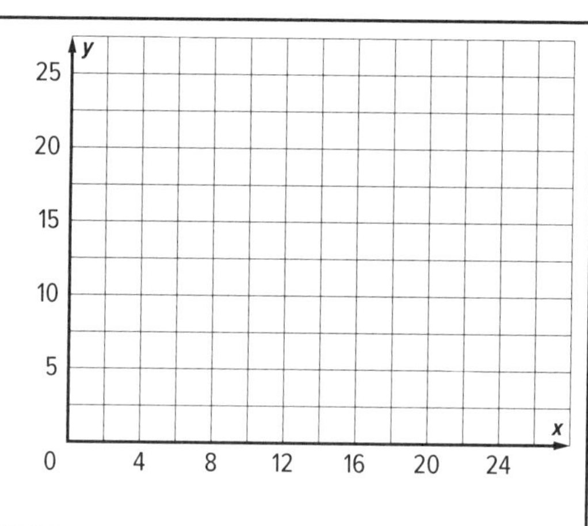

b. **USE A MODEL** If Mary decides to charge \$1.50 for chocolate chip cookies rather than \$1.25, what impact will this have on the graph of the solution set? Give an (x, y) pair that is not in the original solution set, but is in the solution set of the new revised scenario. CCSS SMP 4

c. **USE A MODEL** How does the graph of the inequality change if Mary wants to make at least \$50 each day? How does the graph of the inequality change if Mary wants to make no more than \$25 a day? CCSS SMP 4

EXAMPLE 2 **Absolute Value Inequalities** CCSS A.CED.3

a. **USE A MODEL** A manufacturer of maple syrup sells a product that contains 24 fluid ounces. However, the actual fluid in the bottle varies. The plant currently allows for a maximum difference of 0.5 fluid ounces. If x is the actual number of fluid ounces in the bottle, write an inequality to model the allowable amounts for a 24 ounce bottle. CCSS SMP 4

b. **USE A MODEL** The manager of the plant wants to investigate the changes in the allowable amounts for a 24 ounce bottle of maple syrup that result from increasing or decreasing the maximum allowable difference between the advertised and actual amount. If y represents the maximum allowable difference, then what inequality represents the resultant allowable amounts? Graph the inequality and explain your process. CCSS SMP 4

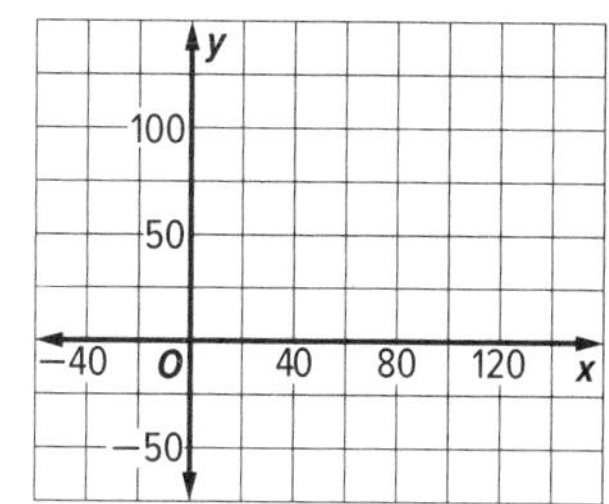

c. **USE A MODEL** The manager now wants to explore the effect of a 0.5 fluid ounce error on different size bottles. Suppose that x still represents the actual volume of syrup in the bottle, but this time y represents the advertised amount on the bottle. If the maximum allowed error is 0.5 fluid ounce, write and graph an inequality to model this situation. Explain your reasoning. CCSS SMP 4

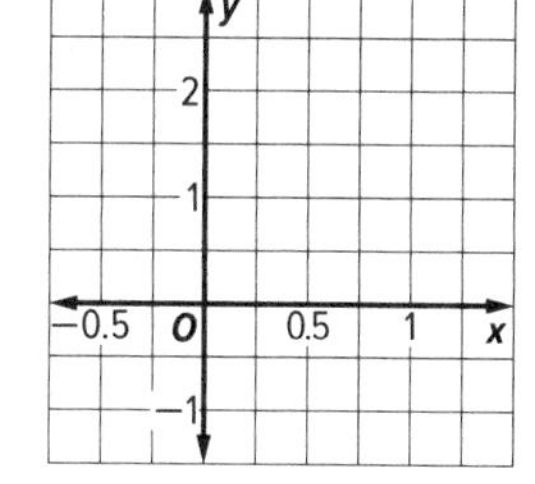

PRACTICE

Use the following scenario to answer Exercises 1–2.

The Math Club at Joe's school is having a fundraiser to earn money for its scholarship fund. The members are selling T-shirts for \$10.50 and sweatshirts for \$20.75. The club needs to raise at least \$2000. CCSS A.CED.3

1. **REASON ABSTRACTLY** Write an inequality describing the number of T-shirts and sweatshirts that need to be sold in order to fund the scholarships. Define the variables and identify any additional constraints on the situation. Graph the inequality. CCSS SMP 2

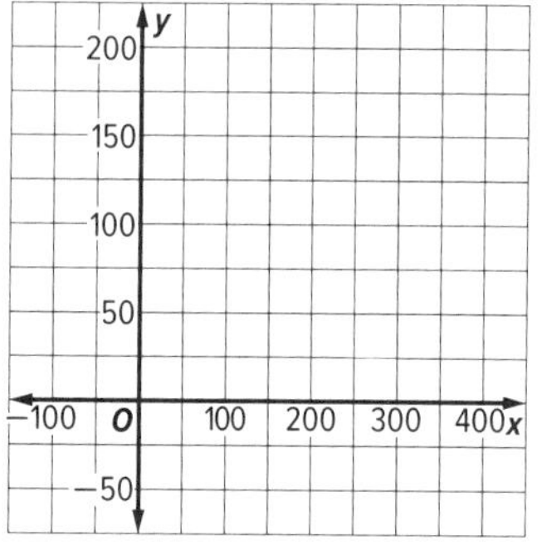

2. a. **REASON ABSTRACTLY** Use your graph to determine if the point (100, 100) is viable or non-viable. Confirm your answer algebraically. CCSS SMP 2

   b. **REASON ABSTRACTLY** Discuss what would happen to the graph if the price of the sweatshirts were raised to \$22.50. How does this affect the required sales? CCSS SMP 2

3. a. **USE A MODEL** A candy company advertises a bag of small candy as containing 1000 pieces. However, because machines are filling the bags, there is always a margin of error one way or another. If x represents the *actual* number of pieces in the bag, and y represents the maximum difference between the count and the advertised 1000 pieces, write and graph an inequality to model this situation. CCSS A.CED.3, SMP 4

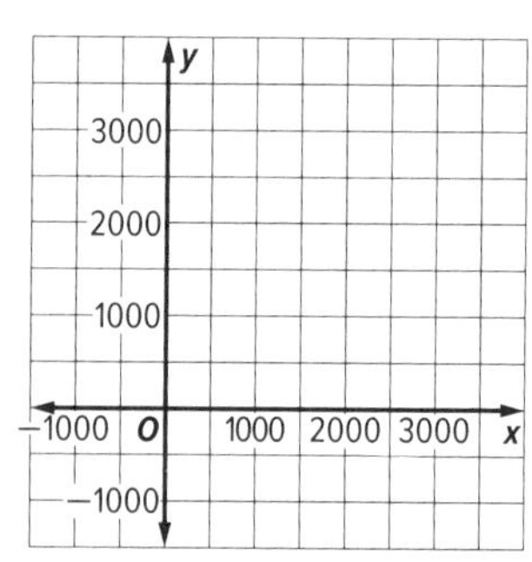

b. **USE A MODEL** Use your graph to determine if (1005, 2) is a solution. Interpret this in the context of the problem. CCSS A.CED.3, SMP 4

c. **REASON ABSTRACTLY** What effect does changing the 1000 to 1500 have on the graph? CCSS A.CED.3, SMP 2

4. a. **USE A MODEL** A nut company is selling bags that contain both peanuts and cashews. They advertise the bag as containing 1 pound of nuts. The maximum allowed error from the advertised weight is 0.1 pounds. Define variables, and write an inequality to model this situation. CCSS A.CED.3, SMP 4

b. **USE A MODEL** Graph the inequality and explain your reasoning. CCSS A.CED.3, SMP 4

y, 2, 1, −1, −0.5, O, 0.5, 1, x, −1

c. **USE A MODEL** Determine if the solution (0.5, 0.65) is a solution. Interpret this in terms of the situation. CCSS A.CED.3, SMP 4

d. **REASON ABSTRACTLY** What impact does changing the allowed error from 0.1 to 0.2 have on the graph? Would the point (0.5, 0.65) from **part c** be viable or not viable? CCSS A.CED.3, SMP 2

Performance Task

Filling a Pool

Provide a clear solution to the problem. Be sure to show all of your work, include all relevant drawings, and justify your answers.

You work at a local swimming pool, and it is time to fill the pool for summer. The shallow end of the pool is 2 feet deep; the pool gradually gets deeper until it is 6 feet deep. Then it drops to 12 feet deep over a 10-ft stretch. Then it is 12 feet deep for the final 20-ft stretch.

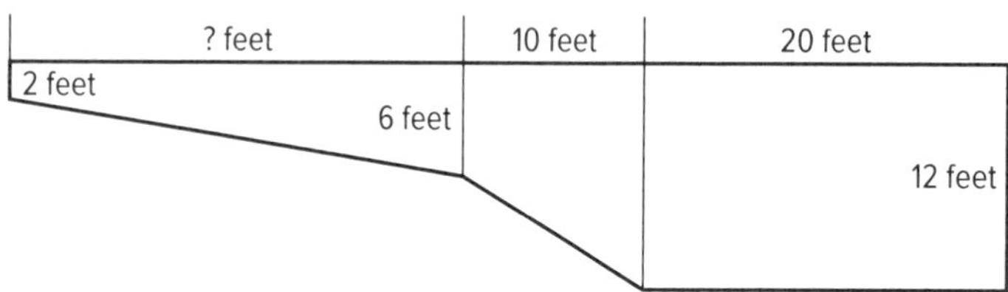

Part A

The depth of the first section of the pool increases at 0.48 in./ft. Find the missing dimension in the diagram. Then determine the volume of the pool if it has a uniform width of 50 feet.

Part B

The pool has jets scattered about its perimeter walls. One jet is able to fill the pool in 146 hours. Alternately, the local fire department will fill the pool if they are available. If the fire department can fill the pool in 5 hours, write an inequality for the number of jets x that need to be turned on while the fire department is filling the pool in order to fill it in less than 4 hours. Solve and explain what the solution means.

Part C

During normal use, water leaves the pool at a rate of about 100 ft^3 every hour due to splashing, evaporation, and filtration.

i) Write an equation for the volume of water in the pool given this loss, t hours after the pool is filled. Assume no additional water is added to the pool.

ii) What percentage of the flow rate needs to be used to constantly replenish this loss?

Performance Task

Weather Prediction

Provide a clear solution to the problem. Be sure to show all of your work, include all relevant drawings, and justify your answers.

A barometer is a device that uses liquid mercury to measure atmospheric air pressure. The height of the mercury column in the barometer gives information about weather conditions. For this task, refer to the following table of barometer readings during a given day.

Time	12:00 P.M.	1:00 P.M.	2:00 P.M.	3:00 P.M.	4:00 P.M.	5:00 P.M.	6:00 P.M.
Barometer Reading (inches of mercury)	33.5	30	31.5	31.5	30	29	28

Part A

Use the data to write an equation to predict the barometer reading at any time during the day. What is a good domain for this equation? What does your equation predict the barometer reading will be at midnight? Is it reasonable to use this equation to predict barometric readings for the next several days? Explain.

Part B

Plot the data from the table on the coordinate grid. Then draw a line of best fit and estimate the rate of change. Compare with the rate of change from **Part A**, and explain the difference.

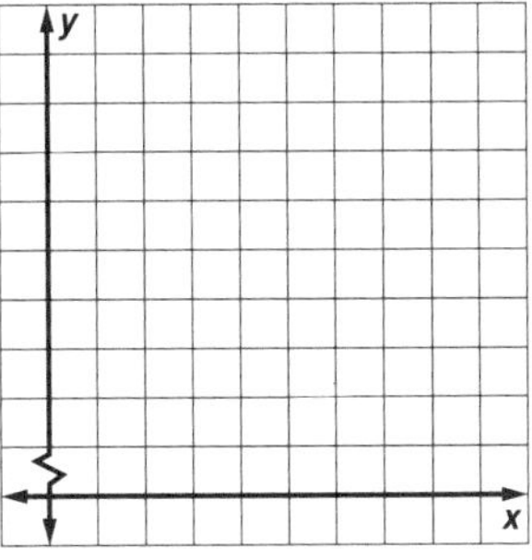

Part C

Average atmospheric pressure at sea level is about 30 inches of mercury. For every foot above sea level, air pressure decreases by about 0.0011 inch of mercury. How far above sea level do you need to be for the air pressure to be 15 inches of mercury?

Standardized Test Practice

1. Samantha and Araina are saving money for college. Both girls started their savings at the same time with an initial deposit. Each girl saves a set amount each month. The graph below shows Samantha's savings. Araina's savings are shown in the table. CCSS F.IF.4, F.IF.9

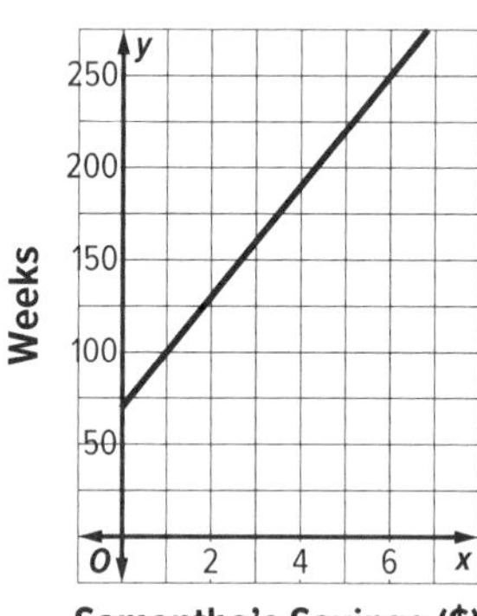

Araina's Savings			
Month	2	5	7
Total Savings ($)	130	250	330

Use the information given above to complete the table.

	Initial Balance ($)	Monthly Savings ($)
Samantha		
Araina		

2. Which of the following best describes the difference between the graphs of $f(x) = -7x + 2$ and $g(x) = -7x - 2$? CCSS F.BF.3

$g(x)$ is a reflection of $f(x)$ in the x-axis

$g(x)$ is translated 4 units left from $f(x)$

$g(x)$ is translated 4 units down from $f(x)$

$g(x)$ is a vertical stretch of $f(x)$ by a factor of 2

3. Consider the expression $-\frac{3}{5}x + \frac{7}{8}$.

The expression is the ☐ of ☐ terms, which are ☐ and ☐. The first term has ☐ factors. The coefficient of the first term is ☐.

CCSS A.SSE.1a, A.SSE.1b

4. Graph the function $y = \left|\frac{1}{2}x + 2\right| - 1$. CCSS F.IF.7b

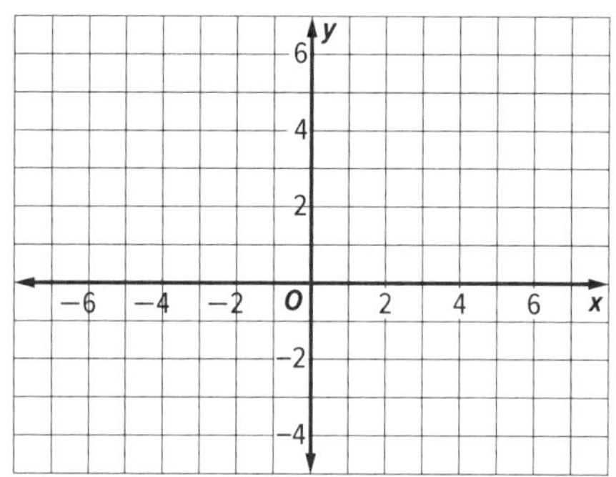

5. Consider the following scatter plot. CCSS F.IF.4

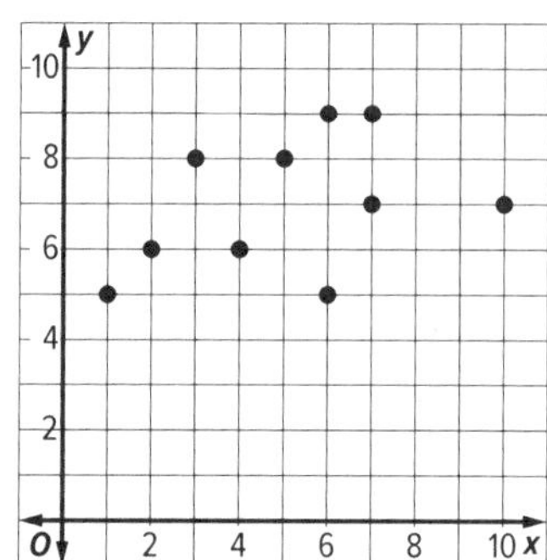

Select the y-intercept and slope that best describe a line of fit for the data.

***y*-intercept**	3	5	7
slope	$\frac{1}{4}$	$\frac{2}{5}$	$\frac{2}{3}$

6. The slope-intercept form of the equation of the line parallel to $y = 3x - 1$ and containing the point $(3, -2)$ is ☐.

The slope-intercept form of the equation of the line perpendicular to $y = 3x - 1$ and containing the point $(3, -2)$ is ☐. CCSS A.CED.2

7. Nathan subscribes to a music download service. He pays a monthly membership fee and a set amount per download. He can spend at most \$10 per month on the music download service. The table below shows the number of songs and total cost for two months.

Month	Number of Downloads	Total Cost (\$)
February	7	5.72
May	11	7.28

Let $f(x)$ represent the total cost for a month, where x is the number of downloads that month.

a. What is the rate of change of $f(x)$? Show your work. Explain what the rate of change represents. CCSS F.IF.6

b. What is a reasonable domain for this situation? Explain your reasoning. CCSS F.IF.5

8. Alonso makes jewelry boxes to sell. It takes Alonso 2 hours to buy materials and set up his workshop, and 3 hours to build each jewelry box. Let $g(x)$ represent the total number of hours needed to build x jewelry boxes. CCSS A.CED.2

a. What is the equation for $g(x)$?

b. Graph $g(x)$.

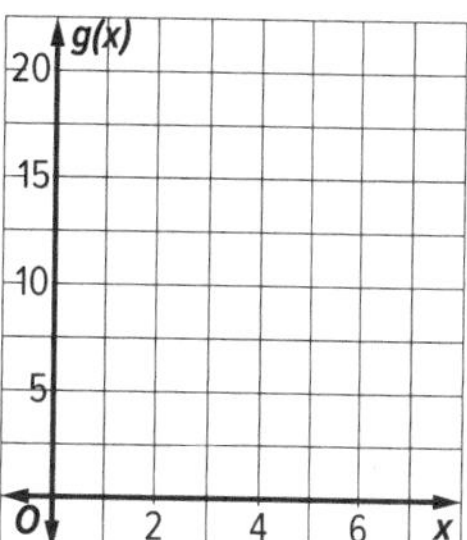

9. Graph the solution set for $|2x - 1| > 7$. CCSS A.CED.3

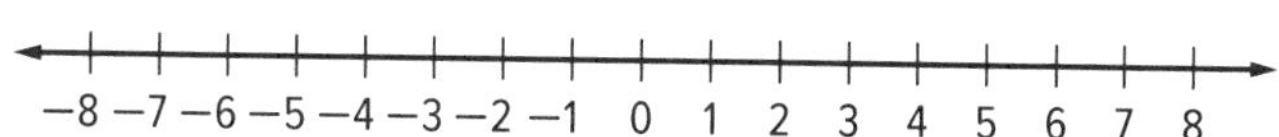

10. If $f(x) = |x|$, write the equation for $g(x)$ if the graph of $g(x)$ is related to the graph of $f(x)$ as follows: translated left 1 unit, translated up 3 units, stretched vertically by a factor of 2. Graph $f(x)$ and $g(x)$. CCSS F.IF.7b, F.BF.3

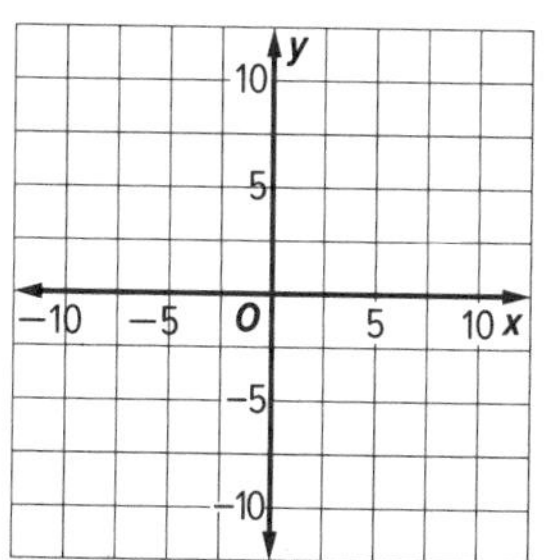

Systems of Equations and Inequalities

CHAPTER FOCUS Learn about some of the Common Core State Standards that you will explore in this chapter. Answer the preview questions. As you complete each lesson, return to these pages to check your work.

What You Will Learn	Preview Question
Lesson 3.1: Solving Systems of Equations	
CCSS A.CED.2 Create equations in two or more variables to represent relationships between quantities; graph equations on coordinate axes with labels and scales. CCSS A.CED.3 Represent constraints by equations or inequalities, and by systems of equations and/or inequalities, and interpret solutions as viable or nonviable options in a modeling context. CCSS A.REI.11 Explain why the *x*-coordinates of the points where the graphs of the equations $y = f(x)$ and $y = g(x)$ intersect are the solutions of the equation $f(x) = g(x)$; find the solutions approximately, e.g., using technology to graph the functions, make tables of values, or find successive approximations. Include cases where $f(x)$ and/or $g(x)$ are linear, polynomial, rational, absolute value, exponential, and logarithmic functions.	CCSS SMP 2 Write a system of equations in two variables that represents that the sum of two numbers is 25 and the difference of the same numbers is 1. Find the solution of the system of equations to determine the two numbers. CCSS SMP 1 A total of 30 quarters and nickels sum to $5.30. Represent this symbolically as a system of equations and solve the system to find the number of quarters and nickels.
Lesson 3.2: Solving Systems of Inequalities by Graphing	
CCSS A.CED.3 Represent constraints by equations or inequalities, and by systems of equations and/or inequalities, and interpret solutions as viable or nonviable options in a modeling context.	CCSS SMP 4 The choir director wants more than 50 boys and girls in the contest choir and for the best balance wants the number of girls to be no more than 10 more than the number of boys. Represent this symbolically as a system of inequalities and graph the system. 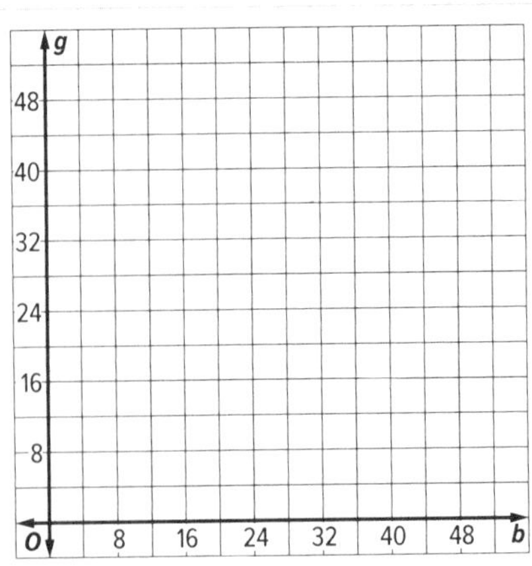

What You Will Learn	Preview Question
Lesson 3.3: Optimization with Linear Programming	
CCSS A.CED.3 Represent constraints by equations or inequalities, and by systems of equations and/or inequalities, and interpret solutions as viable or nonviable options in a modeling context.	CCSS SMP 6 You need to purchase several boxes of pens and pencils. The school store currently has 5 boxes of pens, selling for \$5 per box, and 10 boxes of pencils, selling for \$3 per box. You cannot spend more than \$50. Write a system of inequalities representing the constraints of the situation. Then, graph the system and determine the vertices for the feasible region.
Lesson 3.4: Systems of Equations in Three Variables	
CCSS A.CED.3 Represent constraints by equations or inequalities, and by systems of equations and/or inequalities, and interpret solutions as viable or nonviable options in a modeling context.	CCSS SMP 7 The sum of three numbers is 35. The sum of the first and second number is 30. The sum of the second and third is 15. Write and solve a system of three equations in three variables to find the three numbers. CCSS SMP 2 For a basketball game, children's tickets were \$2, adult tickets were \$5, and senior tickets were \$3. There were 3,000 tickets sold for a total of \$10,903 and there were 200 more children than seniors in attendance. Write a system of equations to represent this information.

3.1 Solving Systems of Equations

Objectives

- Write and solve systems of linear equations.
- Interpret solutions of systems of linear equations as viable or nonviable in modeling contexts.
- Explain why the points where the graphs of systems of equations intersect are the solutions of the systems.

CCSS STANDARDS

Content: A.CED.2, A.CED.3, A.REI.11
Practices: 1, 2, 3, 5, 6, 8
Use with Lesson 3-1

A **system of equations** is two or more equations with the same variables. To solve a system of equations find the ordered pair that satisfies all of the equations.

Solve by Graphing
Step 1 Write each equation in slope-intercept form.
Step 2 Graph each equation and find the point of intersection.
Step 3 Check by substituting into each original equation.

EXAMPLE 1 Solve by Graphing

EXPLORE The high school band was selling ride tickets for the fair. On the first day, 200 child tickets and 100 adult tickets were sold for a total of $400. On the second day, 40 child tickets and 10 adult tickets were sold for a total of $60. What is the price for each child ticket and each adult ticket?

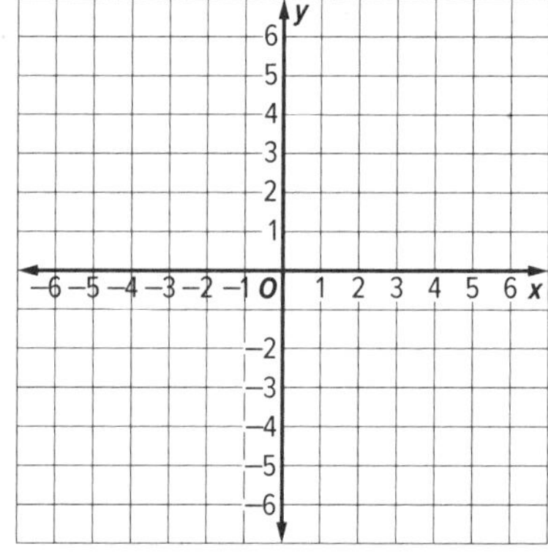

a. **REASON ABSTRACTLY** Write a system of equations to represent this situation. CCSS A.CED.2, SMP 2

b. **USE TOOLS** Graph the system of equations and estimate the point of intersection. CCSS A.REI.11, SMP 5

c. **CALCULATE ACCURATELY** Check your estimate by substituting into the original equations. CCSS A.REI.11, SMP 6

d. **INTERPRET PROBLEMS** What does the point of intersection represent? CCSS A.REI.11, SMP 1

Substitution Method	
Step 1	Solve one equation for one of the variables.
Step 2	Substitute the resulting expression into the other equation to replace the variable, then solve the equation.
Step 3	Substitute to solve for the other variable.

EXAMPLE 2 Solve by Using Substitution

Cassandra and Al are selling pies for a fundraiser. Cassandra sold 3 small pies and 14 large pies for a total of $203. Al sold 11 small pies and 11 large pies for a total of $220. Determine the cost each of one small pie and one large pie.

a. **INTERPRET PROBLEMS** Write a system of equations and solve using substitution. CCSS A.CED.2, SMP 1

b. **EVALUATE REASONABLENESS** Analyze the results. How can you verify that the solution is correct? CCSS A.CED.3, SMP 8

c. **CONSTRUCT ARGUMENTS** Compare your solution with one of your classmates. In **part b** did you both solve for the same variable? If not, were your solutions the same? Was solving for a specific variable easier than the other variable? Why? CCSS A.CED.3, SMP 3

d. **EVALUATE REASONABLENESS** Use a graphing calculator to plot both lines on the same coordinate plane. Does the graph verify your solution? CCSS SMP 8

Elimination Method	
Step 1	Multiply one or both equations by a number to result in two equations that contain opposite terms.
Step 2	Add the equations, eliminating one variable. Then solve the equation.
Step 3	Substitute to solve for the other variable.

EXAMPLE 3 Solve by Using Elimination

Julia has a collection of nickels and dimes. Altogether, she has 17 coins. The total value of the collection is $1.15. How many nickels and how many dimes does she have?

a. Write a system of equations and solve by using elimination. CCSS A.CED.2, SMP 1

b. Analyze the results. Verify the solution. CCSS A.CED.3, SMP 1

c. **USE STRUCTURE** Suppose the same number of coins had a value of $1.20. Without writing or solving a new system of equations, determine how many nickels and dimes there are. Explain your reasoning. CCSS A.CED.3, SMP 7

d. **CRITIQUE REASONING** Describe how to decide which method to use to solve a system of equations. Discuss with your classmates your reasoning. CCSS A.CED.3, SMP 3

PRACTICE

PLAN A SOLUTION **Write and solve a system of equations for each situation. Interpret the solution in terms of the situation.** CCSS A.CED.2, A.CED.3, SMP 1

1. Erasers are on sale for 45 cents each. Pencils are on sale for 40 cents each. Shirley buys 23 pencils and erasers and pays a total of $9.70, not including tax. If Shirley buys only pencils and erasers, how many of each did Shirley buy?

2. A total of \$21,000 is invested in two bonds that pay 4% and 7.5% simple interest. The bond with a 7.5% interest rate earned 6 times as much interest as the bond with a 4% interest rate. How much is invested in each bond? What was the total interest earned? Explain your solution process.

3. Jasmine bought 3 games and 2 DVDs for \$58. John bought 1 game and 4 DVDs for \$46. Determine the price for 1 game and for 1 DVD.

4. Wayne mows his 0.72-acre lawn in 2 hours and 15 minutes. His riding mower cuts grass twice as fast as his push mower. He uses the riding mower 3 times as much as the push mower. How much of the lawn will he mow with the push mower? Use technology to estimate the solution. Explain your solution process. CCSS A.REI.11

5. Mr. Greene invested in two stocks. He purchased shares of a stock priced at \$12 each and shares of a stock priced at \$15 each. Mr. Greene purchased 400 total shares and invested a total of \$5250. How many shares of each of the two stocks did Mr. Greene purchase? How much did Mr. Greene invest in each of the two stocks?

3.2 Solving Systems of Inequalities by Graphing

Objectives

- Use graphs to solve systems of inequalities.
- Use systems of inequalities to model real-world situations.
- Interpret solutions for systems of inequalities and determine their reasonableness.

CCSS STANDARDS

Content: A.CED.3
Practices: 1, 2, 3, 4, 5
Use with Lesson 3-2

A **system of inequalities** is a set of inequalities with the same variables. To solve a system of inequalities, graph the inequalities to determine a common solution region. Ordered pairs within the common region satisfy all of the inequalities in a system.

EXAMPLE 1 Write and Solve a System of Inequalities CCSS A.CED.3

Joanna charges $15 an hour for tutoring and $10 an hour for babysitting. She can work at most 14 hours a week. How many hours should Joanna spend on each job if she wants to earn no less than $125 a week?

a. **PLAN A SOLUTION** Define variables, then write a system of inequalities to represent this situation. State any constraints on the variables and explain them. CCSS SMP 1

b. **USE TOOLS** Graph the system of inequalities. Describe the solution of the system based on the graph. How do the constraints stated in **part a** affect the graph and the solution region? CCSS SMP 5

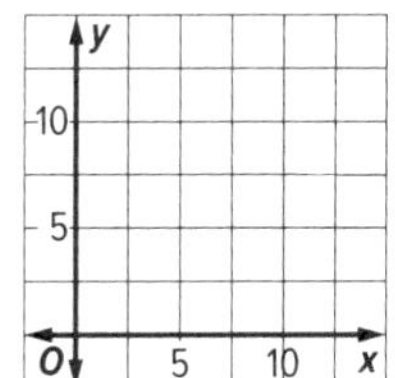

c. **INTERPRET PROBLEMS** Are the points (4, 5), (7, 6), or (5, 10) solutions to the system? Explain in terms of the graph and situation. CCSS SMP 1

EXAMPLE 2 Write a System of Inequalities CCSS A.CED.3

Zinovy created the graph below using an absolute value function and a linear function.

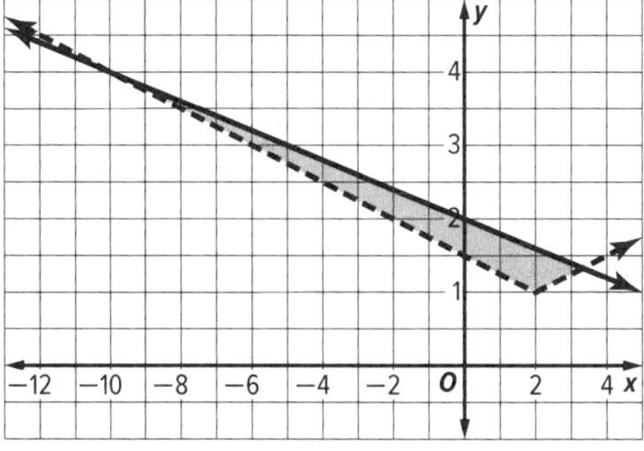

a. **REASON ABSTRACTLY** What are the inequalities that are modeled on the graph? Explain your reasoning. CCSS SMP 2

b. **REASON ABSTRACTLY** In the graph, x represents Zinovy's hours tutoring and y represents his hours babysitting. Zinovy always tutors more hours than he babysits. Write two more inequalities to represent all constraints for the situation. Explain your reasoning. CCSS SMP 2

EXAMPLE 3 Linear Programming CCSS A.CED.3

A manufacturer of refrigerators and dishwashers wants to maximize profits. Use the following table to answer the questions.

	Refrigerators	Dishwashers	Maximum Time Available a Week
Production	2.5 hours	1.75 hours	35 hours
Quality Assurance	1 hour	1 hour	18 hours
Profit per product	$325	$275	

a. **INTERPREPT PROBLEMS** What circumstances will limit the profits made by the manufacturer. How would these limitations be represented algebraically? CCSS SMP 1

b. USE A MODEL Graph a system of inequalities that represents this situation. What constraints have to be placed on the solution to make it feasible? CCSS SMP 4

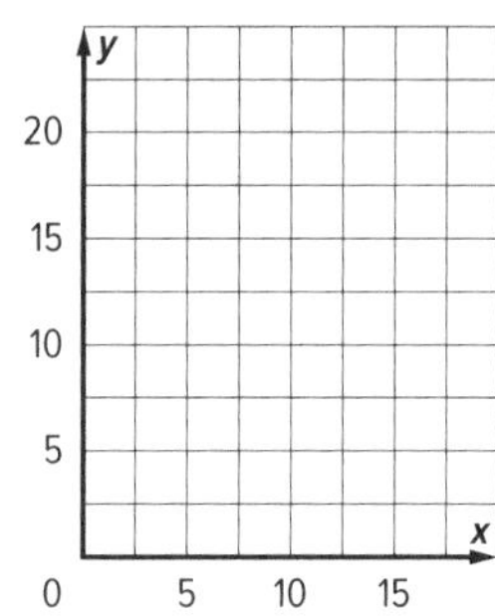

c. MAKE A CONJECTURE Based on the graph, where do you think the maximum profit would be most likely to occur? Why? CCSS SMP 3

d. INTERPRET PROBLEMS Write an equation for the profit and use it to determine how many refrigerators and dishwashers are necessary to maximize profits. CCSS SMP 1

PRACTICE

1. Graph the solution of the system of inequalities. If possible, write an ordered pair solution that lies in the first quadrant. 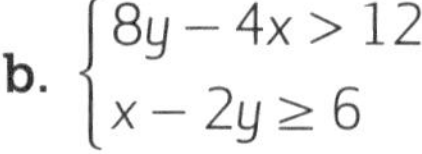

a. $\begin{cases} 2x - y \leq 4 \\ 3x + 2y \geq -6 \end{cases}$

b. $\begin{cases} 8y - 4x > 12 \\ x - 2y \geq 6 \end{cases}$

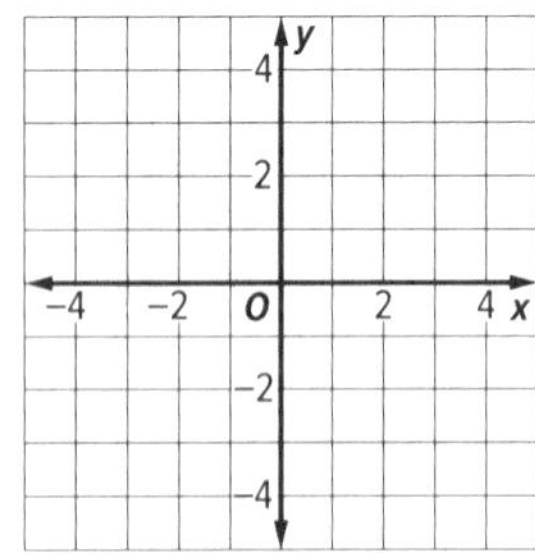

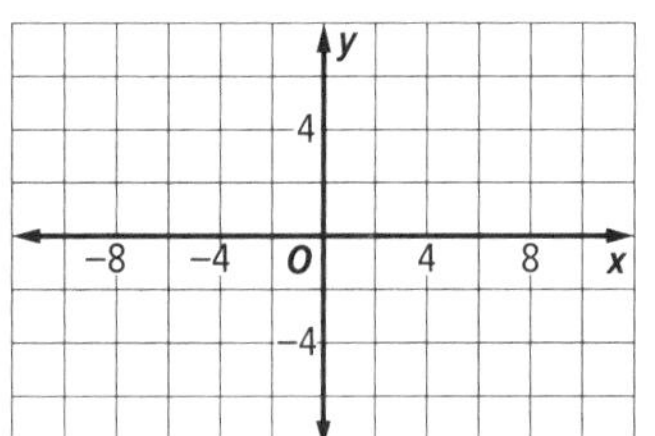

2. **REASON ABSTRACTLY** What system of inequalities satisfies the following solution? CCSS A.CED.3, SMP 2

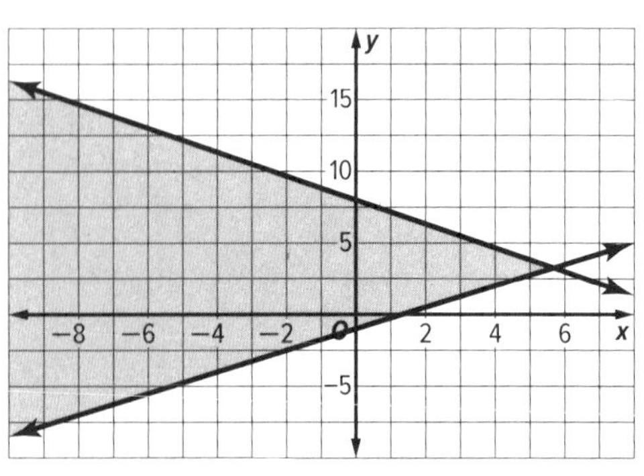

3. **USE A MODEL** To produce a handcrafted rocking chair, a woodworker needs to spend 30 hours on woodworking, and 3 additional hours on painting. The same woodworker can make a simpler kitchen chair in 24 hours, and it takes an additional 3 hours of painting. He has a maximum of 192 hours available for woodworking and 21 hours available for painting in a month. The rocking chair sells for $550, and the kitchen chair sells for $300. Write a system of inequalities to model this situation as well as a profit equation, and find the number of rocking chairs and kitchen chairs the woodworker should make to maximize his revenue for the month. CCSS A.CED.3, SMP 4

4. **CRITIQUE REASONING** Darryl argues that a system of inequalities can have only three types of solutions: infinitely many, a finite number, or no solution. Do you agree? Describe what each of these solutions would look like when graphed. CCSS A.CED.3, SMP 3

5. **USE TOOLS** Use a graphing calculator to solve the following systems of inequalities. List three ordered pairs that satisfy the system. CCSS A.CED.3, SMP 5

a. $\begin{cases} y \le 3x - 4 \\ 2x - 5y \ge 15 \end{cases}$

b. $\begin{cases} 8y - 7x > 4 \\ 14x + 16y \le 6 \end{cases}$

c. $\begin{cases} y \ge \frac{3}{4}|x + 1| \\ 5y - 3x < 6 \\ 2x - 3y > -8 \end{cases}$

6. **REASON ABSTRACTLY** A classmate solves the system $\begin{cases} y \le \frac{1}{3}x + \frac{1}{2} \\ 8x + 5y > 10 \\ 12x - 36y < -24 \end{cases}$ and states that the solution consists of two different regions. Do you agree? Why or why not? CCSS A.CED.3, SMP 2

3.3 Optimization with Linear Programming

Objectives

- Represent constraints by systems of inequalities.
- Interpret solutions as viable or non-viable.

CCSS STANDARDS

Content: A.CED.3
Practices: 1, 2, 4, 5, 6
Use with Lesson 3-3

Linear Programming is a method used to identify optimal **maximum** and **minimum** values. It is used in business to find minimum cost, the maximum profit, or minimum use of resources. Each resource is represented in a system of inequalities with each inequality representing a **constraint**. After the system is graphed and the vertices of the solution set, called the **feasible region**, are substituted into the function, you can determine the maximum and minimum values.

Optimization with Linear Programming
Step 1 Define variables.
Step 2 Write system of inequalities.
Step 3 Graph the system of inequalities.
Step 4 Find the coordinates of the vertices of the feasible region.
Step 5 Write a linear function to be maximized or minimized.
Step 6 Substitute the coordinates of the vertices into the function.
Step 7 Select the greatest or least result. Answer the problem.

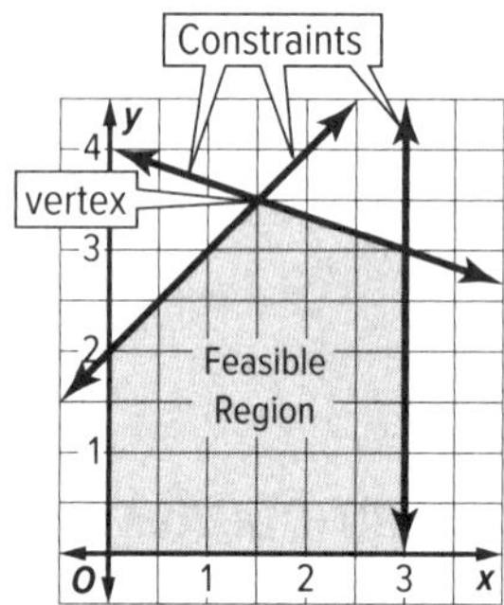

EXAMPLE 1 Solving a Linear Programming Model CCSS A.CED.3

EXPLORE The Jewelry Company makes and sells necklaces. For one type of necklace, the company uses clay beads and glass beads. There are no more than 10 clay beads and at least 4 glass beads per necklace. Four times the number of glass beads is less than or equal to 8 more than twice the number of clay beads. Each clay bead costs \$0.20 and each glass bead costs \$0.40. Find the minimum cost to make a necklace with clay and glass beads and find the combination of clay and glass beads that yields the minimum cost.

a. **REASON ABSTRACTLY** Define variables and write inequalities to represent the constraints. Then write an equation for the cost, C. CCSS SMP 2

b. USE TOOLS Graph the constraints. Is there an area that is bounded by all the constraints? If so, describe it. CCSS SMP 5

c. INTERPRET PROBLEMS Find the coordinates of the vertices of the feasible region. Check that they are correct. CCSS SMP 1

d. USE A MODEL Use the vertices of the feasible region to find the number of clay beads and glass beads that yield the minimum cost. CCSS SMP 4

e. USE A MODEL Suppose that each necklace has between 10 and 12 beads. Describe how you would adjust the system of constraints and describe changes to the graph of the feasible region. CCSS SMP 4

EXAMPLE 2 Maximize Using Linear Programming CCSS A.CED.3

A cell phone company produces basic phones and smartphones. Long-term projections indicate an expected demand of at least 2500 basic phones and 7000 smart phones each day. Because of limitations on production capacity, no more than 4000 basic phones and 9500 smartphones can be made daily. To satisfy a shipping contract, a total of at least 10,000 phones must be shipped each day. If each basic phone sold results in a $2 profit and each smartphone produces a $5 profit, how many of each type should be made daily to maximize net profits?

a. INTERPRET PROBLEMS Define the variables and write the inequalities that represent the constraints. Determine the linear function that represents the net profits. CCSS SMP 1

b. **USE A MODEL** Find the vertices of the feasible region on the graph. CCSS SMP 4

c. **USE A MODEL** Graph the constraints. How many of each type of phone should be made to maximize profits? At which point would the company make the least amount of profit? CCSS SMP 4

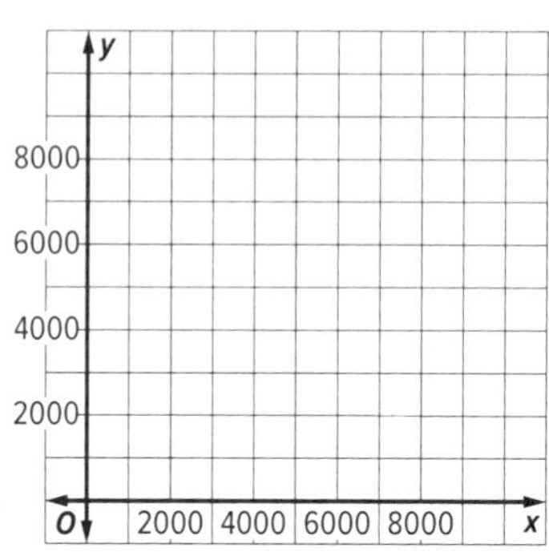

EXAMPLE 3 Unbounded Feasible Region CCSS A.CED.3

A zoo is mixing two types of food for the animals. Each serving is required to have at least 60 grams of protein and 30 grams of fat. Custom Foods has 15 grams of protein and 10 grams of fat and costs 80 cents per unit. Nature's Best contains 20 grams of protein and 5 grams of fat and costs 50 cents per unit. How much of each type should be used to minimize cost to the zoo?

a. **INTERPRET PROBLEMS** Define the variables and write the inequalities that represent the constraints. Determine the linear function to be minimized. CCSS SMP 1

b. **USE A MODEL** Graph the constraints. Find how much of each type of food should be made to minimize costs. What does the unbound region represent? CCSS SMP 4

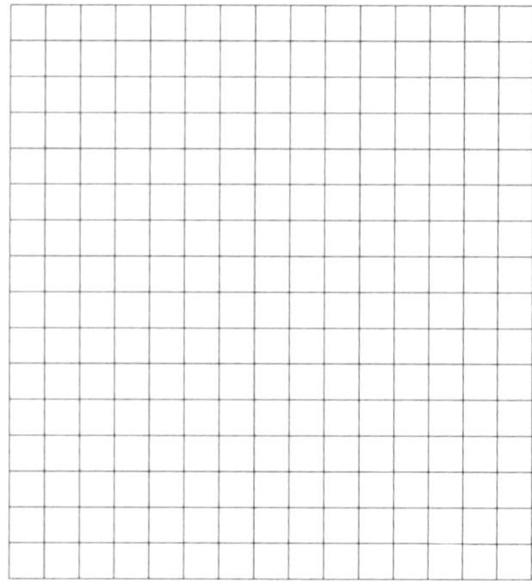

1. **REASON ABSTRACTLY** A baseball team is selling shorts and sweatpants as a fundraiser. They want to sell at least 20 but not more than 100 shorts and at least 20 but not more than 80 sweatpants. The number of shorts they can afford to buy is less than or equal to 60 less than twice the number of sweatpants. If each shorts costs them \$6 and each sweatpants costs them \$14, find the number of shorts and sweatpants they should buy to minimize cost. CCSS A.CED.3, SMP 2

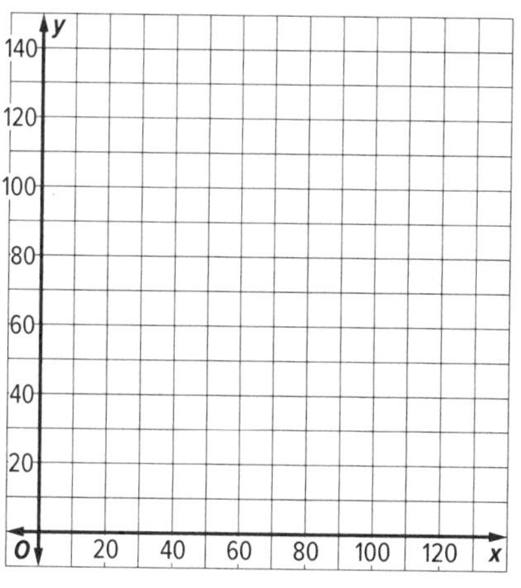

2. **USE A MODEL** A farmer can plant up to 80 acres of land with wheat and barley. He can earn \$5000 for every acre he plants with wheat and \$3000 for every acre he plants with barley. His use of a necessary pesticide is limited by federal regulations to 100 gallons for his entire 80 acres. Wheat requires 2 gallons of pesticide for every acre planted and barley requires just 1 gallon per acre. What is the maximum profit he can make? CCSS A.CED.3, SMP 4

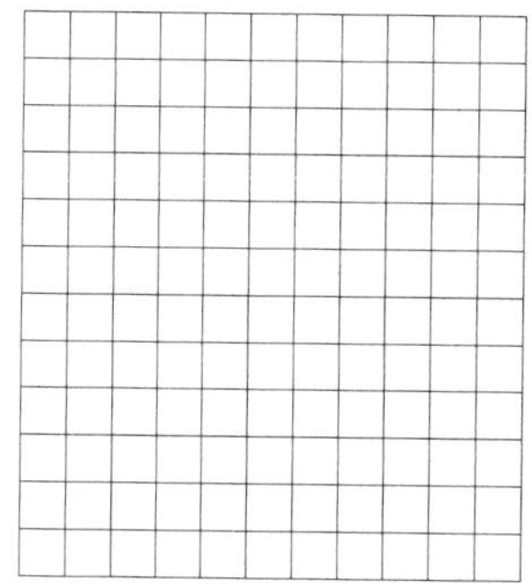

3. **COMMUNICATE PRECISELY** Marcus made a graph to find the feasible region for the following problem. Analyze the graph to determine if his work is correct. Justify your answer. CCSS A.CED.3, SMP 6

The semester test in English consists of short answers and essay questions. Each short answer question is worth 5 points and each essay question is worth 15 points. You may choose up to 20 questions of any type to answer. It takes 2 minutes to answer each short answer question and 12 minutes to answer each essay question. If you have one hour to complete the test and assuming you answer all the questions you attempt correctly, how many of each type of question should you answer to earn the highest score?

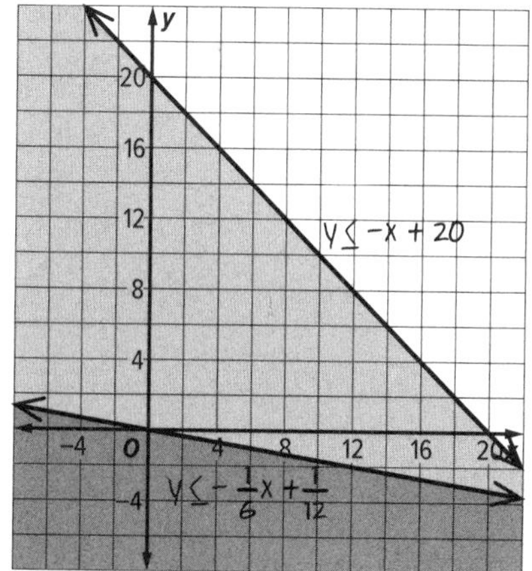

3.4 Systems of Equations in Three Variables

Objectives

- Model real-world problems using systems of linear equations in three variables.
- Determine whether the solution of a system of linear equations in three variables is viable or not viable in the context of the situation.

CCSS STANDARDS

Content: A.CED.3
Practices: 1, 3, 6, 7, 8
Use with Lesson 3-4

Systems of equations in three variables can have one solution, no solution, or infinitely many solutions. A solution of a system in three variables is an ordered triple (x, y, z).

When solving problems involving three variables, first identify the variables and what they represent. Next, write the equations relating the variables using the information in the problem. Then, solve the system. Finally, interpret the solution in the context of the problem.

EXAMPLE 1 Write and Solve a System of Equations CCSS A.CED.3

EXPLORE **A new video game comes in three versions: standard edition, gold edition, and collector's edition. Before taxes, the standard edition costs $50, the gold edition costs $75, and the collector's edition costs $125. A store will have 76 copies of the game available to sell. The number of collector's edition units available will be five fewer than the number of gold edition units, and the number of standard edition units will be twenty more than the number of collector's edition units. The total cost, before tax, of all 76 units is $5625.**

a. **INTERPRET PROBLEMS** Determine how many of each edition will be available. Explain your steps and show your work. CCSS SMP 1

b. **CONSTRUCT ARGUMENTS** A store employee claims that if the total cost of the number of units available is doubled, then there would have to be twice as many of each of the three editions of the game available. He assumes that the number of collector's edition units available will be five fewer than the number of gold edition units, and the number of standard edition units will be twenty more than the number of collector's edition units. Is his reasoning correct? Explain why or why not. CCSS SMP 3

EXAMPLE 2 Reserved Spots in a Parking Garage CCSS A.CED.3

An indoor parking garage reserves 80 spots for certain types of vehicles: SUVs, subcompact cars, and electric-only cars. There are twice as many spaces reserved for subcompact cars as there are for SUVs, and there are 20 more spots reserved for electric-only cars than the number reserved for SUVs.

a. **COMMUNICATE PRECISELY** Determine how many spaces are reserved for each type of vehicle. Explain your process. CCSS SMP 6

b. **COMMUNICATE PRECISELY** Jackson remembers that systems in two linear equations can be solved in three ways: substitution, elimination, and by graphing. He wants to solve this system using as many ways as possible to confirm his answer. Which of the three ways are applicable? Which was used in **part a**? Use the other possibilities to solve the system and confirm the answer. CCSS SMP 6

EXAMPLE 3 How Much Does Food Cost? CCSS A.CED.3

Carmen, Sanford, and Diego go to the cafeteria to buy lunch. The food is separated into three different buffet areas, one devoted to vegetables, one to pasta, and one to gourmet desserts. Each of the three areas has a different price per pound associated with it. Customers fill containers at each of the areas they wish and pay for the containers at the checkout counter. The following are the amounts each of them takes in each category and the prices they pay.

	Vegetable	Pasta	Dessert	Total Price
Carmen		$\frac{1}{2}$ pound	$\frac{1}{4}$ pound	\$7.00
Sanford	$\frac{1}{2}$ pound	$\frac{1}{4}$ pound		\$8.00
Diego	$\frac{1}{4}$ pound	1 pound	$\frac{1}{4}$ pound	\$12.50

a. **INTERPRET PROBLEMS** Determine the cost per pound of each of the three different food types. Use elimination to solve the system and show your work. CCSS SMP 1

b. **EVALUATE REASONABLENESS** A similar buffet exists at another cafeteria. Bart represents the prices per pound at each station with the system $\begin{cases} 2y + z = 25 \\ 2x + y = 34 \\ 2x + 3y + z = 56 \end{cases}$.
Solve the system and evaluate whether Bart's system makes sense for the situation. Justify your answer. CCSS SMP 8

c. **EVALUATE REASONABLENESS** Bart thinks he may have made a mistake in the system he wrote. He thinks that the first equation should be $2y + z = 22$. Solve the system and determine if Bart's new system makes sense. Justify your answer. CCSS SMP 8

PRACTICE

1. **PLAN A SOLUTION** A vendor is having a sale on pool accessories and offers the following combination deals to entice customers to purchase them.

Combo	Price Before 6% Sales Tax
1 Raft and 2 Chlorine Filters	\$220
1 Chlorine Filter and 2 Large Lounge Chairs	\$245
1 Raft and 4 Large Lounge Chairs	\$315

Based on this pricing, if you had \$200, could you buy 1 chlorine filter, 1 raft, and 1 large lounge chair after sales tax is applied? Show your work. CCSS A.CED.3, SMP 1

2. **PLAN A SOLUTION** Three kinds of tickets are available for a concert: orchestra seating, mezzanine seating, and balcony seating. The orchestra tickets cost \$2 more than the mezzanine tickets, while the mezzanine tickets cost \$1 more than the balcony tickets. Twice the cost of an orchestra ticket is \$1 less than 3 times the cost of a balcony ticket. Determine the price of each kind of ticket. CCSS A.CED.3, SMP 1

3. **USE STRUCTURE** When the digits of a 3-digit number are reversed, its value decreases by 198. The tens digit is two more than the hundreds digit, and the sum of the digits is 15. Write and solve a system to find the number. Is the solution unique? Explain. CCSS A.CED.3, SMP 7

Performance Task

Restaurant Planning

Provide a clear solution to the problem. Be sure to show all of your work, include all relevant drawings, and justify your answers.

Congratulations! You are now the head chef at Bob's Bodacious BBQ Bungalow. Friday night is Celebrate Beef Night, when only steak and hamburger dinners are sold. On any night, Bob's can serve a maximum of 300 people. The cost of each burger is $1 and the cost of each steak is $5. Bob's will net a profit of $5 per steak dinner and $3 per burger dinner. You have a meat budget of $1100 for each Friday. What combination of steak and burger dinners would produce the maximum profit?

Part A

Write a system of inequalities to represent the possible numbers of steak and hamburger dinners each Friday. Determine the function for the profit made on the dinners.

Part B

Using the grid provided, graph the system of inequalities and determine the vertices of the feasible region. Be sure to label your axes and scale.

Part C

Determine the number of steak and hamburger dinners that will produce the maximum profit on Bob's Celebrate Beef Night. State the amount of the maximum profit possible.

Part D

One week, Bob's supplier delivers only enough meat to make 140 hamburger dinners. Will this affect the possible maximum profit? Explain your answer.

Performance Task

Art Class

Provide a clear solution to the problem. Be sure to show all of your work, include all relevant drawings, and justify your answers.

Math is an important part of life for everyone, even in Art Class. Sometimes, life presents us with challenges that can only be overcome with math! This performance task will present three different multi-variable problems that apply to Art Class.

Part A

An art supply cabinet contains a total of 75 charcoal pencils, oil pastel crayons, and sketchpads. Twice the number of oil pastel crayons is one more than the number of pencils. Four times the number of oil pastel crayons plus twice the number of pencils equals 8 more than 3 times the number of sketchpads. Write and solve a system of equations to determine the number of each type of these supplies in the cabinet.

Part B

One of your assignments in art class is to create a frame with the material of your choice. You have a 60-inch long strip of braided metal. You will bend it into the shape of a rectangle. You want the width of the rectangle to be $\frac{2}{3}$ its length. Write and solve a system of equations to determine the dimensions of a rectangle that meet this requirement.

Part C

You need to make a 2-gallon mixture of paint. The ratio of the amounts of white paint to blue paint should be 3:5. The amount of red paint in the mixture should be half the amount of the white paint and blue paint combined. Write and solve a system of equations to determine the amount of each type of paint to be used in the mixture.

Standardized Test Practice

1. A math test contains multiple choice and extended response questions. Mizuki had a total score of 87, including 43 for the multiple choice questions. Olivia had a lower overall score, but scored more points on the multiple choice questions than Mizuki. If x represents the points earned from multiple choice questions and y represents points earned from extended response questions, which of the following systems of inequalities represents the possible points that Olivia earned from each type of question? CCSS A.CED.3

$\begin{cases} x < 43 \\ x + y > 87 \end{cases}$ $\begin{cases} y < 87 \\ y - x < 43 \end{cases}$

$\begin{cases} x > 43 \\ x + y < 87 \end{cases}$ $\begin{cases} x > 43 \\ x + 87 > y \end{cases}$

2. Use any method to solve the following system of linear equations. CCSS A.CED.2

$$\begin{cases} y = -2x - 7 \\ 3x + 2y = -9 \end{cases}$$

The solution to this system is

(☐, ☐).

3. Colin, Tia, and Emma bought snacks. The following table shows how many of each item they each bought, and the total cost each spent. CCSS A.CED.2

	Colin	Tia	Emma
Tacos	2	1	2
Yogurts	1	0	2
Juices	1	2	0
Total Cost($)	10.00	8.50	9.00

Fill in the price of each item.

Taco: $ ☐

Yogurt: $ ☐

Juice: $ ☐

4. Antwan wants to buy a total of 10 movies and books at a clearance sale. If each book is \$5 and each movie is \$7 and Antwan wants to spend \$45, the following system of equations represents the number of books and movies he should buy where b and m are nonnegative integers. CCSS A.CED.3

$$\begin{cases} b + m = 10 \\ 5b + 7m = 45 \end{cases}$$

Which of the following best describes the system and the feasibility of Antwan's goals?

One solution, feasible

One solution, not feasible

Many solutions, feasible

5. Latara makes necklaces and bracelets to sell. The graph below shows the feasible region for how many of each she can make, where x is the number of necklaces and y is the number of bracelets. CCSS A.REI.11

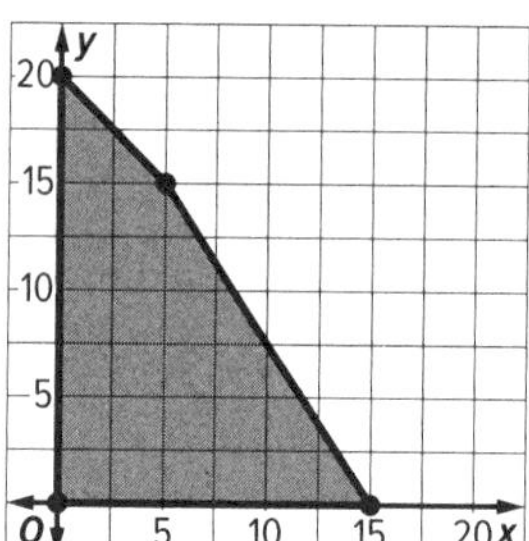

Her profit is \$10 on each bracelet and \$12 on each necklace. She wants to maximize her profit which is represented by the function ☐. She needs to evaluate the function at the vertices ☐, ☐, ☐, and ☐. Based on this, she should make ☐ necklaces and ☐ bracelets.

6. Use a graphing calculator to estimate to the nearest tenth a solution for the following system. CCSS A.REI.11

$$\begin{cases} y = 3.3714x - 2.936 \\ y = 12.0582 - 5.727x \end{cases}$$

(☐, ☐)

7. Andrew is 4 years less than twice the age of his friend, Demetrius. The sum of twice Andrew's age and three times Demetrius's age is 41.

a. If x is Andrew's age and y Demetrius's age, what system of equations represents their ages? CCSS A.CED.2, A.CED.3

b. Graph the system. CCSS A.CED.2

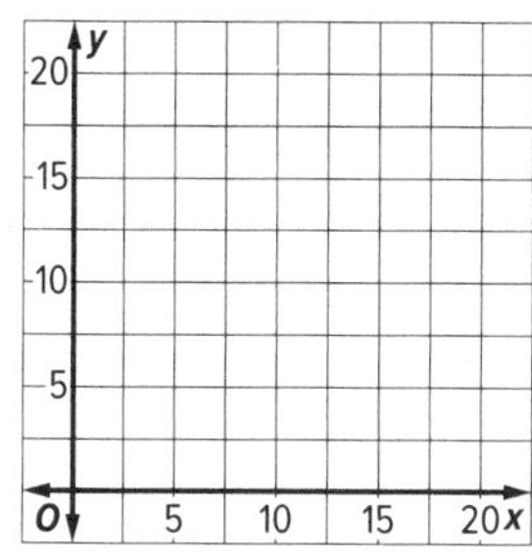

c. How can you identify the solution from the graph? CCSS A.REI.11

d. What is the solution of the system? Explain what the solution means in the context of the problem. CCSS A.CED.3

8. Graph the solution to the following system. CCSS A.REI.11

$$\begin{cases} y < 4x + 3 \\ 5x + 2y \leq 10 \\ y \geq -5 \end{cases}$$

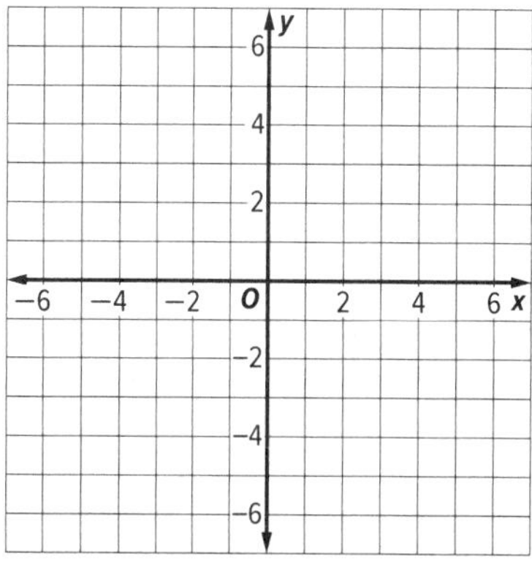

9. A school ski club sold ski hats and gloves with the school logo as a fundraiser. To get the items made, they paid a \$120 service fee as well as \$3.50 per hat and \$4.75 per pair of gloves. They sold the hats for \$6 and gloves for \$8. If they paid \$814.50 to get the items made and made \$1180 in sales, how many of each item did they sell? Explain your reasoning. CCSS A.CED.2, A.CED.3

4 Quadratic Functions

CHAPTER FOCUS Learn about some of the Common Core State Standards that you will explore in this chapter. Answer the preview questions. As you complete each lesson, return to these pages to check your work.

What You Will Learn	Preview Question
Lesson 4.1: Graphing Quadratic Functions	
CCSS F.IF.4 For a function that models a relationship between two quantities, interpret key features of graphs and tables in terms of the quantities, and sketch graphs showing key features given a verbal description of the relationship. CCSS F.IF.9 Compare properties of two functions each represented in a different way (algebraically, graphically, numerically in tables, or by verbal descriptions). **Also addresses:** F.IF.5, A.CED.2	CCSS SMP 7 Graph the quadratic function $f(x) = x^2 - 3x + 2$. 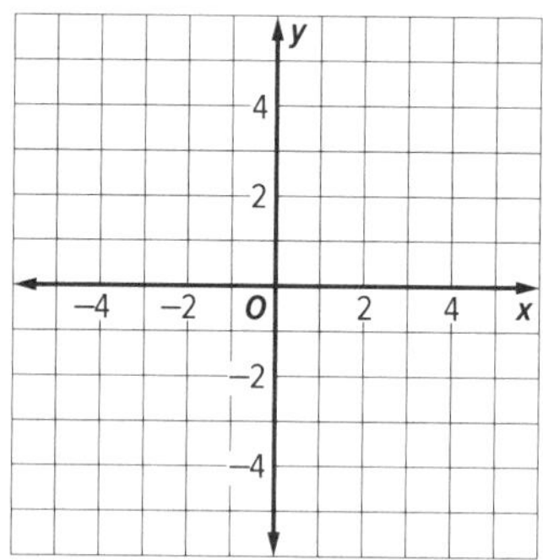
Lesson 4.2: Modeling: Quadratic Functions	
CCSS F.IF.4 For a function that models a relationship between two quantities, interpret key features of graphs and tables in terms of the quantities, and sketch graphs showing key features given a verbal description of the relationship. CCSS F.IF.5 Relate the domain of a function to its graph and, where applicable, to the quantitative relationship it describes. **Also addresses:** A.SSE.1a, A.SSE.1b, A.REI.11, A.CED.4, F.BF.1b	CCSS SMP 4 A bottle rocket launched upwards reaches a height of 25 ft. before returning to the ground 50 ft. from where it was launched. What is the vertex of the parabola that models the flight of the rocket? In what direction does the graph open? What, if any, are the zeros of the equation that represents the graph?
Lesson 4.3: Solving Quadratic Equations by Graphing	
CCSS F.IF.4 For a function that models a relationship between two quantities, interpret key features of graphs and tables in terms of the quantities, and sketch graphs showing key features given a verbal description of the relationship. CCSS A.CED.2 Create equations in two or more variables to represent relationships between quantities; graph equations on coordinate axes with labels and scales.	CCSS SMP 5 Use the graph of $y = x^2 + 4x - 5$ to determine the solutions to the equation $x^2 + 4x - 5 = 0$. 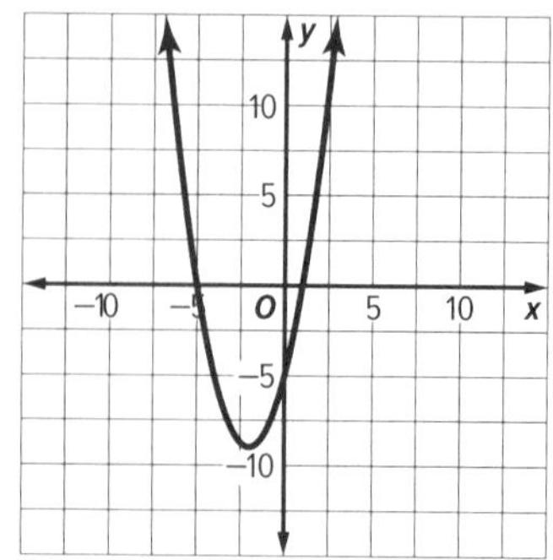

What You Will Learn	Preview Question
Lesson 4.4: Solving by Factoring	
CCSS F.IF.8a Use the process of factoring and completing the square in a quadratic function to show zeros, extreme values, and symmetry of the graph, and interpret these in terms of a context. CCSS A.CED.1 Create equations and inequalities in one variable and use them to solve problems. CCSS A.SSE.2 Use the structure of an expression to identify ways to rewrite it.	CCSS SMP 2 If $-3(2x + 5)(x - 4) = 0$, why are $-\frac{5}{2}$ and 4 the only roots of this equation? Why is -3 not a root?
Lesson 4.5: Solving by Completing the Square	
CCSS A.CED.1 Create equations and inequalities in one variable and use them to solve problems. CCSS F.IF.8a Use the process of factoring and completing the square in a quadratic function to show zeros, extreme values, and symmetry of the graph, and interpret these in terms of a context. **Also addresses:** N.CN.7, F.IF.4	CCSS SMP 6 The dimensions of a rectangular region can be represented by $(x + 2)$ and $(x + 4)$ where x is a number of feet. If the area of the enclosed region is 14, find the value of x and the measure of the dimensions by completing the square.
Lesson 4.6: Solving by Using the Quadratic Formula and the Discriminant	
CCSS N.CN.7 Solve quadratic equations with real coefficients that have complex solutions. CCSS F.IF.4 For a function that models a relationship between two quantities, interpret key features of graphs and tables in terms of the quantities, and sketch graphs showing key features given a verbal description of the relationship. CCSS A.CED.1 Create equations and inequalities in one variable and use them to solve problems. **Also addresses:** A.CED.4, A.APR.4	CCSS SMP 6 Why does a quadratic equation have complex solutions if the discriminant is less than 0?
Lesson 4.7: Transforming Quadratic Functions	
CCSS F.BF.3 Identify the effect on the graph of replacing $f(x)$ by $f(x) + k$, $k\,f(x)$, $f(kx)$, and $f(x + k)$ for specific values of k (both positive and negative); find the value of k given the graphs. Experiment with cases and illustrate an explanation of the effects on the graph using technology. Include recognizing even and odd functions from their graphs and algebraic expressions for them. **Also addresses:** F.IF.8a	CCSS SMP 4 The vertex of $f(x) = x^2$ is located at (0, 0). What are the coordinates of the vertex of $g(x) = (x - 2)^2 + 5$? CCSS SMP 7 What is the equation of a function $g(x)$ whose graph is translated 3 units to the left of and 4 units down from the graph of $f(x) = x^2$?

4.1 Graphing Quadratic Functions

Objectives

- Graph quadratic functions.
- Interpret key features of graphs of quadratic functions.

CCSS STANDARDS

Content: F.IF.4, F.IF.5, F.IF.9, A.CED.2
Practices: 1, 2, 3, 4, 7
Use with Lesson 4-1

The general form of a quadratic function is $f(x) = ax^2 + bx + c$, where $a \neq 0$. The graph of a quadratic function is called a **parabola**, and contains two congruent branches that intersect at the vertex. The graph is also defined by its **axis of symmetry**, a vertical line passing through the vertex of the graph.

KEY CONCEPT Graph of a Quadratic Function

The graph of quadratic function $f(x) = ax^2 + bx + c$ has the following features.

The graph intersects the y-axis at the point ________.

The equation of the ____________________ is $x = -\frac{b}{2a}$.

The x-coordinate of the vertex is ____________ and the y-coordinate is ____________.

EXAMPLE 1 Graph a Quadratic Function

EXPLORE A model rocket is launched from a platform. The height of the rocket after t seconds can be approximated by $h(t) = -16t^2 + 48t + 28$, where $h(t)$ is measured in feet.

a. **USE STRUCTURE** How can you use symmetry to find the coordinates of other points on the graph? Use your method to complete the table. Explain why this works in terms of the context of the problem. CCSS F.IF.4, SMP 7

t	0	1	1.5		
$h(t)$	28				

b. **REASON QUANTITATIVELY** Construct a graph of $h(t)$ and use it to find the domain of the function for this situation. CCSS F.IF.5, SMP 2

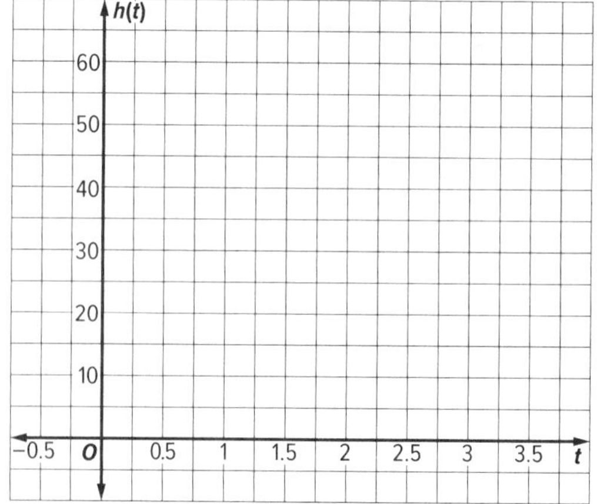

c. **REASON ABSTRACTLY** Determine the value of $h(0)$ and explain its meaning in terms of the function graph and the context of the problem. CCSS F.IF.4, SMP 2

d. **PLAN A SOLUTION** At what time does the rocket begin to fall toward Earth? How high is the rocket at that time? CCSS F.IF.4, SMP 1

KEY CONCEPT Maximum and Minimum Value

Consider the quadratic function $f(x) = ax^2 + bx + c$, where $a \neq 0$.

When $a > 0$, the graph of the function opens ________ and has a ________________.

When $a < 0$, the graph of the function opens ________ and has a ________________.

EXAMPLE 2 Maximum and Minimum Values CCSS F.IF.4

a. **PLAN A SOLUTION** Determine the maximum or minimum value of $f(x) = 9x^2 - 12x + 8$. How can you use a graph to confirm your result? CCSS SMP 1

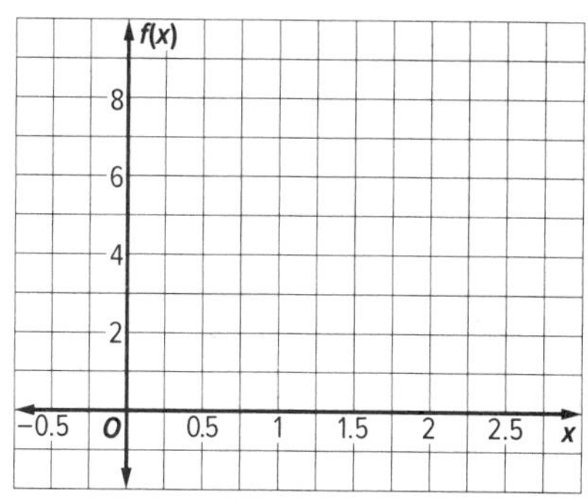

b. **REASON ABSTRACTLY** Find two numbers with a sum of 36 and with a product that is as great as possible. Show your work. CCSS SMP 2

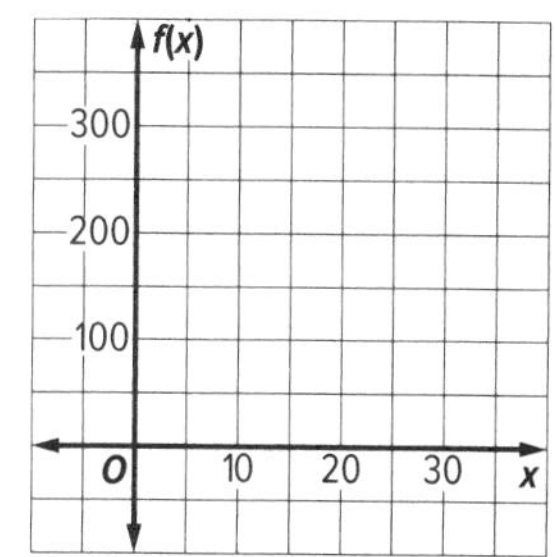

c. **CONSTRUCT ARGUMENTS** What is the least possible product of two numbers with a sum of 24? Use a graph to explain your reasoning. CCSS SMP 3

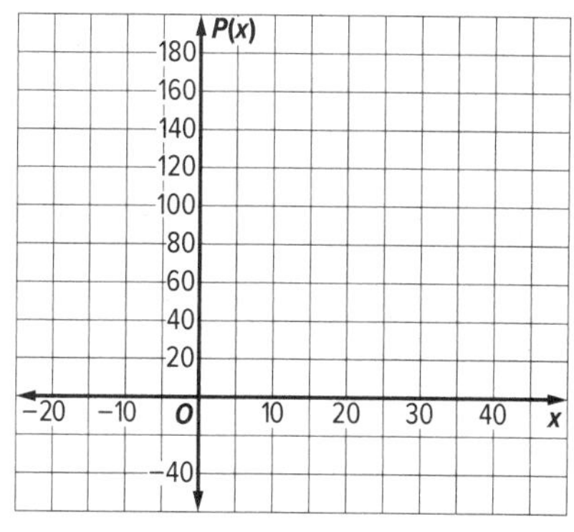

The domain of a quadratic function is the set of all real numbers. The range of a quadratic function is the set of all real numbers less than or equal to the maximum value, or the set of all real numbers greater than or equal to the minimum value. However, both the domain and the range may need to be restricted in the context of a modeling application.

EXAMPLE 3 Maximizing Revenue CCSS F.IF.4

On Friday nights the local cinema typically sells 200 tickets at $6.00 each. The manager estimates that for each $0.50 increase in the ticket price, 10 fewer people will go to the cinema.

a. REASON ABSTRACTLY Write and graph a function to represent the expected revenue, and determine the domain of the function for the situation. Show your work. CCSS A.CED.2, F.IF.5, SMP 2

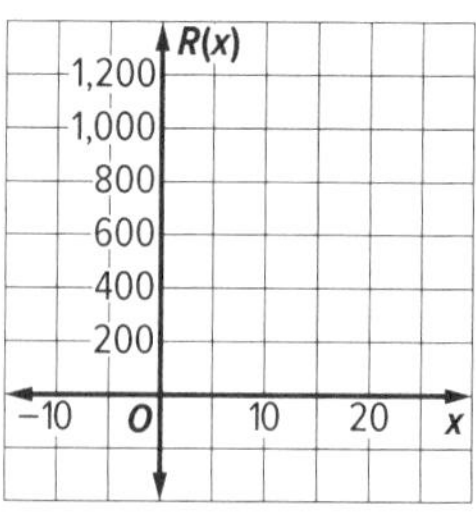

b. CONSTRUCT ARGUMENTS What price should the manager set for a ticket in order to maximize the revenue? What is the maximum possible revenue? Justify your reasoning. CCSS SMP 3

c. INTERPRET PROBLEMS Explain why the graph decreases from $x = 4$ to $x = 20$, and interpret the meaning of the x-intercept of the graph. CCSS SMP 1

PRACTICE

1. **USE A MODEL** A graphic designer wants to know the dimensions of the rectangle with greatest area that is enclosed within the region bounded by the x-axis, the y-axis, and the line $3x + y = 8$. CCSS F.IF.4, SMP 4

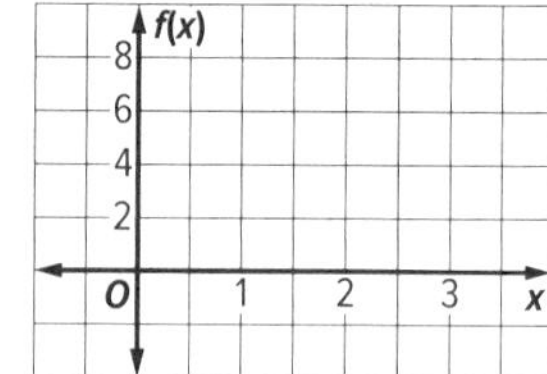

   a. Draw a diagram of the rectangle on a coordinate grid.

   b. Write and graph a function for the rectangle with greatest area. What are the dimensions of the rectangle that give the greatest possible area? CCSS A.CED.2

2. **REASON ABSTRACTLY** Write and graph a function that models the area between the x-axis and the line $y = -4x + 20$, from $x = 0$ to $x = c$. CCSS SMP 2

a. Draw a diagram and write a function that represents this area. What is the domain? CCSS A.CED.2, F.IF.5

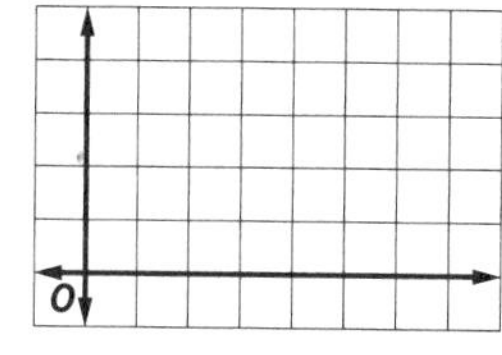

b. What value of c maximizes the area? What is the maximum area? How does a graph of the function confirm your answer? CCSS F.IF.4

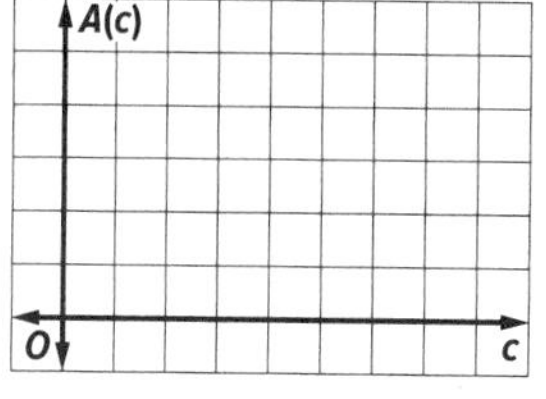

c. **INTERPRET PROBLEMS** Use your diagram in **part a** to explain why your answer in **part b** makes sense in this context. CCSS F.IF.4, SMP 1

3. **CONSTRUCT ARGUMENTS** Which function has a greater maximum: $f(x) = -2x^2 + 6x - 7$ or the function shown in the graph at the right? Explain your reasoning using a graph. CCSS F.IF.4, F.IF.9, SMP 3

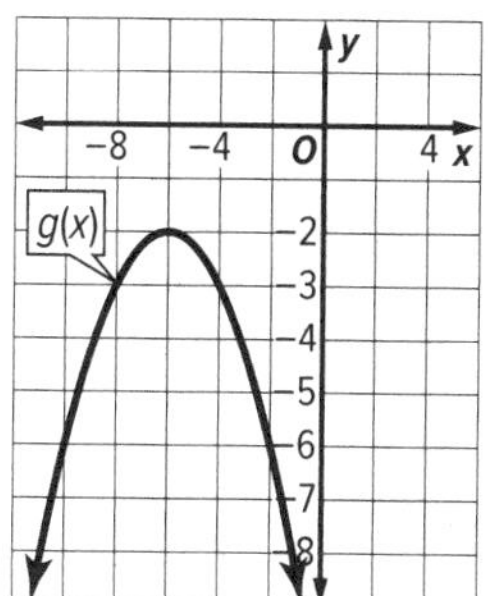

4. **REASON ABSTRACTLY** A farmer has 600 meters of fencing to enclose a field and divide it into three sections. What is the area of the largest field that can be fenced? Explain your reasoning using a graph. CCSS A.CED.2, F.IF.4, SMP 2

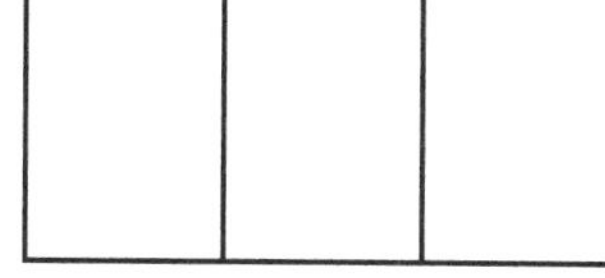

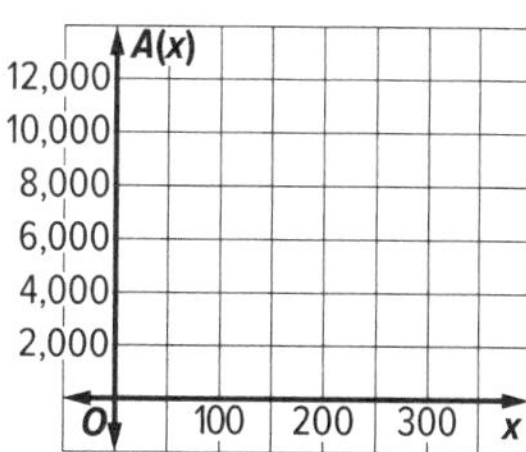

5. **INTERPRET PROBLEMS** The height in feet of a rock thrown into the air is modeled by the function $f(t) = -16t^2 + 58t + 6$. Determine the maximum height of the rock. CCSS F.IF.4, SMP 1

4.2 Modeling: Quadratic Functions

Objectives

- Relate the domain of a quadratic function to the quantitative relationship it describes.
- Interpret parts of an expression that arise as factors and terms of quadratic functions.
- Explain which features of simultaneous graphs of two quadratic functions $f(x)$ and $g(x)$ that intersect are the solutions of the equation $f(x) = g(x)$.

CCSS STANDARDS

Content: A.SSE.1a, A.SSE.1b, A.REI.11, A.CED.4, F.IF.4, F.IF.5, F.BF.1b

Practices: 1, 2, 3, 4, 6, 7, 8

Use with Lesson 4-2

KEY CONCEPT Quadratic Functions

The general form of a quadratic function is ____________, where $a \neq 0$. The quadratic term is ________, the linear term is ________, and the constant term is ________. The graph of a quadratic function is a ____________.

Quadratic functions are used to model and understand problems and scenarios that arise in the physical world and the business world. This lesson addresses several such scenarios.

EXAMPLE 1 Use a Quadratic Model CCSS F.BF.1b

EXPLORE **The school garden club has 60 feet of fencing to use to fence in two adjacent rectangular areas, as shown in the diagram. An existing wall on the school will be one side of the enclosed area.**

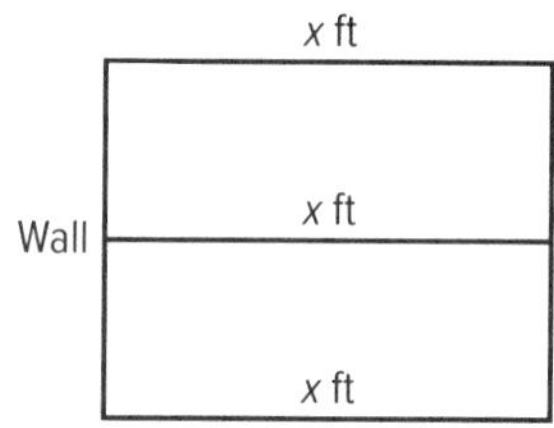

a. **REASON ABSTRACTLY** In the diagram, x represents the width of the three parallel sections of fencing. Write a linear expression for the length of the remaining sections of fencing in terms of x. Explain how you determined the expression. CCSS SMP 2

b. **USE A MODEL** Write a quadratic function that describes the relationship between the dimensions of the enclosure and its area, $A(x)$. State the domain of the function. CCSS F.IF.5, SMP 4

c. **INTERPRET PROBLEMS** Determine the dimensions of the enclosure that will result in the greatest possible area. Then determine the maximum area. Show your work. CCSS SMP 1

EXAMPLE 2 Use a Formula CCSS A.SSE.1a

EXPLORE Objects that are in free fall or are projected straight up with no internal power source are affected by the force of gravity. The formula $h(t) = -\frac{1}{2}gt^2 + v_0t + h_0$ is used to describe the height of an object above Earth.

- g is the acceleration due to gravity.
- $h(t)$ is the height of the object above Earth.
- t is the time, in seconds, that has passed since an object was dropped or projected into the air.
- v_0 is the initial velocity of the object.
- h_0 is the height of the object in its initial position.

a. **USE STRUCTURE** Objects in free fall accelerate toward Earth at a rate of -9.8m/s^2. Derive a general formula for the height, $h(t)$, above Earth, in feet, of a free falling object t seconds after it is dropped or propelled into the air. Assume an initial velocity of v_0m/sec and an initial height of h_0 meters. CCSS A.SSE.1b, SMP 7

b. **CRITIQUE REASONING** Nina states that the formula for an object's height does not make sense because the leading term is negative, and height above Earth cannot be negative. Explain the error in her reasoning. CCSS SMP 2

c. **USE STRUCTURE** A ball is thrown into the air from the ground and reaches a height of $h(t)$ meters after t seconds. Rearrange the formula from **part a** so that it can be used to determine the initial velocity of the ball. Show your work. CCSS A.CED.4, SMP 7

d. **COMMUNICATE PRECISELY** The ball is at a height of 11.4 meters above the ground after 3.3 seconds. Determine the initial velocity of the ball to the nearest tenth. Show your work and label your answer with the appropriate unit(s). CCSS SMP 6

e. **REASON ABSTRACTLY** A rock falls from the top of a cliff that is 25.8 meters high. Write a quadratic function that models the situation. Determine to the nearest tenth of a second the amount of time that it takes the rock to strike the ground. Explain your reasoning. CCSS SMP 2

EXAMPLE 3 Analyze Graphs CCSS A.REI.11

EXPLORE **The diagram shows the height above the ground $h(t)$ feet of two baseballs after t seconds. One ball was thrown straight up by a player on the ground of a stadium. At the same time that the first ball was thrown, the second ball was dropped straight down from a balcony in the stadium.**

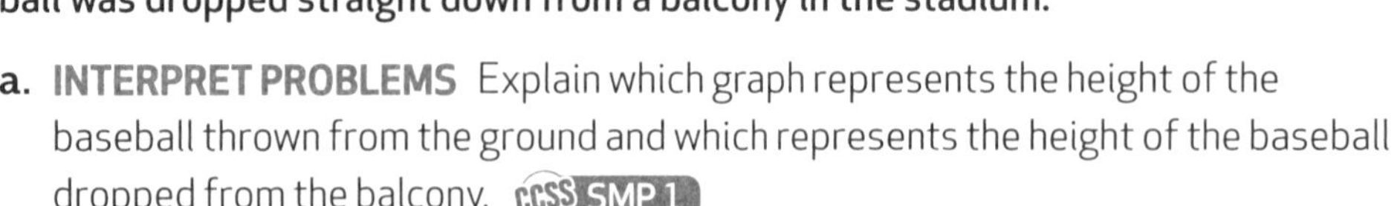

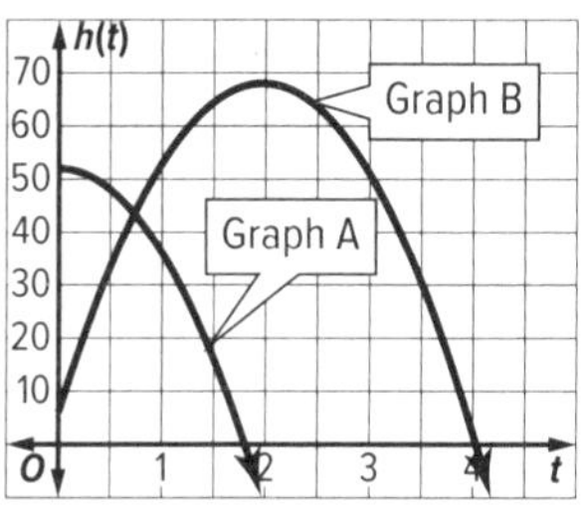

a. **INTERPRET PROBLEMS** Explain which graph represents the height of the baseball thrown from the ground and which represents the height of the baseball dropped from the balcony. CCSS SMP 1

b. **COMMUNICATE PRECISELY** Explain how to use the graph to estimate the length of time for the two balls to reach the same height. Estimate their height at that time. CCSS SMP 6

EXAMPLE 4 Use a Geometric Model CCSS F.BF.1b

EXPLORE **A piece of sheet metal has a length that is three times its width w. It is to be used to make a box with an open top by cutting out 2-inch by 2-inch squares from each corner, then folding up the sides.**

a. **USE A MODEL** Write a quadratic function that represents the volume, $V(w)$, of the box, in square inches. State the domain of the function, including any maximum value of w. CCSS F.IF.5, SMP 4

b. **CRITIQUE REASONING** Timothy states that because all of the terms of the expression for the volume are even, 2 can be factored out of the right side of the function and then divided out of each side to give $V(w) = 3w^2 - 16w + 16$. Is he correct? Explain why or why not. CCSS SMP 3

PRACTICE

1. **DESCRIBE A METHOD** The picture frame shown holds a picture for which the ratio of the length to the width is 3:2. The area of the picture and frame together is 72 square inches. Write a quadratic equation that models the area. Explain how you can use the equation to determine the dimensions of the picture only. CCSS A.SSE.1a, F.BF.1b, SMP 8

3 in.
3 in.

2. **USE STRUCTURE** The function $h(t) = -4.9t^2 + 8.8t + 3$ describes the height of a diver at some point in time after she jumped from a platform above the water. Use the values of the terms and their coefficients to create a scenario in which the height of the diver can be described by the function. Include units and clearly identify any initial values. Explain how you used the quadratic, linear, and constant terms to create your scenario. CCSS A.SSE.1a, A.SSE.1b, SMP 7

3. **USE REASONING** Compare the domains of the area function in **Example 1** and the volume function in **Example 4**. Which is more restrictive than the other? Why? For each function, relate the restrictions to the shape of the graph. CCSS F.IF.4, F.IF.5, SMP 2, SMP 3

4.3 Solving Quadratic Equations by Graphing

Objectives

- Interpret key features of the graphs of quadratic functions in terms of quantities.
- Sketch graphs given a verbal description of the relationship.

CCSS STANDARDS

Content: F.IF.4, F.IF.5, A.CED.2
Practices: 1, 3, 4, 5, 6
Use with Lesson 4-2

Quadratic equations are quadratic functions set equal to a value. The solutions of a quadratic equation are the **roots** of the equation. One method for finding the roots of a quadratic equation is the find the **zeros** of the related quadratic function.

KEY CONCEPT Solutions of a Quadratic Equation

$f(x) = 0$: Solutions are the zeros/roots of the function: $(-2, 0)$ and $(3, 0)$

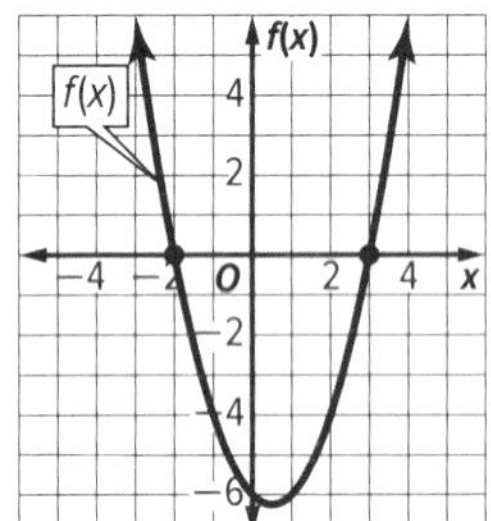

The roots of a quadratic equation correspond to the ________ of the graph of the related quadratic function. These occur where the graph of the quadratic function intersects the graph of the line defined by the function ________.

$g(x) = h(x)$: Solutions are the intersection points of the functions: $(0, 2)$ and $(2, 4)$

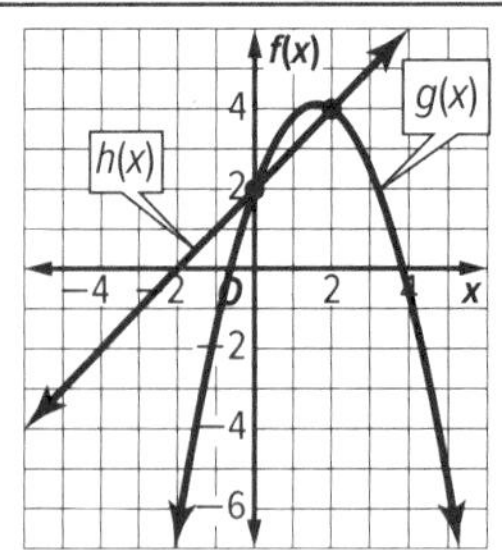

A quadratic equation that is set equal to a nonzero number or expression has solutions where the graph of the ________ intersects the graph of the related function represented by the ________.

EXAMPLE 1 Solve a Quadratic Equation by Graphing

EXPLORE The regular price of a yearbook at a high school is \$30. The school sells an average of 52 yearbooks during the summer after the school year ends. Records for the yearbook publishing company indicate that for every \$3 reduction in the price of a yearbook, the average quantity of yearbooks sold by a school during the summer is 8 more than the average quantity.

a. **USE A MODEL** The yearbook staff plans to hold a summer yearbook sale based on the data provided by the publishing company. Write a function that can be used to model $R(x)$, the revenue (in dollars) that is expected to be generated during the summer sale after x reductions in the original yearbook price. State the domain of the function in the context of the problem. CCSS F.IF.5, A.CED.2, SMP 4

b. INTERPRET PROBLEMS The yearbook staff has a target revenue of \$1200 for the summer sale. Write an equation that can be used to determine the number of \$3 reductions in price that the staff would need to make to reach this target. What type of an equation is this? Explain your reasoning. CCSS A.CED.2, SMP 1

c. COMMUNICATE PRECISELY Graph the function $y = R(x)$ on the coordinate plane. Graph the function $g(x) = 1200$ on the same coordinate plane. Label the functions and their positive intersection point. Interpret the intersection point in the context of the problem in **part b**. Include the yearbook summer sale price as part of your interpretation. CCSS A.CED.2, F.IF.4, SMP 6

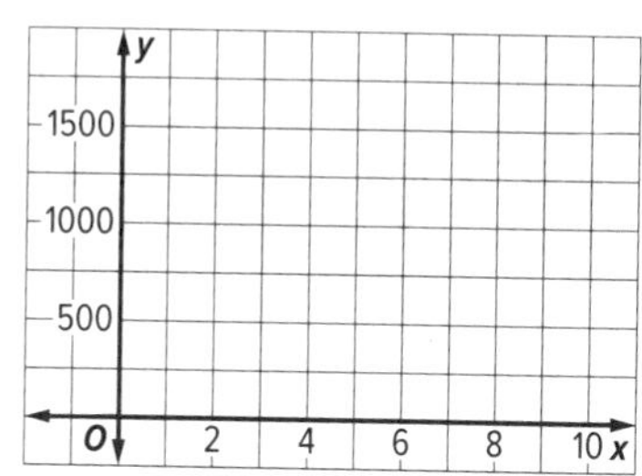

d. USE A MODEL The school must pay the publishing company d dollars for each yearbook. For which values of d and in what situation is the sale price from **part c** unreasonable? Explain your reasoning. CCSS F.IF.4, SMP 4

KEY CONCEPT Height of a Projectile

The height $h(t)$ of an object t seconds after it is dropped straight down or projected straight up is

$$h(t) = -\frac{1}{2}gt^2 + v_0t + h_0,$$

where g is the acceleration due to gravity, v_0 is the initial velocity of the object, and h_0 is the initial height.

EXAMPLE 2 Use Technology to Solve a Quadratic Equation CCSS F.IF.4, A.CED.2

EXPLORE Antoine is hiking and reaches a steep part of the trail that runs along the edge of a cliff. In order to descend more safely, he drops his heavy backpack over the edge of the cliff so that it will land on a lower part of the trail, 38.75 feet below.

a. USE A MODEL Write a quadratic function that can be used to determine the amount of time t that it will take for the backpack to land on the trail below the cliff after Antoine drops it. CCSS SMP 4

b. USE TOOLS Graph the function using a graphing calculator or other graphing utility. Sketch the graph on the coordinate grid. Label the x- and y- intercepts, the axes, and the function. CCSS SMP 5

40
30
20
10
O 1 2 3

c. **USE A MODEL** What do the x- and y-intercepts mean in the context of the problem? Explain. CCSS SMP 4

PRACTICE

1. The owner of a 120-unit storage facility analyzes her records and determines that when she charges \$55 per month, all of her storage units will be rented out. For each increase of \$5 per month, she expects 2 units to be vacated. Let x represent the number of \$5 increases over \$55.

   a. **PLAN A SOLUTION** Write an expression in terms of x that describes the number of units that will be rented after x increases of \$5 are made. Write a second expression in terms of x that describes the amount of rent per unit, in dollars, after x increases of \$5 are made. Then write a quadratic function that represents $R(x)$, the monthly revenue, in dollars, that the owner can expect to make after x increases of \$5 are made. CCSS A.CED.2, SMP 1

   b. **COMMUNICATE PRECISELY** The owner wants to determine the number of \$5 increases in rent that she should make in order to generate a monthly revenue of \$10,000. Explain how the graphs of two functions can be used to determine the number of increases. As part of your explanation, describe the exact functions that will be graphed on the coordinate plane, and what part of the graph you can use to determine whether revenue of \$10,000 is even possible. What would you do if the solution does not yield a whole number? CCSS F.IF.4, SMP 6

c. **COMMUNICATE PRECISELY** Graph the two functions from **part b** on the coordinate grid and determine how many \$5 price increases the owner should make to generate a monthly revenue of \$10,000. CCSS F.IF.4, A.CED.2, SMP 6

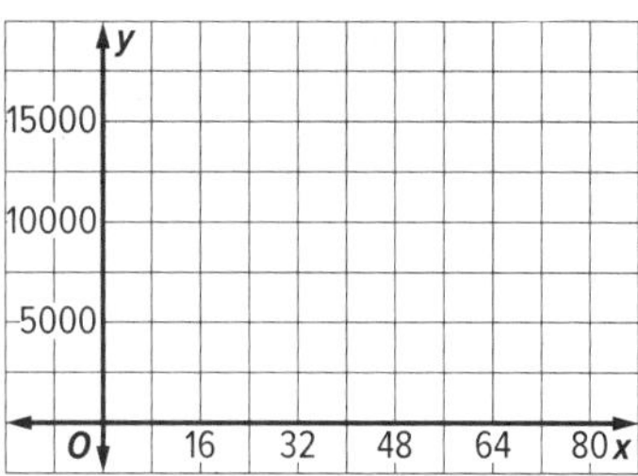

2. **MAKE A CONJECTURE** Complete the following statement that describes a pair of numbers that does not exist in the real number system: "The sum of two numbers is ____________, and their product is ____________." Then draw a graph on the coordinate grid that could represent the scenario. Explain why the graph represents a scenario with no solution in the real number system. CCSS F.IF.4, A.CED.2, SMP 3

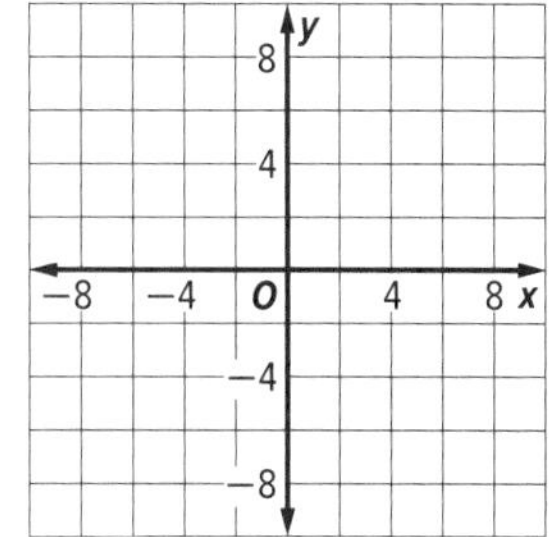

3. **COMMUNICATE PRECISELY** An acrobat springs straight up from a 4-feet high trampoline with a velocity of 28 feet per second. At the same time, a second acrobat directly above the first and 40 feet off the ground releases a baton. Determine how much time will pass until the first acrobat is able to catch the baton. Sketch the graph at the right. Label the functions with their equations set equal to $h_1(t)$ and $h_2(t)$. Label the intersection point. State your answer to the nearest tenth and with the appropriate units. CCSS A.CED.2, F.IF.4, SMP 6

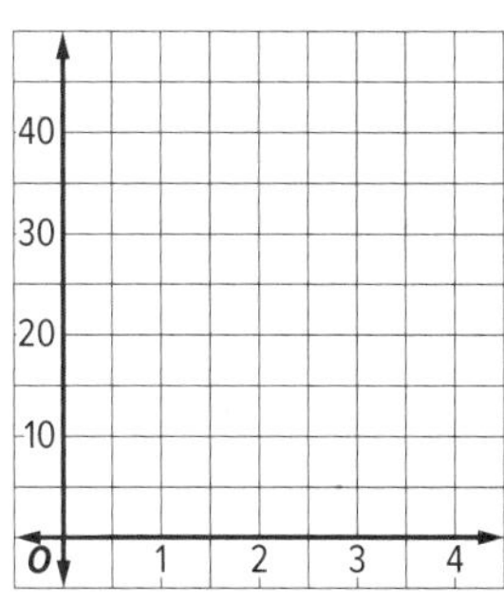

4. **CRITIQUE REASONING** Priya graphs a quadratic function to determine two real numbers with a sum of -10 and a product of -24. She writes an equation to model the situation, representing one number with x and the other with the expression $(-10 - x)$. She plots a number of points and then draws the graph shown to determine the values of the numbers.

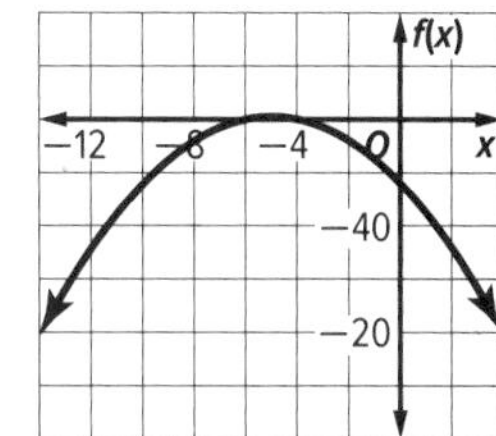

a. Priya states that because the product of the numbers is -24, one must be negative and one must be positive. She indicates that because the graph of the function shows two negative zeros, no such numbers exist to satisfy the conditions. Explain the error that Priya made. CCSS F.IF.4, SMP 3

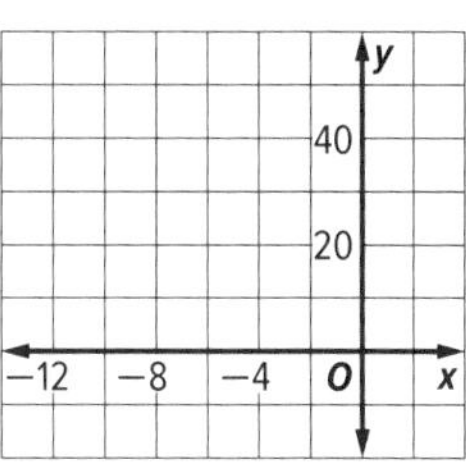

b. Correct Priya's error and solve the problem.

4.4 Solving by Factoring

Objectives

- Solve quadratic equations by factoring.
- Factor quadratic functions to determine key values, and interpret these values in the context of problem situations.

CCSS STANDARDS

Content: F.IF.8a, A.SSE.2, A.CED.1
Practices: 1, 2, 3, 6, 7
Use with Lesson 4-3

The Zero Product Property is used to solve quadratic equations by factoring.

KEY CONCEPT Zero Product Property

For any real numbers a and b, if the product $ab = 0$, then either ________ or ________.

EXAMPLE 1 Solve Equations by Factoring

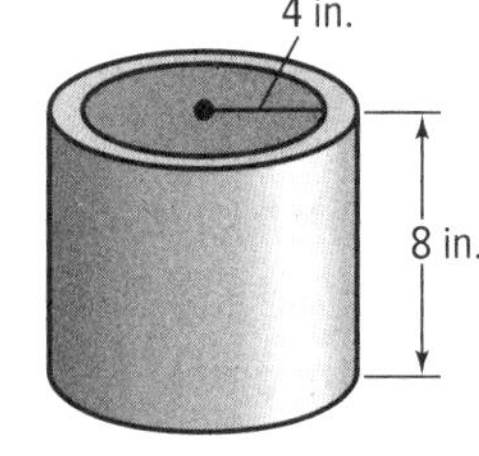

EXPLORE A company manufactures a hollow cylinder component that is closed at one end. It has uniform thickness and the dimensions shown. To make each cylinder 42π cubic inches of material is required.

a. **REASON ABSTRACTLY** Write a function to express the volume of the material that is needed to make the each cylinder. CCSS A.CED.1, A.SSE.2, SMP 2

b. **REASON QUANTITATIVELY** Write an equation that you can solve to find the required thickness of the component. CCSS A.CED.1, SMP 2

c. **USE STRUCTURE** Solve the equation you wrote in **part b**. What property did you use? CCSS F.IF.8a, A.SSE.2, SMP 7

d. **REASON QUANTITATIVELY** A customer wants to purchase a cylinder to fit into a pipe with inside diameter 8.5 inches. Will the standard cylinders produced by this company work for this customer? Explain your reasoning. CCSS A.CED.1, SMP 2

e. **INTERPRET PROBLEMS** If the company decides to make a custom cylinder for the customer in **part d**, how much material should be used to construct it? Explain your reasoning. CCSS A.CED.1, SMP 1

EXAMPLE 2 Use Factoring to Find Intersection Points CCSS A.SSE.2

The graphs of $f(x) = 3x^2 - 2x + 2$ and $g(x) = -x^2 - 7x + 8$ are provided. Determine the points for which the graphs intersect.

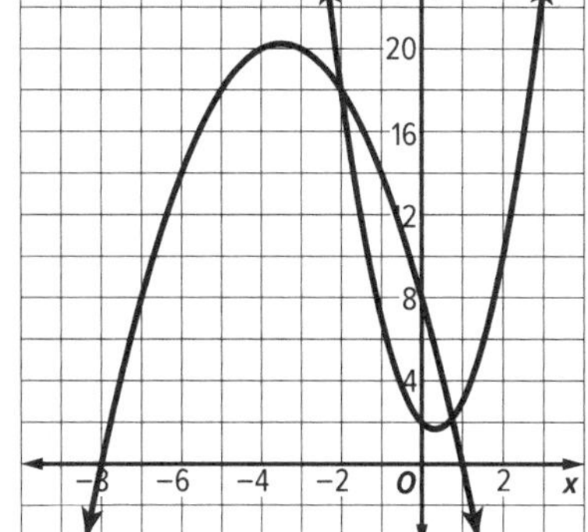

a. **INTERPRET PROBLEMS** If x is a value at which the two graphs intersect, what must be true about $f(x)$ and $g(x)$ at this point? CCSS SMP 1

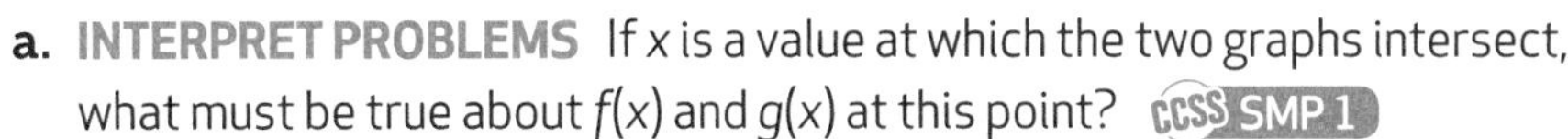

b. **REASON QUANTITATIVELY** Write an equation that can be used to solve for the x-values where the two graphs intersect. CCSS SMP 2

c. **USE STRUCTURE** Solve the equation in **part b** and use substitution to identify the points where the two graphs intersect. CCSS F.IF.8a, SMP 7

EXAMPLE 3 Create Equations and Solve by Factoring CCSS A.CED.1

A box has a base that measures 3 inches by 5 inches, and a height of 11 inches. Increasing each dimension of the base by the same amount causes the volume of the box to increase by 528 cubic inches.

a. **INTERPRET PROBLEMS** Using x as the amount that each dimension of the base is increased by, identify the new dimensions of the box and write an expression for the volume of the box. CCSS SMP 1

b. **REASON QUANTITATIVELY** Find the original volume of the box and the volume after the base increased. Write an equation that can be used to find the amount by which each dimension of the base increased. CCSS SMP 2

c. **USE STRUCTURE** Solve the equation in **part b** and interpret the solutions in the context of the problem. CCSS SMP 7

d. **REASON QUANTITATIVELY** What are the dimensions of the box after the increase? Verify that they produce a volume of 528 cubic inches greater than the original box. CCSS SMP 2

PRACTICE

1. **Carmella is asked to solve the equation $9x^2 - 12x = 5$.**

   a. **USE STRUCTURE** How can you rewrite the equation so that you can use the Zero Product Property? Then factor the expression. CCSS A.SSE.2, F.IF.8a, SMP 7

   b. **COMMUNICATE PRECISELY** Explain how this new equation can be used to determine the maximum or minimum value of $f(x) = 9x^2 - 12x - 5$, and find that value. CCSS F.IF.8a, SMP 6

2. **PLAN A SOLUTION** After being fired straight up into the air, the height of a rocket is approximated by the function $h(t) = -16t^2 + 88t + 168$, where h is measured in feet and t is measured in seconds. How many seconds after launch will the rocket hit the ground? Show your work. CCSS F.IF.8a, SMP 1

3. **PLAN A SOLUTION** How can you find out if $f(x) = 2x^2 - x + 1$ and $g(x) = 2x + 6$ intersect? If so, where do they intersect? How can you verify your answer? CCSS F.IF.8a, A.SSE.2, SMP 1

4. **PLAN A SOLUTION** Find a quadratic equation with roots $\frac{2}{7}$ and $-\frac{5}{2}$. Write the answer in standard form with integer coefficients. Show your work. CCSS A.CED.1, SMP 1

5. **REASON QUANTITATIVELY** The sum of a number and twice its reciprocal is $\frac{11}{3}$. Find the possible values for the number. Show your work. CCSS A.CED.1, F.IF.8a, SMP 2

6. **REASON ABSTRACTLY** A window is designed with a height to width ratio of 1:2. It consists of a center window pane with identical smaller panes on each side. The width of each smaller piece is 18 inches. If the area of the center window pane is 44 square feet, what are the dimensions for the entire window? CCSS A.CED.1, F.IF.8a, SMP 2

7. **CRITIQUE REASONING** Charlie and Paige are solving the equation $2x^2 + 15x + 18 = 0$ and their work is shown. Determine which, if either, of the students is correct. Explain your reasoning. CCSS F.IF.8a, SMP 3

Charlie
$2x^2 + 15x + 18 = 0$
$2x^2 + 4x + 9x + 18 = 0$
$2x(x + 2) + 9(x + 2) = 0$
$(2x + 9)(x + 2) = 0$
$x = -\frac{9}{2}, x = -2$

Paige
$2x^2 + 15x + 18 = 0$
$2x^2 + 12x + 3x + 18 = 0$
$2x(x + 6) + 3(x + 6) = 0$
$(2x + 3)(x + 6) = 0$
$x = -\frac{3}{2}, x = -6$

8. **COMMUNICATE PRECISELY** Explain the differences between the roots of $f(x) = 4x^2 - 20x + 25$ and $g(x) = 4x^2 - 16x + 15$. Discuss the number of roots, using a graph to illustrate your conclusion. CCSS A.SSE.2, SMP 6

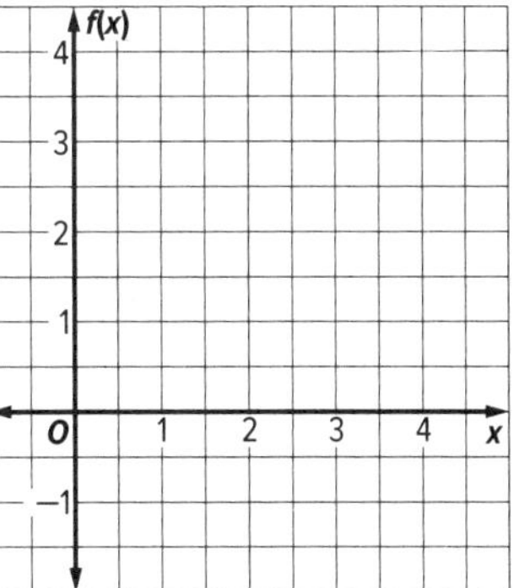

9. **MAKE A CONJECTURE** Find the roots of each of the following functions, $f(x) = 2x^2 - 24x + 70$ and $g(x) = 3x^2 - 9x - 30$. What does a common solution represent with respect to the graphs of $f(x)$ and $g(x)$? Graph the functions to justify your conclusion. CCSS A.SSE.2, SMP 3

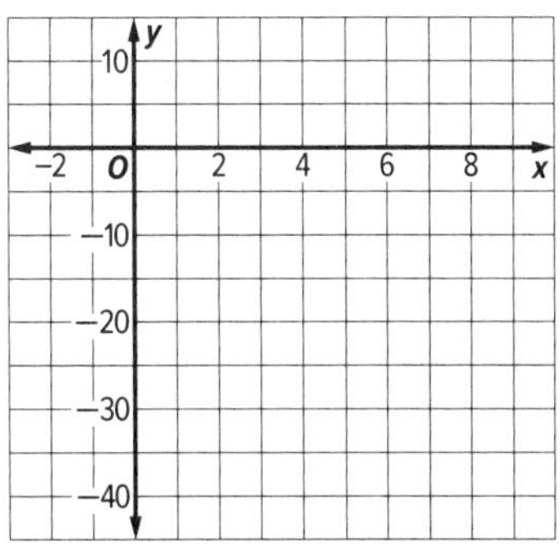

10. **REASON ABSTRACTLY** The Morinho family has three children, Jorge, Isabel, and Miranda, and their current ages are consecutive odd integers. Twice the product of the ages of the oldest and youngest child is 10 more than 16 times the age of the middle child. Write and solve an equation that represents the given information. What are the ages of the three children? CCSS A.CED.1, SMP 2

4.5 Solving by Completing the Square

Objectives

- Solve quadratic equations by completing the square.
- Use the process of completing the square with quadratic functions to determine key values, and interpret these values in the context of problem situations.

CCSS STANDARDS

Content: F.IF.8a, N.CN.7, A.CED.1, F.IF.4

Practices: 1, 2, 3, 6, 7

Use with Lesson 4-5

It is not always possible to solve a quadratic equation by factoring. But it is possible to manipulate the equation until one side is a perfect square, and then take the square root of each side. This process is called **completing the square**.

KEY CONCEPT Completing the Square

To complete the square for the quadratic expression $x^2 + bx$, use the following steps:

Step 1: Find one half of b, the coefficient of x.
Step 2: Square the result from Step 1.
Step 3: Add the result of Step 2 to $x^2 + bx$. Symbols: $x^2 + bx + \left(\frac{b}{2}\right)^2 = \left(x + \frac{b}{2}\right)^2$

EXAMPLE 1 Complete the Square CCSS F.IF.8a

EXPLORE A detective is investigating a traffic accident at an intersection. The car suspected of causing the accident left skid marks stretching 150 feet on the road. The detective uses the following formula for braking distance: $d = \frac{s^2}{20} + 2s$, where d is the distance required to stop on dry pavement and s is the speed of the vehicle.

a. **REASON QUANTITATIVELY** Solve the equation for a car that travels 150 feet before coming to a complete stop. Show your work. CCSS SMP 2

b. **CONSTRUCT ARGUMENTS** The speed limit at the intersection is 25 miles per hour. How fast was the car traveling, and should the driver be ticketed for speeding? CCSS SMP 3

EXAMPLE 2 Complete the Square to Find Zeros CCSS F.IF.8a

Let $f(x) = 4x^2 + 4x - 15$.

a. **PLAN A SOLUTION** Write an equation that can be used to find the zeros of $f(x)$. CCSS SMP 1

b. **CALCULATE ACCURATELY** Isolate the x terms in the equation and complete the square to get an equation of the form $(x - h)^2 = k$. CCSS SMP 6

c. **REASON QUANTITATIVELY** Solve the equation from **part b** and interpret any solutions. CCSS SMP 2

Not all solutions of quadratic equations are real numbers. In some cases, the process of finding a solution requires taking the square root of a negative number, resulting in complex solutions.

Complex Solutions

A complex number is any number that can be written in the form $a + bi$, where a and b are real numbers and $\boldsymbol{i} = \sqrt{-1}$ is the imaginary unit. If the graph of a quadratic function does not have any x-intercepts, the solutions to the related quadratic equation are complex numbers.

EXAMPLE 3 Complex Solutions to Quadratic Equations CCSS N.CN.7

Let $f(x) = 9x^2 - 12x + 148$.

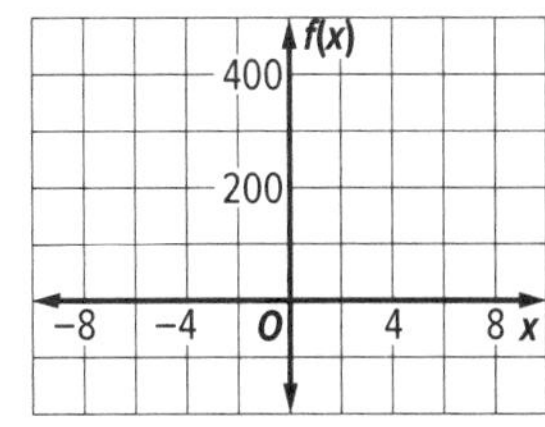

a. **PLAN A SOLUTION** Write an equation that can be used to find the zeros of $f(x)$. CCSS SMP 1

b. **INTERPRET PROBLEMS** Can the equation from **part a** be solved graphically? Can it be solved by factoring? Explain. CCSS SMP 1

c. **CALCULATE ACCURATELY** Isolate the x terms in the equation and complete the square to get an equation of the form $(x - h)^2 = k$. CCSS SMP 7

d. INTERPRET PROBLEMS Solve the equation from **part b**. Write the solutions in the form $a + bi$. CCSS SMP 1

e. REASON QUANTITATIVELY What do the solutions tell you about the zeros of $f(x)$? How do they relate to the graph of $f(x)$? Explain your reasoning. CCSS SMP 2

PRACTICE

1. PLAN A SOLUTION If the length of one side of a square is increased by 4 centimeters and the length of an adjacent side is decreased by 8 centimeters, the area of the figure is decreased by 20%. Find the length of each side of the square to the nearest tenth. CCSS A.CED.1, F.IF.8a, SMP 1

2. PLAN A SOLUTION A rectangle is constructed with one side on the x-axis, one side on the y-axis, and a vertex on the line $4x + y = 16$. If the rectangle is to have an area of 7, what are the dimensions of the rectangle? Show your work and include a diagram. CCSS A.CED.1, F.IF.8a, SMP 1

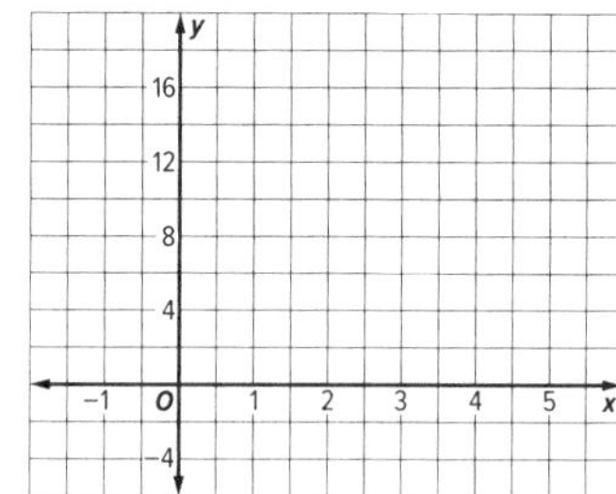

3. USE STRUCTURE Write a quadratic equation with real coefficients and a root of $8 - 2i$. Show your work. CCSS N.CN.7, SMP 7

4. USE STRUCTURE Write a quadratic equation with roots $\frac{1}{2} + \frac{3\sqrt{3}}{2}$ and $\frac{1}{2} - \frac{3\sqrt{3}}{2}$. Show your work. CCSS A.CED.1, SMP 7

5. **CONSTRUCT ARGUMENTS** If a quadratic function has two complex imaginary roots, where are those roots located on the graph of the function? Explain your reasoning. CCSS F.IF.4, SMP 3

6. **PLAN A SOLUTION** Use completing the square to find the solutions to $x^2 + 2x + 11 = 0$. Show your work. CCSS N.CN.7, SMP 1

7. **CONSTRUCT ARGUMENTS** For $x^2 + kx + 9 = 0$, find two integer values of k such that the equation has two real solutions and two integer values of k such that the equation has two solutions that are not real numbers. Show your work. CCSS F.IF.8a, N.CN.7, SMP 3

8. **PLAN A SOLUTION** For which values of k will $f(x) = x^2 - 2kx + 5k$ have a minimum value of -6? Explain your reasoning and verify your answer graphically. CCSS F.IF.8a, SMP 1

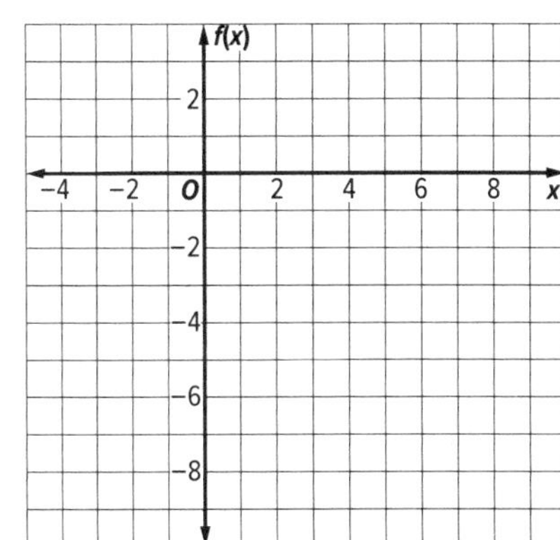

9. **PLAN A SOLUTION** Find the solution set to $x^2 - 6x + 4 = 0$ by completing the square. Show your work. CCSS F.IF.8a, SMP 1

10. **USE STRUCTURE** Write a quadratic equation with roots $4 + \sqrt{2}$ and $4 - \sqrt{2}$. Show your work. CCSS A.CED.1, SMP 7

11. **USE STRUCTURE** Write a quadratic equation with roots $\frac{3}{5} + \frac{\sqrt{2}}{5}\boldsymbol{i}$ and $\frac{3}{5} - \frac{\sqrt{2}}{5}\boldsymbol{i}$. Show your work. CCSS A.CED.1, SMP 7

4.6 Solving by Using the Quadratic Formula and the Discriminant

Objectives

- Use the quadratic formula to solve quadratic equations.
- Analyze quadratic equations that have complex solutions.

CCSS STANDARDS

Content: N.CN.7, N.CN.8, A.CED.1, A.CED.4, A.APR.4, F.IF.4
Practices: 1, 2, 3, 7
Use with Lesson 4-6

You have solved some quadratic equations by graphing, by factoring, and by completing the square. There is also a formula that can be used to solve any quadratic equation.

KEY CONCEPT Quadratic Formula

The solutions of a quadratic equation of the form $ax^2 + bx + c = 0$, where $a \neq 0$, are given by the following formula: $x = \dfrac{-b \pm \sqrt{b^2 - 4ac}}{2a}$.

EXAMPLE 1 Using the Quadratic Formula CCSS A.CED.1

EXPLORE A train hauling coal leaves the train yard traveling west at 32 miles per hour. Five hours later a train carrying appliances leaves the train yard traveling north at 40 miles per hour.

a. **REASON ABSTRACTLY** Construct a diagram of the situation and write a function to express the distance between the two trains x hours after the train carrying appliances has departed. CCSS SMP 2

b. **PLAN A SOLUTION** How long after the train carrying appliances departs will the distance between the two trains be 260 miles? Show your work. CCSS SMP 1

When using the quadratic formula, if the value of the radicand is negative, the solutions will be complex numbers.

EXAMPLE 2 Complex Solutions to Quadratic Equations CCSS N.CN.7

a. **REASON QUANTITATIVELY** Are you able to determine any information about the graph of a quadratic function if you know that the function has complex zeros? Explain your reasoning. CCSS SMP 2

b. CONSTRUCT ARGUMENTS Jamila sells knitted hats to her friends at school, and has determined that the revenue formula is $R(p) = -p^2 + 20p$, where R is the revenue in dollars and p is the selling price of her hats. She wants to make \$200 during her sophomore year. Can she achieve this goal, and what price would she have to charge? Use the quadratic formula and explain your reasoning. CCSS SMP 3

c. PLAN A SOLUTION What are the solutions to the quadratic formula that you found in **part b**? How can you rewrite the equation for the revenue formula with a desired revenue of \$200 in factored form? CCSS N.CN.8, SMP 1

The expression $b^2 - 4ac$, which appears under the radical, is called the discriminant. The value of the discriminant can be used to determine the number and type of roots of a quadratic equation.

KEY CONCEPT Discriminant

The nature of the solutions of $ax^2 + bx + c = 0$ are given as follows:

If $b^2 - 4ac > 0$, the equation has 2 distinct real roots.
If $b^2 - 4ac = 0$, the equation has 1 real root.
If $b^2 - 4ac < 0$, the equation has 2 complex roots.

EXAMPLE 3 Using the Discriminant CCSS N.CN.7

a. CONSTRUCT ARGUMENTS Without finding the roots of $f(x) = 7x^2 - 29x + 31$, determine the number of points of intersection between the graph and the x-axis and explain your reasoning. CCSS SMP 3

b. USE STRUCTURE If $ax^2 + 9x + 1.5$ has two real roots, what are the possible values of a? CCSS SMP 7

c. USE STRUCTURE The graph of a quadratic function $f(x) = ax^2 + bx + c$ is shown. What is the value for c? What are the constraints on a? Explain your reasoning. CCSS F.IF.4, SMP 7

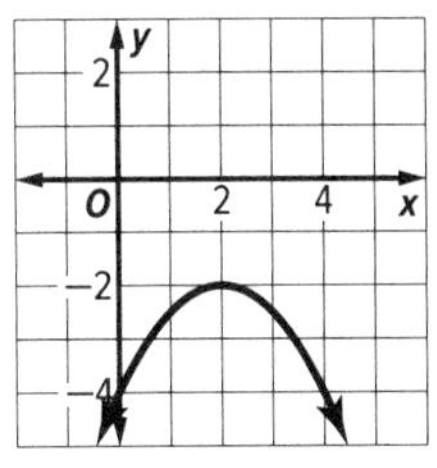

PRACTICE

1. **USE STRUCTURE** Use the quadratic formula to express the roots of $f(x) = ax^2 + bx + c$ in terms of a, c, and h (the x-coordinate of the vertex). CCSS A.CED.4, SMP 7

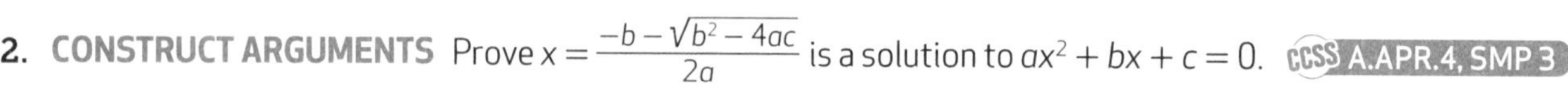

2. **CONSTRUCT ARGUMENTS** Prove $x = \frac{-b - \sqrt{b^2 - 4ac}}{2a}$ is a solution to $ax^2 + bx + c = 0$. CCSS A.APR.4, SMP 3

3. **PLAN A SOLUTION** Construct graphs of the function $f(x) = x^2 + nx + n$, using several different values of n. Explain how to find the values of n for which the graph of $f(x)$ will not intersect the x-axis. Then find the solution set. CCSS N.CN.7, SMP 1

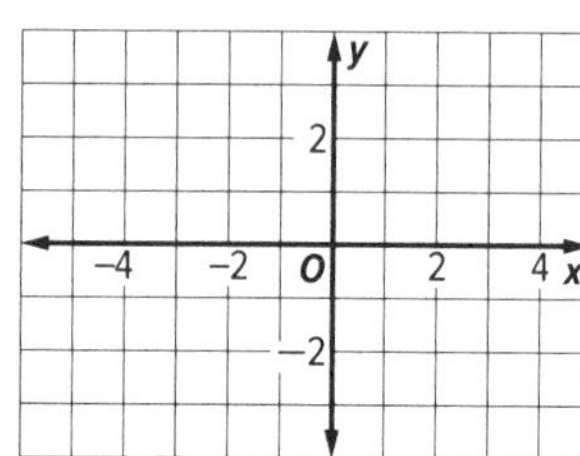

4. **PLAN A SOLUTION** Show how to rewrite the sum of cubes formula as the product of three linear factors. CCSS A.CED.1, SMP 1

5. **CRITIQUE REASONING** Jennifer states that "the product of the roots of the quadratic equation $ax^2 + bx + a = 0$ is 1". Use the quadratic formula to prove or disprove her statement. CCSS A.APR.4, SMP 3

6. **PLAN A SOLUTION** A carnival game has players hit a pad with a large rubber mallet. This fires a ball up a 20-foot vertical chute toward a target at the top. A prize is awarded if the ball hits the target. Explain how to find the initial velocities (in feet per second) for which the ball will fail to hit the target. The height is modeled by the function $h(t) = -16t^2 + vt$, where v is the initial velocity. CCSS A.CED.1, N.CN.7, SMP 1

7. **USE STRUCTURE** The quadratic equation $7x^2 + bx + 5 = 0$ has two complex roots. What are the possible values of b? Show your work. CCSS N.CN.7, SMP 7

8. **PLAN A SOLUTION** A rectangular box has a square base and a height that is one more than 3 times the length of a side of the base. If the sides of the base are each increased by 2 inches and the height is increased by 3 inches, the volume of the box increases by 531 cubic inches. What are the dimensions of the original box? CCSS A.CED.1, SMP 1

9. **PLAN A SOLUTION** What are the roots of the quadratic equation $x^2 + 2x + 5 = 0$? Using these roots, write the equation in factored form. CCSS N.CN.8, SMP 7

10. **CRITIQUE REASONING** Raoul claims that the sum of the roots of a quadratic equation is always equal to $\frac{-b}{a}$. Do you agree? CCSS A.CED.1, SMP 3

11. **REASON QUANTITATIVELY** Let $f(x) = 3x^2 - kx + m$ where k and m are real numbers. CCSS A.CED.1, SMP 2

    a. Find the roots of $f(x)$.

    b. For what values of k and m does $f(x)$ have complex roots?

4.7 Transforming Quadratic Functions

Objectives

- Complete the square to write quadratic functions in vertex form
- Analyze the effects of transformations on the graphs of quadratic functions

Content: F.IF.8a, F.BF.3
Practices: 1, 2, 3, 4, 7
Use with Lesson 4-7

When a quadratic function is written in standard form, you can complete the square to rewrite it in vertex form. The vertex form of a quadratic function is $f(x) = a(x - h)^2 + k$, where (h, k) is the vertex of the parabola, $x = h$ is the axis of symmetry, and a determines the shape of the parabola and the direction in which it opens.
If the coefficient of the quadratic term is not 1, remember to factor the coefficient from the quadratic and linear terms before completing the square.

EXAMPLE 1 Maximizing Revenue CCSS F.IF.8a

EXPLORE **A company has designed a new T-shirt for their line of fitness apparel. To determine a retail price for the T-shirt, the company uses a revenue function defined by $R(p) = 8000p - 200p^2$, where R is the estimated annual revenue and p is the sales price, both measured in dollars.**

a. **REASON QUANTITATIVELY** Use the technique of completing the square to rewrite the function in vertex form. CCSS SMP 2

b. **PLAN A SOLUTION** Construct and describe a graph of the revenue function. CCSS SMP 1

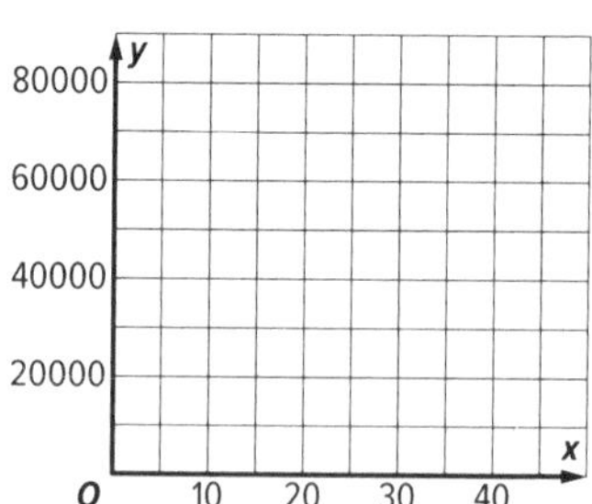

c. **USE A MODEL** What is the retail price the company should select to maximize their potential revenue? At this price, how many T-shirts can the company expect to sell each year? Explain your reasoning. CCSS SMP 4

d. **REASON ABSTRACTLY** Interpret the meaning of the intercepts of the graph, within the context of the problem. CCSS SMP 2

EXAMPLE 2 Height of a Projectile CCSS F.IF.8a

EXPLORE In a popular contest called "Pumpkin Chucking," contestants compete to hurl a pumpkin solely by a catapult. Assume the height of a pumpkin launched by a catapult is modeled by the function $h(t) = -16t^2 + 192t + 8$, where h is measured in feet and t is measure in seconds.

a. **PLAN A SOLUTION** Rewrite the function in vertex form. Describe and sketch the graph of the function. CCSS SMP 1

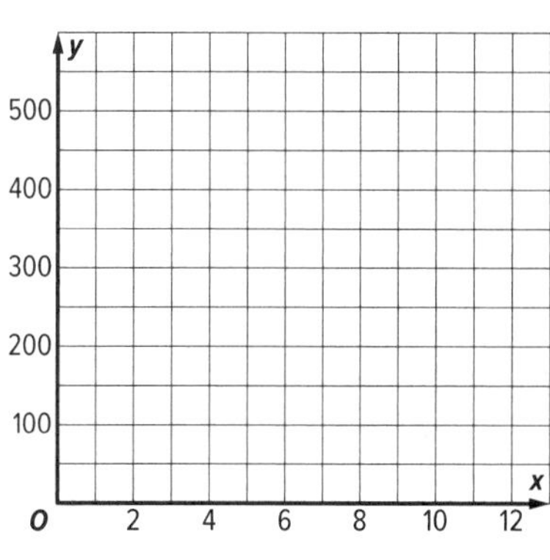

b. **USE A MODEL** What is the maximum height reached by the pumpkin, and how long after launching does this occur? Explain your reasoning. CCSS SMP 4

c. **REASON ABSTRACTLY** Interpret the meaning of the intercepts of the graph, within the context of the problem. CCSS SMP 2

d. **REASON QUANTITATIVELY** Another catapult at the competition launches a pumpkin with height modeled by the function $g(t) = -16t^2 + 200t + 5$. Will this pumpkin reach a height higher than that of the first pumpkin? Which pumpkin is in the air for a longer period of time? Explain your reasoning. CCSS SMP 2

CONCEPT SUMMARY Transformations of Quadratic Functions

The graph of $f(x) = a(x - h)^2 + k$ displays the following transformations:

Horizontal Translation	h units to the ______ if ______, h units to the ______ if ______.
Vertical Translation	k units ______ if ______, k units ______ if ______.
Reflection	Opens up if ______, opens down if ______.
Dilation	Vertical stretch by a factor of ______, for $\|a\| >$ ______. Vertical compression by a factor of ______, for ______ $< \|a\| <$ ______.

EXAMPLE 3 Transformations of Quadratic Graphs CCSS F.BF.3

a. PLAN A SOLUTION Find an equation of the function shown in the graph using vertex form. Show your work. CCSS SMP 1

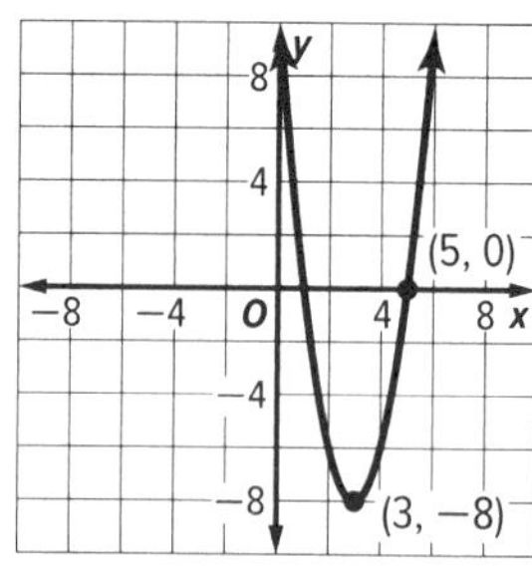

b. PLAN A SOLUTION Describe the transformations that are required to produce the graph of $g(x) = -3x^2 - 6x + 4$ from the graph of $f(x) = x^2$. Show your work. CCSS SMP 1

PRACTICE

1. **PLAN A SOLUTION** Write the function $f(x) = 5x^2 + 25x + 13$ in vertex form. State the coordinates of the vertex. Show your work. CCSS F.IF.8a, SMP 1

2. **PLAN A SOLUTION** Graph the function $f(x) = 8x^2 - 64x + 115$. Show your work. CCSS F.IF.8a, SMP 1

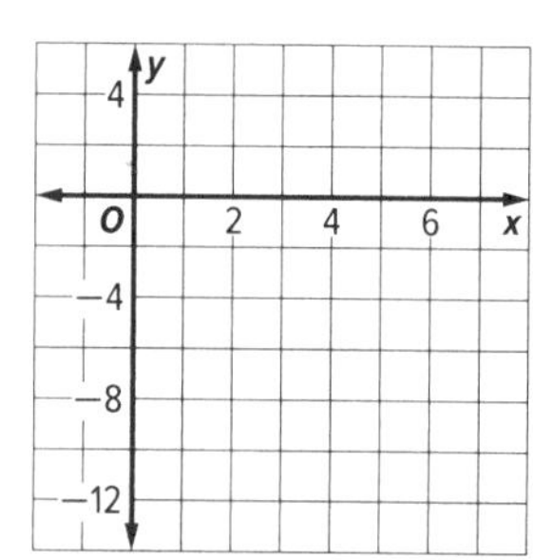

3. The area of a garden enclosed by 60 feet of fencing and divided into two equal plots is modeled by the function $A(x) = -\frac{3}{2}x^2 + 30x$, where A is the area in square feet.

a. **USE A MODEL** Rewrite the function in vertex form and determine the maximum possible area of the garden. Show your work. CCSS F.IF.8a, SMP 4

b. **USE STRUCTURE** Explain how the graph of $A(x)$ is related to the graph of $y = x^2$. Explain in terms of the transformations necessary to produce the graph of $A(x)$ from the graph of $y = x^2$. CCSS F.BF.3, SMP 7

4. **CONSTRUCT ARGUMENTS** Gabrielle examines the two functions represented and states that $g(x)$ has a greater maximum value than $f(x)$. Is she correct? Explain why or why not. CCSS F.IF.8a, SMP 3

$f(x)$

$f(x) = -\frac{1}{2}x^2 - 6x - 10$

$g(x)$

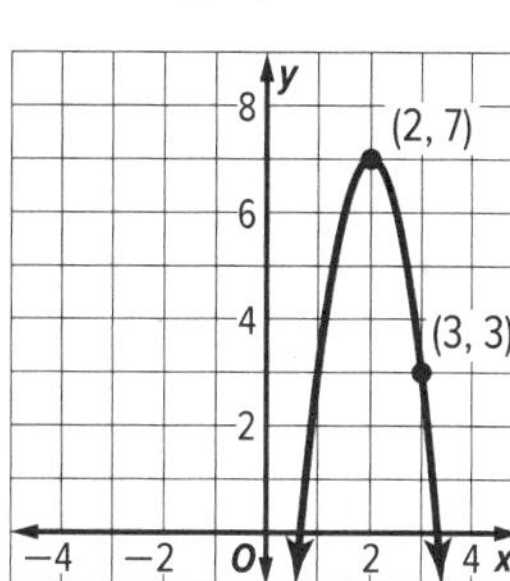

5. **PLAN A SOLUTION** Describe the transformations that are required to produce the graph of $g(x) = x^2 - 10x + 29$ from the graph of $f(x) = -2x^2 - 12x - 19$. Show your work. CCSS F.BF.3, SMP 1

Performance Task

Talking Parabolas

Provide a clear solution to the problem. Be sure to show all of your work, include all relevant drawings, and justify your answers.

A parabola is the set of all points in a plane that are the same distance from a given point known as the focus and a given line known as the directrix. Given a parabola represented by an equation of the form $y = ax^2$, the coordinate of the focus is $\left(0, \frac{1}{4a}\right)$.

Parabolic reflector dishes are formed by "spinning" a parabola around its axis of symmetry to create a bowl. They have the unique property that any wave collected by the dish parallel to the axis of symmetry is reflected to the focus of the parabola (point F in the diagram below). In a similar manner, waves originating at the focus are reflected back off the dish parallel to the axis of symmetry. Reflector dishes are used to collect signals in a wide array of applications. A dish known as the "whisper dish" focuses sound waves and amplifies them to any ear located at the focus.

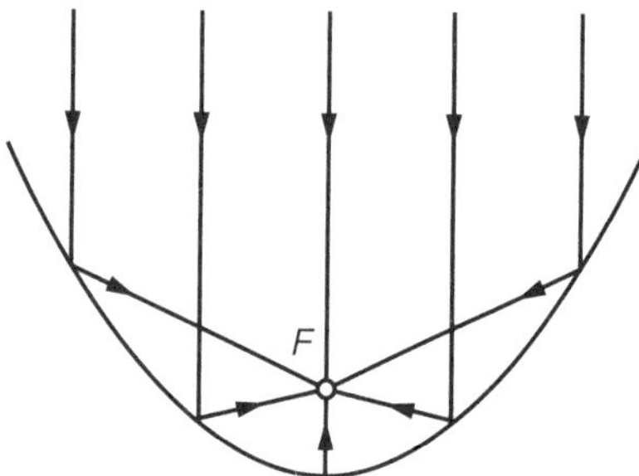

Cross section of the interaction of incoming waves with a parabolic reflector

Part A

The cross section through the center of one such dish, located in Los Angeles, California, can be represented by $y = 0.1x^2$, where both x and y are in feet. How far is the focus from the vertex of the dish?

Part B

If the whisper dish in **Part A** has a circular diameter of 6 feet, with its vertex at (0, 0), determine the dimensions of the smallest rectangular box it can be stored in if the dish faces straight down. (Note that the walls of the box are 0.5 inch thick.)

Part C

Two whisper dishes are often placed directly facing each other, so that one person can whisper into the focus of one dish and another person can hear them at the focus of the other. If a second dish of the same size is placed facing the dish in **Part A**, determine the equation that represents the second dish if their foci are 28 feet apart.

Performance Task

A Backyard Pool

Provide a clear solution to the problem. Be sure to show all of your work, include all relevant drawings, and justify your answers.

A customer of Saul's Swimming Pool Spot has asked that a custom pool be designed for the customer's yard.

Part A

The customer requires that the pool is rectangular and has a perimeter of 100 meters. Write expressions for the length and width of the pool. What is the maximum surface area possible with these dimensions?

Part B

If there is to be a deck 1.5 meters wide around the entire pool, what is the area of the deck, assuming you use the maximized pool area in **Part A**?

Part C

Show that a 1.5 m wide deck will always have an area of 159 square meters, regardless of the value of x, using the expressions for length and width given in **Part A**.

Part D

Write an equation that represents the area A of the deck in terms of its width, q. Use the expressions for length and width given in **Part A**.

1. Graph the function $f(x) = 2x^2 + 4x - 6$.
CCSS A.CED.2

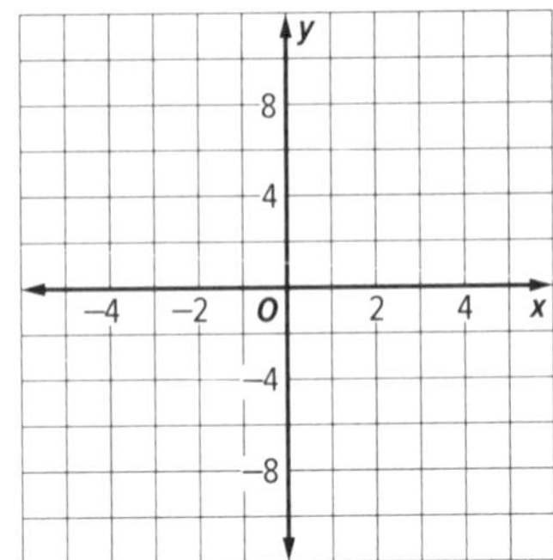

Complete the following: CCSS F.IF.4, F.IF.5

y-intercept: [ ]

x-intercepts: [ ] and [ ]

Minimum value of [ ] occurs at the point [ ].

Graph is symmetric with respect to the line [ ]

Domain: [ ]

2. In factored form, the function $f(x) = 3x^2 + 10x + 8$ is [ ]. Therefore, the zeros of the function are [ ] and [ ].
CCSS F.IF.8a

3. Use the quadratic formula to solve $5x^2 - x + 1 = 0$. CCSS N.CN.7

$x =$ [ ]

4. The expression $49x^2 - 121$ is a difference of [ ]. Therefore, the function $h(x) = 49x^2 - 121$ can be factored as [ ], and the zeros are [ ] and [ ]. CCSS A.SSE.2, F.IF.8a

5. Solve for x. CCSS F.IF.8a

$-5x^2 + 7 = -4$

$x =$ [ ]

6. The quadratic function $g(x)$ passes through the points $(0, 27)$, $(4, 3)$, and $(-2, 51)$. Write $g(x)$ in vertex form. CCSS A.CED.2

[ ]

7. Use the quadratic formula to solve $3x^2 + 4x - 1 = 0$. CCSS N.CN.7

$x =$ [ ]

8. Which of the following equations describe the function shown in the graph? CCSS A.SSE.1a, A.CED.2

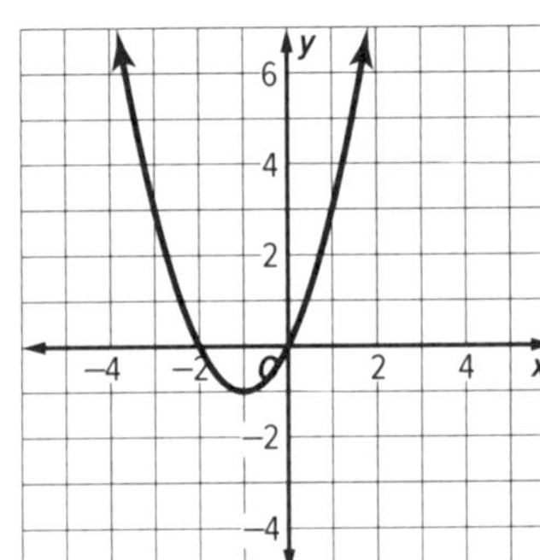

$f(x) = (x - 2)^2$ $\qquad$ $f(x) = x^2 + 2x$

$f(x) = x(x + 2)$ $\qquad$ $f(x) = (x - 1)^2 + 1$

$f(x) = (x + 1)^2 - 1$ $\qquad$ $f(x) = x^2 - 2x$

9. What is the sum of the solutions of the equation $3x^2 - 12x + 5 = 0$? CCSS N.CN.7

-4 $\qquad$ $2\sqrt{\frac{21}{3}}$

$-2\sqrt{\frac{21}{3}}$ $\qquad$ 4

10. Compare each function to $f(x) = x^2$. Indicate how many units the graph is shifted up, down, left, or right from $f(x)$. If the graph is reflected, give the line of reflection. CCSS F.BF.3

Function	Shift Up	Shift Down	Shift Left	Shift Right	Line of Reflection
$g(x) = (x + 2)^2 - 3$					
$h(x) = -3x^2 + 12$					
$j(x) = -(x - 1)^2$					

11. What value of c makes $x^2 + \frac{1}{3}x + c$ a perfect square trinomial? Show your work. CCSS F.IF.8a

12. Check the column that best describes the zero(s) of each function. CCSS N.CN.7

Function	Discriminant	Two Real Solutions	One Real Solution	Two Complex Solutions
$f(x) = 9x^2 - 30x + 25$				
$f(x) = x^2 + 6x + 11$				
$f(x) = 0.5x^2 - 2x - 9$				

13. Pablo stands in the grandstands 8 feet above a soccer field and kicks a ball with an initial vertical velocity of 28 feet per second. The equation $h(t) = -16t^2 + 28t + 8$ gives the height of the ball from the field t seconds after it is kicked.

   a. What equation can you solve to find the time it will take the ball to reach the ground? Explain your answer. CCSS A.REI.11

   b. Solve the equation from **part a** by factoring. Show your work. Are your solutions reasonable in the context of the problem? CCSS A.CED.1, F.IF.8a

14. Consider the function $f(x) = 2x^2 + 4x + 17$.

   a. Transform the function to vertex form, and state the vertex. CCSS F.IF.8a

   b. Is the vertex a maximum or a minimum? Explain how you know. CCSS A.SSE.1a

   c. Compare the graph of $f(x)$ to the graph of $g(x) = x^2$. CCSS F.BF.3

5 Polynomial Functions

CHAPTER FOCUS Learn about some of the Common Core State Standards that you will explore in this chapter. Answer the preview questions. As you complete each lesson, return to these pages to check your work.

What You Will Learn	Preview Question
Lesson 5.1: Operations with Polynomials	
CCSS A.APR.1 Understand that polynomials form a system analogous to the integers, namely, they are closed under the operations of addition, subtraction, and multiplication; add, subtract, and multiply polynomials.	CCSS SMP 7 Which of the following is not a polynomial? Explain. **A.** $(x + 1)(x^2 - 1)$ **B.** $(x^2 + 1) + 3x^3$ **C.** $\frac{x+1}{x-1}$
Lesson 5.2: Dividing Polynomials	
CCSS A.APR.6 Rewrite simple rational expressions in different forms; write $a(x)/b(x)$ in the form $q(x) + r(x)/b(x)$, where $a(x)$, $b(x)$, $q(x)$, and $r(x)$ are polynomials with the degree of $r(x)$ less than the degree of $b(x)$, using inspection, long division, or, for the more complicated examples, a computer algebra system.	CCSS SMP 5 Simplify this expression by factoring the numerator. $\frac{x^2 - 4x - 12}{x + 2}$
Lesson 5.3: Power Functions	
CCSS F.IF.4 For a function that models a relationship between two quantities, interpret key features of graphs and tables in terms of the quantities, and sketch graphs showing key features given a verbal description of the relationship.	CCSS SMP 1 Write a cubic function $f(x)$. Graph the function. Describe the end behavior. 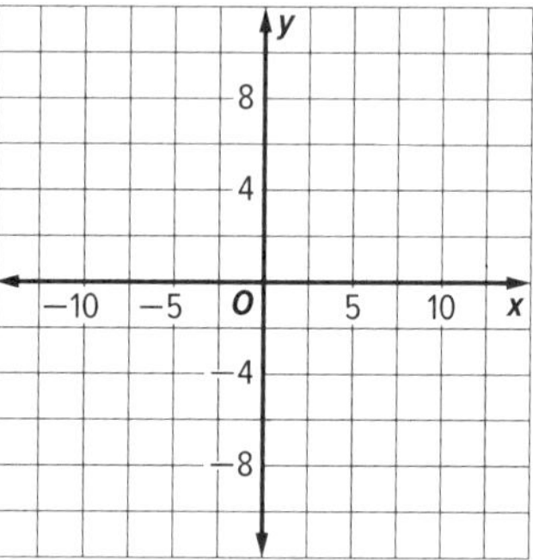
Lesson 5.4: Polynomial Functions	
CCSS F.IF.5 Relate the domain of a function to its graph and, where applicable, to the quantitative relationship it describes. **Also addresses:** F.IF.4, F.IF.7c, F.IF.9, A.CED.2, A.SSE.2	CCSS SMP 8 Suppose $f(x)$ is a power function with an odd-numbered exponent. What is the minimum number of times it will intersect the x-axis?

What You Will Learn	Preview Question
Lesson 5.5: Analyzing Graphs of Polynomial Functions	
CCSS A.CED.2 Create equations in two or more variables to represent relationships between quantities; graph equations on coordinate axes with labels and scales. CCSS F.IF.4 For a function that models a relationship between two quantities, interpret key features of graphs and tables in terms of the quantities, and sketch graphs showing key features given a verbal description of the relationship. CCSS F.IF.7c Graph polynomial functions, identifying zeros when suitable factorizations are available, and showing end behavior.	CCSS SMP 6 A cube x units on each side has the side lengths increased by 1, 2, and 3 units, respectively. Write and graph the corresponding cubic function for the volume of the solid. y; 240; 160; 80; O; 2; 4; 6; 8; x
Lesson 5.6: Modeling: Polynomial Functions	
CCSS F.IF.6 Calculate and interpret the average rate of change of a function (presented symbolically or as a table) over a specified interval. Estimate the rate of change from a graph. **Also addresses:** A.SSE.1a, A.SSE.1b, A.CED.4, F.IF.4, F.IF.5	CCSS SMP 5 The volume of a cube is modeled by the function below. Use a graphing calculator to find the maximum volume of the cube for values of x less than 10. $y = x^3 - 12x^2 + 44x - 48$
Lesson 5.7: Solving Polynomial Equations	
CCSS A.REI.11 Explain why the x-coordinates of the points where the graphs of the equations $y = f(x)$ and $y = g(x)$ intersect are the solutions of the equation $f(x) = g(x)$; find the solutions approximately, e.g., using technology to graph the functions, make tables of values, or find successive approximations. **Also addresses:** A.CED.1	CCSS SMP 8 Suppose a parabola intersects the x-axis at 3 and 5. Identify the general form of a parabola that meets these conditions.
Lesson 5.8: The Remainder and Factor Theorems	
CCSS F.IF.7c Graph polynomial functions, identifying zeros when suitable factorizations are available, and showing end behavior. **Also addresses:** A.APR.2, A.CED.1	CCSS SMP 2 Suppose x is an integer factor of integer y. Is $y \div x$ an integer? Explain.
Lesson 5.9: Roots and Zeros	
CCSS F.IF.4 For a function that models a relationship between two quantities, interpret key features of graphs and tables in terms of the quantities, and sketch graphs showing key features given a verbal description of the relationship. **Also addresses:** N.CN.9, A.APR.3, A.CED.1	CCSS SMP 6 Suppose $f(x)$ is a quadratic function that intersects the x-axis once. Describe the function.

5.1 Operations with Polynomials

Objectives

- Recognize that addition, subtraction, and multiplication of polynomials are closed.
- Add, subtract, and multiply polynomials.

Content: A.APR.1
Practices: 1, 2, 3, 4, 6, 7, 8
Use with Lesson 5-1

A monomial is a number, a variable, or a product of a number and one or more variables and a polynomial is a monomial or a sum of monomials. The **degree of a polynomial** is the greatest degree of any term in the polynomial. To **simplify** a polynomial, it must be rewritten without parentheses or negative exponents. This is achieved by performing the indicated operations and combining like terms.

EXAMPLE 1 Adding and Subtracting Polynomials

EXPLORE **Use the polynomials $f(x) = -8x^4 + 6x^2 - x + 2$ and $g(x) = 8x^4 - 2x^3 + 4x^2 + 3x - 1$ to answer the questions.** CCSS A.APR.1

a. INTERPRET PROBLEMS Evaluate and simplify $f(x) + g(x)$ and $f(x) - g(x)$. Determine whether each expression is a polynomial. If it is a polynomial, state the degree of the polynomial. CCSS SMP 1

b. MAKE A CONJECTURE Make a conjecture about the degree of a sum or difference of two polynomials. Explain your reasoning. CCSS SMP 3

c. CONSTRUCT ARGUMENTS A set is *closed* under a given operation if the result of the operation on any two members of the set is also in the set. Is the set of polynomials closed under addition and subtraction? Justify your answer. CCSS SMP 3

EXAMPLE 2 Multiplying Polynomials

Consider a trapezoid that has one base that measures five more feet than its height. The other base is one foot less than twice its height. Let x represent the height. CCSS A.APR.1

a. **REASON ABSTRACTLY** Write an expression for the area of the trapezoid. CCSS SMP 2

b. **USE STRUCTURE** Show two ways to determine an expression for the area of the trapezoid if its height is changed to $(x + 4)$ feet. CCSS SMP 1

EXAMPLE 3 Multiplying Polynomials

The diagram represents the base area of a circular memorial in a town center. A sidewalk that is 12 feet wide with an area of 384π square feet surrounds the smaller circle. CCSS A.APR.1

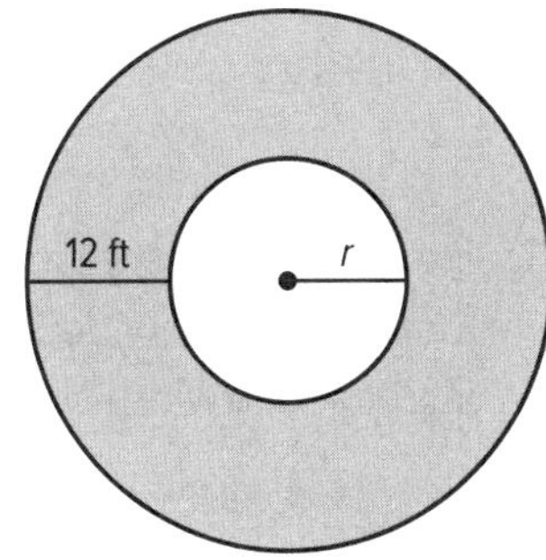

a. **USE A MODEL** Find the radius of the smaller and larger circles. Show your work. CCSS SMP 4

b. **USE A MODEL** A nearby town wants to use the same design concept, but use two squares rather than two circles. Draw and label a diagram with two squares to represent a sidewalk with the same uniform width and area as the circular sidewalk. CCSS SMP 4

c. **DESCRIBE A METHOD** Describe how to find the outer perimeter of the sidewalk. What is the outer perimeter of the sidewalk to the nearest tenth of a foot? CCSS SMP 8

PRACTICE

1. **CALCULATE ACCURATELY** Use the polynomials $f(x) = -6x^3 + 2x^2 + 4$ and $g(x) = x^4 - 6x^3 - 2x$ to evaluate and simplify the given sum or difference. Determine the degree of the resulting polynomial. Show your work. CCSS A.APR.1, SMP 6

   a. $f(x) + g(x)$

   b. $g(x) - f(x)$

2. **CALCULATE ACCURATELY** Use the polynomials $f(x) = 3x^2 - 1$, $g(x) = x + 2$, and $h(x) = -x^2 - x$ to evaluate and simplify the given product. Determine the degree of the resulting polynomial. CCSS A.APR.1, SMP 6

   a. $f(x)g(x)$

   b. $h(x)f(x)$

   c. $[f(x)]^2$

3. Inez wants to increase the size of her rectangular garden. The original garden is 8 feet longer than it is wide. For the new garden, she will increase the length by 25% and increase the width by 5 feet. CCSS A.APR.1

   a. **USE A MODEL** Draw and label a diagram with two rectangles representing the original garden and the new garden. Define a variable and label each dimension with appropriate expressions. CCSS SMP 4

   b. **INTERPRET PROBLEMS** Every foot of the perimeter of the original garden was lined with 7 stones. Write and simplify an expression to represent how many more stones Inez needs for the new garden. CCSS SMP 1

c. **CALCULATE ACCURATELY** Write and simplify an expression for the increase in area of the garden. Find how many square feet the garden's area increased, if the original width of the garden was 10 feet. CCSS SMP 6

4. **CONSTRUCT ARGUMENTS** Given $f(x)$ and $g(x)$ are polynomials, is the product always a polynomial? Explain why or why not. CCSS A.APR.1, SMP 3

5. **USE STRUCTURE** In the box, find the product for $3x^2 - 4x + 1$ and $x^2 + 5x + 6$ using vertical alignment. CCSS A.APR.1, SMP 7

6. **MAKE A CONJECTURE** Use your result to make conjectures about the product of a polynomial with m terms and a polynomial with n terms. Justify your conjecture. CCSS A.APR.1, SMP 3

   a. How many times are two terms multiplied?

   b. What is the least number of terms in the simplified product?

7. **REASON ABSTRACTLY** Complete the table showing closure for the sets shown by writing *yes* or writing *no* and providing a counterexample. You can assume that since division by zero is undefined, it does not affect closure. CCSS A.APR.1, SMP 2

	Addition and Subtraction	Multiplication	Division
Integers			
Rational numbers			
Polynomials			

5.2 Dividing Polynomials

Objectives

- Divide polynomials using long division.

CCSS STANDARDS

Content: A.APR.6
Practices: 1, 2, 3, 4, 5, 6, 7, 8
Use with Lesson 5-2

Division Algorithm: If $f(x)$ and $d(x) \neq 0$ are polynomials, and the degree of $d(x)$ is less than or equal to the degree of $f(x)$, then there exist unique polynomials $q(x)$ and $r(x)$ so that

$$\frac{f(x)}{d(x)} = q(x) + \frac{r(x)}{d(x)}.$$

Note that the degree of $r(x)$ is less than the degree of $d(x)$. In special cases where $r(x) = 0$, $d(x)$ is a factor of $f(x)$.

EXAMPLE 1 Using the Division Algorithm CCSS A.APR.6

EXPLORE Rewrite $\frac{x^2 + 2x - 5}{x - 2}$ in the form $\frac{f(x)}{d(x)} = q(x) + \frac{r(x)}{d(x)}$.

a. **USE STRUCTURE** What value of c and what function $q(x)$ will make the equation $x^2 + 2x + c = (x - 2) \cdot q(x)$ true? Explain your reasoning. CCSS SMP 7

b. **USE STRUCTURE** How can you write $\frac{x^2 + 2x - 5}{x - 2}$ in the form $q(x) + \frac{r}{x - 2}$ where $q(x)$ is a binomial and r is a constant? Simplify and justify your conclusion. CCSS SMP 7

c. **EVALUATE REASONABLENESS** How can you check that the expression you wrote in **part b** is equivalent to $\frac{x^2 + 2x - 5}{x - 2}$? Verify your answer from **part b**. CCSS SMP 8

d. **USE STRUCTURE** Algebraic long division uses the same process as numerical long division. Fill in the missing steps of the long division to find $\frac{x^3 + x^2 - 3x + 1}{x - 1}$. CCSS SMP 7

$$\begin{array}{r} x^2 + \square - \square \\ x - 1 \overline{) x^3 + x^2 - 3x + 1} \\ -(x^3 - x^2) \\ \hline 2x^2 - 3x \\ -(\square) \\ \hline -x + 1 \\ -(\square) \\ \hline 0 \end{array}$$

When dividing polynomials, it is important that the polynomials are in **standard form**. In some cases, not all degrees are represented by terms in the polynomial. If this is the case, place a zero in place of the missing term.

EXAMPLE 2 Dividing Polynomials with Missing Terms CCSS A.APR.6, SMP 6

CALCULATE ACCURATELY Use long division to rewrite $\frac{x^4 + 2x^2 - 3}{x + 4}$ in the form $q(x) + \frac{r(x)}{d(x)}$. Show your work.

EXAMPLE 3 Dividing Polynomials with a Computer Algebra System CCSS A.APR.6

CRITIQUE REASONING Cayla divided $2x^4 + 6x^3 - 7x^2 + 3x + 4$ by $x^2 - 2$. She says that the quotient is $(2x^2 + 6x - 3) \times (x^2 - 2) + 15x - 2$. CCSS SMP 3

a. USE TOOLS Identify the quotient and remainder. Then use a computer algebra system (CAS) to rewrite the quotient in the form $q(x) + \frac{r(x)}{d(x)}$. Do you agree with Cayla? CCSS SMP 5, 3

Input:

$$\frac{2x^4 + 6x^3 - 7x^2 + 3x + 4}{x^2 - 2}$$

Quotient and remainder:

$$2x^4 + 6x^3 - 7x^2 + 3x + 4 = \boxed{(2x^2 + 6x - 3)} \times (x^2 - 2) + \boxed{15x - 2}$$

b. CALCULATE ACCURATELY Verify that your interpretation of the result is correct. CCSS SMP 6

EXAMPLE 4 Using Long Division to Solve Problems CCSS A.APR.6, SMP 4

USE A MODEL **The volume of the cylinder shown is $6x^3 - 10x^2 + 13x - 2$ units^3. If the height is increased by $2x - 3$ and the volume is unchanged, what is an expression for the area of the base? Justify your work.**

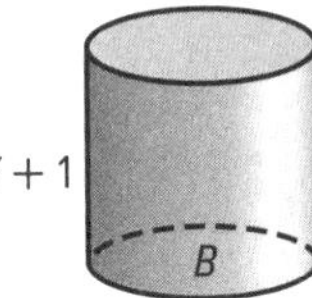

PRACTICE

1. **INTERPRET PROBLEMS** Rewrite $\frac{6x^4 + 2x^3 - 16x^2 + 24x + 32}{2x + 4}$ as $q(x) + \frac{r(x)}{d(x)}$ using long division. What does the remainder indicate in this problem? CCSS A.APR.6, SMP 1

2. **USE STRUCTURE** When a polynomial is divided by $4x - 6$, the quotient is $2x^2 + x + 1$ and the remainder is -4. What is the dividend, $f(x)$? Explain. CCSS A.APR.6, SMP 7

3. **USE STRUCTURE** Determine the constant c in $\frac{3x^5 + 4x^3 - 6x^2 - 15x + c}{3x^2 - 5}$ such that the denominator is a factor of the numerator. Show your work. CCSS A.APR.6, SMP 7

4. **CRITIQUE REASONING** Mariah spilled coffee on her homework. She says that she does not have enough information to recreate the problem. Do you agree? Justify your answer. CCSS A.APR.6, SMP 3

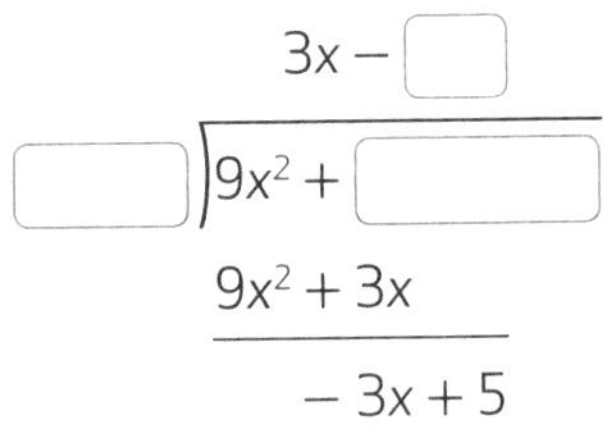

5. **FIND A PATTERN** Rewrite $\frac{x^5 + 2x^2 + x - 2}{2x + 3}$ as $q(x) + \frac{r(x)}{d(x)}$ using long division. CCSS A.APR.6, SMP 7

a. George rewrote the rational expression as: $\frac{1}{4}x^4 - \frac{2}{3}x^3 + \frac{5}{6}x^2 - \frac{7}{8}x + \frac{28}{42} - \frac{\frac{95}{102}}{2x + 3}$. Without using paper and pencil, his friend told him that he had made a mistake. Do you agree? If so, what mistake did George make?

b. How can you use a pattern to tell if the leading coefficient of a quotient is wrong?

c. Simplify $\frac{x^4 + 2x^2 + x - 2}{2x + 4}$.

6. **REASON ABSTRACTLY** Given $\frac{f(x)}{d(x)} = q(x) + \frac{r(x)}{d(x)}$, suppose that you know $q(x) + \frac{r(x)}{d(x)}$. Is it possible to determine $f(x)$? Use an example to illustrate your answer. CCSS A.APR.6, SMP 2

7. **USE A MODEL** Mack has a square garden. A new garden will have the same width and a length that is 3 feet more than twice the width of the original garden. CCSS A.APR.6, SMP 4

Original garden	New garden

a. Define a variable and label each side of the diagrams with an expression for its length.

b. Write a ratio to represent the percent increase in the area of the garden. Use polynomial division to rewrite the expression.

c. Use your expression from **part b** to determine the percent of increase in area if the original garden was a 12-foot square. Check your answer.

8. **CRITIQUE REASONING** Mariella makes claims about the degrees of the polynomials in $\frac{f(x)}{d(x)} = q(x) + \frac{r(x)}{d(x)}$. Do you agree with each claim? Justify your answer and provide examples. CCSS A.APR.6, SMP 3

a. The degree of $d(x)$ must be less than the degree of $f(x)$.

b. The degree of $r(x)$ must be at least 1 less than the degree of $d(x)$.

c. The degree of $q(x)$ must be the degree of $f(x)$ minus the degree of $d(x)$.

9. USE A MODEL The parking lot for a city recreation center needs to be enlarged. The city planner proposes increasing the length and width of the lot by x feet. CCSS A.APR.6, SMP 4

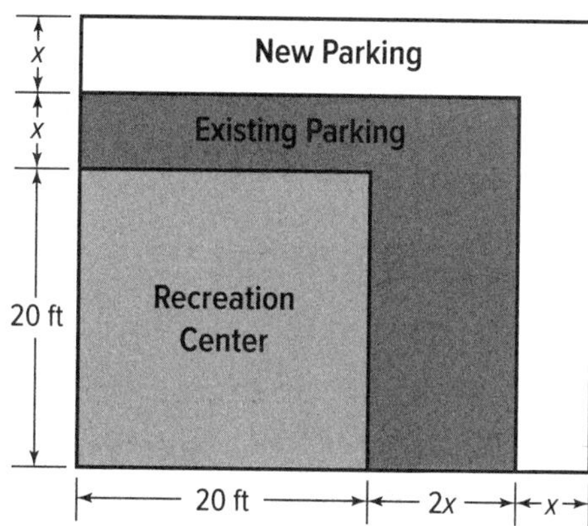

a. USE TOOLS Write functions for the area of the existing parking lot and the area of the new parking lot. Use a CAS to write an equation of the form $\frac{f(x)}{d(x)} = q(x) + \frac{r(x)}{d(x)}$ comparing the area of the new parking lot to the area of the existing parking lot. CCSS A.APR.6, SMP 5

b. REASON QUANTITATIVELY Can the size of the parking area be tripled under this plan? Explain. CCSS A.APR.6, SMP 2

c. DESCRIBE A METHOD Suppose the city planner wants to know how the increase in parking space will affect the area of the entire facility, recreation center and parking lot combined. Explain a method for solving this problem. CCSS A.APR.6, SMP 8

10. USE STRUCTURE A box has length $x + 4$, width $2x - 1$, and height h. If the volume of the box is $2x^3 + x^2 - 25x + 12$, find and simplify an expression for the height h in terms of x. CCSS A.APR.6, SMP 7

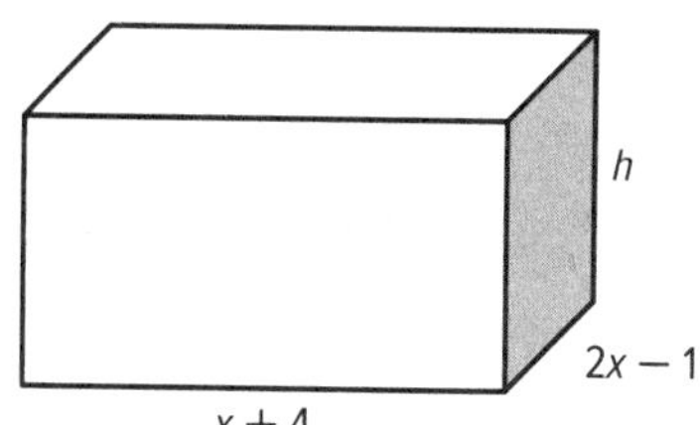

5.3 Power Functions

Objectives

- Interpret key features of power functions in context.
- Sketch graphs of power functions showing key features.

CCSS STANDARDS

Content: F.IF.4
Practices: 1, 2, 3, 4, 5, 6, 7
Use with Explore 5–3

A **power function** has an equation of the form $f(x) = ax^n$, where a and n are any nonzero real number constants. When n is a positive integer, the power function is known as a **monomial function**. Some real-world situations, such as compound interest, can be modeled with a power function.

EXAMPLE 1 Investing in a Certificate of Deposit

EXPLORE Recall that the amount of money earned on an initial deposit a after t years at interest rate r can be represented by $y = a(1 + r)^t$. This can also be written as $y = ax^t$, where x represents the growth rate, $1 + r$.

a. **USE A MODEL** Write an equation that represents the amount of money a that Simone will earn at maturity if she deposits \$100 into a 5 year certificate of deposit (CD) with growth rate x. Then complete the table at right. What happens to the value at maturity of a CD each time the interest rate increases? CCSS SMP 4

$f(x) = 100x^5$	
x	$f(x)$
1	
1.01	
1.02	
1.03	
1.04	
1.05	
1.06	

b. **USE TOOLS** Enter this power function into your graphing calculator and look at the table. What interest rate would Simone need to earn in order to double her money within 5 years? Is this a rate that she will likely be able to procure? CCSS SMP 5

c. **FIND A PATTERN** The table at right shows data for two functions that represent the value at maturity of two other CDs Simone has. What were the initial deposits and interest rates? Compare the value of $f(x)$ with $g(x)$ and $h(x)$. What pattern(s) do you see? CCSS SMP 7

x	$g(x) = 100x^{10}$	$h(x) = 200x^5$
1	\$100	\$200
1.01	\$110.46	\$210.2
1.02	\$121.90	\$220.82
1.03	\$134.39	\$231.85
1.04	\$148.02	\$243.33
1.05	\$162.89	\$255.26
1.06	\$179.08	\$267.65

d. INTERPRET PROBLEMS Use a graphing calculator to determine whether the graphs of the functions intersect over an appropriate domain. What is an appropriate viewing window? What is the meaning of any points of intersection? Sketch and label the graphs at right. CCSS SMP 1

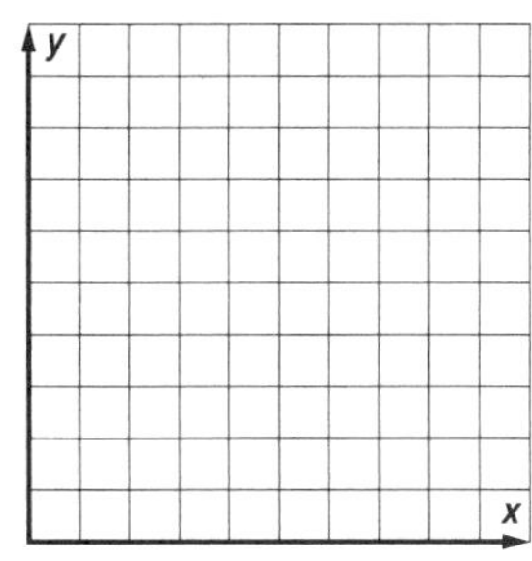

e. REASON QUANTITATIVELY Under what circumstances should Simone consider doubling the investment period? The initial investment? CCSS SMP 2

f. FIND A PATTERN Summarize how the CD model changes for different values of a and n in the power function. CCSS SMP 7

In order to analyze some of the characteristics of the graphs of power functions, refer to the graphs of the three functions from **Example 1** over the domain of all real numbers. These characteristics include symmetry and end behavior, defined as the behavior of a function $f(x)$ as x approaches positive infinity ($+\infty$) and as x approaches negative infinity ($-\infty$).

EXAMPLE 2 Sketch Power Functions over All Real Numbers

a. USE TOOLS Use a graphing calculator to view the graphs of $f(x) = 100x^5$, $g(x) = 100x^{10}$, and $h(x) = 200x^5$ over the domain $[-2, 2]$. Sketch and label the graphs on the coordinate plane at right. CCSS SMP 5

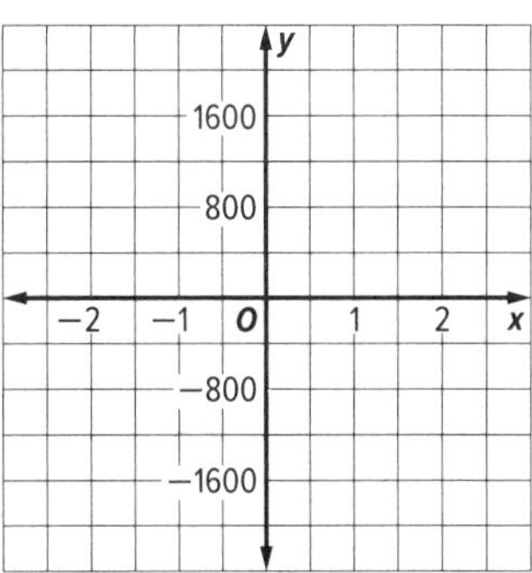

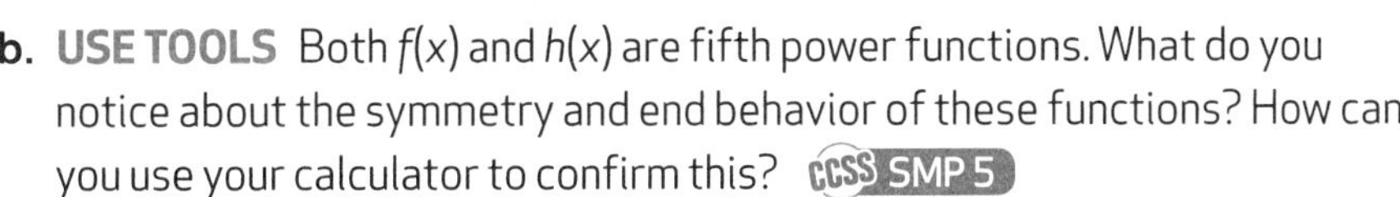

b. USE TOOLS Both $f(x)$ and $h(x)$ are fifth power functions. What do you notice about the symmetry and end behavior of these functions? How can you use your calculator to confirm this? CCSS SMP 5

c. FIND A PATTERN What do you think is true about the symmetry and end behavior of graphs of the power function $f(x) = ax^n$ when n is odd and a is positive? When n is even and a is positive? Justify your answers. CCSS SMP 7

PRACTICE

1. **CONSTRUCT ARGUMENTS** Draw a graph representing the relationship between the length of an edge of a cube and the volume of the cube. Label the function. Why does this graph appear not to have symmetry? CCSS SMP 3

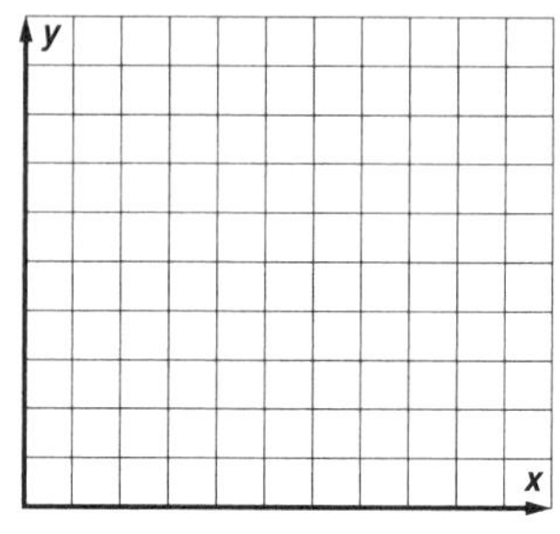

2. The table at right lists some points on the graphs of the power functions $v(x)$ and $w(x)$.

x	$v(x) = -2x^4$	$w(x) = -2x^5$
−3	−162	162
−2	−32	32
−1	−2	2
0	0	0
1	−2	−2
2	−32	−32
3	−162	−162

   a. **COMMUNICATE PRECISELY** Describe the type of symmetry and end behavior displayed by each of the graphs. CCSS SMP 6

   b. **COMMUNICATE PRECISELY** Describe the end behavior of each graph. CCSS SMP 6

   c. **FIND A PATTERN** Compare the graphs of power functions with even exponents to the graphs of power functions with odd exponents where a is any integer. Summarize the relationship between whether the exponent is even or odd, whether the coefficient is positive or negative, and the graphs' symmetry and end behavior. CCSS SMP 7

   d. **USE STRUCTURE** Which power functions are increasing or decreasing over their entire domain? Explain your answer. CCSS SMP 7

3. **REASON QUANTITATIVELY** What point do all power functions contain? Justify your answer algebraically. CCSS SMP 2

4. **USE A MODEL** The parabolic reflector inside a flashlight can be modeled by the function $f(x) = \frac{4}{3}x^2$. CCSS SMP 4

a. Without graphing, what is the minimum point on the graph of this function? Explain.

b. How would the shape of the parabolic reflector change if $a = \frac{3}{4}$ instead? Explain.

5. **INTERPRET PROBLEMS** Without graphing, compare the behavior of the graph of $y_1 = x^{10}$ to the graphs of $y_2 = x^{11}$ and $y_3 = x^{12}$. CCSS SMP 1

6. **USE TOOLS** Use the power functions $f(x) = x^3$, $g(x) = x^5$, and $h(x) = x^7$ to answer the questions. CCSS SMP 5

a. Use a graphing calculator to view and sketch the graphs of $f(x)$, $g(x)$, and $h(x)$ over the domain $[-3, 3]$.

b. Based on the graphs, what three points does each of these functions contain? How could a calculator be used to verify this?

c. Identify any symmetry displayed by the graphs of these functions. How could a calculator confirm your answer?

7. **USE A MODEL** Use the fact that an amount of money earned on an initial deposit a after t years at interest rate r can be represented by $y = ax^t$, where x represents the growth rate, $1 + r$. CCSS SMP 4

a. Consider a 15 year certificate of deposit (CD) with growth rate x. Write an equation that represents the amount of money y after \$15,000 is deposited in the CD and comes to maturity.

b. Use your function from **part a** to complete the table at right. Which growth rate yields approximately twice the amount at maturity than the \$15,000 deposited? What interest rate does this correspond to?

Growth Rate x	Amount at Maturity
1.01	
1.02	
1.03	
1.04	
1.05	
1.06	

5.4 Polynomial Functions

Objectives

- Graph polynomial functions and identify end behavior.
- Interpret key features of graphs of polynomial functions.

STANDARDS

Content: F.IF.4, F.IF.5, F.IF.7c, F.IF.9, A.CED.2, A.SSE.2
Practices: 1, 2, 3, 4, 5, 7
Use with Lesson 5-3

A polynomial function has the form $f(x) = a_nx^n + a_{n-1}x^{n-1} + \cdots + a_2x^2 + a_1x + a_0$, where $a_n \neq 0$; $a_n, a_{n-1}, \ldots, a_2, a_1$, and a_0 are real numbers; and n is a nonnegative integer. The degree of a polynomial function is the ____________________, and represents the ____________________ number of times the graph may intersect the x-axis. The behavior of the graph as $x \to +\infty$ or $x \to -\infty$ will resemble the graph of ____________.

EXAMPLE 1 Volume of a Rectangular Box

EXPLORE **A packaging company is designing a small box with an open top. Starting with a single 12-inch by 8-inch piece of cardboard, equal size squares are removed from each of the corners, and the resulting flaps are folded up to form the sides of the box.**

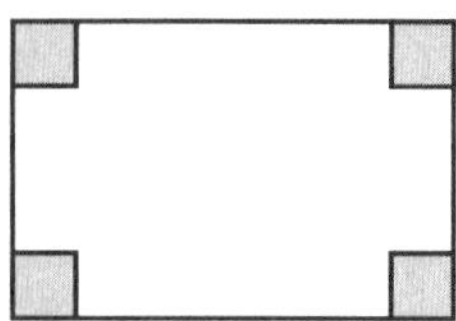

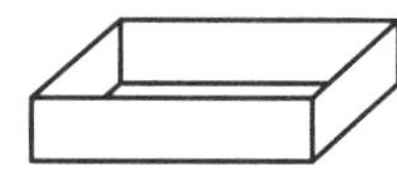

a. **USE A MODEL** Write an expression for the volume of the box and express the result as a polynomial function in terms of the length of an edge of each of the squares removed from each corner. CCSS A.CED.2, A.SSE.2, SMP 4

b. **REASON QUANTITATIVELY** What constraints does the context of the problem place on the variable? CCSS F.IF.5, SMP 2

c. **CONSTRUCT ARGUMENTS** The graph is a graph of the volume function from **part a**. Explain the significance of points A, B, C, D, and E, in the context of the problem. CCSS F.IF.4, F.IF.5, SMP 3

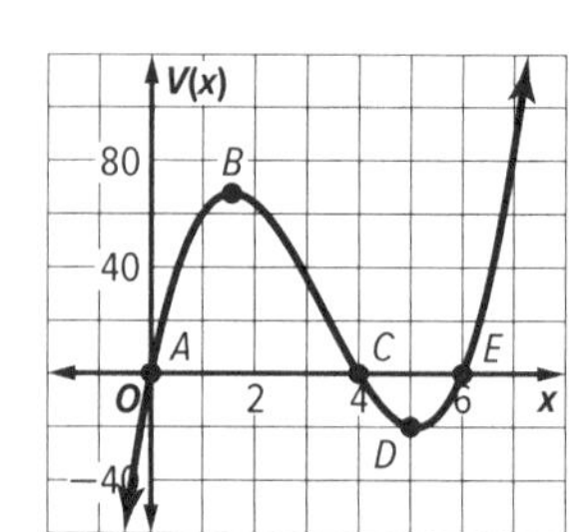

EXAMPLE 2 End Behavior of Polynomial Graphs CCSS F.IF.4, F.IF.9

a. **REASON QUANTITATIVELY** Consider the function $f(x) = -5x^4 + 9x^3 + 7$. Determine the degree and end behavior of the polynomial. Is the end behavior the same as $g(x) = 5x^4 + 9x^3 + 7$? CCSS SMP 2

b. **REASON QUANTITATIVELY** If $f(x) = 3x^3 + 4x^2 - 6x + 2$, determine $f(x - 2)$. Express the result in standard form. How does the end behavior of $f(x - 2)$ compare to $f(x)$? Explain. CCSS SMP 2

c. **USE STRUCTURE** If $f(x) = 6x^3 - x^2 - 12x - 5$, describe the end behavior of $f(-x)$. CCSS SMP 7

EXAMPLE 3 Zeros of Polynomial Graphs CCSS F.IF.4, F.IF.7c

a. **USE STRUCTURE** Determine the number of real zeros for the function shown in the graph. Is the function odd or even? Explain how you know. CCSS SMP 7

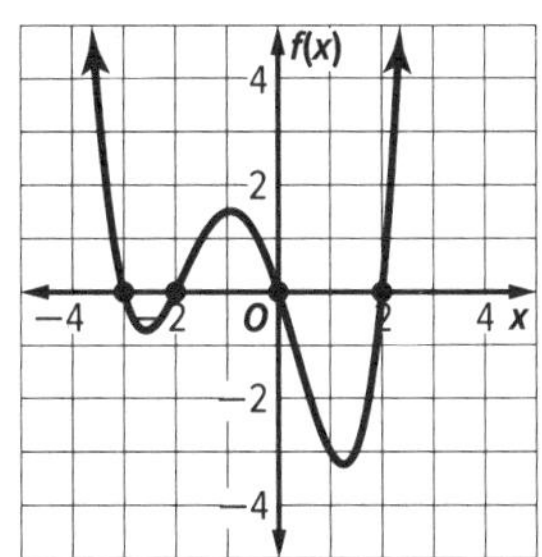

b. **CONSTRUCT ARGUMENTS** Must the graph of a polynomial function cross the x-axis in order for a real zero to exist? Explain your reasoning. CCSS SMP 3

c. **CONSTRUCT ARGUMENTS** Can the graph of a third degree polynomial function intersect the x-axis two times? What about four times? Explain your reasoning. CCSS SMP 3

PRACTICE

1. **PLAN A SOLUTION** For the given function, determine the degree of the polynomial and describe the end behavior. CCSS F.IF.4, SMP 1

   a. $f(x) = 2x^4 - 6x^3 + 7$

   b. $g(x) = 4x - x^2$

   c. $h(x) = 1 - x^3 + x$

2. **CONSTRUCT ARGUMENTS** Explain why a polynomial function with an odd degree must have at least one real zero. CCSS F.IF.4, SMP 3

3. **REASON QUANTITATIVELY** If $f(x) = ax^3 - bx^2 + x$, determine $f(1 - x)$. Express the result in standard form. How does the end behavior of $f(1 - x)$ compare to $f(x)$? CCSS F.IF.4, SMP 2

4. **USE STRUCTURE** Sketch a graph of a polynomial function with the following properties: degree = 3, real zeros = 3, distinct real zeros = 2, $f(x) \to +\infty$ as $x \to -\infty$. CCSS F.IF.7c, SMP 7

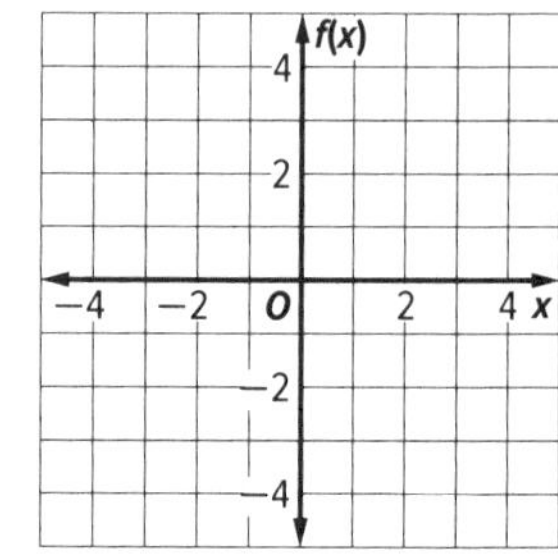

5. **USE STRUCTURE** Sketch a graph of a polynomial function with the following properties: degree = 4, real zeros = 2, distinct real zeros = 2, negative leading coefficient. CCSS F.IF.7c, SMP 7

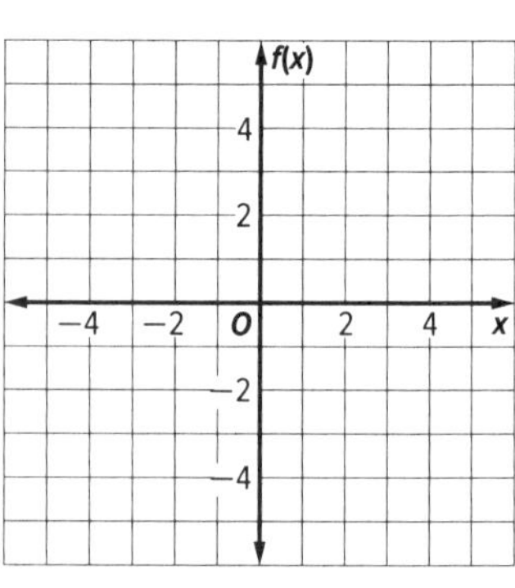

6. **CRITIQUE REASONING** Yuan states that the end behaviors of the functions $g(x) = -3x^4 + 15x^3 - 12x^2 + 3x + 20$ and $h(x) = -3x^4 - 16x - 1$ are exactly the same. Is she correct? Explain why or why not. CCSS F.IF.4, SMP 3

7. **CRITIQUE REASONING** Julia is graphing a function to model the volume of a box with sides x, $6 - x$, and $10 - x$. She recognizes that the graph will have x-intercepts at 0, 6, and 10. Is her graph correct? Explain your reasoning. CCSS F.IF.4, F.IF.5, SMP 3

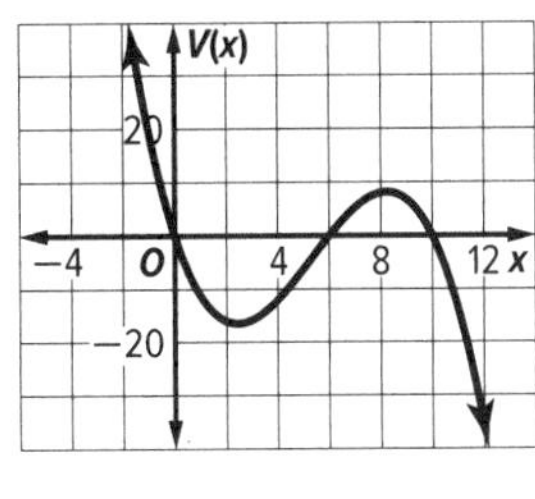

8. **USE STRUCTURE** Consider the function $f(x) = -(3 - x)(4 - x)^2$. Identify the degree of the polynomial and any zeros. CCSS F.IF.7c, SMP 7

9. **USE TOOLS** A company uses the function $f(x) = x^3 + 3x^2 - 18x - 40$ to model the change in efficiency of a machine based on its position with a domain $[-6, 4]$. The graph of $f(x)$ is shown. CCSS F.IF.4, SMP 5

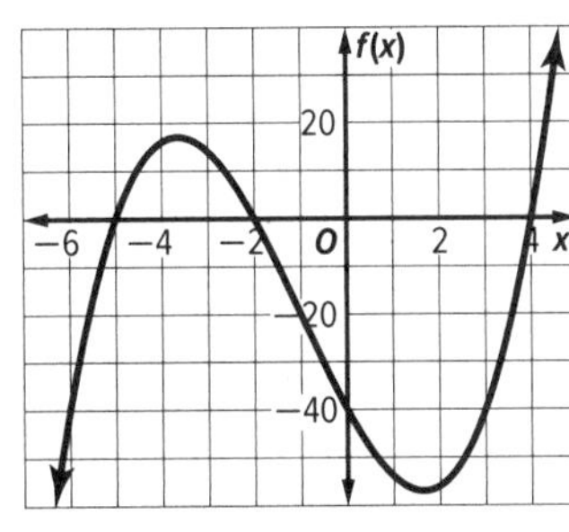

   a. The graph appears to show x-intercepts at $x = -5$, $x = -2$, and $x = 4$. How can you verify that these are x-intercepts?

   b. Identify the intervals on which the function is positive, indicating an increase in efficiency.

5.5 Analyzing Graphs of Polynomial Functions

Objectives

- Interpret key features of graphs of polynomial functions.
- Graph polynomial functions and identify relative extrema.

CCSS STANDARDS

Content: F.IF.4, F.IF.7c, A.CED.2, F.IF.5

Practices: 1, 2, 3, 4, 6, 7

Use with Lesson 5-4

To construct the graph of a polynomial function, you can create a table of several points and then connect them with a smooth continuous curve.

KEY CONCEPT Location Principle

Suppose $f(x)$ represents a polynomial function and a and b are real numbers such that $f(a) < 0$ and $f(b) > 0$. Then the function has ______________ between a and b.

Graphs of polynomial functions may have **extrema**, maximum or minimum values, within certain intervals of the domain. **Relative extrema** consist of relative maximums and relative minimums. A relative maximum has the ______________ when compared to points nearby, and a relative minimum has the ______________ when compared to points nearby.

EXAMPLE 1 Analyzing Graphs of Polynomial Functions CCSS F.IF.4

EXPLORE Use the polynomial function $f(x) = 2x^3 - 7x^2 + 4x + 4$.

a. INTERPRET PROBLEMS Identify points on the graph of $f(x) = 2x^3 - 7x^2 + 4x + 4$ by filling in the table below. Which intervals must contain zeros of the function? Explain your reasoning. CCSS SMP 1

x	−2	−1	0	1	2	3	4
$f(x)$							

b. CONSTRUCT ARGUMENTS Could there be an additional zero somewhere between $x = 3$ and $x = 4$? Explain how to confirm the existence or nonexistence of another zero in that interval. CCSS SMP 3

c. **USE TOOLS** Use a graphing calculator to sketch a graph of the function. CCSS SMP 4

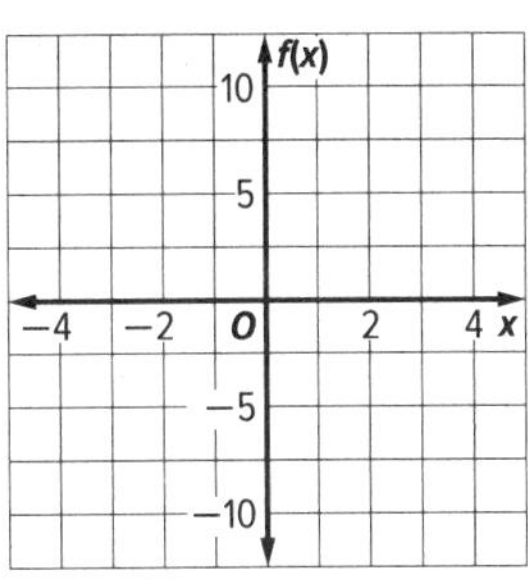

d. **USE STRUCTURE** What does the graph of $f(x)$ tell you regarding the number of zeros between $x = 1$ and $x = 3$? Describe the graph at $x = 2$. CCSS SMP 7

e. **USE STRUCTURE** Identify any relative extrema for the graph of $f(x)$. How can the values of $f(x)$ in the table help you identify relative extrema? CCSS SMP 7

f. **USE A MODEL** Given the graph of $f(x)$ shown at the right, identify any intercepts and/or relative extrema using the labeled points. CCSS SMP 4

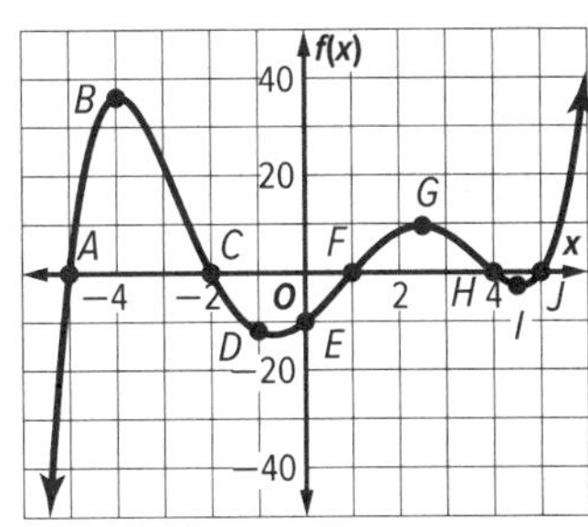

EXAMPLE 2 Constructing Graphs of Polynomial Functions CCSS F.IF.4, F.IF.7c

a. **USE A MODEL** Sketch the graph of a 3rd-degree polynomial function that has zeros at $x = -5$, $x = 1$, and $x = 4$, and a relative maximum at $x = -2.5$. Is the leading coefficient of the function positive or negative? CCSS SMP 4

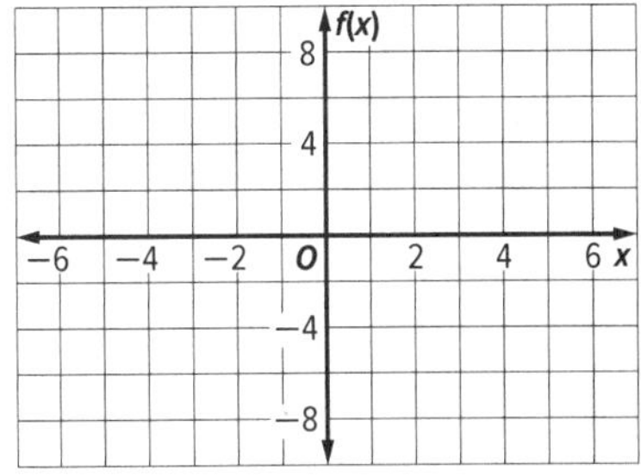

b. **COMMUNICATE PRECISELY** Sketch the graph of a 3rd-degree polynomial function that has zeros only at $x = -2$ and $x = 2$, a y-intercept at $(0, -4)$ and a negative leading coefficient. Describe the graph's behavior at each of the zeros. CCSS SMP 6

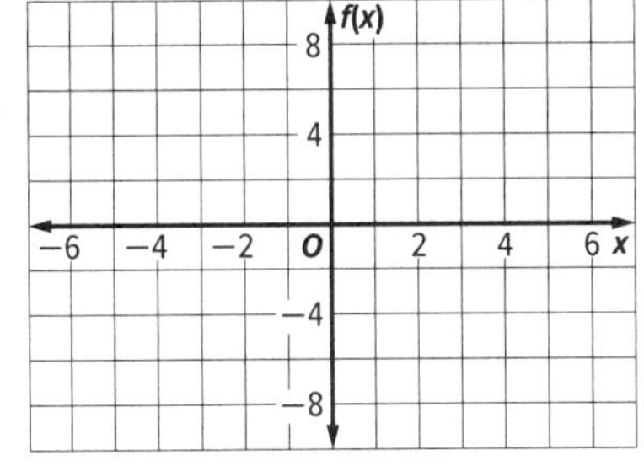

c. **REASON ABSTRACTLY** Explain why knowing the location of the zeros and the leading coefficient of a polynomial function does not allow you to determine the behavior of the graph as $x \rightarrow -\infty$. CCSS SMP 2

EXAMPLE 3 Creating and Analyzing a Polynomial Model

A company packages their drinks in a cylindrical can with a radius *r* of 2 inches and a height *h* of 6 inches. They are considering changing the dimensions of the can so that the height is decreased by the same amount as the radius is increased.

a. **REASON QUANTATIVELY** Let x represent the change in the dimensions. Write a function to model the volume $V(x)$ of the new can design. What is the appropriate domain for this situation? CCSS A.CED.2, SMP 2

b. **USE A MODEL** What will be the new dimensions of the can if the dimensions are changed by 2 inches? Graph the function from **part a**. Use the graph to approximate the new volume. Explain how you determined each quantity. CCSS F.IF.4, SMP 4

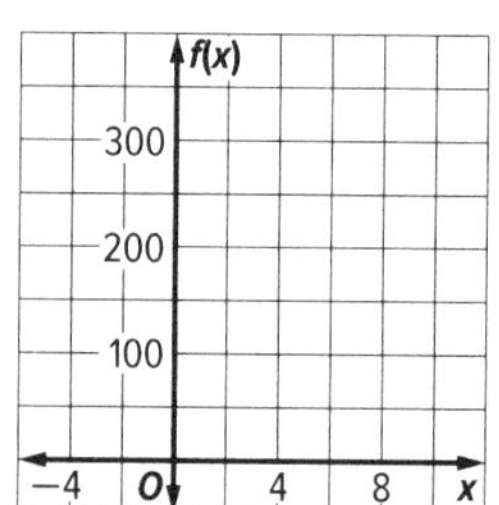

c. **CALCULATE ACCURATELY** Find $V(2)$. Compare your results to your estimate from the graph in **part b**. CCSS F.IF.7c, SMP 6

d. **REASON QUANTITATIVELY** Use the graph from **part b** to determine the approximate dimensions of the can that will maximize the volume. Use a graphing calculator to verify your approximation. CCSS F.IF.4, SMP 2

PRACTICE

1. **INTERPRET PROBLEMS** Identify points on the graph of $f(x) = -2x^3 - 6x^2 - x + 1$ by filling in the table below.

x	-3	-2	-1	0	1	2	3
$f(x)$							

a. Determine which intervals must contain zeros of the function. Explain your reasoning. CCSS F.IF.4, SMP 1

b. With the aid of a graphing calculator, sketch the graph of $f(x)$ on the grid provided. Are zeros present in the intervals determined from **part a**? Are there any zeros other than those mentioned in **part a**?

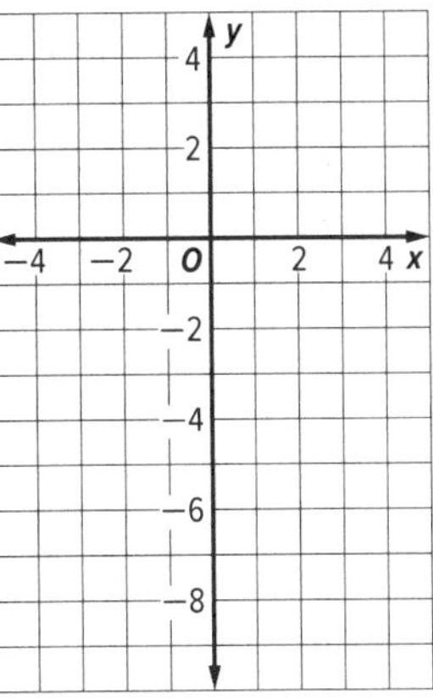

2. **USE STRUCTURE** Sketch two possible graphs of a 4th-degree polynomial function that has zeros only at $x = -3$ and $x = 3$, and a y-intercept at $(0, -9)$ or $(0, 9)$. Compare the possible end behavior of the functions. Determine if the leading coefficient of the function must be positive or negative and explain your reasoning. CCSS F.IF.7c, SMP 7

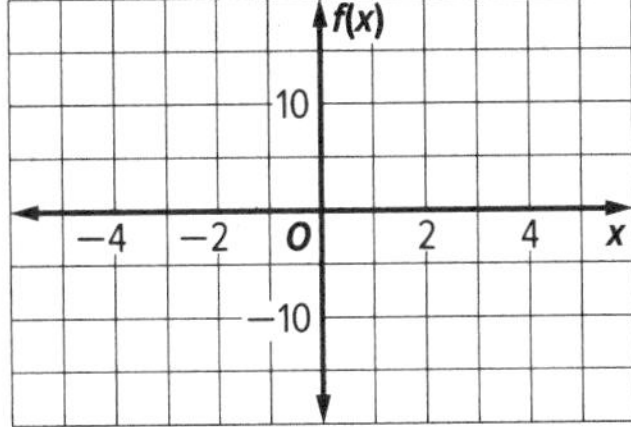

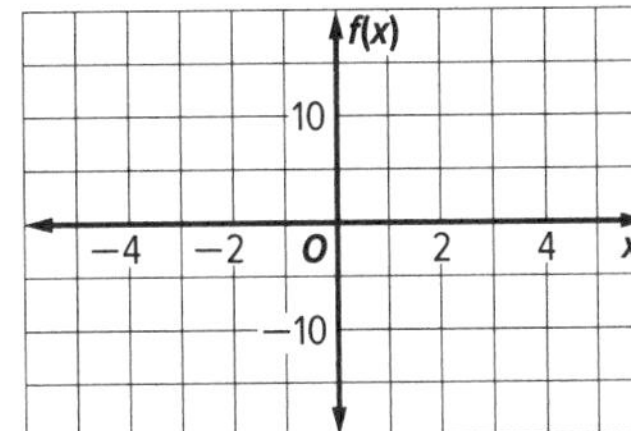

3. **USE STRUCTURE** Sketch the graph of a 3rd degree polynomial function that has a relative minimum at $x = -3$, passes through the origin, and has a relative maximum at $x = 2$. Describe the end behavior of the graph, if possible. Based on the sketch, determine whether the leading coefficient is negative or positive. CCSS F.IF.7c, SMP 7

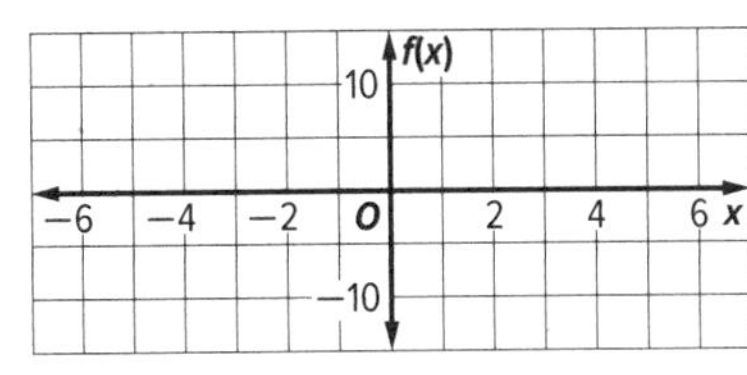

4. USE A MODEL A box has a square base with sides of 10 centimeters and a height of 4 centimeters. For a new box, the height is increased by twice the same amount as the lengths of the sides of the base are decreased. Graph the function. What new dimensions will produce a box with the greatest volume? Describe your solution process. CCSS A.CED.2, F.IF.7c, SMP 4

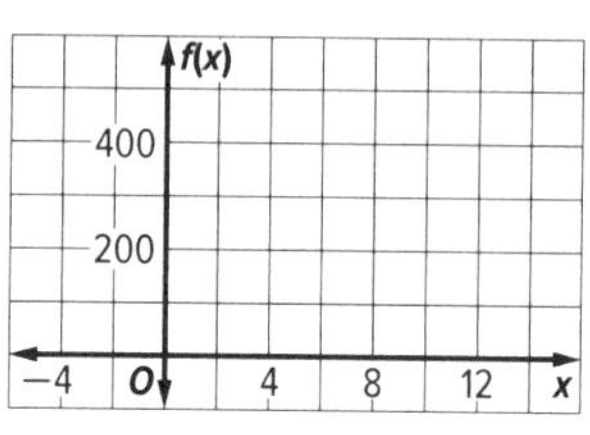

5. USE A MODEL A company wants to maximize the profits from manufacturing and selling a chemical solution. Currently, when priced at one dollar per ounce, the company sells an average of 40,000 ounces of the solution per month. They want to discover if changing the price can increase profits. CCSS F.IF.4, F.IF.5, SMP 4

a. Let x represent the selling price of the chemical solution. Market research shows that the function $S(x) = 110{,}000 - 70{,}000x$ can be used to model the number of ounces sold per month at the price x. Write a function to model the total profit per month using the fact that profit equals the number of ounces sold times the selling price.

b. Graph the function for profit from **part a**. Interpret the meaning of negative y-values on the graph. What is the appropriate domain for this situation? Find the selling price that maximizes profit.

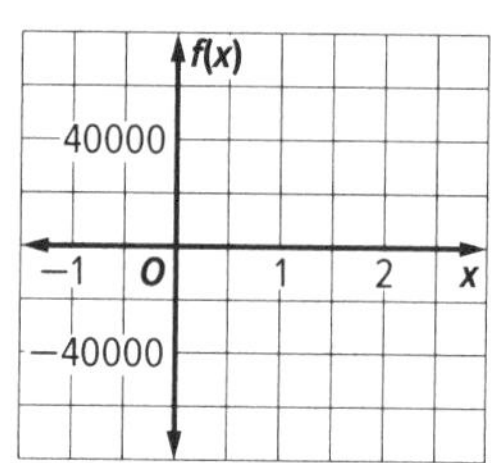

6. Use the function $f(x) = x^3 - 2x^2 - 4x + 1$ to answer the questions.

a. CALCULATE ACCURATELY Identify points on the graph of $f(x)$ by filling in the table below. Determine which intervals must contain zeros of the function. Explain your reasoning. CCSS F.IF.4, SMP 6

x	−3	−2	−1	0	1	2	3
$f(x)$							

b. **USE TOOLS** Use a graphing calculator to sketch a graph of the function. Are there any zeros outside the intervals identified in **part a**? CCSS SMP 4

c. **USE STRUCTURE** Identify any relative extrema for the graph of $f(x)$. How can the values of $f(x)$ in the table help you identify relative extrema? CCSS SMP 7

7. **USE A MODEL** Several engineering students have built a catapult for a class project. They test the catapult by launching a watermelon and model the height of the watermelon in feet by the function $h(t) = -16t^2 + 40t$ where t is time in seconds after the launch. CCSS SMP 4

   a. Sketch the graph of the function $h(t)$ and find the zeros. CCSS F.IF.7c

   b. Considering the context of the problem, what is an appropriate domain for $h(t)$? Explain your reasoning. CCSS F.IF.5

   c. How many seconds after the watermelon is launched does it reach the ground? Explain your reasoning. CCSS F.IF.4

   d. Use the graph of $h(t)$ to find the maximum height of the watermelon. When does the watermelon reach the maximum height? Explain your reasoning. CCSS F.IF.4

5.6 Modeling: Polynomial Functions

Objectives

- Interpret the graphs of polynomial functions in a real-world context.
- Define the domain of polynomial functions suitable for a real-world context.

CCSS STANDARDS

Content: A.SSE.1a, A.CED.4, F.IF.4, F.IF.5

Practices: 1, 2, 3, 4, 5, 6

Use with Extend 5-4

Recall that a polynomial function is a function in which the coefficients are real numbers with the leading coefficient non-zero and the exponents of the variable are whole numbers. You will use these types of functions to model situations that occur in the real world.

EXAMPLE 1 Predicting Revenue CCSS A.SSE.1a, F.IF.4

EXPLORE Stephanie is given the following information about her company's revenue over the last 15 years. Her supervisor asks her to project the company's revenue over the next five years.

Year	0	1	2	3	4	5	6	7	8	9	10	11	12	13	14
Revenue (in billions)	23.4	25.3	25.2	25.3	27.1	30.8	31.9	34.3	35.5	37.8	36.1	38.1	40.9	42.3	45.0

a. **USE A MODEL** Make a scatter plot of the data that models the relationship between the year and the revenue. Label the axes appropriately. CCSS SMP 4

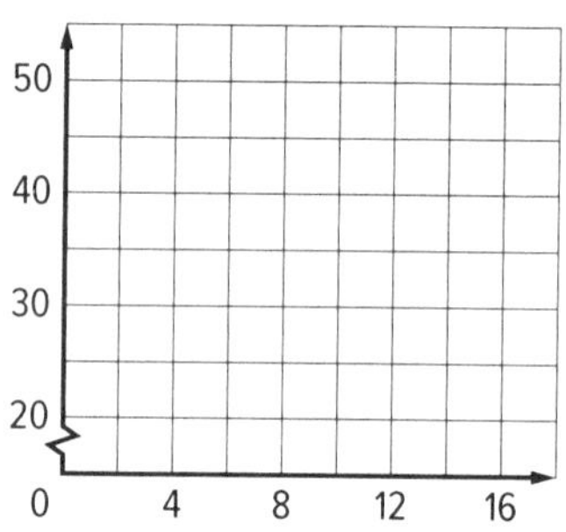

b. **MAKE A CONJECTURE** Analyze the graph you created in **part a**. What type of polynomial function would you choose to represent the data? Why? CCSS SMP 3

c. **INTERPRET PROBLEMS** A coworker gives Stephanie the polynomial function $f(x) = -0.0020x^3 + 0.0532x^2 + 1.1389x + 23.1485$ as a potential model, in billions. Calculate the revenue for the next five years, rounding to the tenths place. Are the results reasonable? CCSS SMP 1

d. CRITIQUE REASONING Stephanie analyzes the function her coworker presented more carefully and finds a flaw. What is the flaw in the function? Why? CCSS SMP 3

EXAMPLE 2 Storing Grain CCSS A.SSE. 1a, F.IF.4, F.IF.5

A town uses a silo in the shape of a cylinder to store grain. The volume of the silo is represented by $V = \pi(x^3 - 6x^2 + 9x)$, where x is the height of the silo in feet.

a. REASON ABSTRACTLY Determine the radius of the silo as a function of x. How does the radius compare to the height? CCSS A.CED.4, SMP 2

b. USE TOOLS Use a graphing calculator to sketch a model that represents the volume of the silo. Label axes appropriately. CCSS SMP 5

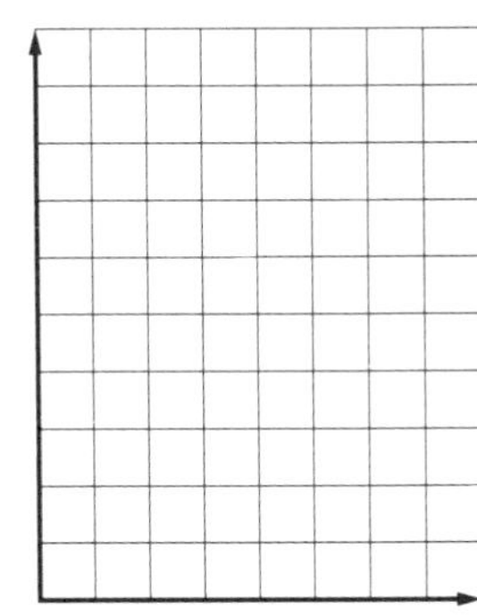

c. COMMUNICATE PRECISELY What is the domain of the model? Explain any restrictions that apply. CCSS SMP 6

d. INTERPRET PROBLEMS Use your answer from **part a** to write an expression for the volume of the silo in terms of the radius r. What are the domain restrictions for this model? Which model do you prefer to use to represent the volume of the silo? Explain your reasoning. CCSS SMP 1

e. INTERPRET PROBLEMS Use a graphing calculator to sketch the volume of the silo in terms of the radius r found in **part d**. How does this graph compare to the graph from **part b**?

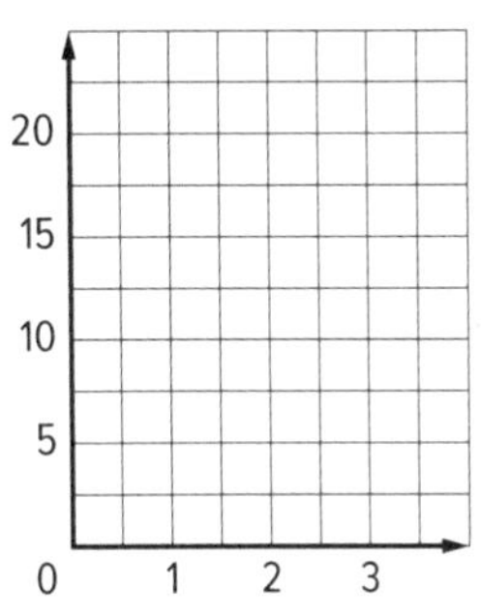

PRACTICE

1. The diagrams show patterns for making two boxes from rectangular pieces of cardboard. A square of side length x inches has been cut from the corners of each box.

Box 1:

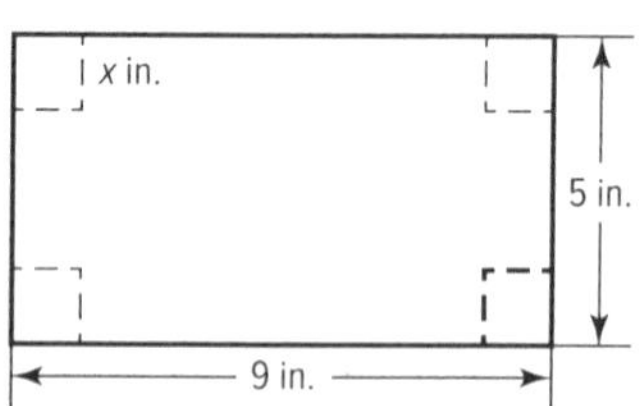

Box 2:

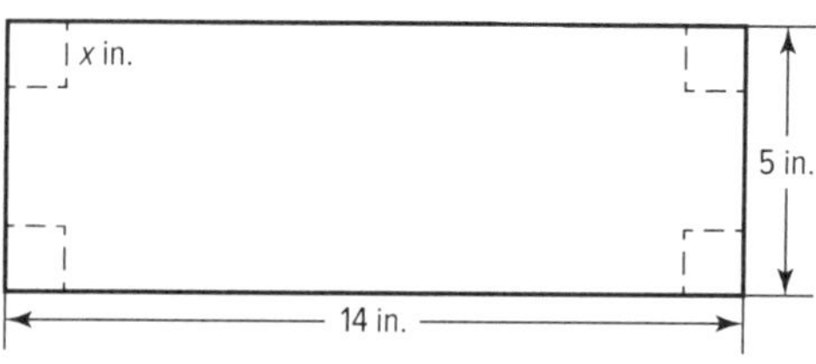

a. **REASON ABSTRACTLY** Write the polynomial function that models the volume of each box. CCSS A.SSE.1a, SMP2

b. **USE TOOLS** Use a graphing calculator to graph the equation of the volume for each box. Label axes appropriately. CCSS F.IF.4, SMP 5

Box 1:

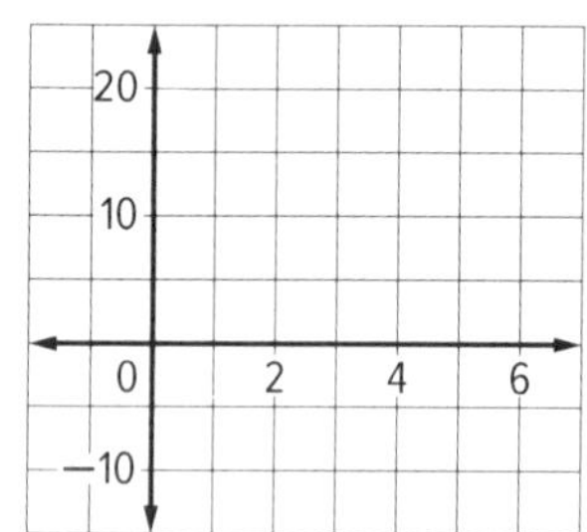

Box 2:

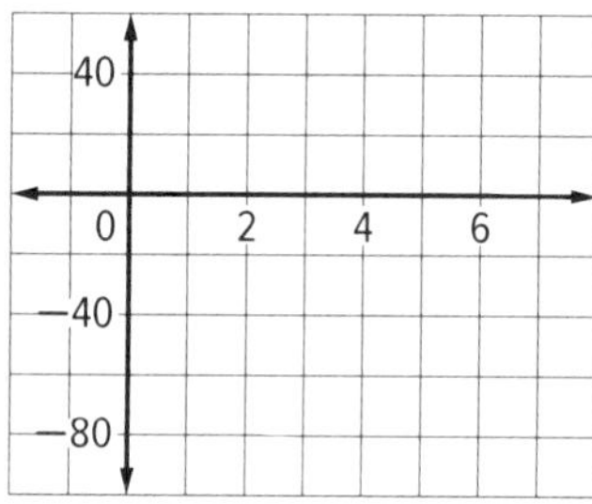

c. **INTERPRET PROBLEMS** In the context of this situation, explain the significance of x–intercepts on the graphs of these functions. CCSS F.IF.4, SMP 1

d. **USE A MODEL** Use the graphs from **part b** to determine which box has the greater maximum volume. CCSS F.IF.4, SMP 4

e. **INTERPRET PROBLEMS** Use the graphs from **part b** to estimate the side length of the cut-out square that will make Box 2 have the same volume as the maximum volume of Box 1. CCSS F.IF.4, SMP 1

2. **MAKE A CONJECTURE** Look at the following graph, what type of polynomial function should be used to model it? Why? CCSS F.IF.4, SMP 3

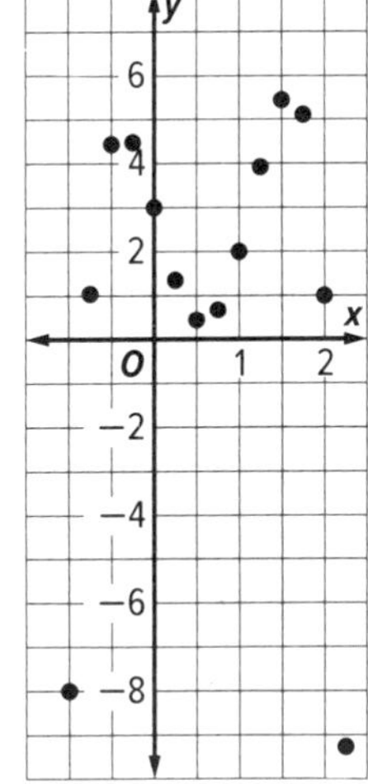

3. **COMMUNICATE PRECISELY** Juan wants to create a polynomial function that models the average value of stocks for a petroleum company in years. What would be an appropriate domain for Juan to use if he starts after the stock market crash in 1929? CCSS F.IF.5, SMP 6

4. A canister has the shape of a cylinder with spherical caps on either end. The volume of the canister in cubic millimeters is modeled by the function $V(x) = \pi(x^3 + 3x^2) + \frac{4}{3}\pi x^3$ where x represents the radius of the canister in millimeters.

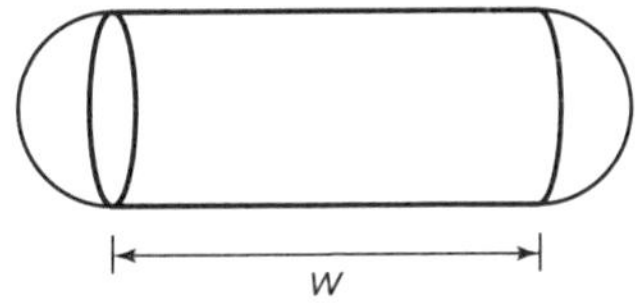

a. **INTERPRET PROBLEMS** Find an expression for the width of the cylindrical portion of the canister w in terms of the radius x. Interpret the expression in the context of the problem. *Hint: the volume of a sphere is* $\frac{4}{3}\pi r^3$. CCSS A.CED.4, SMP 1

b. **USE TOOLS** Use a graphing calculator to sketch the model that represents the volume of the canister. Label axes appropriately. CCSS F.IF.4, SMP 5

c. **COMMUNICATE PRECISELY** What is the domain of the model? Explain any restrictions that apply. CCSS F.IF.5, SMP 6

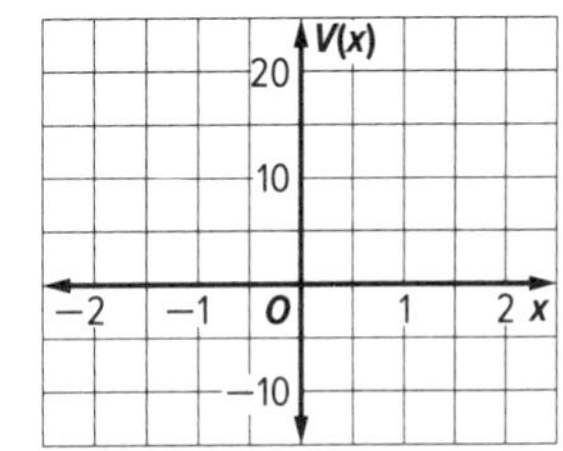

5.7 Solving Polynomial Equations

Objectives

- Create and solve polynomial equations.
- Use graphs to solve polynomial equations.

CCSS STANDARDS

Content: A.CED.1, A.REI.11
Practices: 1, 2, 3, 4, 5, 7
Use with Lesson 5-5

There are many different methods for solving polynomial equations. They include graphing, factoring, and substitution.

EXAMPLE 1 Modeling with Polynomial Functions CCSS A.CED.1, A.REI.11

EXPLORE An air show features demonstrations by vintage aircraft. A function to model the altitude of one of the airplanes is $h(t) = 10t^4 - 50t^3 - 50t^2 + 450t$, where altitude h is measured in feet and time t is measured in minutes. Part of the graph of $h(t)$ is shown.

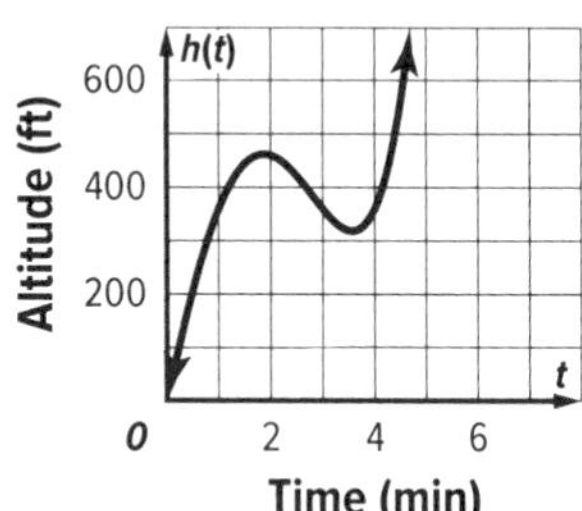

a. **REASON QUANTITATIVELY** Explain how you can determine the amount of time the plane spends below a given altitude. CCSS SMP 2

b. **REASON ABSTRACTLY** How many minutes will the plane spend below an altitude of 360 feet? Explain your reasoning. CCSS SMP 2

c. **USE A MODEL** Explain how you could solve the problem graphically. CCSS SMP 4

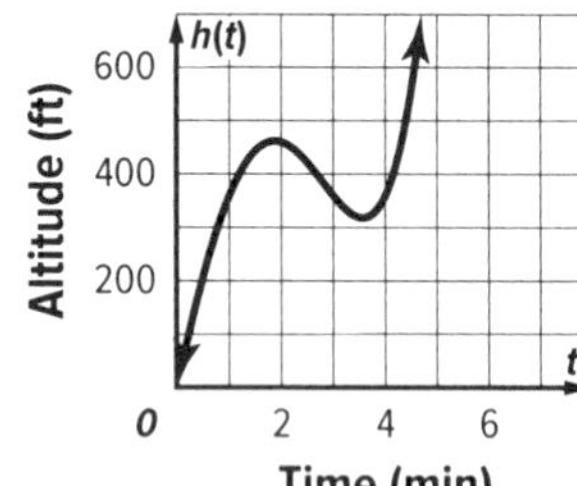

EXAMPLE 2 Solve Polynomial Equations by Factoring CCSS A.CED.1

a. **PLAN A SOLUTION** Describe a plan for solving the equation $3x^3 + 375 = 0$. Then solve and check your solutions. CCSS SMP 1

b. PLAN A SOLUTION Describe a plan for solving the equation $x^3 + 5x^2 + 9x + 45 = 0$. Then solve the equation. CCSS SMP 1

EXAMPLE 3 Solve Polynomial Equations in Quadratic Form

a. PLAN A SOLUTION Describe a plan for solving the equation $8x^4 - 22x^2 + 12 = 0$. Then solve the equation. CCSS SMP 1

b. PLAN A SOLUTION Describe a plan for solving the equation $25x^4 + 130x^2 + 25 = 0$. Then solve the equation. CCSS SMP 1

EXAMPLE 4 Solve Polynomial Equations Graphically CCSS A.REI.11

a. USE TOOLS Use a graphing calculator to sketch the graphs of $f(x) = x^4 - 2x - 5$ and $g(x) = -x^6 + 3x^2 + 3$ and approximate any points of intersection. CCSS SMP 5

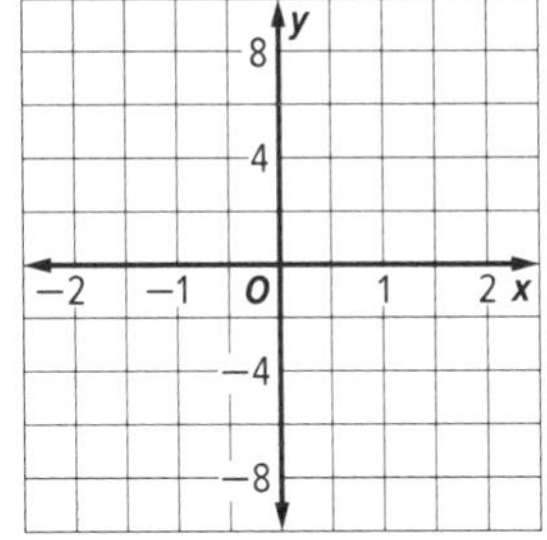

b. REASON QUANTITATIVELY Explain the algebraic significance of a point of intersection on the graphs of $f(x)$ and $g(x)$. How does this relate to solving the equation $f(x) = g(x)$? CCSS SMP 2

c. CRITIQUE REASONING A student claims the equation $x^4 - 64x^2 + 5 = 5x + 7$ has no solutions because there are no points of intersection on the graphs of $f(x) = x^4 - 64x^2 + 5$ and $g(x) = 5x + 7$. The student's graph is shown at right. Do you agree with the student? Explain your reasoning. CCSS SMP 3

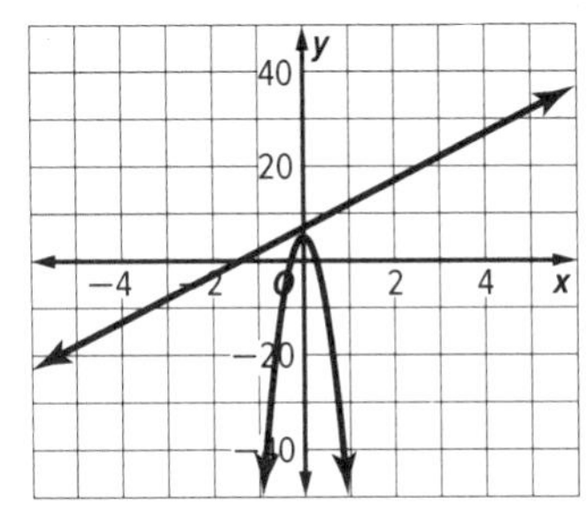

PRACTICE

1. **USE STRUCTURE** If the equation $ax^2 + bx + c = 0$ has solutions $x = m$ and $x = n$, what are the solutions to $ax^4 + bx^2 + c = 0$. Explain your reasoning. CCSS A.CED.1, SMP 7

2. **REASON QUANTITATIVELY** Explain why the x-coordinates of the points of intersection of $f(x) = x^3 + x^2 - 14x - 4$ and $g(x) = x^3 - 3x^2 - 6x + 28$ represent the solutions to $x^3 + x^2 - 14x - 4 = x^3 - 3x^2 - 6x + 28$. Then use the graphs to solve the equation. CCSS A.REI.11, SMP 2

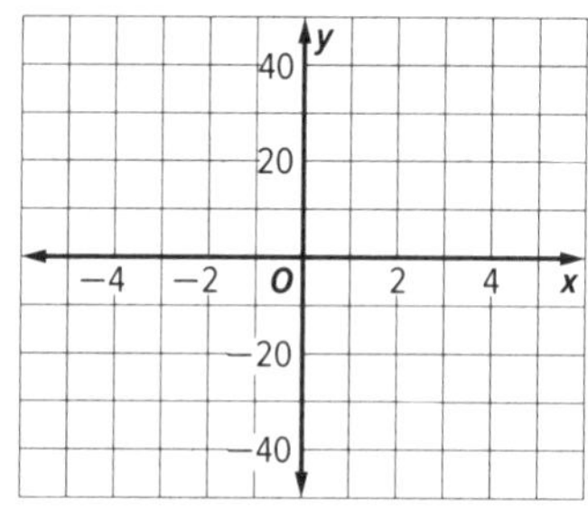

3. **REASON QUANTITATIVELY** A rectangular box has dimensions of x inches, $(x + 5)$ inches, and $(x - 2)$ inches. If the volume of the box is $30x$ cubic inches, explain how to find the dimensions of the box. Show your work. CCSS A.CED.1, SMP 2

4. **PLAN A SOLUTION** Explain two different methods for solving the equation $2x^3 - 3x^2 + 7x + 29 = x^3 + 2x^2 + 19x - 7$. Solve the equation using each method and verify the results are the same. CCSS A.REI.11, SMP 1

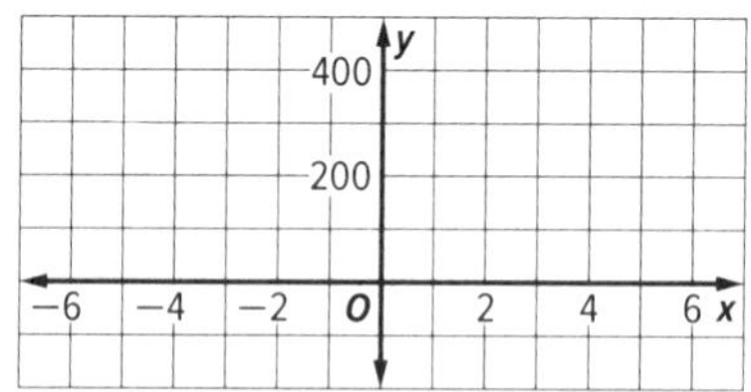

5. **USE STRUCTURE** Find the solutions to $(a + 3)^4 - 2(a + 3)^2 - 8 = 0$. Show your work. CCSS A.CED.1, SMP 7

6. **USE STRUCTURE** What are the factors of the polynomial function shown in the graph? Find two possible equations for the function. CCSS A.CED.1, SMP 7

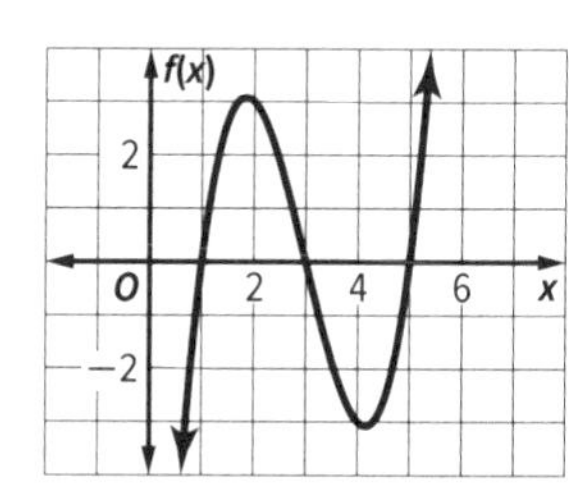

7. **USE TOOLS** A company models its profit in dollars using the function $P(x) = 70{,}000(x - x^4)$ on the domain $(0, 1)$ where x is the price at which they sell their product in dollars. Use a graphing calculator to sketch a graph and find the price at which their product should be sold to make a profit of \$20,000. CCSS A.CED.1, SMP 5

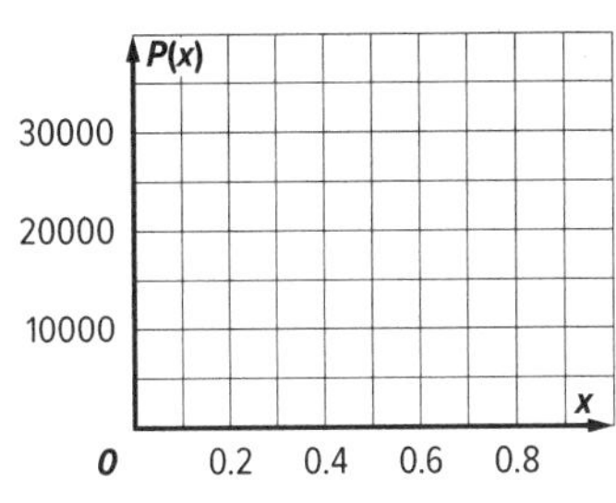

8. **USE STRUCTURE** The combined volume of a cube and a cylinder is 1,000 cubic inches. If the height of the cylinder is twice the radius and the side of the cube is four times the radius, then what would be the radius of the cylinder to the nearest tenth of an inch? CCSS A.CED.1, SMP 7

9. **USE A MODEL** The height of a passenger car for two roller coasters can be modeled by the functions $f(x) = \frac{1}{20}(x^3 - 60x^2 + 900x)$ and $g(x) = \frac{1}{12{,}000}(x^5 - 144x^4 + 7384x^3 - 158{,}400x^2 + 1{,}210{,}000x)$ where x is time in seconds for the first 35 seconds of the ride. CCSS SMP 4

   a. **INTERPRET PROBLEMS** If the two roller coasters start at the same time, then what equation would determine the times when the passenger cars of each roller coaster are at the same height? Can you determine any solutions by inspecting the equation? CCSS A.CED.1, SMP 1

   b. **USE TOOLS** Use a graphing calculator to sketch a graph of the functions $f(x)$ and $g(x)$ and solve the equation from **part a**. CCSS A.REI.11, SMP 5

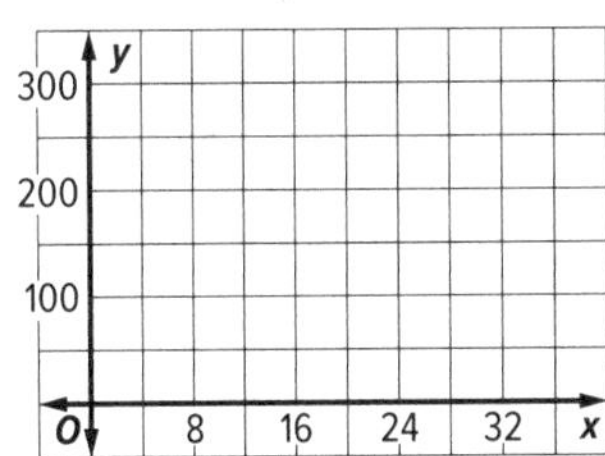

   c. **INTERPRET PROBLEMS** Write an equation to determine the times for which the passenger car modeled by $f(x)$ was at a height of 150 feet. Use a graphing calculator to solve the equation. CCSS A.CED.1, SMP 1

5.8 The Remainder and Factor Theorems

Objectives

- Apply the Remainder Theorem to evaluate polynomial functions.
- Use the Factor Theorem to identify factors of polynomial functions.

CCSS STANDARDS

Content: A.APR.2, F.IF.7c, A.CED.1

Practices: 1, 2, 3, 4, 6, 7

Use with Lesson 5-6

The Factor Theorem provides a connection between the zeros of a polynomial function and its factors. This aids in factoring, graphing, and especially writing polynomial functions when given a zeros along with points on a graph.

KEY CONCEPT Factor Theorem

The binomial $(x - r)$ is a factor of the polynomial function $P(x)$ if and only if $P(r) = 0$.

EXAMPLE 1 Connecting the Factor Theorem with Graphs of Polynomial Functions

EXPLORE Consider the graph of the polynomial function $P(x)$ as shown.

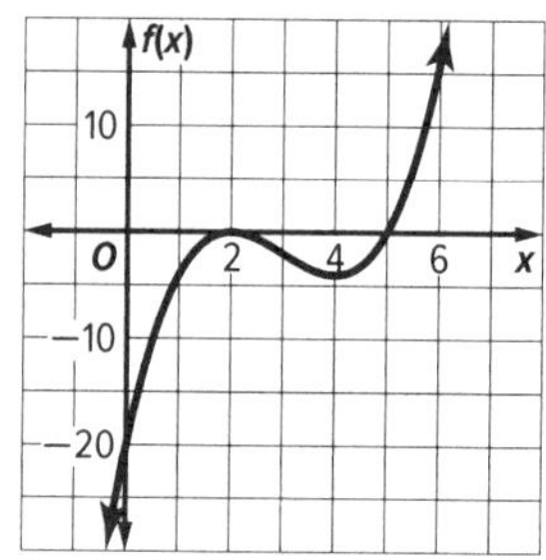

a. **USE STRUCTURE** Given the graph of $P(x)$, evaluate $P(5)$. What does this tell you about $(x - 5)$? Explain your answer. CCSS A.APR.2, SMP 7

b. **CONSTRUCT ARGUMENTS** Give the factors of $P(x)$. Do you have enough information to determine $P(x)$? Explain your answer. CCSS A.APR.2, SMP 3

c. **USE STRUCTURE** If $P(x)$ is a third degree polynomial function, can you determine $P(x)$? Explain your answer. CCSS A.CED.1, SMP 7

The Remainder Theorem introduces a result regarding the division of a polynomial by a binomial. In fact, the Factor Theorem stated above is a special case of the Remainder Theorem, a statement relating division to function evaluation.

KEY CONCEPT Remainder Theorem

If a polynomial function $P(x)$ is divided by $(x - r)$, the remainder is $P(r)$.

KEY CONCEPT Synthetic Substitution and Depressed Polynomials

For a given polynomial function $f(x)$ and value a, **synthetic substitution** is the process of evaluating $f(a)$ by using synthetic division to divide $f(x)$ by $x - a$.

When a polynomial is divided by one of its binomial factors, the quotient is called the **depressed polynomial**.

EXAMPLE 2 Using the Remainder Theorem

The percent change in a stock price over the course of a year is modeled by $f(x) = -0.5x^3 + 10x^2 - 58x + 80$, where x is the time in months.

a. **INTERPRET PROBLEMS** The stock price showed no gain or loss at the end of February. During which other months was there no change in price? Explain your reasoning. CCSS A.APR.2, SMP 1

b. **USE A MODEL** Graph the function on the coordinate plane. During which months was there a positive change in the stock price? CCSS F.IF.7c, SMP 4

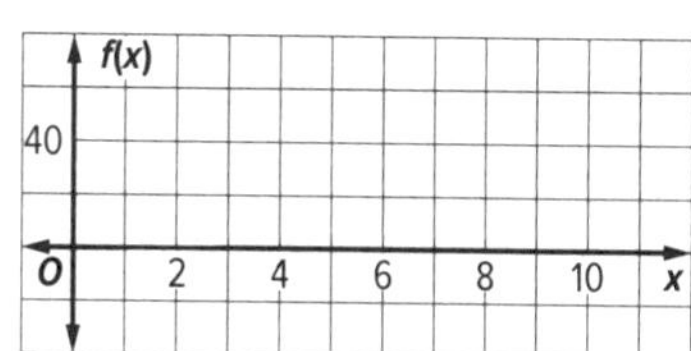

c. **USE STRUCTURE** What is the percent change in the stock price on the last day of June? Explain how you found your answer. CCSS A.APR.2, SMP 7

EXAMPLE 3 Apply the Remainder and Factor Theorems to Polynomials CCSS A.APR.2

a. **REASON ABSTRACTLY** If $(x - 2)$ is a factor of $f(x) = 3x^3 - 11x^2 + kx + 4$, determine the value of k and express the polynomial function in factored form. Explain your reasoning. CCSS SMP 2

b. **REASON QUANTITATIVELY** When the polynomial function $f(x) = 7x^3 - x^2 + kx + 4$ is divided by $(x + 2)$, the remainder is 2. Determine the value of k. Explain your reasoning. CCSS SMP 2

c. **CRITIQUE REASONING** Pablo divides $3x^4 - 4x^3 + x - 6$ by $x - 1$ and finds that

$$\frac{3x^4 - 4x^3 + x - 6}{x - 1} = 3x^3 - x^2 - x - \frac{6}{x - 1}$$

Is Pablo correct? Explain using the Remainder Theorem. CCSS SMP 3

PRACTICE

1. **USE STRUCTURE** Given that $f(-8) = 0$ and $f(x) = x^3 - x^2 - 58x + 112$, find all the factors of the polynomial and use the result to construct a graph. Explain your reasoning. CCSS A.APR.2, F.IF.7c, SMP 7

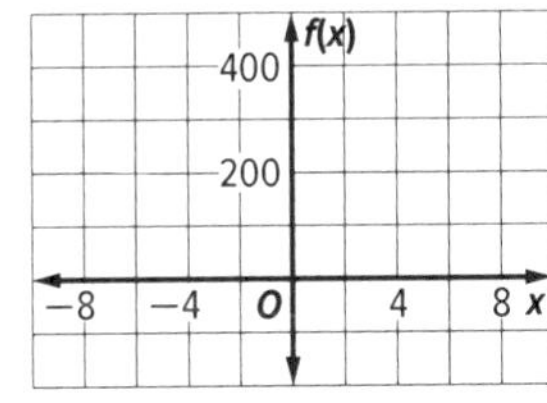

2. **REASON ABSTRACTLY** If $P(1) = 0$ and $P(x) = 10x^3 + kx^2 - 16x + 3$, find all the factors of $P(x)$ and use them to graph the function. Explain your reasoning. CCSS A.APR.2, F.IF.7c, SMP 2

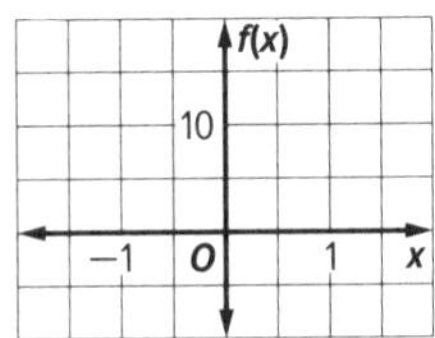

3. **USE STRUCTURE** The graph of a polynomial function is shown. What are the factors of the function? What is the equation of the function? Show your work. CCSS A.APR.2, A.CED.1, SMP 7

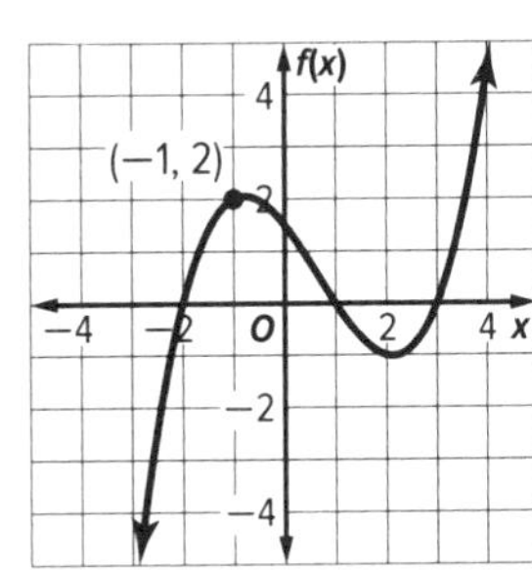

4. **COMMUNICATE PRECISELY** Divide the polynomial function $f(x) = 4x^3 - 10x + 8$ by the factor $(x + 5)$. Then state and confirm the Remainder Theorem for this particular polynomial function and factor. CCSS A.APR.2, SMP 6

5. **USE STRUCTURE** The graph of a polynomial function is shown. What are the factors of the function? What is the equation of a function of least degree to match the graph? CCSS A.APR.2, A.CED.1, SMP 7

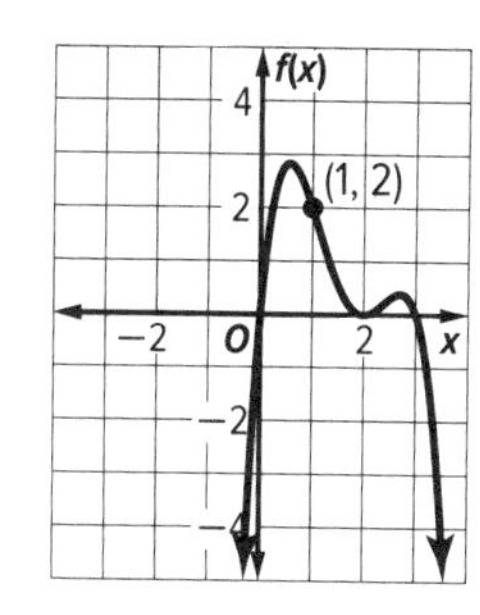

6. **REASON ABSTRACTLY** The volume of a box with a square base is $V(x) = 2x^3 + 15x^2 + 36x + 27$. If the height of the box is $(2x + 3)$, what are the sides of the base in terms of x? CCSS A.APR.2, SMP 2

7. **USE STRUCTURE** The polynomial function $P(x)$ is symmetric in the y-axis and contains the point $(2, -5)$. What is the remainder when $P(x)$ is divided by $(x + 2)$? Explain your reasoning. CCSS A.APR.2, SMP 7

8. **CALCULATE ACCURATELY** Verify the Remainder Theorem for the polynomial $x^2 + 3x + 5$ and the factor $(x - \sqrt{3})$ by first using synthetic division. CCSS A.APR.2, SMP 6

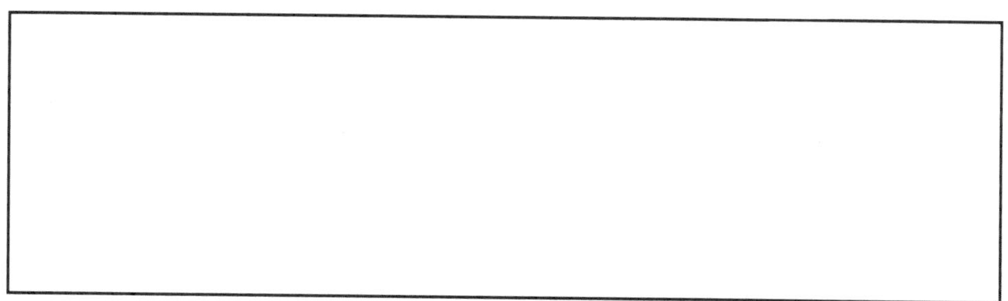

9. **USE A MODEL** If $(x + 6)$ is a factor of $kx^3 + 15x^2 + 13x - 30$, determine the value of k, factor the polynomial and confirm the result graphically. CCSS A.APR.2, F.IF.7c, A.CED.1, SMP 4

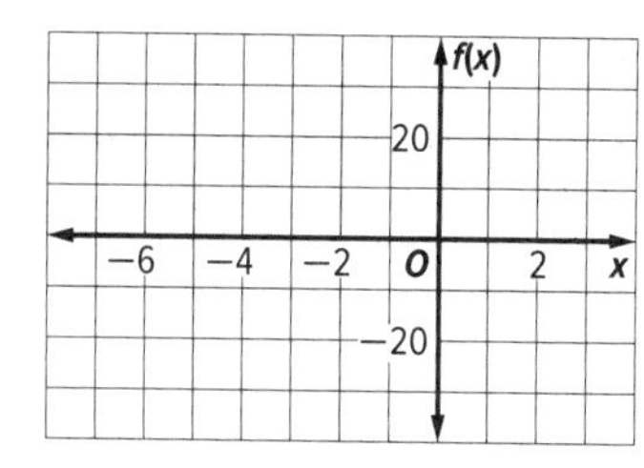

10. **COMMUNICATE PRECISELY** When the function $P(x) = 2x^4 + 3x^3 - 24x^2 - 13x + 12$ is divided by $x^2 - 2x - 3$, the remainder is zero. Explain how you can find the zeros of $P(x)$. Determine the zeros and use them to sketch a graph of the function. CCSS A.APR.2, F.IF.7c, SMP 6

11. **USE STRUCTURE** The points $(-1, 4)$, $(5, 10)$, $(3, 0)$, and $(0, 0)$ are on the graph of a 4th degree polynomial function $P(x)$. Give two possible equations for $P(x)$. Show your work. CCSS A.APR.2., SMP 7

12. **PLAN A SOLUTION** For a cubic function $P(x)$, if $P(-2) = -12$, $P(1) = -15$, $P(2) = 0$, and $P\left(-\frac{3}{2}\right) = 0$, write the equation for $P(x)$. Explain your answer. CCSS A.APR.2, A.CED.1, SMP 1

13. **USE STRUCTURE** For a cubic function $P(x)$, $P(2) = -90$, $P(-8) = 0$, and $P(5) = 0$. CCSS SMP 1

    a. Write two possible equations for $P(x)$. Explain your answer. CCSS A.CED.1, A.APR.2

    b. Graph your equations from **part a**. What three points do these graphs have in common? CCSS F.IF.7c

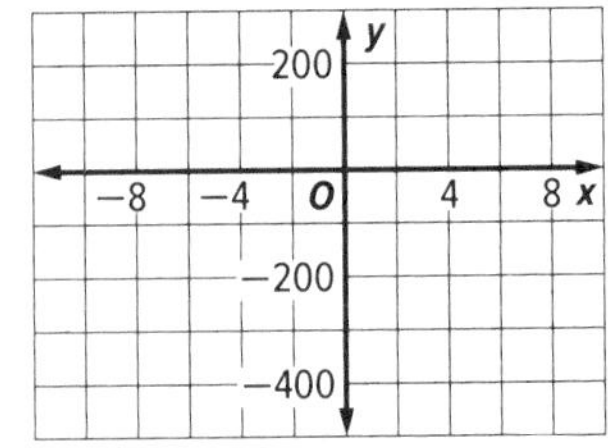

    c. If $P(4) = 60$, write the equation for $P(x)$. CCSS A.CED.1

14. **INTERPRET PROBLEMS** Tyrone made a table of x values and their corresponding $P(x)$ values for a polynomial function. CCSS A.APR.2, SMP 1

x	−3	−1	0	1	2	4
$P(x)$	−18	0	6	2	0	122

a. Use the Factor Theorem to name all of the factors of the polynomial shown in the table. Explain your answer.

b. What is the remainder of the polynomial when divided by $(x - 1)$? Explain.

15. **CONSTRUCT ARGUMENTS** A third degree polynomial function has factors $(x + 2)$, $(x - 2)$, and $(x - 3)$. Sketch three possible graphs of the function. Explain how there can be more than one graph with these factors. CCSS A.APR.2, F.IF.7c, SMP 3

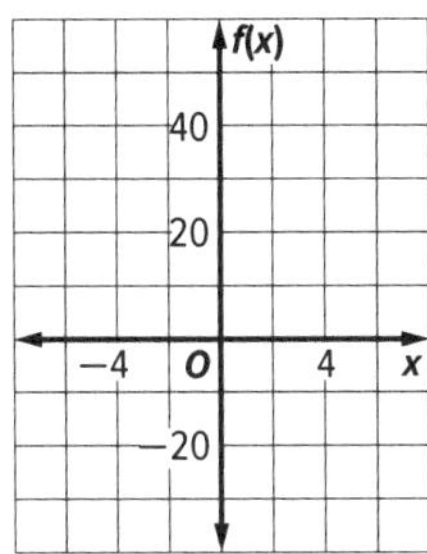

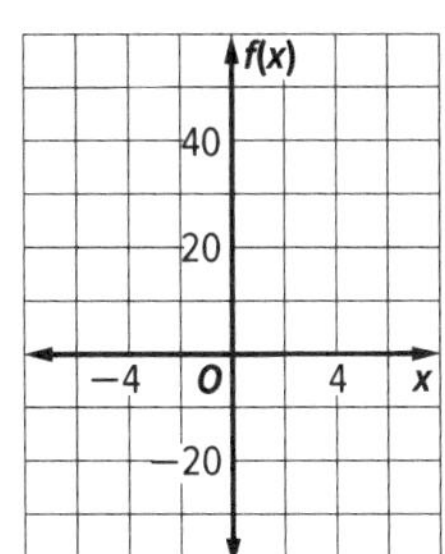

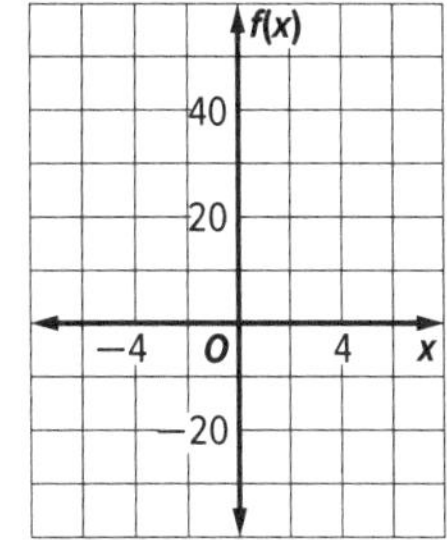

16. **INTERPRET PROBLEMS** Use the graph of polynomial function $P(x)$ to approximate k if $P(x) = q(x)(x - 3) + k$ for some polynomial function $q(x)$. Explain your reasoning. CCSS A.APR.2, SMP 1

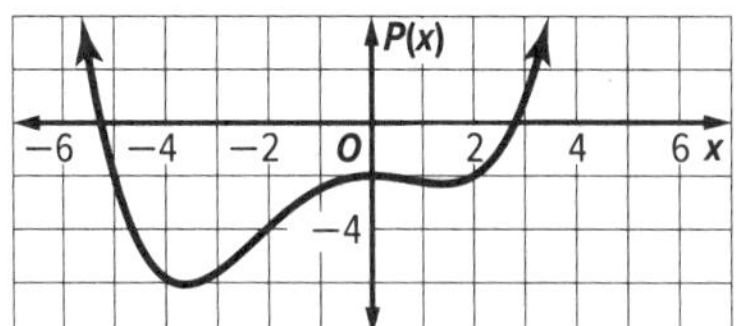

17. **INTERPRET PROBLEMS** The graph of a cubic polynomial function $P(x) = x^3 - 4.1x^2 - 31.3x + 71$ is shown at right. There appear to be zeros near $x = -5$, $x = 2$, and $x = 7$. Explain one way to determine if each value is a zero and identify any factors using the Factor Theorem. CCSS A.APR.2, SMP 1

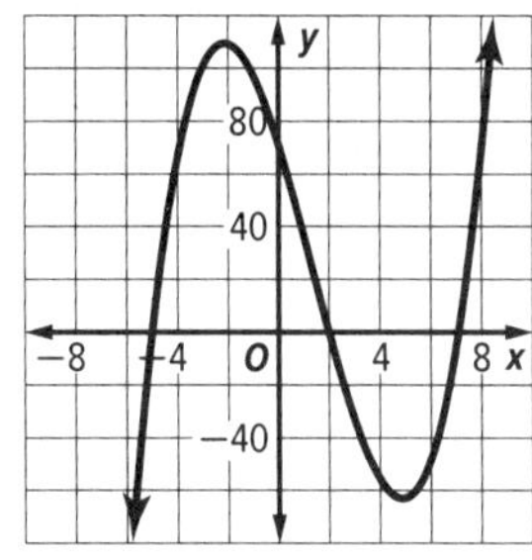

5.9 Roots and Zeros

Objectives

- Use the Fundamental Theorem of Algebra to analyze roots of polynomial equations and zeros of related polynomial functions.
- Use factors and zeros to write and graph polynomial functions.

Content: N.CN.9, A.APR.3, F.IF.4, A.CED.1
Practices: 1, 3, 6, 7
Use with Lesson 5-7

The real zeros of a polynomial function are the *x*-intercepts of the graph. The factor $(x - r)$ corresponds to the *x*-intercept $(r, 0)$, the zero *r*, and the root $x = r$. When the zeros of a polynomial function are known, the associated factors can be used to determine the function and its graph. If the coefficients of the polynomial function are real, any complex zeros will appear as conjugate pairs $a + bi$ and $a - bi$.

EXAMPLE 1 Finding Polynomials from *x*-Intercepts CCSS N.CN.9, F.IF.4

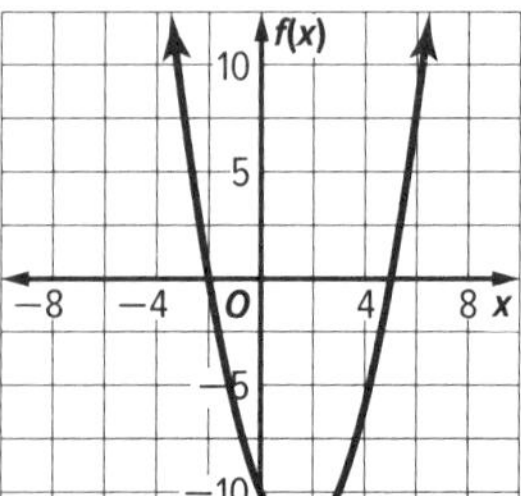

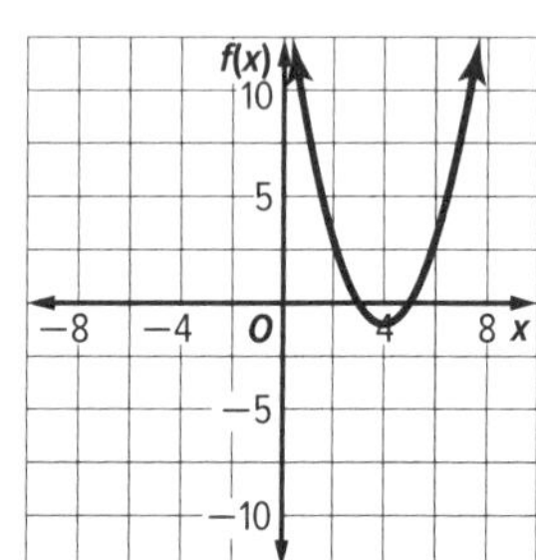

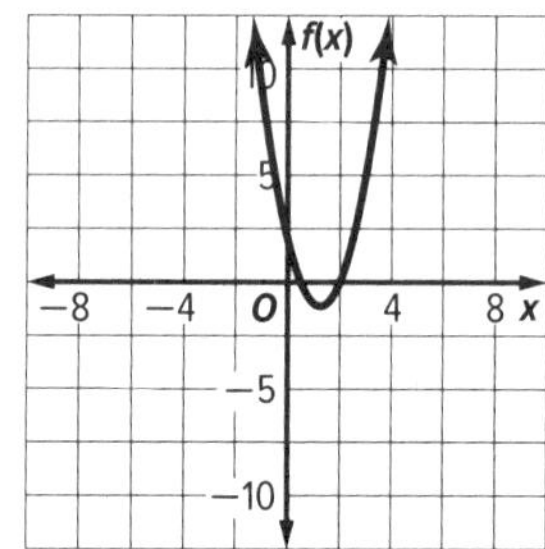

a. COMMUNICATE PRECISELY Determine the zeros and the function shown in each graph. CCSS SMP 6

b. USE STRUCTURE Is there a relationship between the zeros of the function and the coefficients and constants in the polynomial? When does this relationship exist? CCSS SMP 7

c. COMMUNICATE PRECISELY Suppose a polynomial function has complex zeros $2 + 3i$ and $2 - 3i$. Explain the challenges in finding a quadratic polynomial function associated with these zeros. CCSS SMP 6

d. USE STRUCTURE Without multiplying the factors, find a quadratic polynomial equation with roots $2 + 3i$ and $2 - 3i$. Explain your reasoning. CCSS SMP 7

KEY CONCEPT Complex Conjugates Theorem

Let $P(x)$ be a polynomial function with real coefficients. If $a + bi$ is a complex zero of $P(x)$, then so is its complex conjugate $a - bi$.

EXAMPLE 2 Investigate the Complex Conjugate Theorem CCSS N.CN.9

a. **CALCULATE ACCURATELY** The polynomial function $P(x) = x^2 - 4x + 5$ has the complex zero $2 + i$. Verify that the complex conjugate $2 - i$ is also a zero of $P(x)$. CCSS SMP 6

b. **USE STRUCTURE** Let $f(x)$ be a polynomial function with real coefficients and degree greater than zero. Regarding zeros, what does the Fundamental Theorem of Algebra guarantee about $f(x)$? CCSS SMP 7

c. **CONSTRUCT ARGUMENTS** If the polynomial function $f(x)$ has real coefficients and zero $a + bi$, can you find another zero of $f(x)$? What if $f(x)$ has degree 1? Explain your reasoning. CCSS SMP 3

KEY CONCEPT Fundamental Theorem of Algebra

Every polynomial equation with degree greater than zero has ________ in the set of complex numbers.
A polynomial equation of degree n has ________ in the set of complex numbers, including repeated roots.

EXAMPLE 3 Determine a Polynomial Function CCSS A.CED.1

a. **USE STRUCTURE** Write a polynomial function of least degree with integral coefficients and zeros that include -3 and $3 + 5i$. Explain your reasoning. CCSS SMP 7

b. **USE STRUCTURE** Write a polynomial function of least degree with integral coefficients and zeros that include $2, \frac{1}{4}$, and $-1 - i$. Explain your reasoning. CCSS SMP 7

c. **USE STRUCTURE** Write a polynomial function of least degree with integral coefficients and zeros that include $4-\sqrt{3}$ and $4+3i$. Explain your reasoning. 

KEY CONCEPT Descartes Rule of Signs

For a polynomial function $P(x)$ that has real coefficients,

the number of positive real zeros = ______________________,
or is less than this by an even number.

the number of negative real zeros = ______________________,
or is less than this by an even number.

EXAMPLE 4 Determine the Zeros of a Polynomial Function CCSS N.CN.9, F.IF.4

a. **CONSTRUCT ARGUMENTS** How many zeros does $P(x) = x^3 - 14x^2 + 64x - 80$ have? State the possible number of positive real zeros, negative real zeros, and imaginary zeros. Explain your reasoning. CCSS SMP 3

b. **CONSTRUCT ARGUMENTS** How many times could the graph of $P(x) = x^3 - 14x^2 + 64x - 80$ intersect the x-axis? Explain your reasoning. CCSS SMP 3

c. **PLAN A SOLUTION** Determine all of the zeros of $P(x) = x^3 - 14x^2 + 64x - 80$. Where does the graph intersect the x-axis? Explain. CCSS SMP 1

d. **USE STRUCTURE** Graph $P(x)$ using a graphing calculator and provide a sketch of the results to verify the zeros found in **part c**. CCSS SMP 7

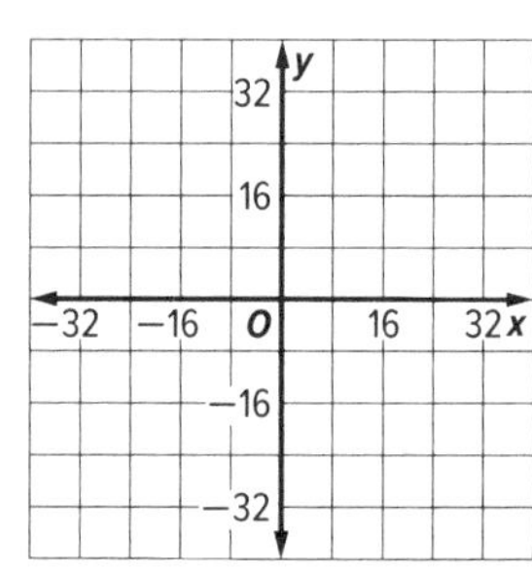

PRACTICE

1. **USE STRUCTURE** Write a polynomial function of least degree with integral coefficients and zeros that include $-\frac{3}{2}$ and $2 - 2\sqrt{2}$. Explain your reasoning. CCSS N.CN.9, A.CED.1, SMP 7

2. **USE STRUCTURE** Write a polynomial function of least degree with integral coefficients and zeros that include 4 and $3 - 4\boldsymbol{i}$. Explain your reasoning. CCSS N.CN.9, A.CED.1, SMP 7

3. **PLAN A SOLUTION** Determine all of the zeros of $P(x) = x^3 - x^2 - 15x - 9$ and construct a graph of the function. CCSS N.CN.9, A.APR.3, SMP 1

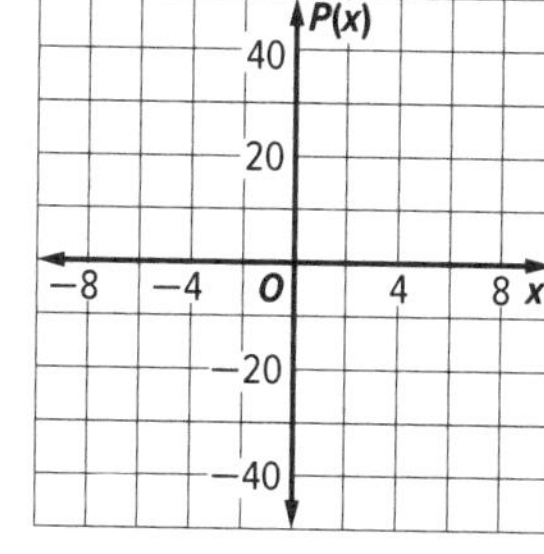

4. a. **USE STRUCTURE** Write a polynomial function of least degree with integral coefficients and zeros that include $-1 - 4\boldsymbol{i}$ and $\frac{2}{3} + \frac{1}{3}\boldsymbol{i}$. Explain your reasoning. CCSS N.CN.9, A.CED.1, SMP 7

   b. **USE STRUCTURE** Use your answer from **part a** to write another polynomial function with integral coefficients having the same degree and zeros. Explain your reasoning. CCSS N.CN.9, A.CED.1, SMP 7

   c. **USE STRUCTURE** Are you able to sketch these graphs based on the zeros? Explain your reasoning. Use a calculator to graph the functions from **parts a** and **b**. CCSS N.NC.9, SMP 7

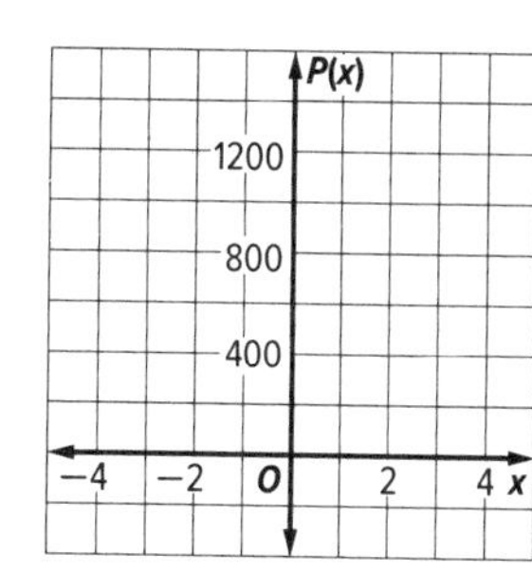

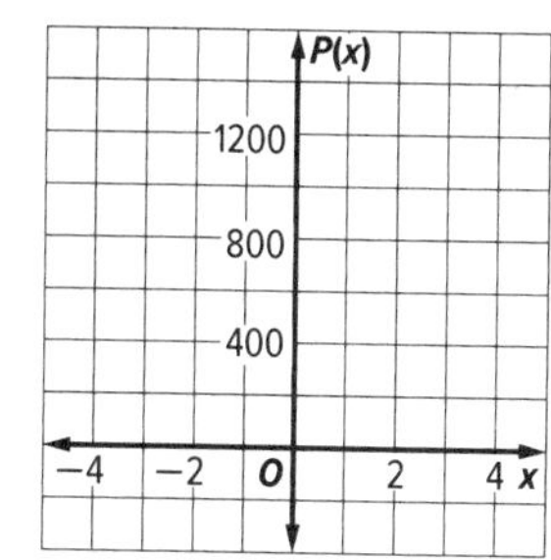

5. **USE STRUCTURE** The graph shows a function of degree 4. State the number of positive real zeros, negative real zeros, and imaginary zeros of $P(x)$. Explain your reasoning. CCSS N.CN.9, F.IF.4, SMP 7

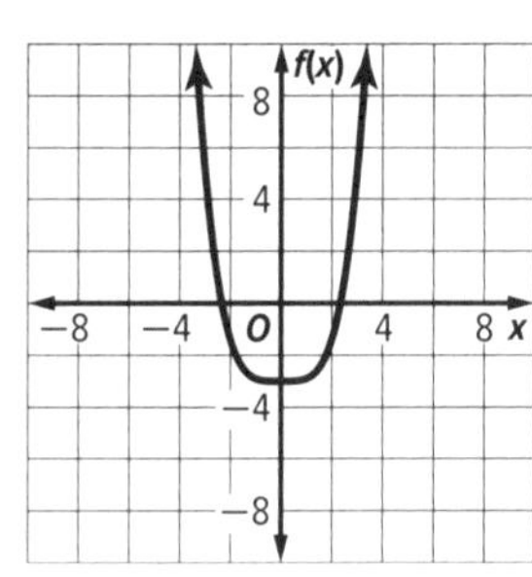

6. **CONSTRUCT ARGUMENTS** Explain why the graph of $P(x) = -4(x + 2)(x - 3)^3$ crosses the x-axis at $x = 3$, while the graph of $R(x) = 2(x - 1)(x - 3)^4$ does not cross the x-axis at $x = 3$. CCSS F.IF.4, SMP 3

7. **PLAN A SOLUTION** During the first 6 years after introducing a new product, a company's revenue is modeled by $R(x) = -x^4 + 11x^3 - 43x^2 + 75x + 125$, where $R(x)$ is measured in thousands of dollars and x is measured in years. For what period of time were the profits greater than \$175,000? Explain your answer. CCSS N.CN.9, F.IF.4, SMP 1

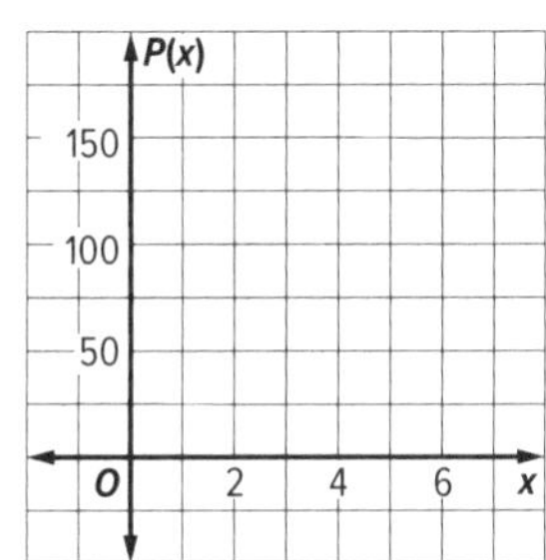

8. **USE STRUCTURE** The graph models is of a function with degree 3. State the number of positive real zeros, negative real zeros, and imaginary zeros of $P(x)$. Explain your reasoning. CCSS N.CN.9, F.IF.4, SMP 7

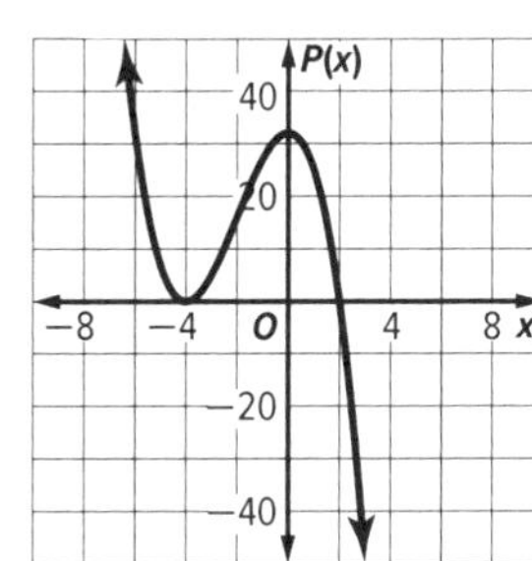

9. **CONSTRUCT ARGUMENTS** Can you verify that a complex number is a zero of a polynomial function by looking at its graph? How can you verify that $3 + 2i$ is a zero of $f(x) = x^3 - 8x^2 + 25x - 26$. Explain your reasoning. CCSS N.CN.9, A.APR.2, SMP 3

10. PLAN A SOLUTION Sketch the graph of two functions of different degree that only have zeros −3, 1, and 2. CCSS F.IF.7c, SMP 1

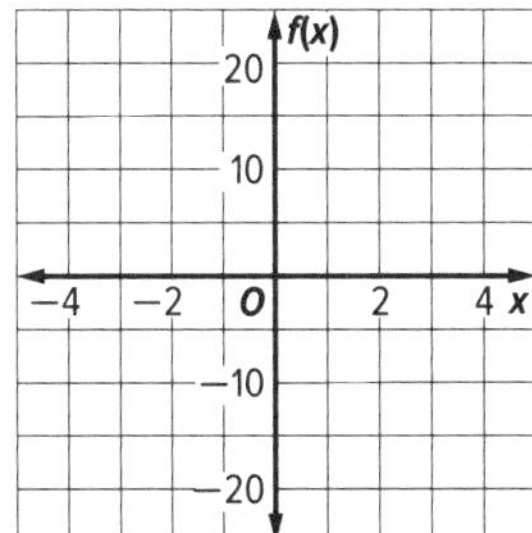

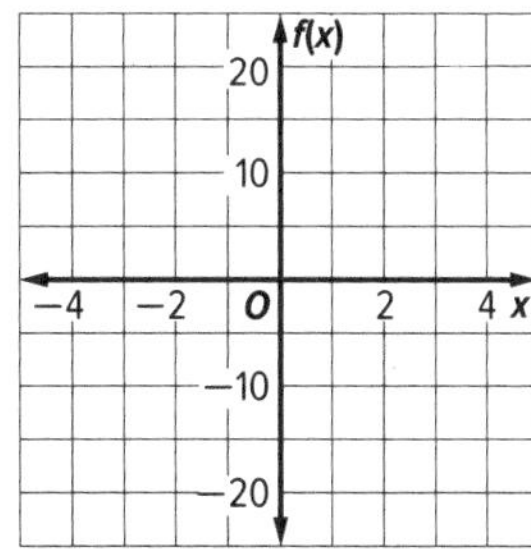

11. CONSTRUCT ARGUMENTS How many cube roots of 8 exist in the set of complex numbers? Explain your reasoning. Then find the cube roots of 8.

12. CONSTRUCT ARGUMENTS Use the zeros to construct a hand-drawn graph of $P(x) = x^3 - 7x^2 + 7x + 15$. Discuss the accuracy of your graph, and what could be done to improve the accuracy. CCSS A.APR.3, SMP 3

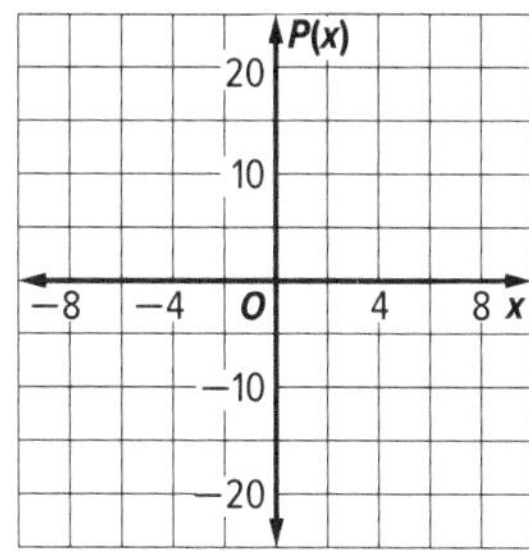

13. USE STRUCTURE A fourth degree polynomial function $P(x)$ with real coefficients contains the factor $(x - 2\boldsymbol{i})^2$. Write a possible equation for $P(x)$. Explain your reasoning. CCSS N.CN.9, A.CED.1, SMP 7

14. Let the polynomial function $f(x)$ have real coefficients, degree 5, and zeros $4 + 3\boldsymbol{i}$, $2 - 7\boldsymbol{i}$, and $6 + b\boldsymbol{i}$ where b is some real number constant. CCSS A.CED.1

a. CONSTRUCT ARGUMENTS What can be determined about the constant b? Explain your reasoning. CCSS SMP 3

b. USE STRUCTURE Using your answer from **part a**, write a possible equation for $f(x)$. CCSS SMP 7

Performance Task

Polynomial Design

Provide a clear solution to the problem. Be sure to show all of your work, include all relevant drawings, and justify your answers.

A graphic artist is creating a figure like the one shown. She is using a graphing program and working with two polynomial functions to draw the curves.

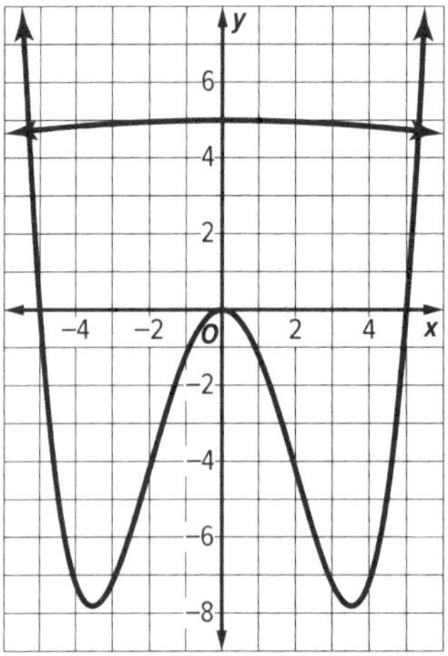

Part A

Do the graphs exhibit any symmetry? What are the simplest possible degrees of the two polynomial functions? Write a possible equation for the lesser degree polynomial function. Justify your answer.

Part B

Devise a plan to find an equation for the greater degree polynomial function. Include any theorems you use, and explain your reasoning.

Part C

Create your own design that satisfies the following conditions.

- You must use polynomial functions to define the boundaries.
- At least one of the polynomials must have degree greater than 2.
- One polynomial cannot be negative of the other.
- The region bounded by the polynomials you choose must contain points in each of the four quadrants.

Performance Task

Painting Polynomials

Provide a clear solution to the problem. Be sure to show all of your work, include all relevant drawings, and justify your answers.

The local rec center has to paint several walls of similar shape. The figure represents the shape of the walls.

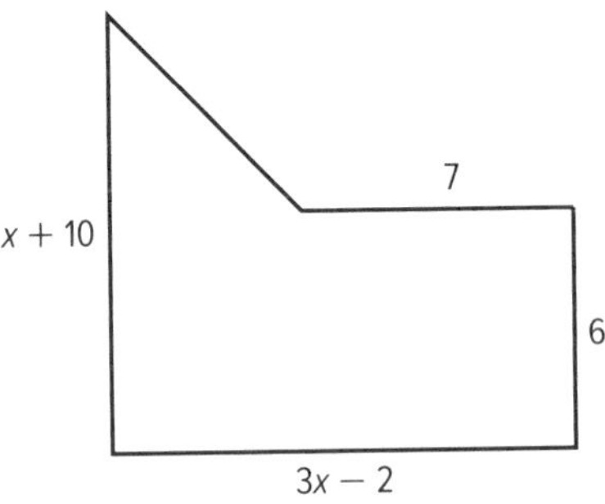

Part A

Write an expression for the area in square units. Justify your expression.

Part B

What are the limits of the variable x in this context? Explain your reasoning.

Part C

Use technology to graph the function. Sketch the graph below. Describe any relative extrema, any zeros of the function, and the end behavior.

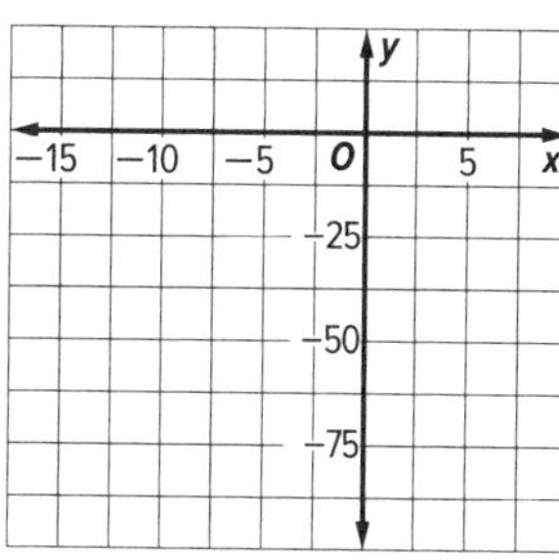

Standardized Test Practice

1. Write the polynomial $f(x)$ of lowest degree that has roots -2, 3, and 5 and has leading coefficient 1. CCSS N.CN.9

2. Jeff buys a box of pencils. The box is a rectangular prism with dimensions as shown.

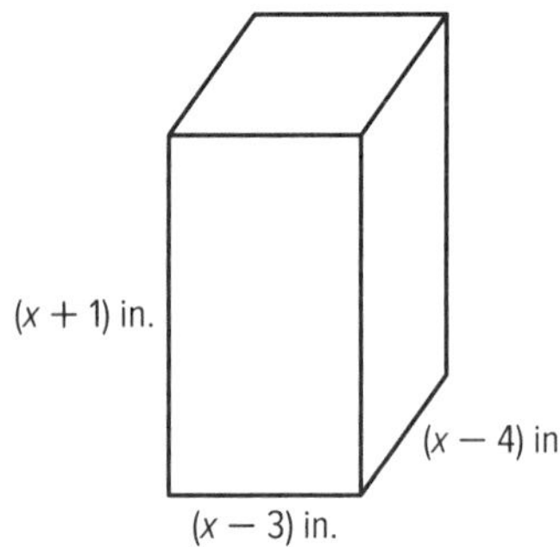

If the volume of the box is 42 cubic inches, then the dimensions of the box are ______, ______, and ______. CCSS A.CED.1

3. Find all roots of the following equation. CCSS A.APR.3, N.CN.9

$x^4 - 2x^3 - 16x^2 + 82x - 65 = 0$

4. Which function has the end behavior described below? CCSS F.IF.4

As $x \to -\infty$, $f(x) \to +\infty$, and as $x \to +\infty$, $f(x) \to -\infty$.

$f(x) = 4x + 5$

$f(x) = 4x^2 + 5x - 1$

$f(x) = -4x^2 + 5x - 1$

$f(x) = -4x^3 + 5x - 1$

5. What is the sum $(2x^4 - 3x^3 + 5x^2 - x + 11) + (3x^3 - x^2 + 6x - 4)$? CCSS A.APR.1

6. The value of g is proportional to the 5th power of d. When d is 20, g is 14,400. CCSS F.IF.4, A.APR.3

What is the function?

Graph the function, showing intercepts and end behavior.

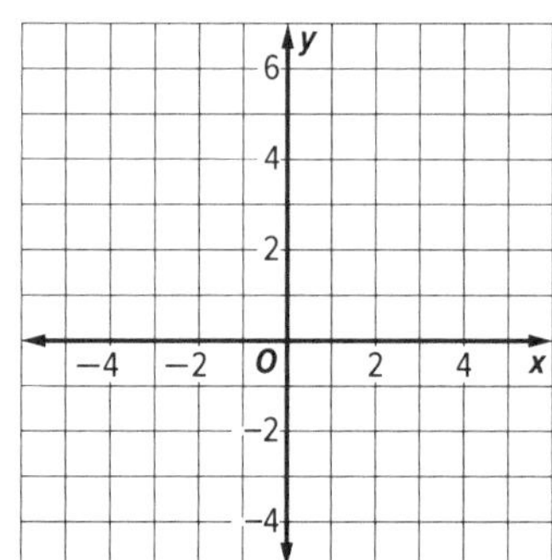

7. What are the roots of $f(x) = 2x^3 + 7x^2 + 2x - 3$? CCSS A.APR.3, N.CN.9

Graph the function. CCSS A.APR.3

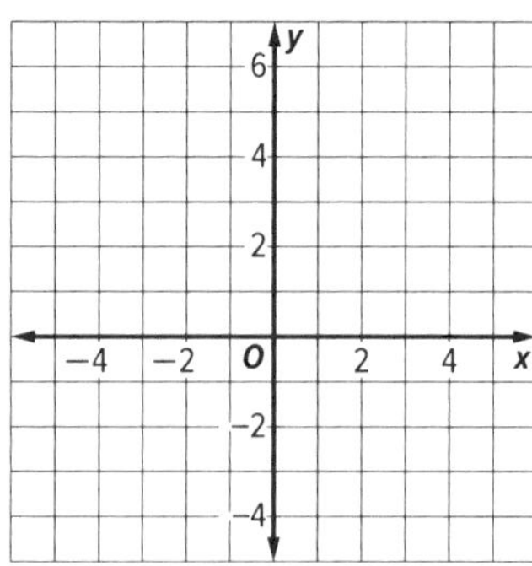

8. What is the quotient $(2x^4 - 3x^3 + 5x^2 - x + 11) \div (x - 3)$? CCSS A.APR.6

9. For each of the following graphs, identify whether the degree of the function shown is odd or even, and whether it has a positive or negative leading coefficient. CCSS F.IF.7c

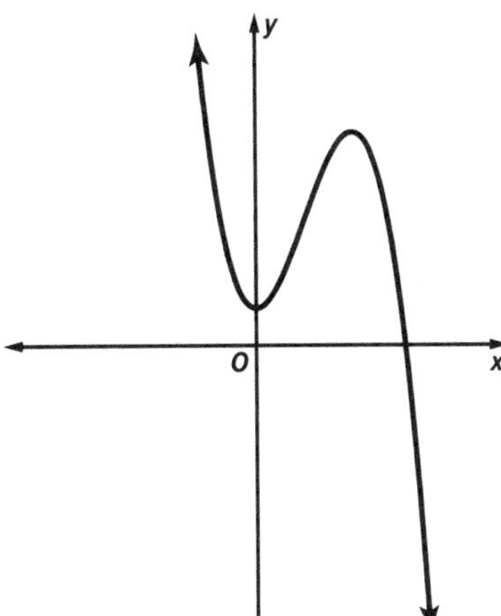

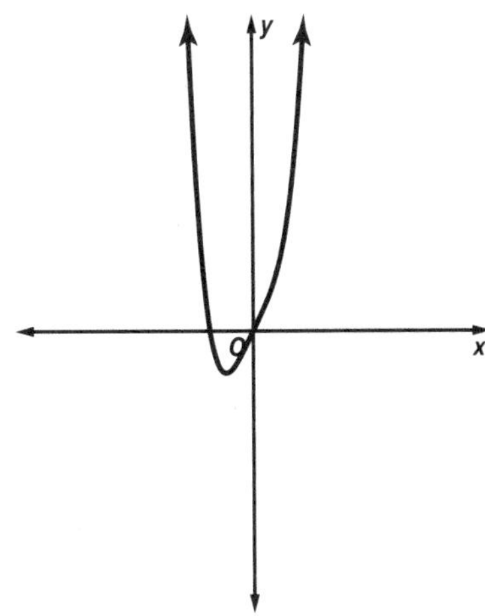

10. A cubic function $f(x)$ has zeros 1 and $1 + 2i$. CCSS A.APR.3

a. Write the function as a product of an unknown constant, a linear factor, and a quadratic factor.

b. Use the fact that $f(0) = 10$ to determine the value of the constant. Write the function using the value of the constant.

c. Describe the end behavior of the graph of the function.

d. Graph the function to check **part b**. Does your graph show the function value given in **part b**? Does your graph show the real zero that was given? Does your graph match the end behavior you described in **part c**?

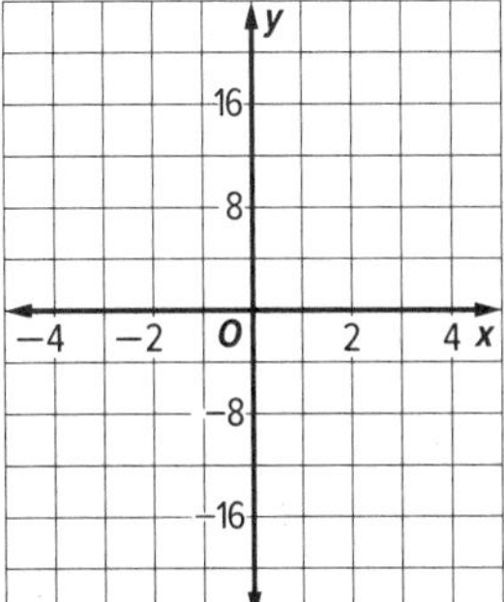

11. Complete the synthetic division started below to determine if $x = -3$ is a zero of $6x^6 + 13x^5 + 9x^4 - 11x^3 - 8x + 4$. CCSS A.APR.2

$$\begin{array}{r|l} -3 & 6 \\ \hline & 6 \end{array}$$

Is $x = -3$ a zero of this polynomial? Explain how you know.

12. a. Is it possible for a polynomial of degree 5 to have 7 unique roots? Explain. CCSS N.CN.9

b. Is it possible for a polynomial of degree 5 to have 4 unique roots? Explain. CCSS N.CN.9

6 Radical Functions

CHAPTER FOCUS Learn about some of the Common Core State Standards that you will explore in this chapter. Answer the preview questions. As you complete each lesson, return to these pages to check your work.

What You Will Learn	Preview Question
Lesson 6.1: Operations on Functions	
CCSS F.BF.1b Combine standard function types using arithmetic operations. CCSS F.IF.9 Compare properties of two functions each represented in a different way (algebraically, graphically, numerically in tables, or by verbal descriptions.	CCSS SMP 1 For the given right triangle, write a function, $f(x)$, for finding the hypotenuse. 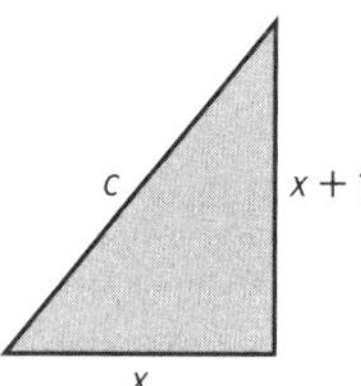
Lesson 6.2: Inverse Functions and Relations	
CCSS F.BF.4a Solve an equation of the form $f(x) = c$ for a simple function f that has an inverse and write an expression for the inverse. CCSS F.IF.5 Relate the domain of a function to its graph and, where applicable, to the quantitative relationship it describes. **Also addresses:** F.IF.4	CCSS SMP 5 The graph of a function's inverse is symmetric about the line $y = x$. Sketch the graph of the function $y = 3x - 2$ on the grid provided and then sketch its inverse. 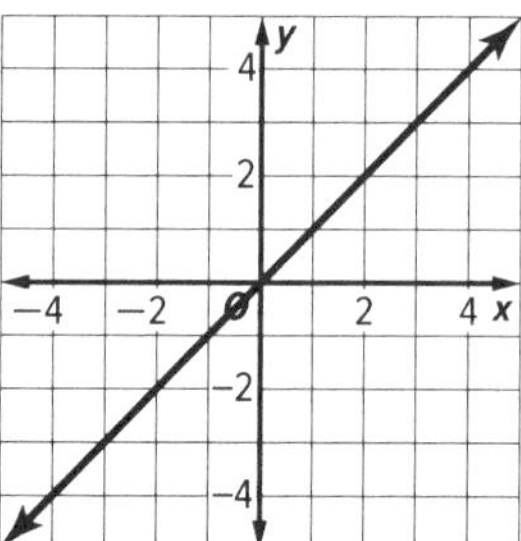
Lesson 6.3: Square Root Functions & Inequalities	
CCSS F.BF.3 Identify the effect on the graph of replacing $f(x)$ by $f(x) + k$, $k\,f(x)$, $f(kx)$, and $f(x + k)$ for specific values of k (both positive and negative); find the value of k given the graphs. Experiment with cases and illustrate an explanation of the effects on the graph using technology. CCSS F.IF.7b Graph square root, cube root, and piecewise-defined functions, including step functions and absolute value functions. **Also addresses:** A.CED.2, F.IF.4, F.IF.5, F.IF.9, A.SSE.2	CCSS SMP 3 Identify the inequality shown. Explain your reasoning. 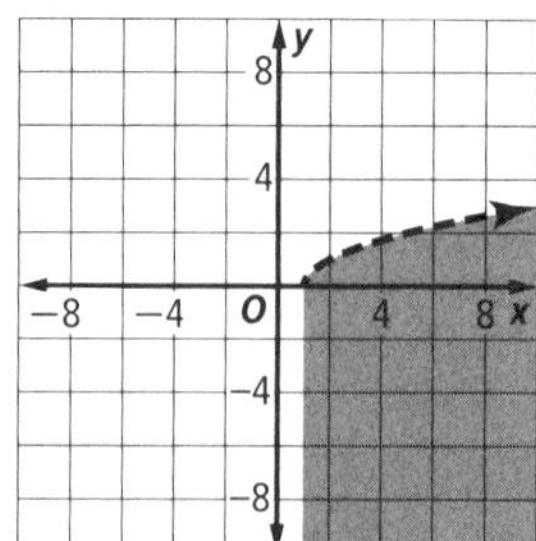 **A.** $y < \sqrt{x - 1}$ **B.** $y < \sqrt{x} - 1$ **C.** $y < \sqrt{x} + 1$

What You Will Learn	Preview Question
Lesson 6.4: Modeling: Square Root Functions	
CCSS F.IF.5 Relate the domain of a function to its graph and, where applicable, to the quantitative relationship it describes. CCSS F.IF.4 For a function that models a relationship between two quantities, interpret key features of graphs and tables in terms of the quantities, and sketch graphs showing key features given a verbal description of the relationship. CCSS A.SSE.1a Interpret parts of an expression, such as terms, factors, and coefficients. CCSS A.SSE.1b Interpret complicated expressions by viewing one or more of their parts as a single entity.	CCSS SMP 2 The area of a circle with radius r is $A = \pi r^2$. A company manufactures circular metal discs of different sizes. Write a function $r(a)$ for the radius of a disc given its area a and interpret the domain in the context of the problem.
Lesson 6.5: *n*th Roots	
CCSS A.SSE.2 Use the structure of an expression to identify ways to rewrite it.	CCSS SMP 6 The volume of a cube is modeled by this function: $f(x) = (x + 2)^3$. Solve for x if the volume is 125 cubic units.
Lesson 6.6: Operations with Radical Expressions	
CCSS A.SSE.2 Use the structure of an expression to identify ways to rewrite it.	CCSS SMP 8 Suppose $f(x)$ is a radical function. Describe the graph of $g(x) = f(x + 1) + 1$. CCSS SMP 7 Rewrite the following expression in the form ax^m for appropriate choices of a and m. $\sqrt{4x}\,(\sqrt{16x} + \sqrt{x})$
Lesson 6.7: Solving Radical Equations and Inequalities	
CCSS A.REI.11 Explain why the x-coordinates of the points where the graphs of the equations $y = f(x)$ and $y = g(x)$ intersect are the solutions of the equation $f(x) = g(x)$; find the solutions approximately, e.g., using technology to graph the functions, make tables of values, or find successive approximations. Include cases where $f(x)$ and/or $g(x)$ are linear, polynomial, rational, absolute value, exponential, and logarithmic functions. **Also addresses:** A.REI.2	CCSS SMP 2 Suppose $f(x) = 0$ is a radical equation. What can you conclude about real number a, if $f(a) = 0$?

6.1 Operations on Functions

Objectives

- Model relationships by adding, subtracting, multiplying, and dividing functions.
- Compare properties of functions that are presented in different formats, such as in tables, verbal descriptions, algebraic equations, or graphs.

CCSS STANDARDS

Content: F.IF.9, F.BF.1b
Practices: 2, 3, 4, 6, 7, 8
Use with Lesson 6-1

EXAMPLE 1 Model Areas with Functions CCSS F.BF.1b

EXPLORE An architect is given specific directions for planning the dimensions of a new art exhibit showroom.

a. **USE A MODEL** The length of one room is to be 4 feet longer than the width. The adjacent room is to have the same length, but its width is to be twice the width of the first room. The diagram at right shows the dimensions in terms of the width of the first room, x.

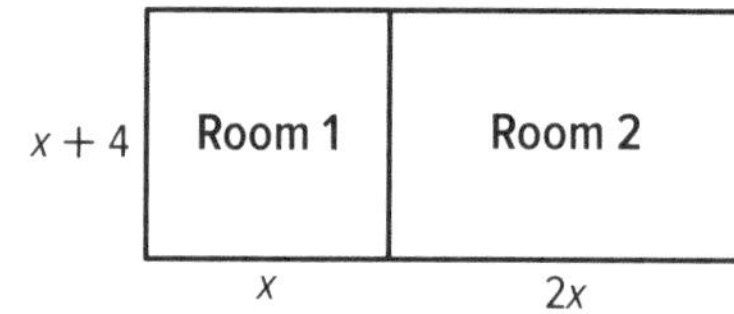

Write functions $A_1(x)$ and $A_2(x)$ to represent the areas of the rooms 1 and 2, respectively. Explain how you combined the given expressions to find the area of each room. CCSS SMP 4

b. **USE A MODEL** Write a new function $A(x)$ that models the simplified area of the full floor plan. CCSS SMP 4

c. **EVALUATE REASONABLENESS** If the artists decide the width of the first room should be 10 feet, evaluate your three functions from **parts a** and **b** for $x = 10$. Describe the relationship you find among the functions. CCSS SMP 8

d. **USE A MODEL** When the walls are constructed, their height will be half of the length of the rooms. Write a function $H(x)$ representing the height; $V_1(x)$ and $V_2(x)$, the volumes of each room; and $V(x)$, the volume of the entire space, in terms of x. Show your work. CCSS SMP 4

e. REASON ABSTRACTLY The artists tell you that the function $6x + 16$ can be used to model a different attribute of this space. What is this attribute? What did the artist do to come up with this function? How do you know? CCSS SMP 2

f. USE A MODEL The artists decide to change the dimensions of the floor plan such that the total area is modeled by $A(x) = 2x^2 + 14x + 24$. If the area for the first room is unchanged, then what function models the new area of the second room? How do you know? CCSS SMP 4

g. REASON ABSTRACTLY If the two rooms share the length of the adjacent wall, use the function from **part d** to show that the width of the second room can be modeled by the function $w_2(x) = x + 6$. Explain. CCSS SMP 2

h. EVALUATE REASONABLENESS Evaluate $A(x) = 2x^2 + 14x + 24$, $B(x) = x + 4$, and $C(x) = 2x + 6$ for $x = 15$. Describe the relationship you find among the functions in terms of the context. CCSS SMP 8

KEY CONCEPT

Functions can be added, subtracted, multiplied and divided just like operations on real numbers. Complete each statement with the appropriate rule showing how to combine functions using arithmetic operations.

Operation	Definition
Addition	$(f + g)(x) =$ ______
Subtraction	$(f - g)(x) =$ ______
Multiplication	$(f \cdot g)(x) =$ ______
Division	$\left(\frac{f}{g}\right)(x) = \frac{f(x)}{g(x)}$ ______

EXAMPLE 2 Compare Equations and Graphs of Functions CCSS F.BF.1b, F.IF.9

The graphs represent the four arithmetic operations on the functions $f(x) = x^2 + 1$ and $g(x) = 2x$.

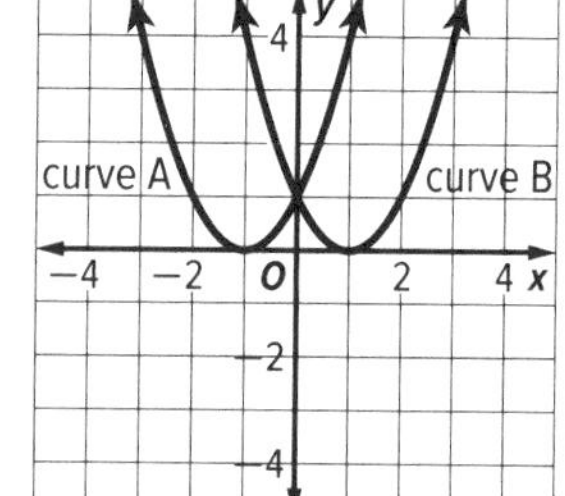

a. REASON QUANTITATIVELY Which graph represents $(f + g)(x)$? Explain your reasoning. CCSS SMP 2

b. REASON QUANTITATIVELY Which graph represents $(f - g)(x)$? Explain your reasoning. CCSS SMP 2

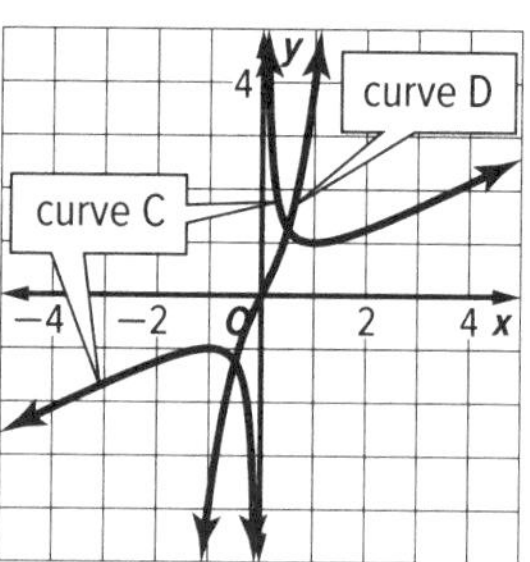

c. USE STRUCTURE Of the two graphs that remain, which represents $(f \cdot g)(x)$ and which represents $\left(\frac{f}{g}\right)(x)$? Explain your reasoning. CCSS SMP 7

d. REASON ABSTRACTLY The table below represents one of four functions derived from doing the four arithmetic operations on the functions above. Determine its maximum or minimum and use that information to determine which operation it represents. CCSS SMP 2

x	3	2	1	0	−1	−2	−3	−4
y	16	9	4	1	0	1	4	9

PRACTICE

1. **REASON ABSTRACTLY** The National Center for Education Statistics reports data showing that college enrollment for men since 2006 (in thousands) can be modeled by the function $f(x) = 389x + 7500$, where x represents the number of years since 2006. Similarly, enrollment for women can be modeled by the function $g(x) = 480x + 10075$. Write a function for $(f + g)(x)$ and interpret what it represents. CCSS F.BF.1b, SMP 2

2. **REASON ABSTRACTLY** Jasper is raising money for a local charity by making and selling cupcakes. The cost of making x cupcakes can be represented by the function $C(x) = 0.625x$, and his revenue from selling x cupcakes can be represented by the function $R(x) = 3.45x$. Explain what these functions must mean about Jasper's cupcakes. Then write a function that represents his profit. Explain your reasoning. CCSS F.BF.1b, SMP 2

3. **CRITIQUE REASONING** The area of the base of a rectangular prism is given by $A(x) = x^2 - 7x - 30$ and its height is given by $h(x) = 2x + 5$. Deepti says that the volume of the prism would be given by $V(x) = 2x^3 - 9x^2 - 95x - 150$. Do you agree or disagree and why? CCSS F.BF.1b, SMP 3

4. **USE A MODEL** The total cost of producing x photo greeting cards is given by the function $f(x) = 0.8x + 25$. Write a function showing $\left(\frac{f}{g}\right)(x)$, where $g(x) = x$. What does this new function model? CCSS F.BF.1b, SMP 4

5. a. **CALCULATE ACCURATELY** Given $f(x) = -2x + 5$ and $g(x) = x + 1$, determine whether $(f \cdot g)(x)$ or $h(x)$ has a higher y-intercept. Explain your reasoning. CCSS F.BF.1b, F.IF.9, SMP 6

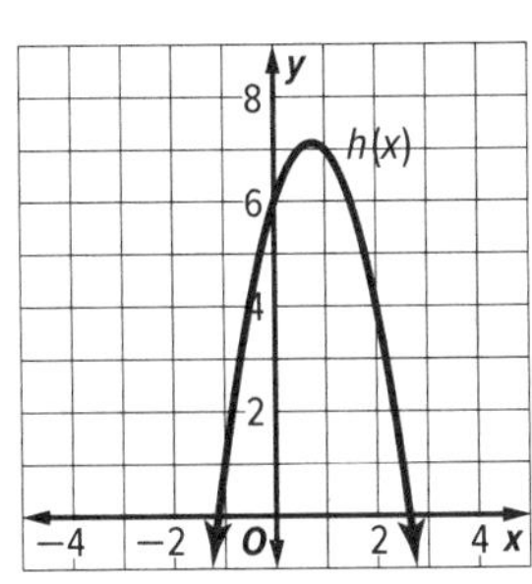

b. **CALCULATE ACCURATELY** Consider the two functions from **part a**. Which function has a larger spread between its x-intercepts? Explain CCSS F.BF.1b, SMP 3

6. **CALCULATE ACCURATELY** The following table shows various values of functions $f(x)$, $g(x)$, and $h(x)$. CCSS F.BF.1b, SMP 6

x	−1	0	1	2	3	4
f(x)	7	−2	0	2	4	1
g(x)	−3	−4	−5	0	1	1
h(x)	0	4	1	1	5	5

Use the table to find the following values:

$(f + g)(-1) =$ ______ $(h - g)(0) =$ ______ $(f \cdot h)(4) =$ ______

$\left(\frac{f}{g}\right)(3) =$ ______ $\left(\frac{g}{h}\right)(2) =$ ______ $\left(\frac{g}{f}\right)(1) =$ ______

7. **REASON ABSTRACTLY** If $(f + g)(3) = 5$ and $(f \cdot g)(3) = 6$, find $f(3)$ and $g(3)$. Explain your reasoning. CCSS F.BF.1b, SMP 2

6.2 Inverse Functions and Relations

Objectives

- Find the inverse of a function.
- Compare properties of functions and their inverses.
- Restrict the domain of a function so it has an inverse function.

CCSS STANDARDS

Content: F.IF.4, F.IF.5, F.BF.4a
Practices: 1, 2, 3, 4, 6, 7, 8
Use with Lesson 6-2

The **inverse relation of a function** reverses the inputs and the outputs of a function. For example, if (a, b) is an element of a function, then (b, a) must be an element of its inverse. The inverse of a function, however, may or may not be a function itself. Many real-world situations can be modeled by considering a function and its inverse, such as calculating original and sale prices, finding side lengths and areas, or converting units of measure. To find the inverse of a function given as an equation in the form $f(x) = c$ where c represents a simple expression in terms of x, replace $f(x)$ with y, then interchange x and y. Solve the equation for y and the result will be the inverse of $f(x)$. This is usually denoted by replacing y with $f^{-1}(x)$.

EXAMPLE 1 Find the Inverse of a Function CCSS F.BF.4a

a. **FIND A PATTERN** Find the inverse of the functions $A = \{(2, 3), (-1, 4), (0, -2), (1, 5)\}$ and $B = \{(-3, -2), (0, -3), (1, -2), (4, 1)\}$. CCSS SMP 7

b. **REASON ABSTRACTLY** Are both the inverses of the functions in **part a** functions? Explain. What has to be true about a function if its inverse is going to be a function? CCSS SMP 2

Determine the inverse relation for the function $f(x) = \frac{3}{2}x - 2$. Name the inverse $f^{-1}(x)$.

c. **PLAN A SOLUTION** Replace $f(x)$ with y in the function rule and then interchange x and y. CCSS SMP 1

d. **PLAN A SOLUTION** Solve the resulting equation for y and then replace y with $f^{-1}(x)$. CCSS SMP 1

e. **FIND A PATTERN** Evaluate $f(2)$ and $f^{-1}(1)$. What do you notice? CCSS SMP 7

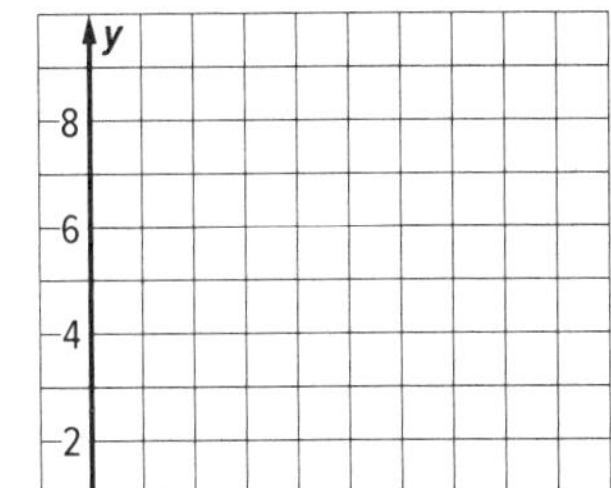

f. **FIND A PATTERN** Graph $f(x)$, $f^{-1}(x)$, and the line $y = x$. What do you notice? CCSS SMP 7

KEY CONCEPT

Complete the statement about inverses.

If two functions f and g are inverse functions, then $f(a) = b$ if and only if $g(\_\_\_) = \_\_\_$.

EXAMPLE 2 Model Garden Planning

EXPLORE The table gives an example of a function and its inverse. The function models the number of feet, $f(x)$, you can alter the side length of a square garden in order to achieve a planting area of x square feet.

Function	Inverse
$f(x) = -4 + \sqrt{x}$	$g(x) = (x + 4)^2$

a. **USE A MODEL** Evaluate $f(100)$ and explain what the calculation means in the context of this scenario. Then describe what must be true about the dimensions of the original square garden. CCSS F.IF.4, SMP 4

b. **REASON QUANTITATIVELY** Based on the context of the model, how must the domain and range be restricted for $f(x)$? Explain. CCSS F.IF.5, SMP 2

c. **REASON ABSTRACTLY** Describe how $g(x)$ relates to the context of this model and why there is a restriction on the domain. CCSS F.IF.5, SMP 2

d. **FIND A PATTERN** Sketch graphs for $f(x)$ and $g(x)$ on the coordinate plane on the right. Describe the relationships between the graphs. CCSS F.IF.4, SMP 8

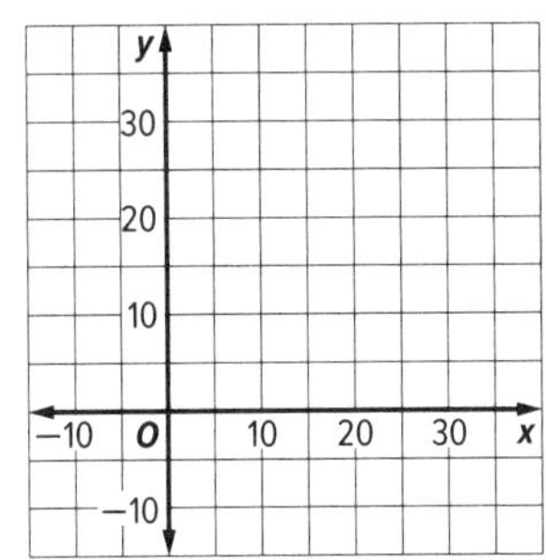

e. USE A MODEL Determine the x- and y-intercepts of $f(x)$ and $g(x)$. What do you notice? What do these points represent in the context of the model? CCSS F.IF.4, SMP 4

EXAMPLE 3 When is the Inverse of a Function a Function?

Body Mass Index (BMI) is a scale used to indicate whether an adult over age 20 is at a healthy weight. The formula for calculating Body Mass Index is BMI $= \frac{\text{weight(lb)}}{[\text{height(in.)}]^2} \times 703$. The function that generates weight as a function of height corresponding to an average healthy BMI of 21.75 is $w = f(h) = 0.031h^2$.

a. CONSTRUCT ARGUMENTS The graph of $f(x) = 0.031x^2$ and its inverse are shown on the right. Are $f(x)$ and its inverse both functions? Explain. CCSS F.IF.4, SMP 3

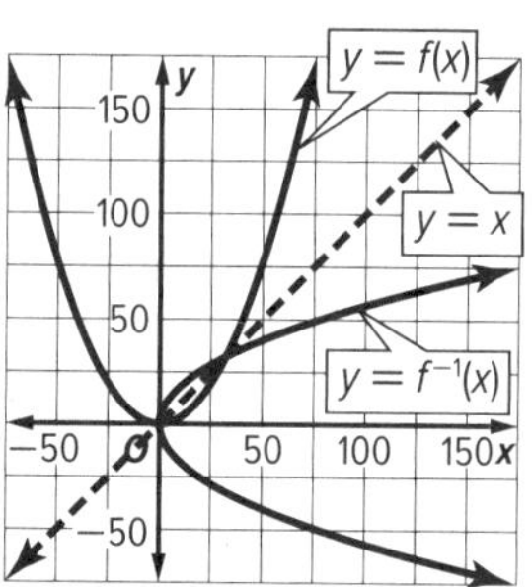

b. DESCRIBE A METHOD What can you look for in the graph of a function to see if its inverse will also be a function? CCSS F.IF.4, F.BF.4a, SMP 8

c. REASON ABSTRACTLY You can often limit the domain of a function to insure that its inverse is also a function. How could you limit the domain of the graph of the original function so that its inverse would be a function? CCSS F.BF.4a, SMP 2

d. INTERPRET PROBLEMS In the context of the Body Mass Index function, what are appropriate limits on the domain and range of the function $w = f(h) = 0.031h^2$? Explain your reasoning. CCSS F.IF.5, SMP 1

e. USE A MODEL Write an expression for the inverse of the function $w = f(h) = 0.031h^2$, $h = g(w)$. Is the inverse also a function? How do you know? CCSS F.BF.4a, SMP 4

EXAMPLE 4 The Meaning of the Inverse in the Context of a Problem

a. USE STRUCTURE The graph at right is of the function $f(x) = -0.025x^2 + 5x - 50$. This function can be used to model the profit of a coffee shop depending on the number of cups of coffee, x, that are sold each morning. Identify 5 points that must be on the graph of the inverse of this function. Explain your reasoning. CCSS F.IF.4, SMP 7

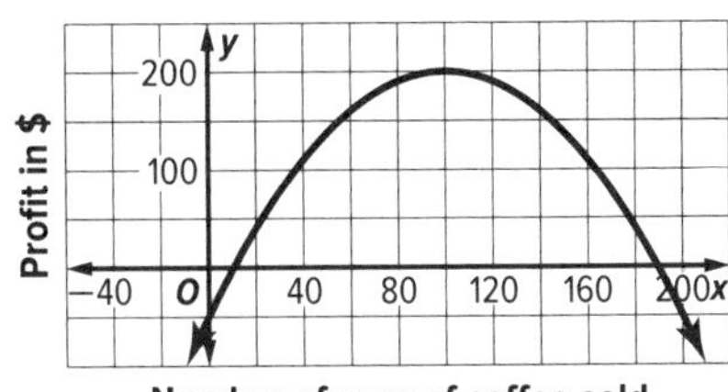

b. COMMUNICATE PRECISELY Plot and then connect the points you found in **part a** to sketch the inverse on the plane at right. How are the graphs of $f(x)$ and its inverse related? Discuss any symmetry of the lines and the location of key points in your explanation. CCSS F.IF.4, F.BF.4a, SMP 6

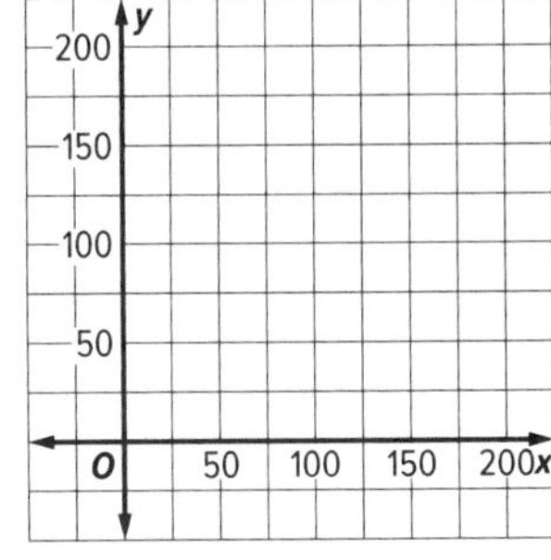

c. CONSTRUCT ARGUMENTS Is the inverse of $f(x)$ a function? Explain your reasoning. CCSS F.IF.5, F.BF.4a, SMP 3

d. COMMUNICATE PRECISELY What information do you get from the inverse of $f(x)$? CCSS F.IF.4, SMP 6

e. REASON ABSTRACTLY How could the domain of the function be restricted so that the inverse was a function? Would this make sense in the context of the problem? Explain. CCSS SMP 2

1. a. **USE A MODEL** Write and graph a function, $V(x)$, which gives the resulting volume of an open box made from a sheet of metal with squares of side length x removed from each corner, as shown in the diagram below. Explain how the domain of this function must be restricted to make sense within the context of this scenario. CCSS F.IF.5, SMP 4

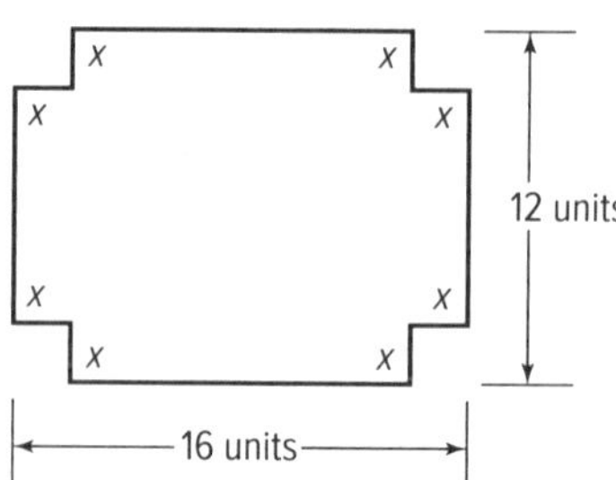

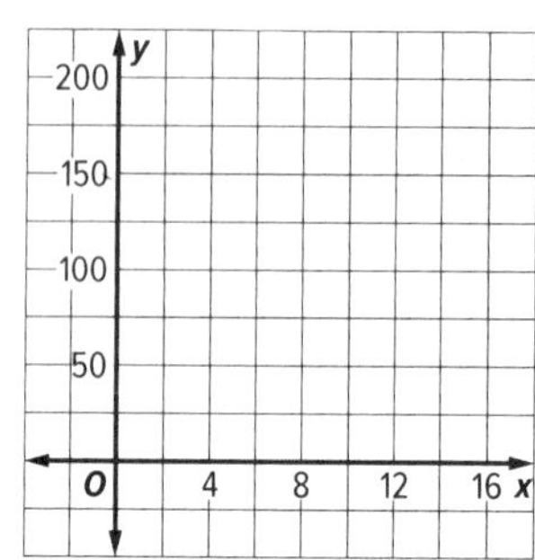

b. **INTERPRET PROBLEMS** Does restricting the domain to $0 \leq x \leq 6$ in **part a** above allow for the inverse to also be a function? How do you know? CCSS SMP 1

2. **CONSTRUCT ARGUMENTS** We have seen that functions with restricted domains can have inverses that are functions and yet without the restriction fail to be functions. Consider a real-world example in which an object is thrown upward with an initial velocity of v_0. Its height above the ground can be expressed as a functions of time using $h(t) = v_0t - 16t^2$. Explain why this real-world function cannot have an inverse function and what restrictions would need to be placed on it to ensure that an inverse function exists. CCSS F.IF.5, SMP 3

3. **USE STRUCTURE** Show that $f(x) = 4x^3$ and $g(x) = \sqrt[3]{\frac{x}{4}}$ are inverses of each other. CCSS F.BF.4a, SMP 7

4. USE A MODEL Graph the function $f(x) = -\frac{3}{4}x + 5$, its inverse, and $y = x$. Write the equation of the inverse. Compare the slopes and y-intercepts. Can you come to a conclusion regarding the relationship between the slope and y-intercept of a linear function and its inverse? CCSS F.IF.4, SMP 4

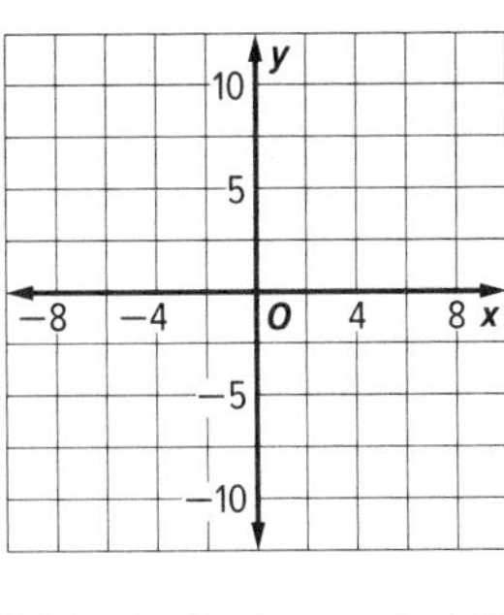

5. USE A MODEL The function $f(x) = 0.8x$ gives the discounted price for any original price x. Determine the inverse for this function. Then interpret what the function and its inverse mean in the context of this situation. What is the discount percentage? CCSS F.BF.4a, SMP 4

6. REASON ABSTRACTLY Explore the relationship between $(f + g)^{-1}(x)$ and $f^{-1}(x) + g^{-1}(x)$. CCSS F.BF.4, SMP 2

a. Suppose that functions $f(x)$, $g(x)$, and $(f + g)(x)$ all have inverse functions on the domain [0, 3]. Consider the following table of values.

x	0	1	2	3
$f(x)$	0	3	1	4
$g(x)$	1	0	4	3

Calculate the following values.

$f^{-1}(3) + g^{-1}(3) =$ ______ $f^{-1}(1) + g^{-1}(1) =$ ______

b. Find $(f + g)(1)$. Use this to find $(f + g)^{-1}(3)$. Find $(f + g)(0)$. Use this to find $(f + g)^{-1}(1)$.

c. Jonathan claims that $(f + g)^{-1}(x) = f^{-1}(x) + g^{-1}(x)$ and $(f + g)^{-1}(x) = f^{-1}(x) + g^{-1}(x)$. Use **part b** to explain why he cannot be correct.

d. Consider the functions $f(x) = 2x + 1$ and $g(x) = 2x - 1$. Find $(f + g)^{-1}(x)$ and $f^{-1}(x) + g^{-1}(x)$. Are they the same?

6.3 Square Root Functions and Inequalities

Objectives

- Graph square root functions using translations and reflections of a parent function.
- Create square root functions to model a real-world situation.
- Identify the region associated with a given radical inequality in an applied context.

CCSS STANDARDS

Content: F.IF.4, F.IF.5, F.IF.7b, F.IF.9, A.CED.2, F.BF.3, A.SSE.2
Practices: 1, 2, 3, 4, 6
Use with Lesson 6-3

A **square root function** is a function that contains the square root of a variable. The parent function of a square root function is $f(x) = \sqrt{x}$.

EXAMPLE 1 Dimensions of a Triangular Flower Garden

EXPLORE You are preparing to plant a triangular flower garden in your backyard, as shown. The side of the triangle that runs alongside the house is fixed at 15 feet. You can choose the length ℓ of the second leg that runs along the fence. The length of the hypotenuse is labeled b.

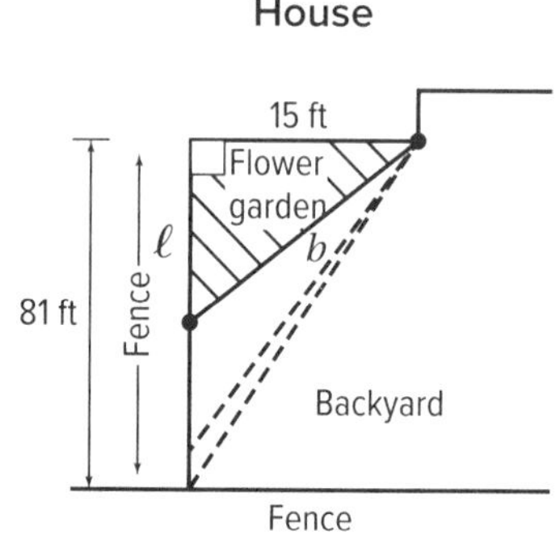

a. **USE A MODEL** Express b as a function of ℓ. CCSS A.SSE.2, A.CED.2, SMP 4

b. **REASON ABSTRACTLY** Express the area of the triangular flower garden, $A(b)$, as a function of b. Describe your solution process. CCSS A.SSE.2, A.CED.2, SMP 2

c. **INTERPRET PROBLEMS** What is the domain of the function $A(b)$ in the context of this situation? Explain the limitations. CCSS F.IF.5, SMP 1

d. **USE A MODEL** Sketch a graph of $A(b)$, labeling all axes appropriately. What is the area of the largest such flower garden that can be planted? CCSS A.CED.2, F.IF.4, SMP 4

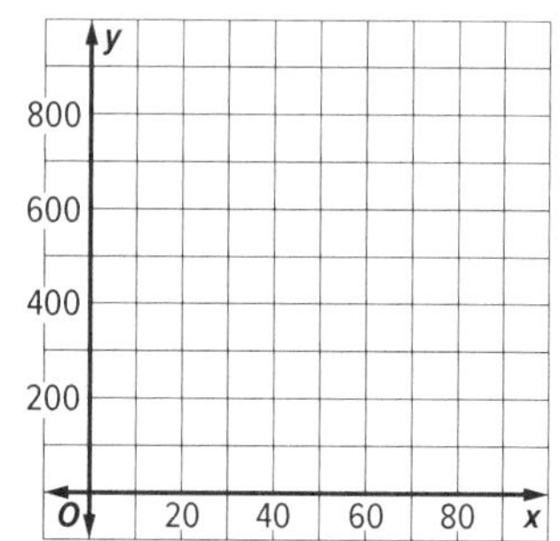

e. COMMUNICATE PRECISELY Describe the graph of the square root function by noting any intercepts, how it increases, and what would happen if the domain was infinite. CCSS F.IF.4, SMP 6

A **square root inequality** is an inequality involving square roots. They are graphed in a manner similar to the way other inequalities have been graphed.

EXAMPLE 2 Square Root Inequalities CCSS F.BF.3, F.IF.7b

a. REASON ABSTRACTLY Let $f(x)$ be the parent function of the square root function, so $f(x) = \sqrt{x}$. If $g(x)$ represents the new function after the following transformations are performed on $f(x)$, what is the new function? CCSS SMP 2

i. Translated left 1 unit

ii. Reflected in the line $x = -1$

iii. Reflected in the x-axis

iv. Translated down 5 units

b. CRITIQUE REASONING Alejandro is asked to verbally describe how $g(x) \leq -\sqrt{-x-1} - 5$ would look graphed. He states that the graph would be a solid curve located in the third quadrant with a minimum at $(-1, -5)$ that continues up to infinity as x decreases to negative infinity with shading under the curve. Do you agree? Explain. CCSS SMP 3

c. INTERPRET PROBLEMS Sketch a graph of the inequality in **part b**. What is different about the graph of the inverse of the graph from **part b**? CCSS SMP 1

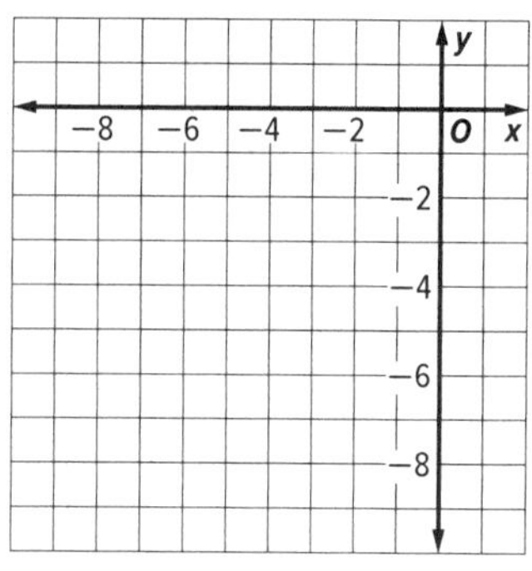

PRACTICE

1. Consider the function $f(x) = -\sqrt{3 - x} + \frac{13}{2}$ and the function $g(x)$ whose graph is shown at right. CCSS F.IF.9

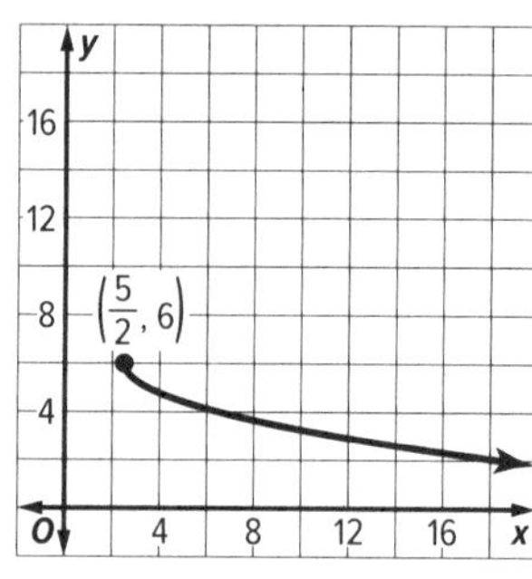

a. **INTERPRET PROBLEMS** Determine which function has the greater maximum value. Explain your reasoning. CCSS SMP 1

b. **INTERPRET PROBLEMS** Compare the domains of the two functions. CCSS F.IF.5, SMP 1

c. **INTERPRET PROBLEMS** Compare the rates of change of the two functions. CCSS SMP 1

2. A silo is in the shape of a right circular cone sitting on the ground, as shown. The cost per square foot to apply an environmentally-friendly UV protective paint to the surface of the silo not on the ground is \$1.25. Suppose the radius is $r = 60$ feet. The surface to be painted can be described by the following function of height, h: $S(h) = 60\pi\sqrt{3600 + h^2}$. Let $f = h^2$ in this formula to define the new function $S(f) = 60\pi\sqrt{3600 + f}$.

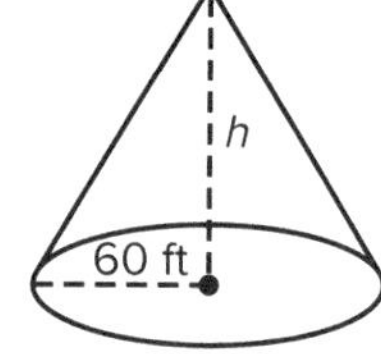

a. **INTERPRET PROBLEMS** Graph $S(f)$ and determine its domain in this context. CCSS F.IF.7b, F.IF.5, SMP 1

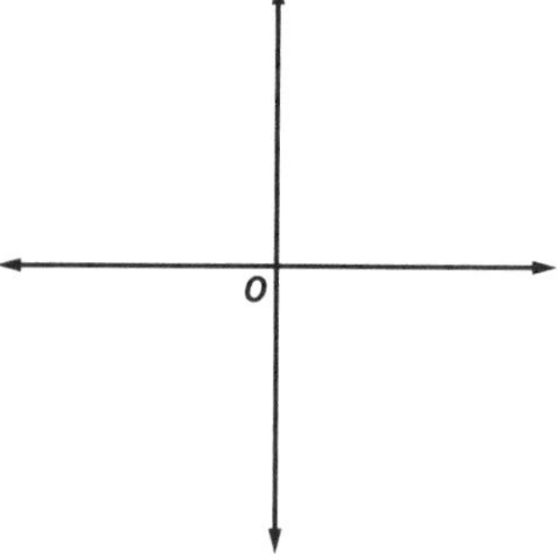

b. **REASON ABSTRACTLY** Construct a function of f that describes the cost of applying N coats of this paint to the surface of the silo. Explain your reasoning. CCSS A.CED.2, SMP 2

c. **REASON QUANTITATIVELY** Determine the range of the cost that will be incurred by applying either 2 or 3 coats of this paint to the entire surface of the silo if the height can vary between 100 feet and 150 feet. Express your answer as an inequality where the endpoints are rounded to the nearest dollar. Explain your reasoning. CCSS F.IF.5, SMP 2

3. **USE A MODEL** Pita produces the following graph for the inequality $f(x) > \sqrt{-x + 2} - 3$. Graph its inverse on the same coordinate plane along with the line of symmetry. Give the inequality that defines the graph of the inverse. Are there any restrictions that must be placed on the domain of the inverse? CCSS F.IF.5, SMP 4

4. The velocity of a dropped object just before it strikes the ground can be modeled using the equation $v = \sqrt{2gh}$ where v is the velocity in feet per second, g is the acceleration of gravity, 32 ft/sec^2, and h is the initial height from which the object is dropped, in feet. CCSS F.IF.4, F.IF.7b

a. **CRITIQUE REASONING** Angelo says that in order to double the final velocity of the dropped object one should double the drop height. Do you agree? Sketch the graph of the function and use your sketch to explain your answer. CCSS SMP 3

b. **COMMUNICATE PRECISELY** Discuss the attributes of the graph, including intercepts, rate of increase, and end behavior. How could this information help you determine the validity of Angelo's conjecture? CCSS SMP 6

6.4 Modeling: Square Root Functions

Objectives

- Interpret square root functions found in real-world applications in terms of the context.
- Define an appropriate domain for square root functions in a real-world context.

CCSS STANDARDS

Content: A.SSE.1a, A.SSE.1b, F.IF.4, F.IF.5

Practices: 1, 2, 3, 4, 5, 6, 7

Use with Lesson 6-3

You learned that a square root function is a type of radical function that contains a square root of a variable. In this lesson, you will discover how this type of function is applied to real-world situations.

EXAMPLE 1 Investigating Speed

EXPLORE **Martin is training to become a traffic accident investigator. He is currently studying skid marks created on dry asphalt and how to use them to determine the speed of a vehicle with the help of the formula, $s = \sqrt{21d}$, where s is the speed of the vehicle in miles per hour and d is the length of the skid mark in feet.**

a. **CALCULATE ACCURATELY** Complete the table to show the missing length of the skid mark or the speed of a vehicle. Round to the nearest whole number. CCSS F.IF.4, SMP 6

Length of Skid Mark (feet)	Speed of Vehicle (mph)
0	0
50	
	46
	56
200	
	72

b. **CONSTRUCT ARGUMENTS** What is the domain of the function $s = \sqrt{21d}$? Given the table in **part a**, is the graph of $s = \sqrt{21d}$ increasing or decreasing on its domain? What are the s- and d-intercept? CCSS F.IF.4, F.IF.5, SMP 3

c. **USE TOOLS** Use a graphing calculator to graph the function, and then sketch a graph that Martin can use to quickly determine the approximate speed of a vehicle. Label the axes appropriately. CCSS F.IF.4, SMP 5

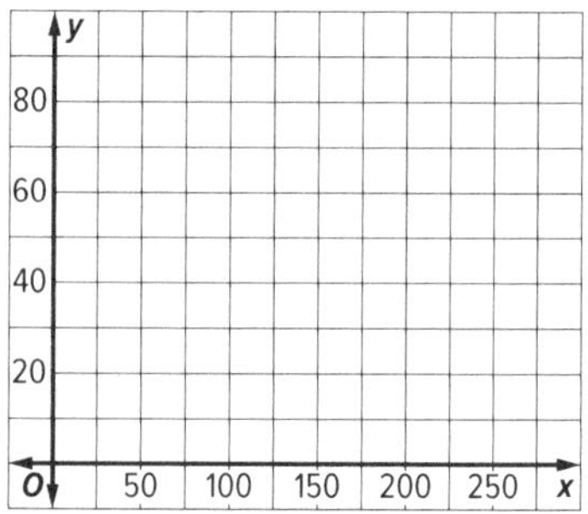

d. **REASON ABSTRACTLY** Rewrite the formula showing d as a function of s. Describe the new formula as a sentence. CCSS A.SSE.1a, SMP 2

e. **REASON ABSTRACTLY** What is the parent function of the function you wrote in **part d**? Find its domain and range and state their meaning in terms of the graph of the function. What does this mean in terms of the context of this situation? CCSS F.IF.4, F.IF.5, SMP 2

f. **INTERPRET PROBLEMS** A vehicle is unknowingly traveling on a road that has just collapsed in a small sinkhole. If the driver sees the sinkhole 65 feet ahead and is traveling at a speed of 35 mph, will there be enough road surface available to come to a complete stop before reaching the sinkhole? CCSS F.IF.4, SMP 1

EXAMPLE 2 Making Woven Rugs

A rug maker wants to add circular rugs to his inventory. The rugs will be of various sizes.

a. **REASON ABSTRACTLY** Write a function, $A(r)$, that models the relationship between the radius and the area of a rug. What function can be used to find the radius of each circular rug as a function of its area? Describe both functions in words. CCSS A.SSE.1a, SMP 2

b. **REASON ABSTRACTLY** Find the domain and range of the functions. Based on the domain and range of the functions, which quadrant will they be graphed in? Explain. 

c. **USE A MODEL** Sketch a graph that models $r(A)$ if the rugs can have an area of at most 40 square feet. CCSS F.IF.4, SMP 4

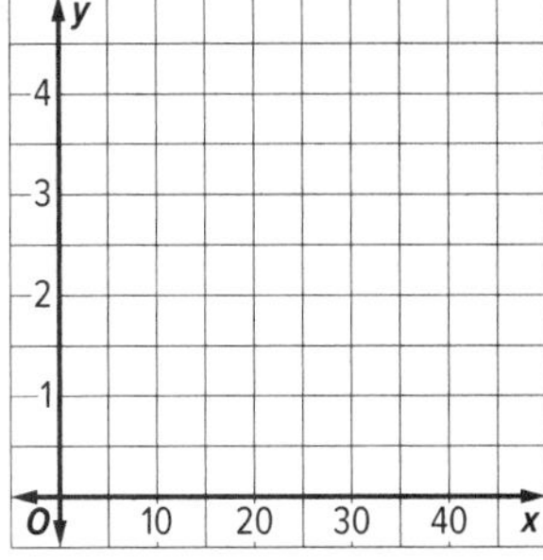

d. **USE STRUCTURE** Is the graph of $r(A)$ increasing or decreasing from zero to 40? CCSS F.IF.4, SMP 7

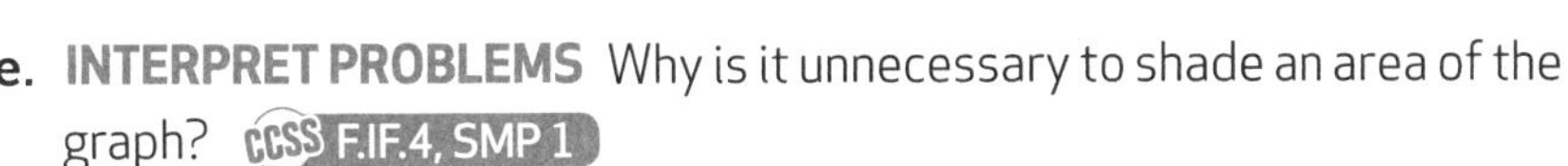

e. **INTERPRET PROBLEMS** Why is it unnecessary to shade an area of the graph? CCSS F.IF.4, SMP 1

PRACTICE

In Example 1, calculations were computed for a dry asphalt road. The formula changes to $s = \sqrt{9d}$ when the road is wet.

1. a. **FIND A PATTERN** Fill in the following features of the function $s = \sqrt{9d}$. CCSS F.IF.5, SMP 7

Domain: ________

Range: ________

s-intercept: ________

d-intercept: ________

The graph of $s = \sqrt{9d}$ is ________ on the domain.

b. **USE A MODEL** Using your answers from **part a**, make a rough sketch of the graph of $s = \sqrt{9d}$. CCSS F.IF.4, SMP 4

c. **USE TOOLS** Use a graphing calculator to sketch a graph that models the function under the new road conditions. Label axes appropriately. CCSS SMP 5

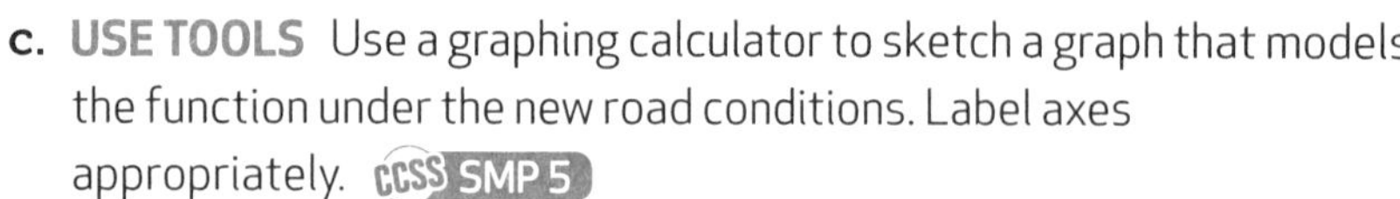

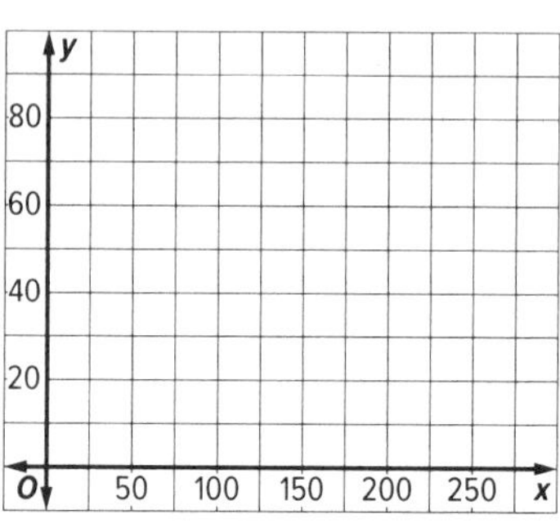

2. Compare the models for the dry and wet roads from **Example 1** and **Exercise 1**.

a. **COMMUNICATE PRECISELY** Which aspects of functions $s = \sqrt{21d}$ and $s = \sqrt{9d}$ are similar, and which are different? How do these similarities and differences impact the graphs of the functions? CCSS F.IF.4, SMP 6

b. **FIND A PATTERN** The skid mark function in general is $s = \sqrt{kd}$, where s is the speed of the car, d is the length of the skid mark, and k is a constant. In **Example 1**, $k = 21$. In **Exercise 1**, $k = 9$. Why might this constant be different in different situations? What factors might contribute to this constant? CCSS A.SSE.1b, SMP 7

c. **COMMUNICATE PRECISELY** Both graphs are increasing on their domains. What does this mean in the in terms of the relationship between speed and skid mark distance? Describe what the difference between the two graphs means in the context of the situation. CCSS F.IF.4, SMP 6

3. **REASON QUANTITATIVELY** The general formula for calculating the speed of a vehicle is $s = \sqrt{30fd}$, where f is the coefficient of friction for the road surface the vehicle is traveling.

a. How does the value of f affect the speed of the vehicle in relation to the length of the skid mark produced? Why? CCSS A.SSE.1a, SMP 2

b. **FIND A PATTERN** Suppose d was some unknown constant. Describe the general formula in terms of s and f. CCSS A.SSE.1b, SMP 7

4. **REASON ABSTRACTLY** Rewrite the general formula showing f as a function of s and d. Using words, describe the function in terms of s and d. What is the coefficient of friction if a driver is travelling at 50 mph and skids 175 feet? CCSS A.SSE.1a, SMP 2

Electrical engineers use the formulas $P = VI$ and $V = IR$ when working with circuits, where P is the electrical power in watts, V is the voltage in volts, I is the current in amperes, and R is the resistance in ohms.

5. **INTERPRET PROBLEMS** Given that voltage (V) measures the strength of an electrical current and amperes (I) measures how *much* current there is, interpret the equation $P = VI$ by explaining what happens to the electrical power as both V and I change. Explain how to derive the function that calculates voltage in terms of power and resistance. Give the function, and interpret the relationship between V and P given that R is usually constant. CCSS A.SSE.1a, SMP 1

6. **FIND A PATTERN** Given that the resistance is 240 ohms, complete the table using the function derived in **Exercise 5**. Round values to the nearest hundred. Is the function increasing or decreasing over the interval $375 \le P \le 37{,}500$? Do you notice a pattern with the increase or decrease? CCSS F.IF.4, SMP 7

Power (watts)	Voltage (volts)
375	
	600
3375	
	1200
9375	
37,500	

7. **CRITIQUE REASONING** The students in a mathematics class are given the graph shown at right and are asked to determine the domain of the function. Erin states that the domain is $\{x \mid -4 \le x \le 10$, where x is an integer$\}$. James disagrees and says that the domain is $\{x \mid x \ge -4\}$. Is either student correct? Explain. CCSS F.IF.5, SMP 3

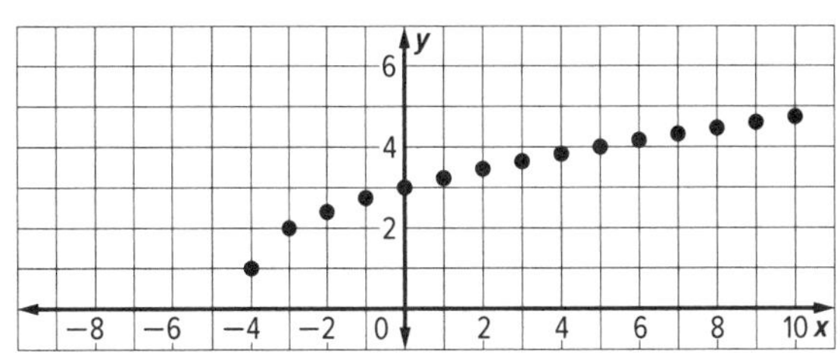

6.5 nth Roots

Objectives

- Simplify radical expressions.

CCSS STANDARDS

Content: A.SSE.2, A.SSE.3c
Practices: 1, 2, 3, 6, 7
Use with Lesson 6-4

The inverse of raising a number to the *n*th power is finding the *n*th root of a number.

EXAMPLE 1 Determine Roots CCSS A.SSE.2, A.SSE.3c

EXPLORE Use the expressions $\sqrt[5]{x^5}$ and $\sqrt[4]{x^4}$.

a. REASON QUANTITATIVELY Evaluate $\sqrt[5]{x^5}$ when $x = 3$ and when $x = -3$. Does $\sqrt[5]{x^5} = x$ for both values of x? Explain. CCSS SMP 2

b. REASON QUANTITATIVELY Evaluate $\sqrt[4]{x^4}$ when $x = 3$ and when $x = -3$. Does $\sqrt[4]{x^4} = x$ for both values of x? Explain. CCSS SMP 2

c. COMMUNICATE PRECISELY What can you conclude about the root of an expression with an odd index and an odd exponent? What can you conclude about the root of an expression with an even index and an even exponent? CCSS SMP 6

d. MAKE A CONJECTURE If you do not know whether x is positive or negative, then what rule can be followed to simplify $\sqrt[n]{x^n}$? CCSS SMP 3

e. REASON ABSTRACTLY Explain why using absolute value is necessary to simplify $\sqrt{x^2}$, but it is not necessary for $\sqrt{x^4}$. CCSS SMP 2

f. REASON QUANTITATIVELY You know from previous work that $\sqrt[n]{x} = x^{\frac{1}{n}}$. Use this information and your understanding of the rules for raising a power to a power to prove that $\sqrt[n]{x^n} = x$ when $x \geq 0$. CCSS SMP 2

Definition of *n*th roots: For any real numbers a and b, and any positive integer n, if $a^n = b$, then a is an nth root of b.

Real *n*th Roots

Suppose n is an integer greater than 1, and a is a real number.

a	n is even.	n is odd.
$a > 0$	1 unique positive and 1 unique negative real root $\pm\sqrt[n]{a}$; positive root is principal root.	1 unique positive and 0 negative real roots.
$a < 0$	0 real roots	0 positive and 1 negative real root: $\sqrt[n]{a}$
$a = 0$	1 real root: $\sqrt[n]{0} = 0$	1 real root: $\sqrt[n]{0} = 0$

EXAMPLE 2 Determining *n*th roots CCSS A.SSE.2

USE STRUCTURE Rewrite each expression using the indicated method. Then simplify the expression. CCSS SMP 7

a. $\sqrt[4]{81x^4}$; Prime factorization

b. $\sqrt[3]{-125x^{12}y^6}$; Power of a Product Property and Power of a Power Property

c. $\sqrt[5]{(x^5 + 32)^{10}}$; Power of a Power Property

d. $\sqrt{(x^4 - 14x^2 + 49)}$; Factor

e. **CONSTRUCT ARGUMENTS** For what values of x is $\sqrt[6]{64x^5}$ undefined in the set of real numbers? Explain your reasoning. CCSS SMP 3

EXAMPLE 3 Graphing *n*th Roots CCSS A.SSE.2

CRITIQUE REASONING Seamus claims that the function $f(x) = \sqrt{x^6}$ is the same as the function $g(x) = x^3$. CCSS SMP 6

a. Is Seamus correct? Simplify $f(x)$ to prove your result.

b. Graph $g(x)$ and $f(x)$ to confirm your answer from **part a**.

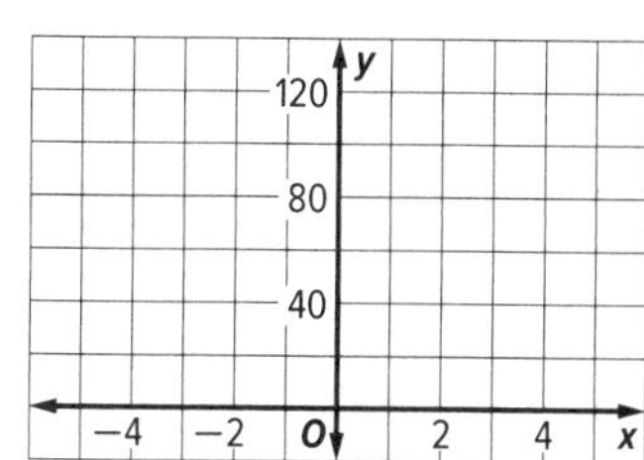

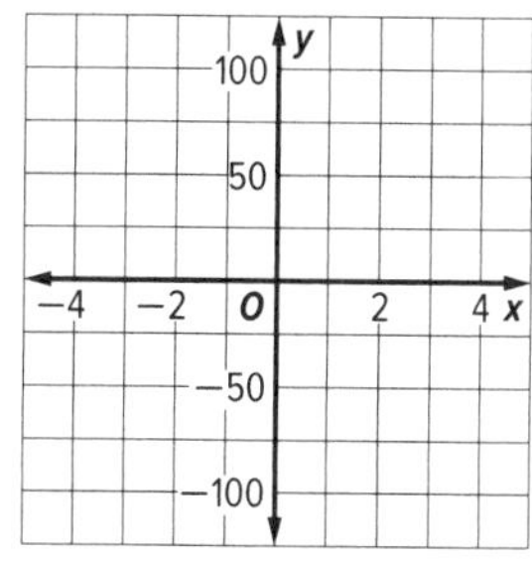

PRACTICE

1. **CALCULATE ACCURATELY** A cube has a volume of $512x^3$ cubic inches. Find the measure of each side of the cube. Explain your solution process. CCSS A.SSE.2, SMP 6

2. **USE STRUCTURE** Simplify $(\sqrt[4]{16})^2$. Explain how you can use the properties of radical expressions to rewrite the expression to make calculations easier. CCSS A.SSE.2, SMP 7

3. **REASON ABSTRACTLY** Simplify $\sqrt[b]{m^{3b}}$, where $b > 0$. Explain your reasoning. CCSS A.SSE.2, SMP 2

4. **COMMUNICATE PRECISELY** Does $\sqrt{x^2 - 9} = x - 3$ for all values of x? Explain your reasoning. CCSS A.SSE.2, SMP 6

5. **INTERPRET PROBLEMS** There are no real nth roots of a number w. What can you conclude about the index and the number w? CCSS A.SSE.2, SMP 1

6. **REASON ABSTRACTLY** Determine the values of x for which $\sqrt{x^2} \neq x$. Explain your answer. CCSS A.SSE.2, SMP 2

7. The volume V of a sphere can be found using the formula $V = \frac{4}{3}\pi r^3$. CCSS A.SSE.2, SMP 6

   a. **CALCULATE ACCURATELY** Using the formula for volume given above, determine the formula for the radius of a sphere in terms of volume. Show your work. CCSS SMP 6

   b. **CALCULATE ACCURATELY** What is the radius of a sphere with a volume of 288π cubic inches? CCSS SMP 6

   c. **CALCULATE ACCURATELY** Describe how decreasing the volume of the sphere by a factor of 8 will change its radius. What is the new radius? Explain. CCSS SMP 6

8. The depreciation rate is calculated by the following formula: $r = 1 - \sqrt[n]{\frac{T}{P}}$, where r is the depreciation rate, n is the age of the item in years, T is the resale price in dollars, and P is the original price in dollars. CCSS A.SSE.2

a. **CALCULATE ACCURATELY** Find the depreciation rate, to the nearest hundredth, of a car originally purchased for $52,425 that has depreciated over an eight-year period to $9856. Describe your solution process. CCSS SMP 6

b. **REASON QUANTITATIVELY** Solve for P in terms of T, n, and r. CCSS SMP 2

9. The formula $d = \sqrt[3]{6t^2}$ represents the distance d in millions of miles a planet is from the Sun in terms of t, the number of Earth-days it takes for the planet to orbit to the Sun. It takes Jupiter 4332 Earth-days to complete one orbit. CCSS A.SSE.2

a. **CALCULATE ACCURATELY** How many millions of miles is Jupiter from the Sun? Round to the nearest million. Describe your solution process. CCSS SMP 6

b. **CALCULATE ACCURATELY** Solve for t in terms of d. Determine how many Earth days it would take a planet that was twice as far as Jupiter is from the sun to orbit the sun. Compare the result with the time it takes Jupiter to orbit the sun. CCSS SMP 6

10. **REASON ABSTRACTLY** Construct an example where you raise a constant to an nth power and then take its nth root. Then, use the law of exponents to show that this process is commutative. CCSS A.SSE.3c, SMP 2

11. **USE STRUCTURE** Which of the following functions are equivalent? Justify your answer. CCSS A.SSE.2, SMP 7

$f(x) = \sqrt[3]{x^9}$ $\quad g(x) = \sqrt{x^6}$ $\quad r(x) = (\sqrt[3]{x})^9$ $\quad s(x) = (\sqrt{x})^6$

6.6 Operations with Radical Expressions

Objectives

- Simplify radical expressions.
- Perform arithmetic operations to simplify radical expressions arising in applied contexts.

CCSS STANDARDS

Content: A.SSE.2
Practices: 1, 2, 3, 6, 7
Use with Lesson 6-5

The properties of square roots all work for expressions involving n^{th} roots.

KEY CONCEPT Product and Quotient Properties of Radicals

Product Property: For any integer $n > 1$, and any real numbers a and b,
$\sqrt[n]{a \cdot b} = \sqrt[n]{a} \cdot \sqrt[n]{b}$.
If n is odd, then a and b can be any real numbers.
If n is even, then a and b must be nonnegative real numbers.
Quotient Property: For any integer $n > 1$ and any real numbers a and b where $b \neq 0$,
$\sqrt[n]{\frac{a}{b}} = \frac{\sqrt[n]{a}}{\sqrt[n]{b}}$.
If n is odd, then a and b can be any real numbers with $b \neq 0$.
If n is even, then a and b must be nonnegative real numbers with $b \neq 0$.

EXAMPLE 1 Properties of Radical Expressions CCSS A.SSE.2

EXPLORE An artist is designing a sculpture involving cubes and rectangular prisms. Part of the sculpture will feature the group of figures shown below. Figure A is a cube with volume 384 in^3 and Figure B is a cube with volume 864 in^3. One side of Figure C has the same length as a side of Figure A and another side of Figure C has the same length as a side of Figure B.

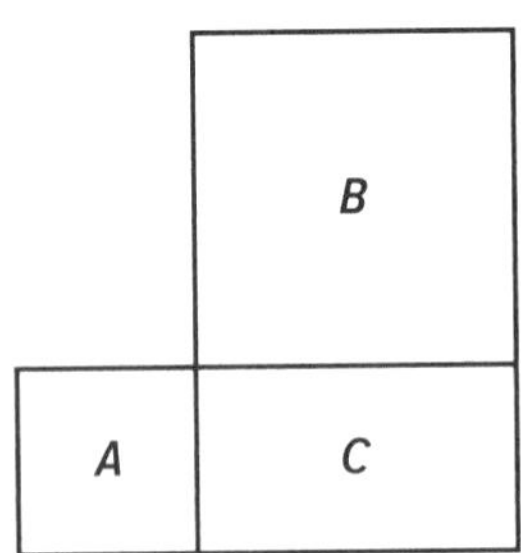

a. **CALCULATE ACCURATELY** Find the length of the sides of Figure A and Figure B. CCSS SMP 6

Figure A:

Figure B:

b. **CALCULATE ACCURATELY** Find the area of the rectangle formed by the face of Figure C bordering the edges of Figure A and Figure B. CCSS SMP 6

c. **INTERPRET PROBLEMS** He decides to enlarge Figure B so that it has a new side of $12\sqrt[3]{6}$ inches. If he wants the new face of the face of Figure C bordering the edges of Figure A and Figure B and the old face to be similar rectangles, what is the length of the new side of Figure A? Describe your solution process. CCSS SMP 1

d. **CRITIQUE REASONING** Antwan says that it is possible to write an expression like $\frac{4\sqrt[3]{6}}{6\sqrt[3]{4}}$ as an equivalent expression with no radical in the denominator. He says that he can do so by multiplying the numerator and denominator by $\sqrt[3]{4}$. Do you agree with Antwan? Explain. CCSS SMP 3

CONCEPT SUMMARY Simplifying Radical Expressions

A radical expression, $\sqrt[n]{p(x)}$, is in simplified form when the following conditions are met:

- The index n is as small as possible.
- The radicand contains no factors (other than 1) that are nth powers of a number or polynomial.
- No radicals appear in a denominator of a fraction.
- The radicand contains no fractions.

EXAMPLE 2 Average Rate of Change of Concentration CCSS A.SSE.2

A drug dose is administered to a patient intravenously. The percent of the dose that is expected to remain in the patient's bloodstream t hours after starting the IV is modeled by the function $f(t) = \frac{100}{\sqrt{t-1}}$, where $t \geq 2$. The average rate of change of this percent over the time interval $a \leq t \leq b$ is defined as $\frac{f(b) - f(a)}{b - a}$ percent per hour.

a. **REASON QUANTITATIVELY** What is the average rate of change of $f(t)$ over the interval $2 \leq t \leq 3$? Express the answer as a simplified radical expression. Then express the value of the expression to the nearest hundredth. Show your work. CCSS SMP 2

b. **REASON ABSTRACTLY** Write an expression in simplest square root form for the average rate of change of $f(t)$ over the interval $2 \leq t \leq 2 + h$, where h can be any positive real number. Show your work. CCSS SMP 2

c. **REASON ABSTRACTLY** We can estimate the exact rate at $t = 2$ hours by selecting a value of h that is very close to zero in the expression from **part b**. Use a calculator and $h = 0.00001$ to find an estimate for the rate of change at the two hour mark. CCSS SMP 2

PRACTICE

1. **USE STRUCTURE** Write, in simplest form, the ratio of the sides of the two cubes described. CCSS A.SSE.2, SMP 7

   a. The volumes of the two cubes are $270x$ cubic inches and $32x^2$ cubic inches.

   b. The surface areas of the two cubes are $6x^4$ square feet and $6(x + 1)$ square feet.

2. **a.** **REASON ABSTRACTLY** Rewrite each of the following expressions as a single expression in the form ax^m for appropriate choices of a and m. Show your work. CCSS A.SSE.2, SMP 2

   i. $\sqrt{x}(\sqrt{x} + \sqrt{4x})$

   ii. $\sqrt{x}(\sqrt{x} + \sqrt{4x} + \sqrt{9x})$

   iii. $\sqrt{x}(\sqrt{x} + \sqrt{4x} + \sqrt{9x} + \sqrt{16x})$

   b. **FIND A PATTERN** More generally, simplify $\sqrt{x}\left(\sqrt{x} + \sqrt{4x} + \ldots + \sqrt{n^2x}\right)$ for any positive integer n. Use the fact that $1 + 2 + \ldots + n = \frac{n(n+1)}{2}$ CCSS A.SSE.2, SMP 7

3. **CALCULATE ACCURATELY** If the area of the trapezoid shown is 200 square feet, what is the height h of the trapezoid? CCSS A.SSE.2, SMP 6

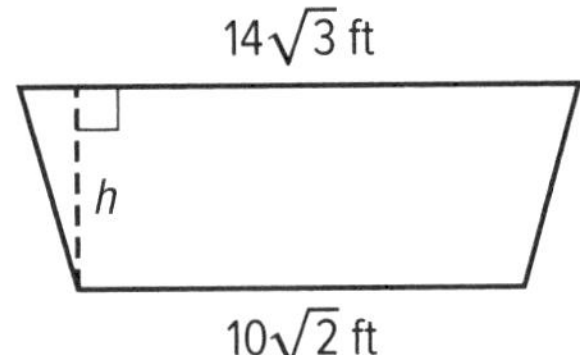

4. A spherical paperweight with a volume of 72π cubic centimeters is to be packaged in a gift box that is a cube. There must be at least 2 centimeters of packing material around the paperweight to protect it during shipping. The formula for the volume of a sphere is $V = \frac{4}{3}\pi r^3$.

   a. **USE STRUCTURE** Write an expression for the minimum length of a side of the gift box. Show your work. CCSS A.SSE.2, SMP 7

   b. **CONSTRUCT ARGUMENTS** The shipper wants to use a box with a volume of 384 cubic centimeters that they already have in inventory. Is this box suitable? CCSS A.SSE.2, SMP 3

5. **CRITIQUE REASONING** Kendra, Micha, and Ruiz simplify the radical expression $\sqrt[n]{\frac{2^{n^2}}{8^{2n}}}$. They know that at most only one person's work is correct. Identify the errors, if any, in the students' work. Is there one of these students with whom you agree? CCSS A.SSE.2, SMP 3

Kendra	*Micha*	*Ruiz*
$\sqrt[n]{\frac{2^{n^2}}{8^{2n}}} \underset{\text{Step}(1)}{=} \frac{\sqrt[n]{2^{n^2}}}{\sqrt[n]{8^{2n}}}$	$\sqrt[n]{\frac{2^{n^2}}{8^{2n}}} \underset{\text{Step}(1)}{=} \frac{\sqrt[n]{2^{n^2}}}{\sqrt[n]{8^{2n}}}$	$\sqrt[n]{\frac{2^{n^2}}{8^{2n}}} \underset{\text{Step}(1)}{=} \frac{\sqrt[n]{2^{n^2}}}{\sqrt[n]{8^{2n}}}$
$\underset{\text{Step}(2)}{=} \frac{2^{\sqrt[n]{n^2}}}{8^{\sqrt[n]{2n}}}$	$\underset{\text{Step}(2)}{=} \frac{\sqrt[n]{2^n \cdot 2^n}}{\sqrt[n]{8^{2n}}}$	$\underset{\text{Step}(2)}{=} \frac{\sqrt[n]{(2^n)^n}}{\sqrt[n]{(8^2)^n}}$
	$\underset{\text{Step}(3)}{=} \frac{\sqrt[n]{2^n} \cdot \sqrt[n]{2^n}}{\sqrt[n]{(8^2)^n}}$	$\underset{\text{Step}(3)}{=} \frac{2^n}{8^2}$
	$\underset{\text{Step}(4)}{=} \frac{2 \cdot 2}{8^2}$	$\underset{\text{Step}(4)}{=} \frac{2^n}{64}$
	$\underset{\text{Step}(5)}{=} \frac{1}{16}$	

6. **INTERPRET PROBLEMS** Graph the function in **Example 2**, $f(t) = \frac{100}{\sqrt{t-1}}$. Draw a line segment from $(2, f(2))$ to $(3, f(3))$ and another line segment from $(2, f(2))$ to $(5, f(5))$. CCSS A.SSE.2, SMP 1

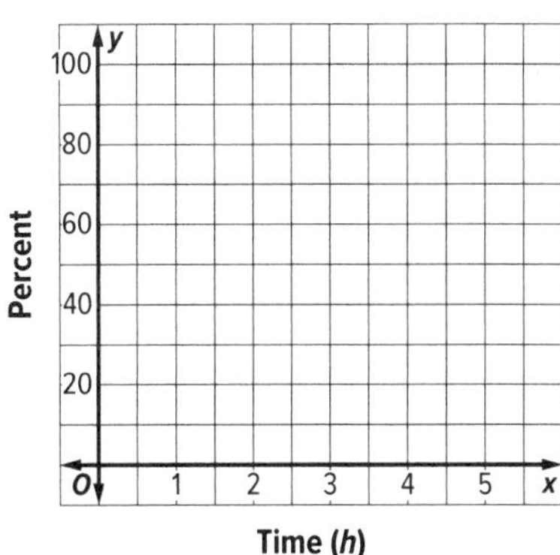

   a. What does the slope of each line segment represent?

   b. As t increases, does the slope between two points increase, decrease, or stay about the same? What does this mean? Explain your reasoning.

6.7 Solving Radical Equations and Inequalities

Objectives

- Solve radical equations and inequalities and determine when a solution of an equation is not viable in the applied context.
- Use the graphs of radical functions to find the solutions of radical equations depending on parameters.

CCSS STANDARDS

Content: A.REI.2, A.REI.11
Practices: 1, 2, 3, 5, 6, 7, 8
Use with Lesson 6-7

Radical equations are equations that include radical expressions.

KEY CONCEPT Solving Radical Equations

STEP 1: Isolate the radical expression on one side of the equation.

STEP 2: Raise each side of the equation to the power equal to the index of the radical to eliminate the radical.

STEP 3: Solve the resulting polynomial equation. Check your results.

When solving radical equations, the result may be a number that does not satisfy the original equation. Such a number is called an extraneous solution.

EXAMPLE 1 Hanging a Dartboard CCSS A.REI.2

EXPLORE A circular dartboard has an area of 8 square feet.

a. **INTERPRET PROBLEMS** Can the dartboard be hung at the end of a 3-foot wide hallway? Explain your reasoning. CCSS SMP 1

b. **REASON ABSTRACTLY** More generally, a circular dartboard has an area of *A* square feet. What is the minimum width of the end of a hallway where this dartboard can be hung? Explain your reasoning. CCSS SMP 2

c. **REASON ABSTRACTLY** Find the circumference of the dartboard from **part b** in terms of *A*, the area. Simplify your answer. CCSS SMP 2

EXAMPLE 2 Solving Radical Equations CCSS A.REI.2

a. **CONSTRUCT ARGUMENTS** Compare and contrast the solutions of the equations $\sqrt{x+3} = x - 3$ and $\sqrt{x+3} = 3 - x$. What was the effect of multiplying the right side by -1? Use graphs of the functions to illustrate your answer. CCSS A.REI.11, SMP 3

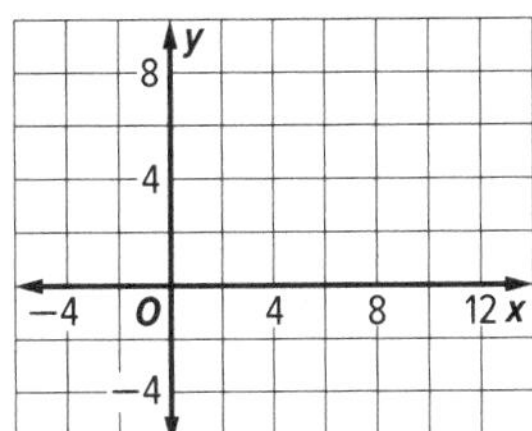

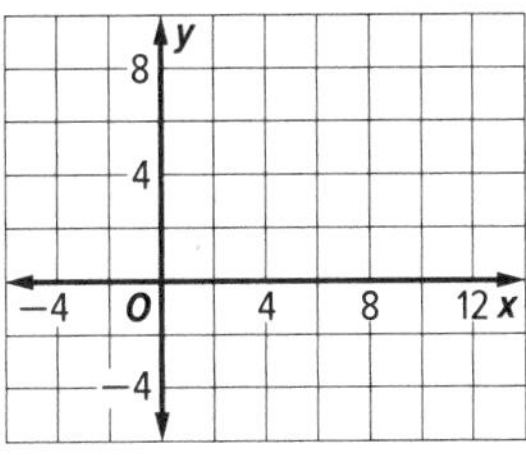

b. **COMMUNICATE PRECISELY** Explain how the process of solving a square root radical equation can produce an extraneous solution. How can you determine if solutions are extraneous? CCSS SMP 6

c. **REASON QUANTITATIVELY** If you want the radical equation $\sqrt{x-4} = a$ to have an extraneous solution of $x = 13$, what could you choose to replace a in the equation? Explain your answer, and find any true solutions to your equation. CCSS SMP 2

A **radical inequality** has a variable in the radicand. The following steps are used to solve such an inequality.

KEY CONCEPT Solving Radical Inequalities

STEP 1: If the index of the root is even, identify the values of the variable for which the radicand is nonnegative.

STEP 2: Solve the inequality algebraically.

STEP 3: Test values to check your solution.

EXAMPLE 3 Solving Radical Inequalities CCSS A.REI.2

a. PLAN A SOLUTION Describe how to solve the radical inequality $\sqrt{x+4} > x - 2$ algebraically and confirm your results graphically. CCSS A.REI.11, SMP 1

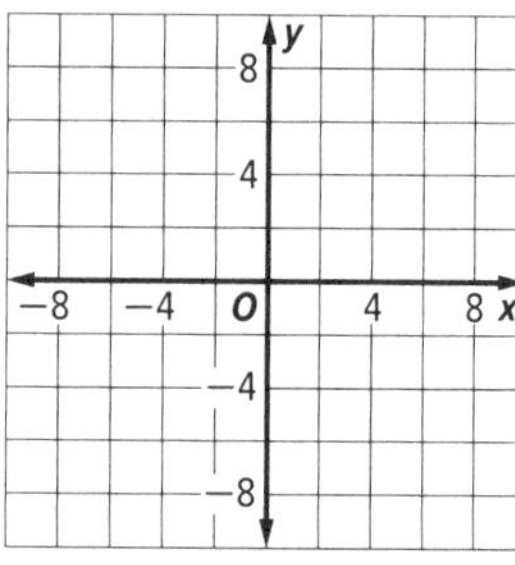

b. USE STRUCTURE For what values of b does the inequality $\sqrt{x} > \sqrt{2x} + b$ have a positive solution? Explain your answer. CCSS SMP 7

EXAMPLE 4 Supply and Demand

A company produces personalized cell phone cases. The number of cases produced can be expressed as $S = \sqrt{4p}$, where S is the supply (in thousands) and p is the price of the case (in dollars). The demand for cell phone cases (in thousands) can be expressed as $D = \sqrt{165 - p^2}$.

a. USE TOOLS Show how a graph can be used to estimate the price at which the supply is equal to demand. How many cases can they expect to sell at this price? CCSS A.REI.11, SMP 5

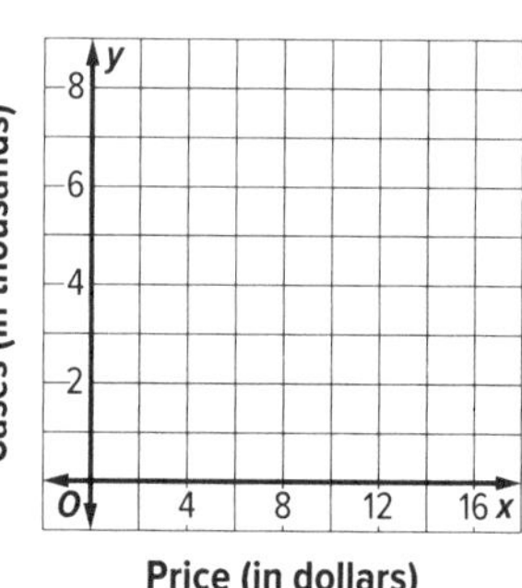

b. CALCULATE ACCURATELY Verify your answer algebraically. CCSS SMP 6

EXAMPLE 5 Solving Radical Equalities and Inequalities with Two Radical Expressions CCSS A.REI.2

a. **REASON QUANTITATIVELY** Consider the equation $\sqrt{x-2}+1=\sqrt{x+1}$. Square both sides of the equation and simplify. CCSS SMP 2

b. **REASON QUANTITATIVELY** In your equation from **part a**, isolate the radical on one side of the equation and square both sides. Simplify, and solve for x. CCSS SMP 2

c. **REASON QUANTITATIVELY** How many solutions did you get in **part b**? Substitute the values to check to see if any are extraneous. Graph both sides of the equation to verify your answer. CCSS SMP 2

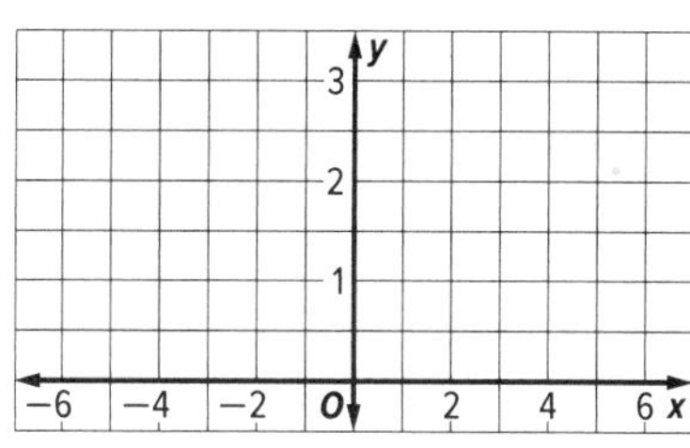

d. **REASON QUANTITATIVELY** Use the work from **parts a–c** to solve $\sqrt{x-2}+1<\sqrt{x+1}$. CCSS SMP 2

e. **CRITIQUE REASONING** Katy claims that the equation $\sqrt{x+2}+1=\sqrt{-x-1}$ has two real solutions, whereas the equation in **part a** had only one. Solve this equation to determine if Katy is correct. CCSS SMP 6

f. **USE TOOLS** Use a calculator to graph the equation and confirm your answer from **part e**. CCSS SMP 5

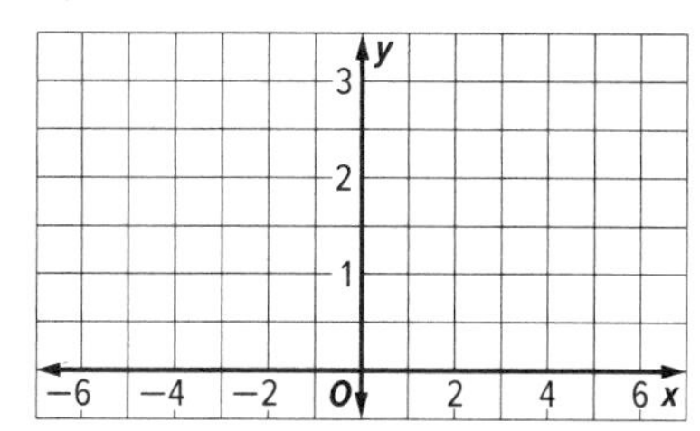

PRACTICE

1. **REASON QUANTITATIVELY** If you want the radical equation $\sqrt{a} = x + 2$ to have an extraneous solution of $x = -5$, what could you choose to replace a in the equation? Explain your answer, and find any true solutions to your equation. CCSS A.REI.2, SMP 2

2. **REASON QUANTITATIVELY** If you want the radical equation $\sqrt{x+7} = \sqrt{x-5} + c$ to have an extraneous solution of $x = 9$, what number could you choose to replace c in the equation? Explain your answer, and find any true solutions to your equation. CCSS A.REI.2, SMP 2

3. **REASON ABSTRACTLY** For what non-negative values of a will the inequality $\sqrt{ax} < \sqrt{x-a}$ have a solution? Explain your answer. CCSS A.REI.2, SMP 2

4. **DESCRIBE A METHOD** Explain how to find the solutions to $\sqrt[4]{10x+11} - \sqrt{x+2} = 0$ graphically and confirm your results algebraically. What are the solutions? Construct the graph. CCSS A.REI.2, A.REI.11, SMP 8

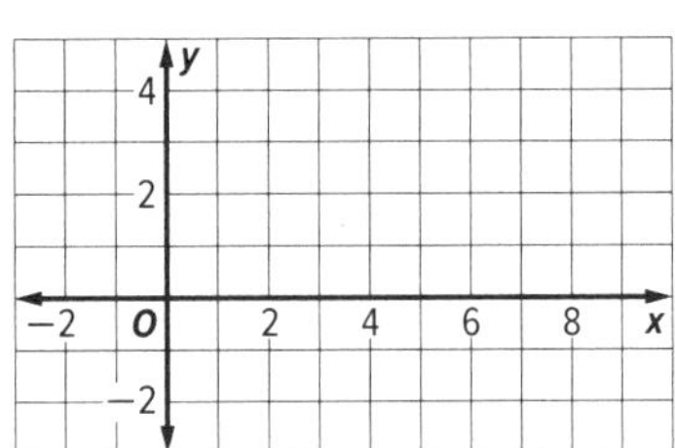

5. **DESCRIBE A METHOD** Explain how to find the solutions to $\sqrt{x-5} - \sqrt[4]{x+7} = 0$, both algebraically and graphically. Describe how to find any extraneous solutions. Illustrate your answer. CCSS A.REI.2, A.REI.11, SMP 8

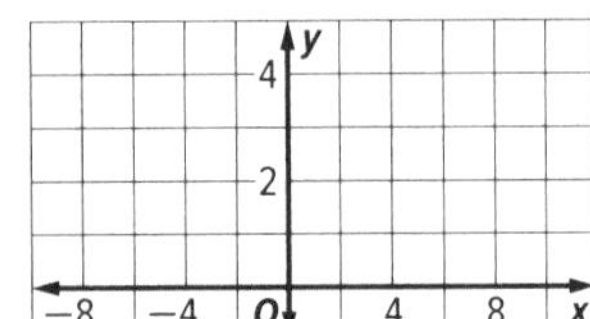

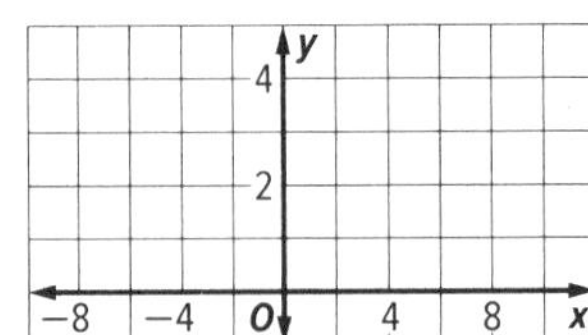

6. **USE TOOLS** The surface area of a sphere is 20 cm^2 greater than the surface area of a cube. Find functions to represent the radius of the sphere and the side length of the cube, in terms of the surface area of the cube. Describe how a graphing calculator can be used to find the surface area of each object, if the radius of the sphere equals the side length of cube? Sketch the graph. Find the surface area of the cube and the sphere. CCSS A.REI.11, SMP 5

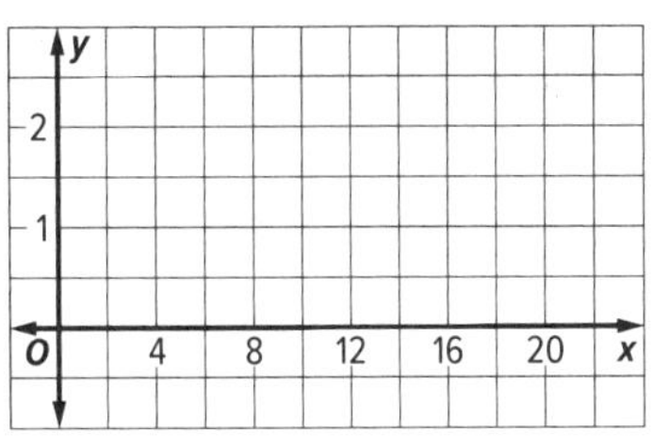

7. a. **REASON ABSTRACTLY** Explain how we know that the equation $\sqrt{x-5}+1=\sqrt{(2-x)}$ has no solutions without having to actually solve it. Confirm this by graphing the two sides of the equation. CCSS A.REI.11, SMP 2

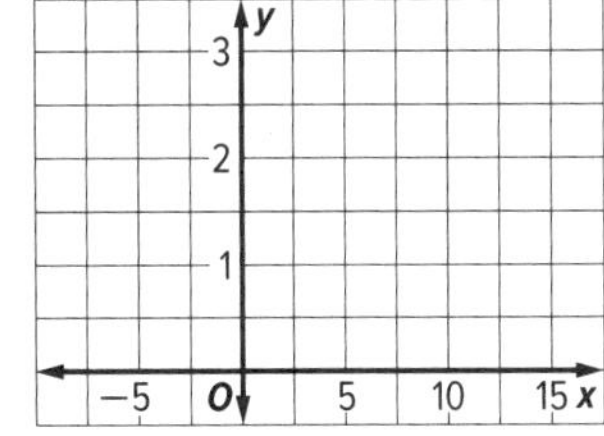

b. **REASON ABSTRACTLY** If the equation in **part a** becomes an inequality, $\sqrt{x-5}+1<\sqrt{(2-x)}$, will we then get real solutions? CCSS A.REI.11, SMP 2

8. a. **REASON QUANTITATIVELY** Solve the equation $\sqrt{x-1}+1=\sqrt{x-2}$, and graph the two sides of the equation to confirm your answer. CCSS A.REI.11, SMP 2

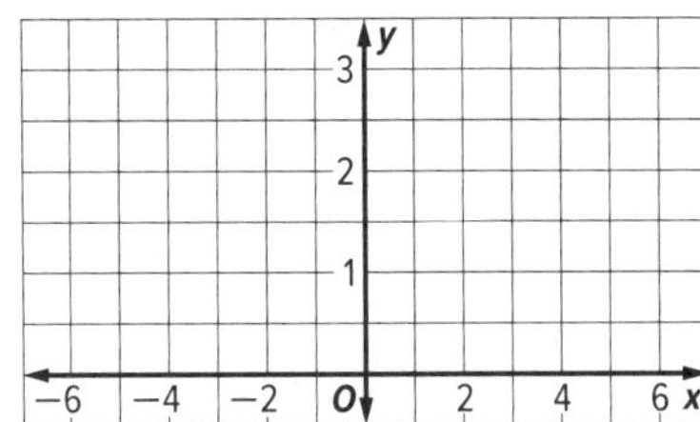

b. **REASON QUANTITATIVELY** Solve the equation $\sqrt{x-1}+1=\sqrt{x+2}$, and graph the two sides of the equation to confirm your answer. CCSS A.REI.11, SMP 2

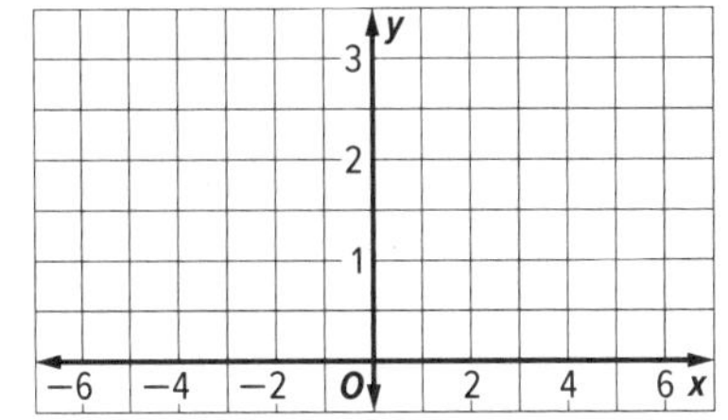

Performance Task

Speed Can Be a Drag

Provide a clear solution to the problem. Be sure to show all of your work, include all relevant drawings, and justify your answers.

Drag force is a force due to air resistance that an object moving through the air feels. It is proportional to the square of the object's velocity. A graph for the drag force on a certain car is given below, where v is in meters per second.

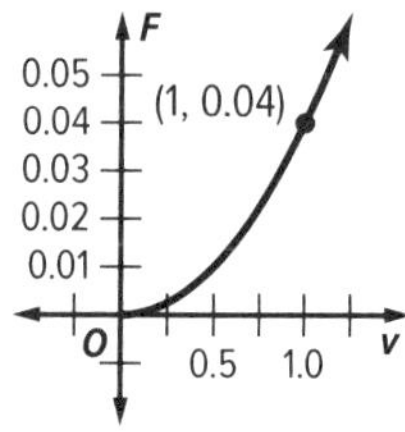

Part A

Find $F(v)$, then find the inverse of $F(v)$. If the function $F(v)$ has a domain of $0 \leq v \leq 75$, then what are the domain and range of the inverse function $v(F)$?

Part B

Suppose the car is traveling 5 m/s, and then begins accelerating at 3 m/s^2. Express the force as a function of time *t* since the car began to accelerate. (Hint: Use unit analysis.)

Part C

The total force on an object is the product of its mass and acceleration. That is, $F_{total} = ma$. The car has a mass of 1300 kg. Using your answers and the information given above, determine the following:

i) The force that is being applied to the car to accelerate it (do not include the drag force)

ii) The time at which the drag force will equal the force applied to accelerate the car

Performance Task

Torricelli's Law

Provide a clear solution to the problem. Be sure to show all of your work, include all relevant drawings, and justify your answers.

The diagram shows a water tank, with a small drainage hole near the base. Let h_0 represent the initial depth of the water in the tank at $t = 0$.

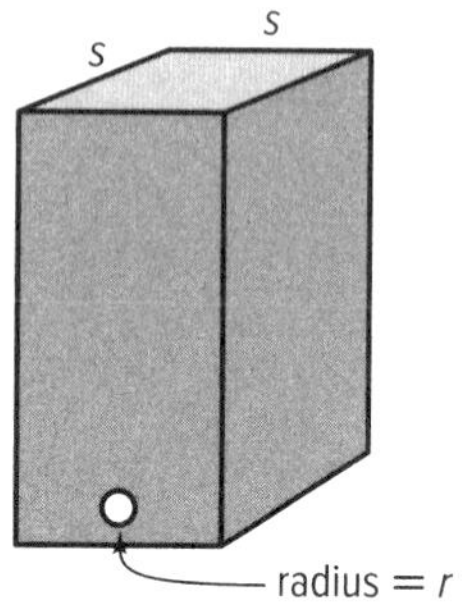

Part A

Torricelli's Law states that the velocity v of the water coming out of the drainage hole is equal to the square root of twice the product of the depth of the water and the acceleration due to gravity g. Write an equation representing Torricelli's Law. Then, write an equation for the depth of the water in terms of the velocity and the acceleration due to gravity.

Part B

The constant g is about 9.81 N/kg, where N stands for Newtons. Look up the units for Newtons, and then show that the units of the expression in your equation from **Part A** are the expected units for a velocity.

Part C

If h is the depth of the water at time t as the tank drains, it can be shown that $2\sqrt{h} = \frac{-r^2}{s^2}\sqrt{2g} \cdot t + C$, where C is a constant. Use the initial state of the water in the tank to determine the value of C. Substitute the value of C into the original equation.

Part D

Solve your equation from **Part C** for t. How long would it take the tank to drain if it is full and contains 90 cubic meters of water? Use $s = 3$ m and $r = 15$ cm.

Standardized Test Practice

1. Given $f(x) = x^2 + 4x - 5$ and $g(x) = 2x^2 + 13x + 15$, complete the following. Include any restrictions on the domain. CCSS F.BF.1b

$(f + g)(x) =$ ____

$(f - g)(x) =$ ____

$(fg)(x) =$ ____

$\left(\frac{f}{g}\right)(x) =$ ____

2. Simplify each expression. Assume all variables are positive. CCSS A.SSE.2

$\sqrt[3]{\frac{16x^{12}}{216}} =$ ____

$\sqrt[4]{81x^8y^2} =$ ____

3. Consider the function $f(x) = \frac{\sqrt{x-2}}{3}$. CCSS F.IF.5

Graph $f(x)$.

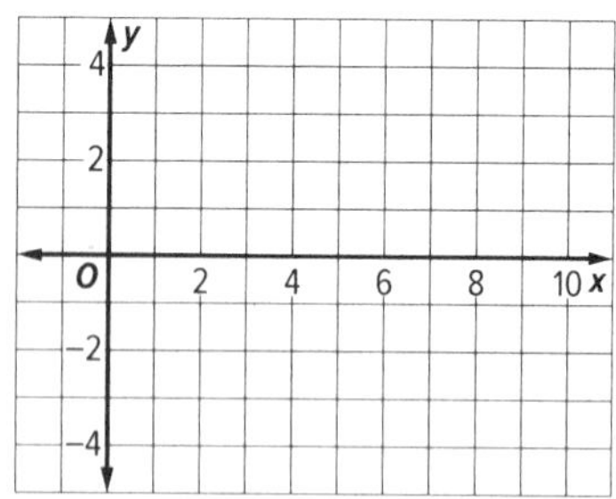

What are the domain and range of $f(x)$?

Domain: ____

Range: ____

4. Solve $\sqrt{3x - 1} \geq 7$ for x. CCSS A.REI.2

____

5. The diagonal length of a right rectangular prism is given by the formula $d = \sqrt{x^2 + y^2 + z^2}$, where x, y, and z are the dimensions of the prism.

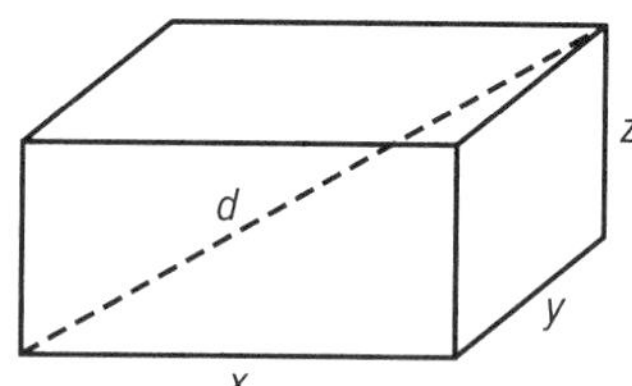

Write the formula that gives the height of the prism, z, in terms of d, x, and y and use it to find the height when the length and width are 2 and 4 and the diagonal is 7. CCSS A.CED.2

____

6. If $f(x) = 2x^2 - 4x$, $g(x) = 5x + 3$, and $h(x) = -x^2 + 4$, complete the following. CCSS F.BF.1b

$(f + g + h)(x) =$ ____

$(f - g - h)(x) =$ ____

7. Simplify the following expressions. CCSS A.SSE.2

$\sqrt{35}(3\sqrt{2} + \sqrt{7})^2 =$ ____

$4\sqrt{11} - 2\sqrt{2}(5\sqrt{6} - 3\sqrt{22}) =$ ____

8. Graph the function $f(x) = \sqrt{2x - 1} + 3$. CCSS F.IF.7b

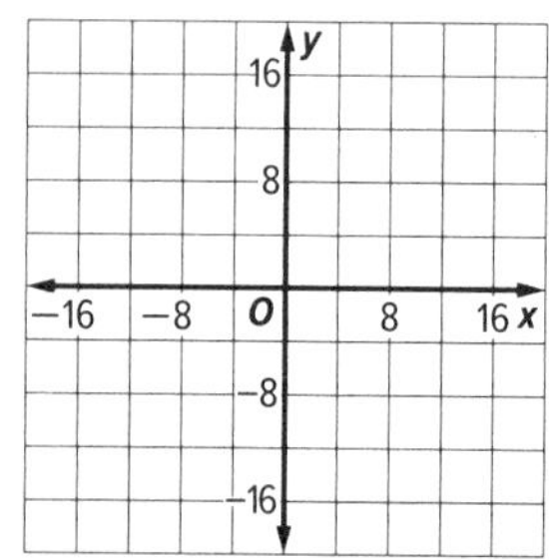

9. The graphs of $f(x) = 4x^2 - 1$ with domain $x \geq 0$ and $g(x) = x^3 + 1$ are given on a square scale. On the same axes sketch the graphs of $f^{-1}(x) = \frac{\sqrt{x+1}}{2}$ and $g^{-1}(x) = \sqrt[3]{x-1}$, the inverses of $f(x)$ and $g(x)$. CCSS F.IF.7b

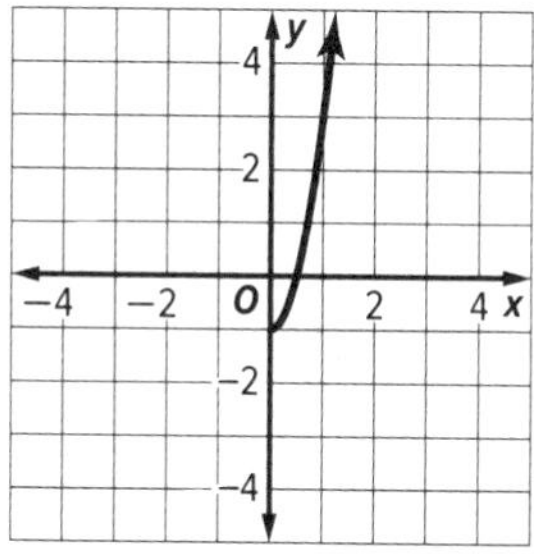

$f(x) = 4x^2 - 1$ on domain $x \geq 0$

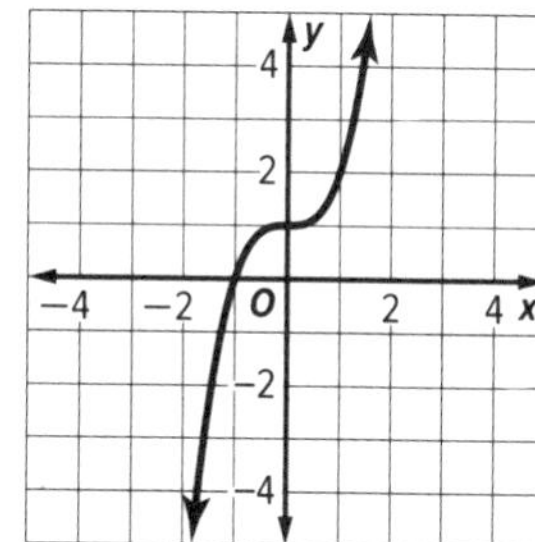

$g(x) = x^3 + 1$

10. The following table shows the age of a tree and the diameter of its trunk.

Age (yrs)	4	6	8	10	12	14	16
Diameter (cm)	3.9	4.7	5.4	6.1	6.7	7.2	7.7

What type of model best fits the data? Explain by comparing properties of the table to properties of the model. CCSS F.IF.9

11. The function $f(x) = 2\pi(x + 1)$ gives the circumference of a target in a computer game, x seconds after the player clicks on the target. What is the inverse of this function? Explain what it represents. CCSS F.BF.4a

12. Solve $4 + \sqrt{x + 8} = x$ for x. Describe your solution process. CCSS A.REI.2

7 Exponential and Logarithmic Functions

CHAPTER FOCUS Learn about some of the Common Core State Standards that you will explore in this chapter. Answer the preview questions. As you complete each lesson, return to these pages to check your work.

What You Will Learn	Preview Question
Lesson 7.1: Graphing Exponential Functions	
CCSS F.IF.7e Graph exponential and logarithmic functions, showing intercepts and end behavior, and trigonometric functions, showing period, midline, and amplitude. **Also addresses:** F.IF.4, F.IF.5, F.IF.8b, F.BF.3, A.CED.1, F.IF.6, A.REI.11	CCSS SMP 5 Use a calculator to generate a table of values for these functions. Which function increases at a greater rate? Try to account for this greater rate. A. $y = x^2$ B. $y = 2^x$
Lesson 7.2: Solving Exponential Equations and Inequalities	
CCSS A.CED.1 Create equations and inequalities in one variable and use them to solve problems. Include equations arising from linear and quadratic functions, and simple rational and exponential functions. CCSS A.SSE.4 Derive the formula for the sum of a finite geometric series (when the common ratio is not 1), and use the formula to solve problems.	CCSS SMP 7 This is the graph of an exponential function. Sketch the inverse of this function in the line $y = x$. 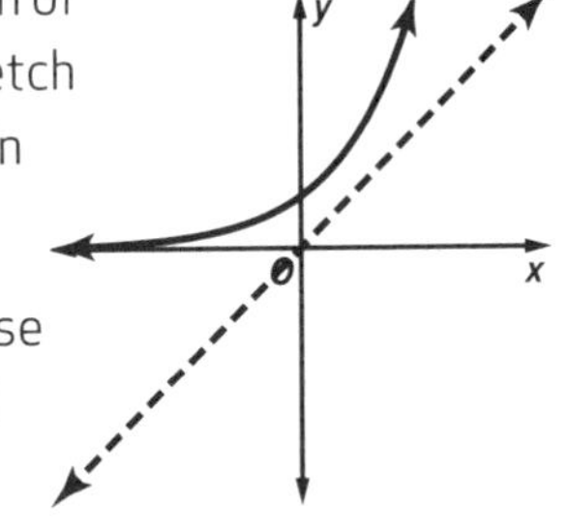 CCSS SMP 7 State the inverse of the function $f(x) = 2^x$ and use it to solve the equation $2^{3x} = 72$.
Lesson 7.3: Logarithms and Logarithmic Functions	
CCSS F.BF.3 Identify the effect on the graph of replacing $f(x)$ by $f(x) + k$, $kf(x)$, $f(kx)$, and $f(x + k)$ for specific values of k (both positive and negative); find the value of k given the graphs. Experiment with cases and illustrate an explanation of the effects on the graph using technology. Include recognizing even and odd functions from their graphs and algebraic expressions for them. CCSS F.IF.7e Graph exponential and logarithmic functions, showing intercepts and end behavior, and trigonometric functions, showing period, midline, and amplitude. **Also addresses:** F.IF.4, F.IF.5, A.CED.2, F.LE.4	CCSS SMP 2 Suppose $f(x)$ is a logarithmic function. Which function results in a vertical translation of $f(x)$? Explain. $g(x) = f(x) + 1, \quad h(x) = f(x + 1)$ CCSS SMP 2 If (1, 0) is a point on the graph of $f(x)$, find a point on the graph of $d(x) = -2f(x + 2) + 1$. Explain.

What You Will Learn	Preview Question
Lesson 7.4: Solving Logarithmic Equations and Inequalities	
CCSS A.REI.11 Explain why the *x*-coordinates of the points where the graphs of the equations $y = f(x)$ and $y = g(x)$ intersect are the solutions of the equation $f(x) = g(x)$; find the solutions approximately, e.g., using technology to graph the functions, make tables of values, or find successive approximations. **Also addresses:** A.CED.1	CCSS SMP 8 Describe a method for solving for *x* in this equation. Solve for *x*. $\log_{10}[\log_{10}(x + 1)] = 1$
Lesson 7.5: Properties of Logarithms	
CCSS A.CED.1 Create equations and inequalities in one variable and use them to solve problems. Include equations arising from linear and quadratic functions, and simple rational and exponential functions.	CCSS SMP 7 What property of exponents allows you to simplify this expression? What is the simplified form? $2^3 \cdot 2^x \cdot 2^{-4}$ CCSS SMP 7 Use properties of logarithms to rewrite the expression $\log_2(3) + 3\log_2(x) - \log_2(x + 1)$ as a single logarithm.
Lesson 7.6: Common Logarithms	
CCSS A.SSE.2 Use the structure of an expression to identify ways to rewrite it. For example, see $x^4 - y^4$ as $(x^2)^2 - (y^2)^2$, thus recognizing it as a difference of squares that can be factored as $(x^2 - y^2)(x^2 + y^2)$. **Also addresses:** A.CED.1	CCSS SMP 6 How could you use the properties of logarithms to solve this equation? What is *x*? $10^{x+1} = 25$
Lesson 7.7: Base *e* and Natural Logarithms	
CCSS A.SSE.2 Use the structure of an expression to identify ways to rewrite it. For example, see $x^4 - y^4$ as $(x^2)^2 - (y^2)^2$, thus recognizing it as a difference of squares that can be factored as $(x^2 - y^2)(x^2 + y^2)$.	CCSS SMP 3 For what value of *n* is *x* equal to 1? $\log_n(5^{7x+1}) = 8$ CCSS SMP 7 Write a logarithmic equation equivalent to the exponential equation $e^{3-x} = 17$.
Lesson 7.8: Modeling: Exponential and Logarithmic Functions	
CCSS F.IF.8b Use the properties of exponents to interpret expressions for exponential functions. CCSS F.LE.4 For exponential models, express as a logarithm the solution to $ab^{ct} = d$ where *a*, *c*, and *d* are numbers and the base *b* is 2, 10, or *e*; evaluate the logarithm using technology. **Also addresses:** F.IF.4, A.CED.1, A.CED.2, A.CED.3, A.SSE.1a, A.SSE.1b, A.SSE.2	CCSS SMP 4 A bank account earns 5% interest compounded annually. How much will a $1500 deposit be worth after *n* years?

7.1 Graphing Exponential Functions

Objectives

- Use exponential functions to model real-life growth and decay.
- Identify and implement transformations to exponential functions.

CCSS STANDARDS

Content: F.IF.7e, F.IF.8b, F.BF.3, A.CED.1, F.IF.4, F.IF.5, F.IF.6 A.REI.11

Practices: 1, 2, 3, 4, 5, 6, 7

Use with Lesson 7-1

A function in which the base is a constant and the exponent is the independent variable is an **exponential function**.

KEY CONCEPT Exponential Function

The function $f(x) = b^x$ is an exponential function. This function represents exponential growth if $b > 1$ and exponential decay if $0 < b < 1$. Characteristics of $f(x)$:

- Domain = {all real numbers}
- Range = {positive real numbers}
- y-intercept: $(0, 1)$
- asymptote : x axis, $y = 0$

EXAMPLE 1 Exponential Graphs and Transformations CCSS F.BF.3, F.IF.4

EXPLORE In this exploration, you will explore transformations of the function $f(x) = b^x$.

a. USE STRUCTURE Let $f(x) = 2^x$ and $g(x) = -2^{x-3}$. What transformations of $f(x)$ will result in the graph of $g(x)$? Construct the graphs of both functions. How are the y-intercept and asymptote of $f(x)$ transformed? CCSS F.IF.7e, SMP 7

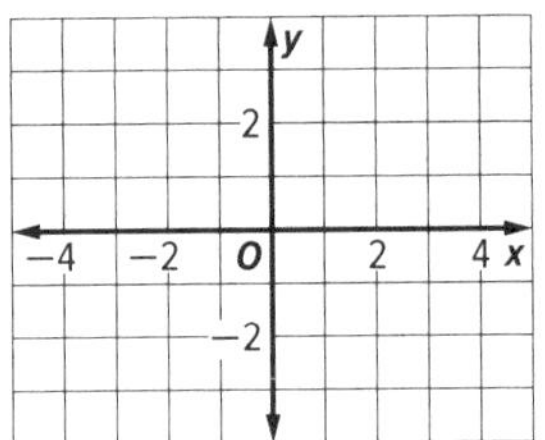

b. USE STRUCTURE Let $f(x) = 0.5^x$ and $g(x) = \frac{1}{2}(0.5)^x + 4$. State the domain and range for each function. What transformations of the graph of $f(x)$ will create the graph of $g(x)$? What happens to the asymptote and y-intercept of $f(x)$ under the transformations? CCSS SMP 7

c. PLAN A SOLUTION Consider the function $f(x) = (3)^x$. If the graph of $f(x)$ is reflected in the x-axis and translated 2 units to the left and 3 units down to create $g(x)$, find the equation for $g(x)$. What are the domain and range, and the equation for the asymptote of $g(x)$? Graph the function $g(x)$ and label any intercepts. CCSS F.IF.7e, A.CED.1, SMP 1

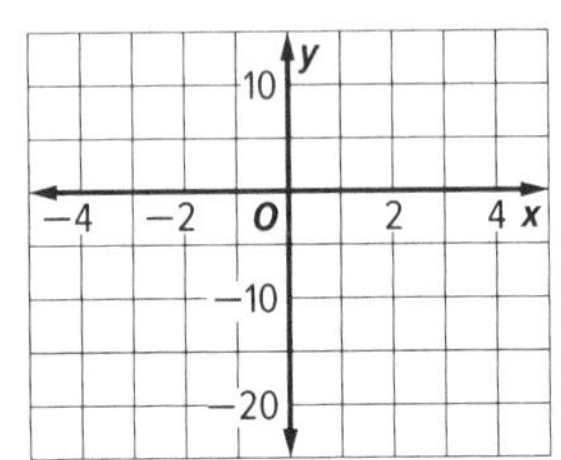

d. **COMMUNICATE PRECISELY** Let $f(x) = (5)^x$ and $g(x) = (5)^{-x}$. How are the graphs of the functions similar and how are they different? Where do the graphs intersect? CCSS SMP 6

KEY CONCEPT Growth and Decay

Exponential functions may be used to model constant percent changes over repeated and regular time intervals within a specific duration of time.

- A constant percent increase is modeled by the growth function $A(t) = a(1 + r)^t$ where $A(t)$ is the amount after t time periods, a is the initial amount and r is the percent increase per time period.
- A constant percent decrease is modeled by the decay function $A(t) = a(1 - r)^t$, where $A(t)$ is the amount after t time periods, a is the initial amount and r is the percent decrease per time period.

EXAMPLE 2 Exponential Decay

Radiocarbon dating is a technique used to estimate the age of organic material found at archaeological sites. When a living organism dies, the carbon-14 isotope, C_{14}, begins to decay according to an exponential model. The ratio of the amount of the isotope in a decaying organism to the amount of the isotope that existed in the living organism decreases by 50% of its previous value every 5730 years, an interval referred to as the half-life of carbon-14.

a. **PLAN A SOLUTION** How could you calculate the number of half-life cycles of the C_{14} isotope over an interval of t years? How can you know the C_{14} ratio at the time of an organism's death? How can you use this information to write a function $C(t)$ for the C_{14} ratio, where t represents the years since the organism died? CCSS F.IF.8b, A.CED.1, SMP 1

b. **USE A MODEL** Use a graphing calculator to graph $C(t)$. Sketch the graph on the grid provided. What viewing window and scale did you choose and why? State the domain and range in the context of the problem. If there is an asymptote, what does it represent in the problem? CCSS F.IF.5, F.IF.7e, SMP 4

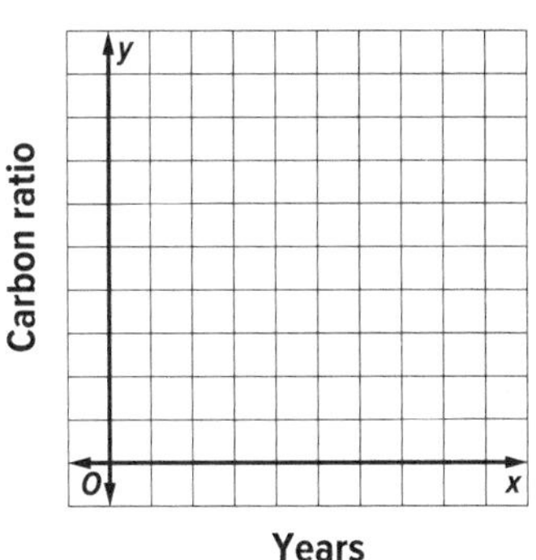

c. **CALCULATE ACCURATELY** If an organism died 30,000 years ago, what carbon ratio would you expect to find? Show your work. CCSS A.CED.1, SMP 6

d. **MAKE A CONJECTURE** Carbon dating is used for objects up to about 50,000 years old; other techniques are used for older items. Use the graph to make a conjecture about why carbon dating might not be effective for older items. Explain your reasoning. CCSS F.IF.4, SMP 3

EXAMPLE 3 Exponential Growth

Regina recently inherited an oil painting that was purchased by her great-grandfather for $100 in 1938. In 2013 she took the painting to an appraiser, who estimated its value at $300,000.

a. **REASON ABSTRACTLY** Write a function $A(t)$ that models the change in value of the painting over t years. What was the annual percent of increase in the value of the painting? Show your work. CCSS F.IF.8b, A.CED.1, SMP 2

b. **USE A MODEL** Using a graphing calculator, graph the function in **part a** and sketch the graph on the grid provided. On the graph, mark the point that represents the value of the painting in 2013. If the painting were to continue appreciating at the same rate, estimate the value in 2028 to the nearest $100,000. Show your work. CCSS F.IF.7e, SMP 4

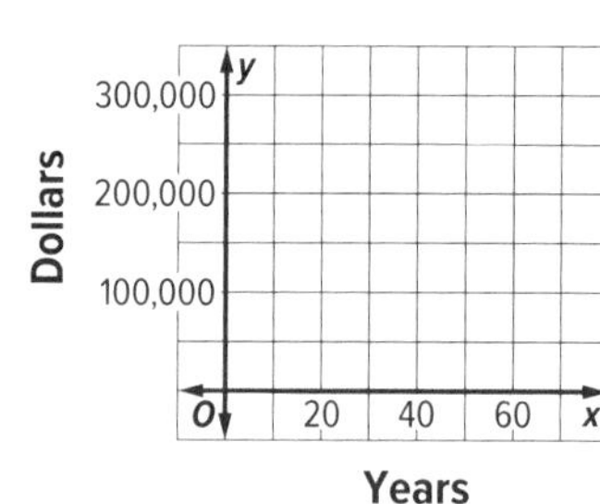

c. **USE TOOLS** How can you determine during which year the value of the painting will reach $750,000? Use a calculator and explain your process. CCSS A.REI.11, SMP 5

PRACTICE

1. **REASON ABSTRACTLY** Polonium is an element discovered in 1898 by Marie and Pierre Curie. If a sample of radioactive polonium decays to 12.5% of its original value after 414 days, what is the half-life of polonium? Explain your reasoning. CCSS F.IF.8b, SMP 2

2. **CRITIQUE REASONING** While analyzing the function $f(x) = 3^x$, Vanita says that applying a vertical stretch by a factor of 9 is exactly the same as translating the graph 2 units to the left. Is she correct? Explain your reasoning. CCSS F.BF.3, SMP 3

3. At age 28, Susan makes a single $22,000 investment that earns 5% interest each year. CCSS F.IF.8b, A.CED.1

   a. **CALCULATE ACCURATELY** If Susan leaves the investment untouched until she turns 65, how much will the investment be worth at that time? CCSS SMP 6

   b. **INTERPRET PROBLEMS** Susan's twin brother, Samuel, waits until he is 45 years old to invest money for retirement. At that time, he can only find an investment that earns 3% interest per year. If he wishes to have the same amount of money as Susan when they turn 65, how much does he need to invest? Show your work. CCSS SMP 1

4. **REASON ABSTRACTLY** The value of an automobile depreciates by approximately 15% each year after purchase. Jayden paid $28,000 when he bought his car 15 years ago. CCSS F.IF.6, F.IF.7e, F.IF.8b, A.CED.1, SMP 2

   a. Write and graph a function that models how the value of the car depreciated during the time that Jayden has owned it.

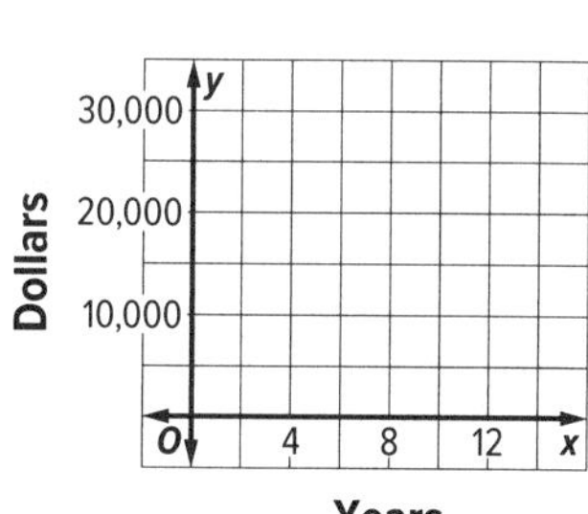

   b. How does the average decrease in value during the first five years of ownership compare to the last five years of ownership?

5. **REASON QUANTITATIVELY** Over 7 years, an investment grows from \$25,000 to \$38,849.66. By what percent did the investment grow each year? Show your work.

CCSS F.IF.8b, A.CED.1, SMP 2

6. **USE STRUCTURE** Let $f(x) = 6^x$; $g(x) = -\frac{1}{2}(6)^{x+3} - 1$ and $h(x) = \frac{3}{2}(6)^{x-5}$.

CCSS F.BF.3, F.IF.4, SMP 7

   a. How could you transform $f(x)$ to create the graphs of $g(x)$ and $h(x)$?

   b. How could you directly transform $g(x)$ to create the graph of $h(x)$? What are the asymptotes of each function?

7. Use the given functions to answer each question.

   a. **USE STRUCTURE** Let $f(x) = (4)^x$ and $g(x) = (4)^{-x} + 1$. What transformations of $f(x)$ will result in the graph of $g(x)$? Construct the graphs of both functions. How are the y-intercept and asymptote of $f(x)$ transformed? CCSS F.IF.7e, SMP 7

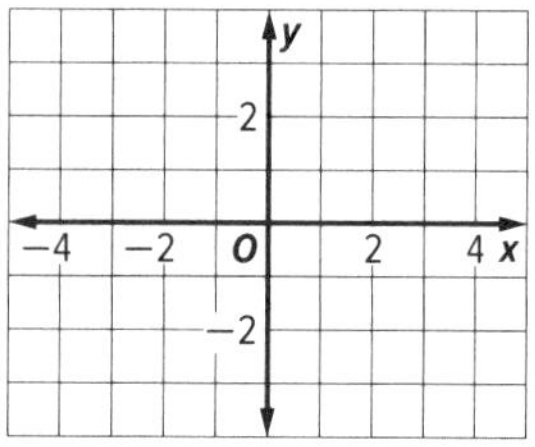

   b. **PLAN A SOLUTION** Consider the function $f(x) = (0.6)^x$. If the graph of $f(x)$ is reflected in the y-axis, stretched vertically by a factor of 2, and translated 1 unit down to create $g(x)$, find the equation for $g(x)$. What are the domain, range, and y-intercept of $g(x)$? Graph the function $g(x)$ and label any intercepts.

   CCSS F.IF.7e, A.CED.1, SMP 1

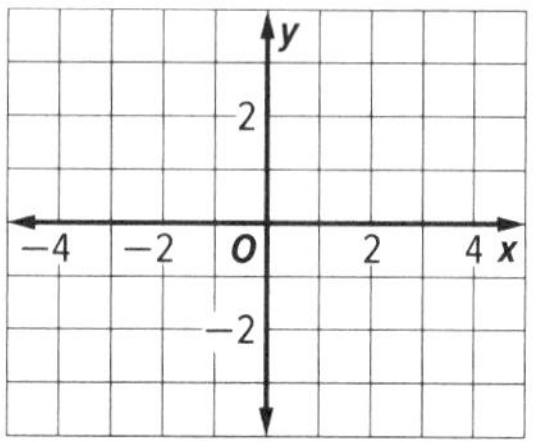

8. INTERPRET PROBLEMS A compound undergoes exponential decay with initial amount 27.3 grams.

a. If the amount of the compound decreases by ten percent each year, write a function $A(t)$ modeling the amount of the compound remaining after t years.

b. Use your function from **part a** to find the number of grams to the nearest tenth of the compound remaining after 3 years.

9. USE A MODEL A population of bacteria grows exponentially with initial population 20,000. CCSS SMP 4

a. After one day the bacteria population grows to 30,000. Write an exponential function $P(t)$ to model the bacteria population after t days. CCSS A.CED.1

b. Use a graphing calculator to graph $P(t)$ and sketch the graph on the grid provided. Identify any intercepts and the end behavior as $t \to +\infty$ and explain these features in the context of the problem. CCSS F.IF.7e

y
250,000
150,000
50,000
O
2
4
6
x

c. Based on the context, what is an appropriate domain for $P(t)$? Explain your reasoning. CCSS F.IF.5

d. Use your graph and the aid of a graphing calculator to find the number of days it takes for the population to reach 100,000. Explain your reasoning. CCSS A.REI.11

7.2 Solving Exponential Equations and Inequalities

Objectives

- Derive the formula for a finite geometric series.
- Use properties of exponents to solve equations and inequalities.

STANDARDS

Content: A.SSE.4, A.CED.1
Practices: 1, 2, 3, 5, 6, 7
Use with Lesson 7-2

A **geometric sequence** is a sequence of terms in which the ratio between any two consecutive terms is constant.
A **geometric series** is the sum of the terms of a geometric sequence.

EXAMPLE 1 Sum of a Geometric Series CCSS A.SSE.4

EXPLORE After graduating from college, Sofia is ready to start a career in finance. She is offered a starting salary of $40,000, and two choices for how to receive an annual raise.

	Starting Salary	Annual Increase
Option 1	$40,000	$1,400
Option 2	$40,000	3%

a. **REASON ABSTRACTLY** Would Sofia's salaries under Option 2 form a geometric sequence? Explain your answer. CCSS SMP 2

b. **REASON QUANTITATIVELY** For Option 2, write an equation to calculate the sum of her salaries for her first six years working for the company. CCSS SMP 2

c. **PLAN A SOLUTION** How does this compare to her total earnings under Option 1 for the same time period? CCSS SMP 1

d. **USE STRUCTURE** Find an expression for her total earnings from **part b** using the fact that $1 + x + x^2 + x^3 + \cdots + x^n = \frac{x^{n+1} - 1}{x - 1}$ for $x \neq 1$. Generalize this result to express the sum of n terms, if the first term is represented by a, and the ratio between consecutive terms is r. CCSS SMP 7

e. CONSTRUCT ARGUMENTS If Sofia stays with the same company for 30 years, which salary option is a better choice? Explain your answer. CCSS SMP 3

KEY CONCEPT Compound Interest

You can calculate compound interest using the following formula:

$$A = P\left(1 + \frac{r}{n}\right)^{nt}$$

where A is the amount in the account after t years, P is the principal amount invested, r is the annual interest rate, and n is the number of compounding periods per year.

EXAMPLE 2 Compound Interest CCSS A.CED.1

a. PLAN A SOLUTION Suppose $15,000 is deposited in an account that offers 3.5% annual interest, compounded monthly. How can you find the balance after 1 year and the percent increase from the initial deposit? Show your work. CCSS SMP 1

b. MAKE A CONJECTURE In **part a**, why was the percent increase greater than the annual interest rate? CCSS SMP 3

EXAMPLE 3 Exponential Equations and Inequalities CCSS A.CED.1

Scientists estimate the rat population on an island is currently 10,000 and will triple every 120 days.

a. PLAN A SOLUTION For how many days will the population remain less than 100,000? Show how to determine the answer graphically. CCSS SMP 1

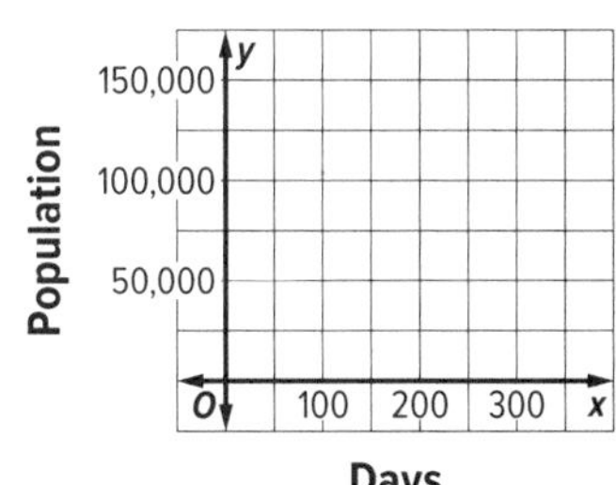

b. USE TOOLS The local government is working to obtain a poison to eradicate the rats. The total number of doses available is 200 and is increasing by 3% every day. Write a function to model the amount of rat poison available each day. Use a graph to determine when the number of doses will be greater than the rat population. CCSS SMP 5

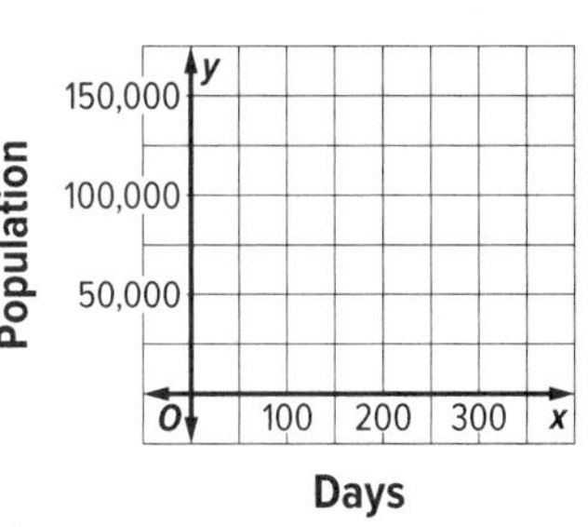

PRACTICE

1. **PLAN A SOLUTION** Lottery winners have two choices for receiving their winnings. They can choose an \$80 million dollar immediate payout, or 20 annual payments that begin at \$4 million dollars and increase 2% each year. Which option results in the larger prize? Show your work. CCSS A.SSE.4, SMP 1

2. A geometric series can be combined with compound interest to find a formula for payments on a loan or mortgage. For a mortgage with a principal *P*, a monthly interest rate *i*, and consisting of *n* monthly payments, each monthly payment is $\frac{Pi}{1-(1+i)^{-n}}$. CCSS A.SSE.4

   a. **REASON ABSTRACTLY** For a 30 year \$300,000 mortgage with a 6% annual interest rate, how much interest is collected over the life of the loan? Show your work. CCSS SMP 2

   b. **REASON QUANTITATIVELY** Suppose a family can afford to make monthly payments of \$1000 to \$1250. If annual interest rates are 4%, what is the greatest 20-year mortgage they can afford? Round your answer to the nearest \$1000. CCSS SMP 2

3. **USE TOOLS** After a patient is given a dose of medicine, the concentration in the bloodstream is 3.0 mg/mL. The concentration decays exponentially, and drops to 1.5 mg/mL after 2 hours. The medicine is ineffective at concentrations less than 0.6 mg/mL. If the patient is given a dose at 10 A.M., could the next dose be given at 3 P.M. and without the level of medication dropping below the effective concentration? Use a graph to justify your answer. CCSS A.CED.1, SMP 5

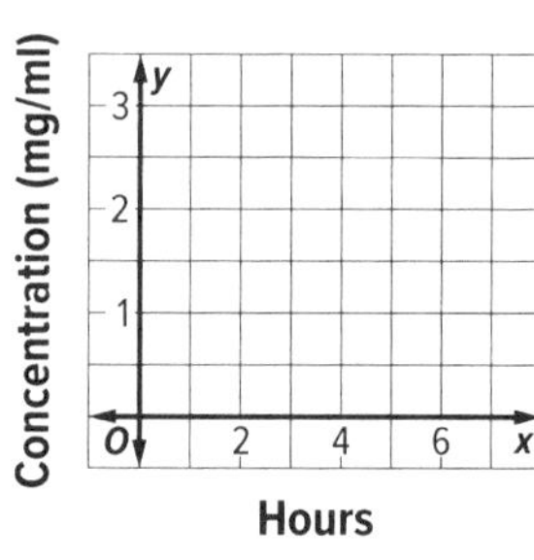

4. **CALCULATE ACCURATELY** Suppose you have the option to get a lump sum of \$5,000,000 at the end of one year or receive one penny when the year starts, two pennies after the first week, four pennies after the second week, eight pennies after the third week, and so forth until the year ends. Using the fact that there are about 52 weeks in a year, which option would result in more money? CCSS A.SSE.4, SMP 6

5. **CRITIQUE REASONING** Tom uses a graphing calculator to graph $f(x) = 5(1.25)^x$ and $g(x) = 1.5^x$. He notices that the gap between the curves increases as x increases, and concludes that $f(x) > g(x)$ for $x > 0$. Is he correct? Use a graph to support your reasoning. Could you have written and solved an equation to come to the same conclusion? CCSS A.CED.1, SMP 3, SMP 5

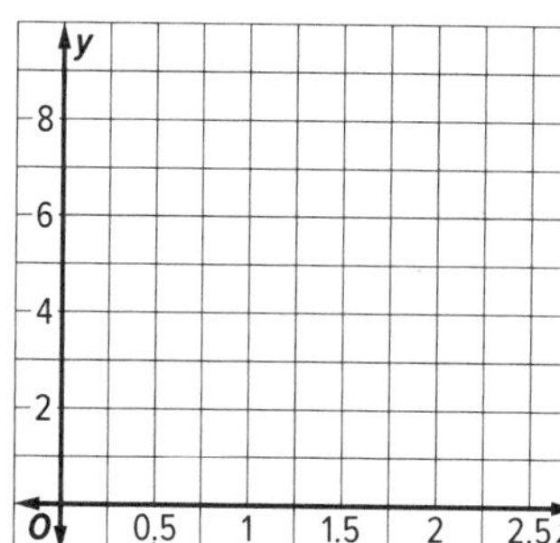

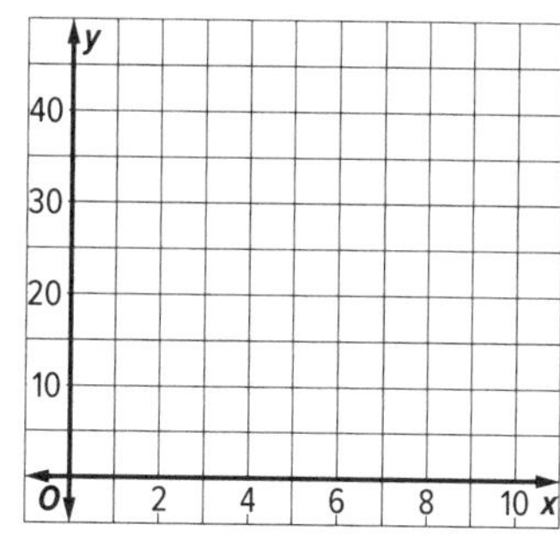

6. **CALCULATE ACCURATELY** Use the information given to answer the questions. CCSS A.SSE.4, SMP 6

   a. A college professor was hired with a starting salary of $44,000. If she receives a 2% raise each year, how much money will she have earned after working 25 years?

   b. Considering the total amount earned in 25 years, would it have been better to start at $35,000 and receive a 3% raise each year? Explain your reasoning.

7. **USE STRUCTURE** An athlete plans on running 20 miles for the first week of training and increasing that mileage by 10% each week. CCSS A.SSE.4, SMP 7

   a. How many total miles will the athlete have run after training for 10 weeks? Explain your reasoning.

   b. The athlete hopes to run a total of 1,000 miles in preparation for a race at the end of 20 weeks of training. Will the athlete reach this goal? Explain your reasoning.

   c. Another athlete also increased running mileage by 10% each week but reached 1000 miles in 15 weeks. Approximately how many miles did the athlete run for the first week of training? Explain your reasoning.

7.3 Logarithms and Logarithmic Functions

Objectives

- Graph logarithmic functions.
- Interpret key features of the graphs of logarithmic functions.
- Identify effects of transformations on a graph of a logarithmic function.

Content: F.IF.4, F.IF.5, F.IF.7e, F.BF.3, A.CED.2, F.LE.4
Practices: 1, 2, 5, 6, 7, 8
Use with Lesson 7-3

The definition of logarithms can be used to express logarithms in exponential form.

Logarithms with Base *b*

Let b and x be positive numbers, $b \neq 1$. The *logarithm with base b of x* is denoted $\log_b x$ and is defined as the exponent y that makes the equation $b^y = x$ true.

Suppose $b > 0$ and $b \neq 1$ for $x > 0$. Then there is a number y such that

$$\log_b x = y, \text{ if and only if, } b^y = x.$$

EXAMPLE 1 Graphing Logarithmic Functions CCSS F.IF.4

EXPLORE **Use a graph to explore the function $y = \log_2 x$.**

a. **USE STRUCTURE** Rewrite $y = \log_2 x$ as an exponential equation. Create a table of values, substitute values for x and compute the corresponding y value. What are the constraints on the domain? CCSS F.IF.4, F.LE.4, SMP 7

x	Exponential Form: $2^y = x$	Logarithmic Form: $y = \log_2 x$	(x, y)
1	If $2^y = 1$, then $y = 0$	$0 = \log_2 1$	$(1, 0)$
2	If , then $y = 1$		
4	If $2^y = 4$, then $y =$		
8	If , then $y = 3$		
16	If $2^y = 16$, then $y =$		

b. **REASON QUANTITATIVELY** The x-values in **part a** do not indicate what happens when $0 < x < 1$. Complete the table. CCSS F.IF.4, SMP 2

x	Exponential Form: $2^y = x$	Logarithmic Form: $y = \log_2 x$	(x, y)
$\frac{1}{2}$	If $2^y = \frac{1}{2}$, then $y = -1$	$-1 = \log_2 \frac{1}{2}$	$\left(\frac{1}{2}, -1\right)$
$\frac{1}{4}$	If , then $y = -2$		
$\frac{1}{8}$	If $2^y = \frac{1}{8}$, then $y =$		

c. **USE TOOLS** Use the data from the tables to graph the function $y = \log_2 x$. CCSS F.IF.7e, SMP 5

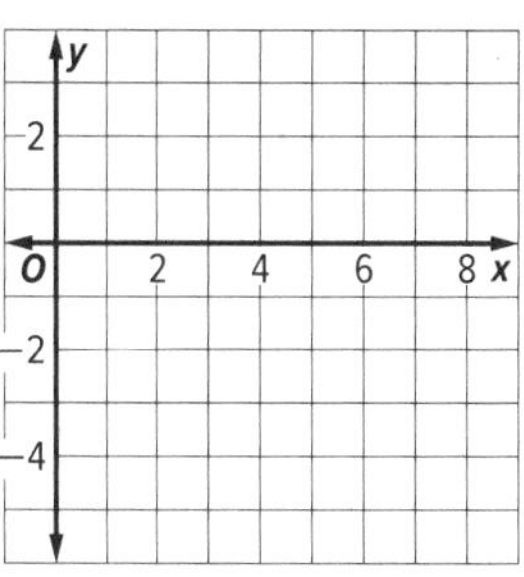

d. **COMMUNICATE PRECISELY** Describe the domain and range of the function $y = \log_2 x$. How do the domain and range compare to the domain and range of the function $y = 2^x$? CCSS F.IF.4, F.IF.5, SMP 6

e. **COMMUNICATE PRECISELY** Describe the end behavior of the function $y = \log_2 x$ and how it relates to the asymptote. CCSS F.IF.4, F.IF.5, SMP 6

f. **COMMUNICATE PRECISELY** Over what interval of the domain is the function increasing and where is it decreasing? Explain how you know. CCSS F.IF.4, F.IF.5, SMP 6

The function $f(x) = \log_b x$ is the parent logarithmic function.

Parent Logarithmic Function

Parent function: $f(x) = \log_b x$

Domain: all positive real numbers

Asymptote: $x = 0$

Type of Graph: Continuous, one-to-one

Range: all real numbers

x-intercept: $(1, 0)$

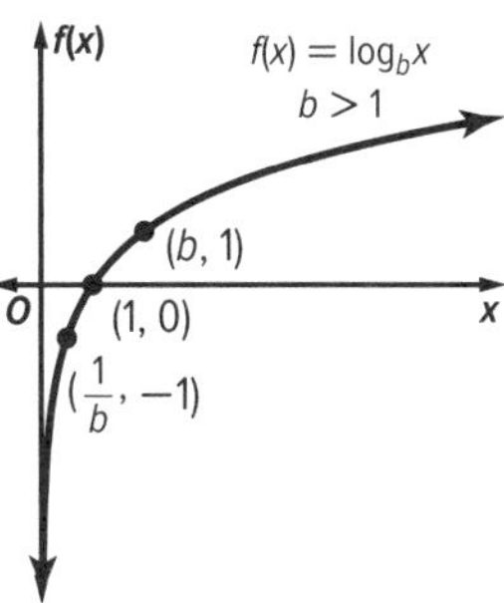

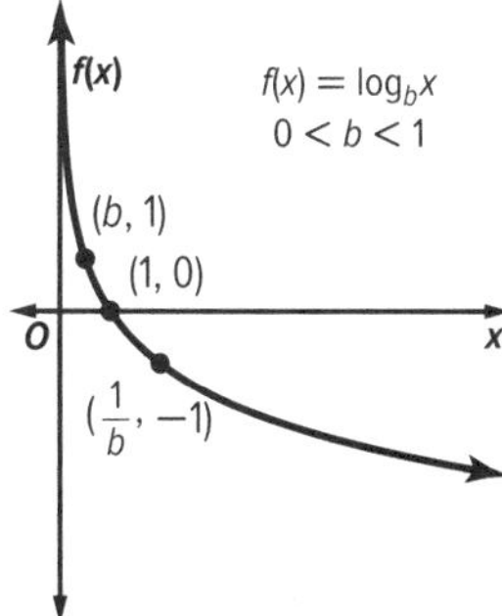

EXAMPLE 2 Transformations of Logarithmic Functions CCSS F.BF.3

a. **CALCULATE ACCURATELY** Graph $y = \log_2(x + 2)$. Compare the graph to the graph of $y = \log_2 x$ in **Example 1c**. Describe the transformation. What is the asymptote for the function $y = \log_2(x + 2)$? Why is it different than the asymptote of the function $y = \log_2 x$? CCSS SMP 6

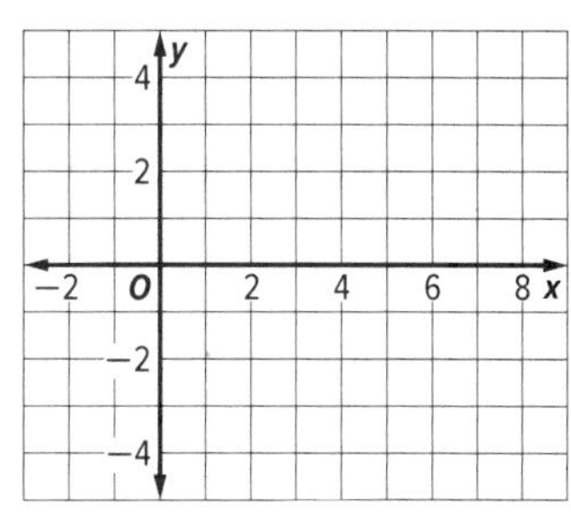

b. CALCULATE ACCURATELY Graph $y = \log_2 x + 2$. Describe the transformation from the parent function. How does the transformation differ from the transformation $y = \log_2(x + 2)$? CCSS SMP 6

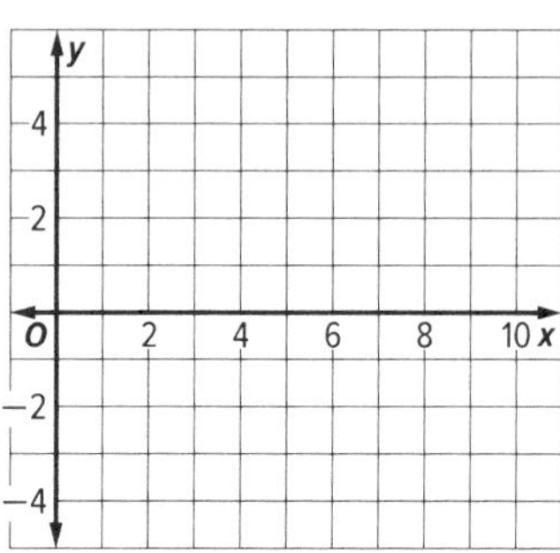

c. DESCRIBE A METHOD Describe the transformation $y = a\log_2 x$, where $a \neq 1$. Consider negative values of a, positive values of a, and fractional values of a. CCSS SMP 8

d. DESCRIBE A METHOD Describe the relationship between the graph of the function $f(x) = \frac{1}{2}\log_2 x$ and the graph of the function $g(x) = -\frac{1}{2}\log_2 x$. Compare their domains and ranges, their asymptotes and their end behavior. CCSS SMP 8

EXAMPLE 3 Analyzing Logarithmic Functions CCSS F.IF.5

The loudness of sound is measured in a logarithmic scale using a unit called decibels (dB) $d = 10 \log_{10} \frac{I}{I_0}$, where d is the decibel rating, I is the power or intensity of the sound, and I_0 is the softest sound (the threshold sound) that the human ear can hear. Noise-induced hearing loss can be caused by prolonged exposure to any loud noise over 85 (dB).

a. PLAN A SOLUTION Determine the decibel rating of sound with an intensity of $10{,}000\, I_0$. Would prolonged exposure to this intensity lead to noise-induced hearing loss? Explain. CCSS SMP 1

b. CALCULATE ACCURATELY If a sound has an intensity of 85 decibels, how much more intense is it than the threshold of sound? CCSS SMP 6

c. REASON QUANTITATIVELY If sound intensity doubles, how does the decibel rating change? Justify your answer. CCSS SMP 2

d. REASON QUANTITATIVELY Can decibel ratings be negative? Explain. CCSS SMP 2

PRACTICE

1. **CALCULATE ACCURATELY** Nine rabbits stage a daring escape in broad daylight from an animal shelter, eventually establishing a colony behind the shelter. After one month, the population has risen to 27 rabbits. After two months the population is 81; after three months, 243, and so on. Write a logarithmic function where y represents months and x represents rabbits in the colony. How many will be in the colony at 5 months? At what month will the colony reach 729 rabbits? CCSS A.CED.2, F.LE.4, SMP 6

2. **COMMUNICATE PRECISELY** Graph the logarithmic function $f(x) = \log_5(x - 3) + 4$. Identify the domain, range, asymptotes, intercepts, and end behavior. CCSS F.BF.3, SMP 6

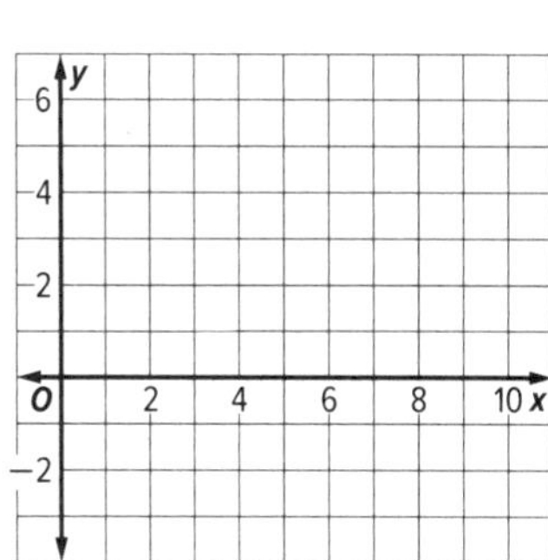

3. **COMMUNICATE PRECISELY** Sketch the graph of $f(x) = 3^x$ and $g(x) = \log_3 x$ on the same coordinate plane. Define the domain, range, intercepts, and asymptotes of each. Compare the two graphs. CCSS F.IF.7e, SMP 6

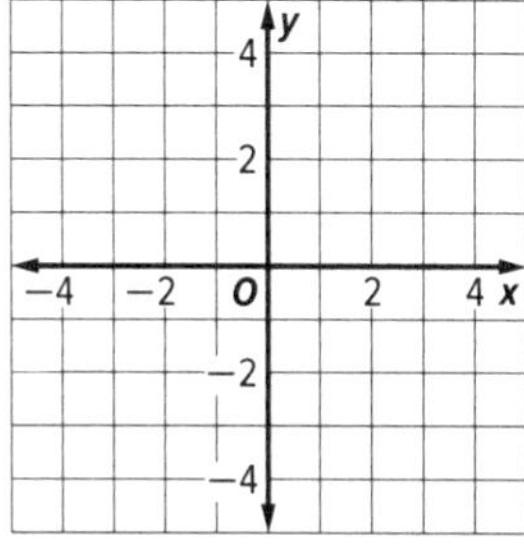

4. **USE STRUCTURE** A normal conversation has a noise rating of 60 decibels. A digital audio player at full volume has a noise rating of 100 decibels. How many times more intense is the digital audio player than a normal conversation? CCSS F.IF.5, SMP 7

5. COMMUNICATE PRECISELY Without using a calculator, answer the questions for the function $f(x) = \log_{10} x$. CCSS F.LE.4, SMP 6

a. Describe the values of $f(x)$ for $1000 \leq x \leq 10{,}000$.

b. Describe the values of x, given that $f(x)$ is negative.

c. By what amount will x increase given that $f(x)$ is increased by one? Explain how you know.

6. INTERPRET PROBLEMS For an American Elk, the antler spread a inches and the shoulder height h inches of an adult male elk are related by the function $h = 116 \log_{10}(a + 40) - 176$. Approximate the shoulder height of an adult male elk with an antler spread of 55 inches. Choose other values for a and discuss the solution with your classmates. Are there any values for a that are not reasonable? CCSS F.IF.5, SMP 1

7. CALCULATE ACCURATELY Compare $\log_3 10$ and $\log_7 40$ by using the definition of logarithms. Which quantity is greater? Do not use a calculator. CCSS F.LE.4, SMP 6

8. REASON QUANTITATIVELY Find the domain of the function $f(x) = \log_2(12 - 4x)$. CCSS F.IF.5, SMP 2

9. COMMUNICATE PRECISELY Graph the logarithmic function $f(x) = -\log_7(x + 3)$. Identify the domain, range, asymptotes, intercepts, and end behavior. CCSS F.BF.3, SMP 6

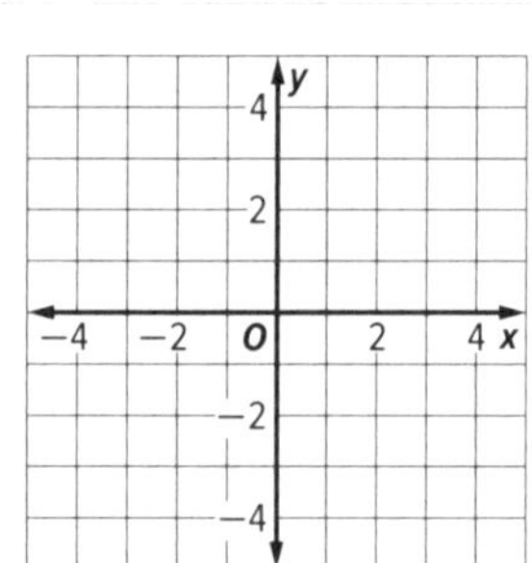

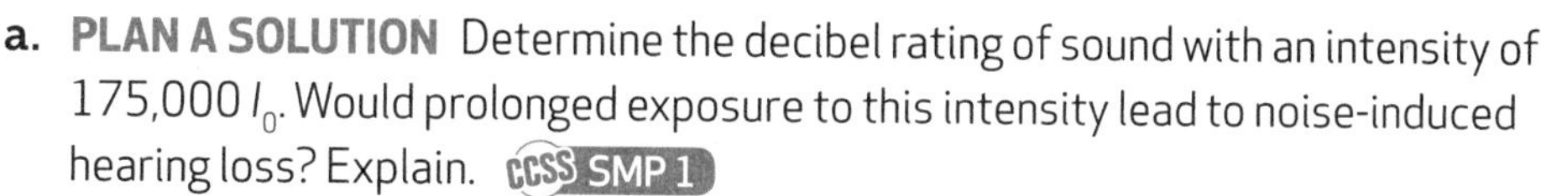

10. Answer the following questions concerning the Decibel scale. CCSS F.IF.5

a. PLAN A SOLUTION Determine the decibel rating of sound with an intensity of 175,000 I_0. Would prolonged exposure to this intensity lead to noise-induced hearing loss? Explain. CCSS SMP 1

b. CALCULATE ACCURATELY If a sound has an intensity of 36 decibels, how much more intense is it than the threshold of sound? CCSS SMP 6

c. REASON QUANTITATIVELY If sound intensity increases by a factor of 3.5, how does the decibel rating change? Justify your answer. CCSS SMP 2

11. USE A MODEL A compound undergoes exponential decay with an annual decay rate of 50%. CCSS SMP 4

a. If the initial amount of the compound is 50 grams, write a logarithmic function $T(x)$ for the time it takes for the compound to decay to x grams. CCSS A.CED.2

b. Use the aid of a calculator to complete the table of values, identifying points on the graph of $T(x)$. How do you expect the values of x to behave as $t \rightarrow +\infty$? CCSS F.IF.4

t	Exponential Form: $x = 50(0.5)^t$	Logarithmic Form: $t = \log_{0.5}\left(\frac{x}{50}\right)$	(x, t)
0	If $t = 0$, then $x = 50(0.5)^0 = 50$	$0 = 0 = \log_{0.5}\left(\frac{50}{50}\right) = \log_{0.5}(1)$	(50, 0)
1			
2			
3			
4			
5			

c. Use the table of values from **part b** to sketch a graph of the function $T(x)$ on the grid provided. Identify any intercepts and explain their significance in the context of the problem. CCSS F.IF.7e

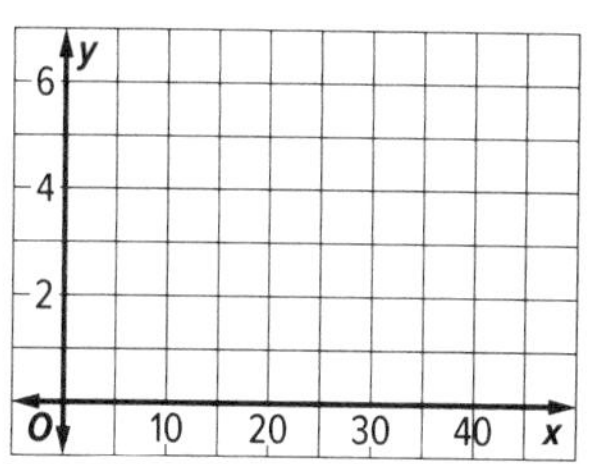

7.4 Solving Logarithmic Equations and Inequalities

Objectives

- Write and solve logarithmic equations and inequalities.
- Use graphs to solve logarithmic equations and inequalities.

Content: A.CED.1, A.REI.11
Practices: 1, 2, 3, 5, 6
Use with Lesson 7-4

KEY CONCEPT Properties of Equality and Inequality for Logarithmic Functions

Property of Equality: If b is a positive number other than 1, then $\log_b x = \log_b y$ if and only if $x = y$

Properties of Inequality: Assume $b > 1$ and $x > 0$.

If $\log_b x > y$, then $x > b^y$	If $\log_b x < y$, then $0 < x < b^y$
$\log_b x > \log_b y$ if and only if $x > y$	$\log_b x < \log_b y$ if and only if $x < y$

When working with equations and inequalities involving logarithmic expressions, you must exclude any value from your solution set that makes an input of a logarithmic expression negative .

EXAMPLE 1 Modeling with Logarithmic Equations

Maria learns that the percentage of a girl's adult height at x years old can be modeled by the equation $f(x) = 62 + 35 \log_{10} (x - 4)$, where x represents the girl's age (from ages 5 to 15) and $f(x)$ represents the percentage of her adult height that she has grown so far.

a. **INTERPRET PROBLEMS** Maria's sister Danielle is 6 years old. Approximately what percentage of her adult height has Danielle grown so far? Explain how you found your answer. CCSS A.CED.1, SMP 1

b. **COMMUNICATE PRECISELY** Between which ages does a girl reach between 60–80% of her adult height? Write and solve a compound inequality to find the ages. Round to the nearest tenth of a year. Explain your answer. CCSS A.CED.1, SMP 6

c. **USE TOOLS** Explain how you can obtain the solution in **part b** graphically. Then sketch a graph of the solution. CCSS A.REI.11, SMP 5

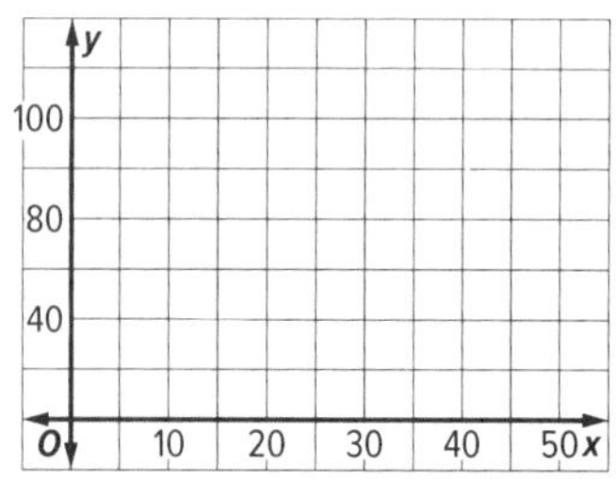

d. **CALCULATE ACCURATELY** Maria's other sister Carla is 5 feet 3 inches tall. This is 90% of her adult height. How old is she now? About how tall will she be as an adult? Explain how you found her adult height. CCSS A.CED.11, SMP 6

EXAMPLE 2 Modeling a Bird Population

The population of a certain species of bird is modeled by the function $f(t) = 2.2 + A\log_3 t$, where $t \geq 1$ is measured in years and $f(t)$ is measured in tens of thousands of birds. Assume that $t = 1$ corresponds to the year 2003.

a. **REASON ABSTRACTLY** If it is known that $f(4) = 5.2$, determine the exact value of A. Show your work. CCSS A.CED.1, SMP 2

b. **REASON ABSTRACTLY** The environmental committee for a town is interested in the time-span during which the population of this species of bird is between 80,000 and 100,000. Write a logarithmic inequality that can be used to find this time period. CCSS A.CED.1, SMP 2

c. **PLAN A SOLUTION** Solve your inequality from **part b** algebraically. Show your work. Round the highest and lowest value of the time-span to the nearest thousandth. CCSS A.CED.1, SMP 1

d. **CALCULATE ACCURATELY** Convert the highest and lowest value of the time-span in **part c** to year and month. Round each value to the nearest month. Explain your reasoning. CCSS A.CED.1, SMP 6

e. **COMMUNICATE PRECISELY** Explain how you could obtain the solution of the inequality in **part a** graphically. Sketch the graph. CCSS A.REI.11, SMP 6

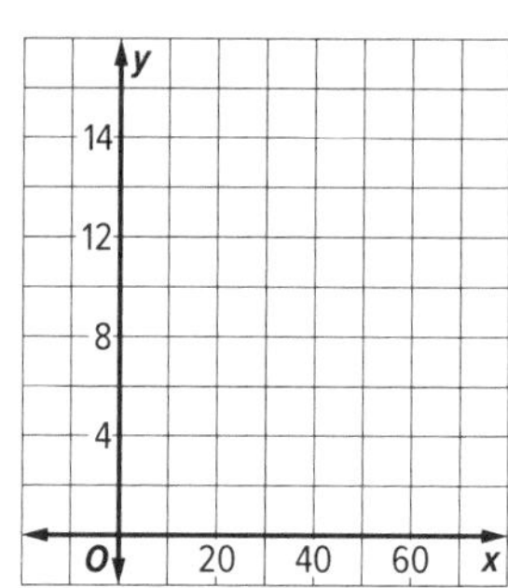

PRACTICE

1. The percentage of a boy's adult height at x years old can be modeled by the function $f(x) = 29 + 48.8 \log_{10}(x + 1)$, where x represents the boy's age and $f(x)$ represents the percentage of his adult height.

   a. **INTERPRET PROBLEMS** At what age does a boy reach 65% of his adult height? Round the age to the nearest tenth year. Explain your answer. CCSS A.CED.1, SMP 1

   b. **USE TOOLS** Explain how you can find the age of a boy when he has reached 75% of his adult height or greater using a graph. Write the inequality to represent the age range. Round to the nearest tenth of a year. CCSS A.REI.11, SMP 5

2. **CONSTRUCT ARGUMENTS** Consider the inequality $\log_4(x + b) \geq \log_4(x + 1)$, where b is a real number. Determine the solution set for this inequality in terms of b. Reason graphically showing all steps. CCSS A.REI.11, SMP 3

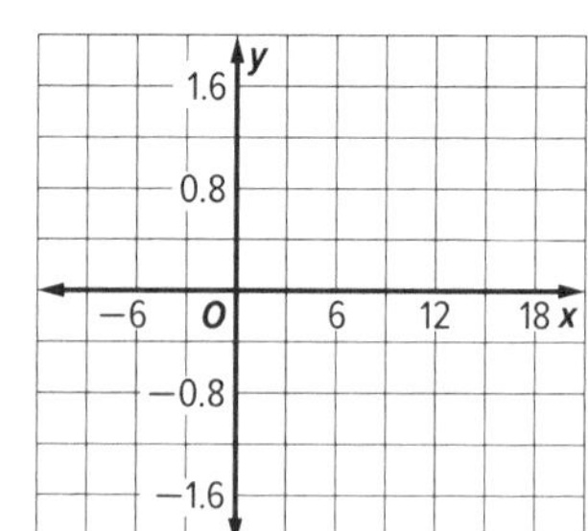

3. **CRITIQUE REASONING** Anvi and David solve the inequality $25 \log_{10}(x-9) \leq 38$. Is either of them correct? Explain your reasoning. CCSS A.CED.1, SMP 3

Anvi

$$25 \log_{10}(x-9) \leq 38$$
$$\frac{25 \log_{10}(x-9)}{25} \leq \frac{38}{25}$$
$$\log_{10}(x-9) \leq 1.52$$
$$x - 9 \leq 10^{1.52}$$
$$x - 9 \leq 33.11$$
$$x \leq 42.11$$

David

$$25 \log_{10}(x-9) \leq 38$$
$$\frac{25 \log_{10}(x-9)}{25} \leq \frac{38}{25}$$
$$\log_{10}(x-9) \leq 1.52$$
$$x - 9 \leq 10^{1.52}$$
$$x - 9 \leq 33.11$$
$$x \geq 24.11$$

4. **PLAN A SOLUTION** Explain the difference between solving the equations $\log_2(x+3) = 5$ and $\log_2(x+3) = \log_2 5$ and then solve each for x. CCSS A.CED.1, SMP 1

5. **INTERPRET PROBLEMS** As mentioned in the previous lesson, the loudness of sound is measured in a logarithmic scale using a unit called decibels (dB) $d = 10 \log_{10} \frac{I}{I_0}$, where d is the decibel rating, I is the power or intensity of the sound, and I_0 is the softest sound (the threshold sound) that the human ear can hear. CCSS A.CED.1, SMP 1

   a. Sound A has a decibel value of 25 and sound B has a decibel value of 68. How does the intensity of sound B compare to the intensity of sound A?

   b. If sound A is 5,000 times the intensity of sound B, what is their difference in decibel values?

7.5 Properties of Logarithms

Objectives

- Create and solve logarithmic equations.
- Apply properties of logarithms to solve equations.
- Use technology to find approximations of solutions to logarithmic equations.

CCSS STANDARDS

Content: A.CED.1
Practices: 1, 3, 4, 5, 7
Use with Lesson 7-5

KEY CONCEPT Properties of Logarithms

For all positive numbers a, b, c, and x, where $x \neq 1$:

$$\log_x ab = \log_x a + \log_x b$$

$$\log_x \frac{a}{b} = \log_x a - \log_x b$$

$$\log_x a^c = c \log_x a$$

EXAMPLE 1 Use Properties of Logarithms to Solve Equations CCSS A.CED.1

a. PLAN A SOLUTION Explain how to determine the values of a and b if $\log_2 a^3b^2 = 19$ and $\log_2 \frac{a^4}{b^5} = 10$. CCSS SMP 1

b. PLAN A SOLUTION Use a calculator to estimate the solutions to $3^{\log_{10} 2x} = x$. Explain how to use the properties of logarithms to solve the equation algebraically. CCSS SMP 1

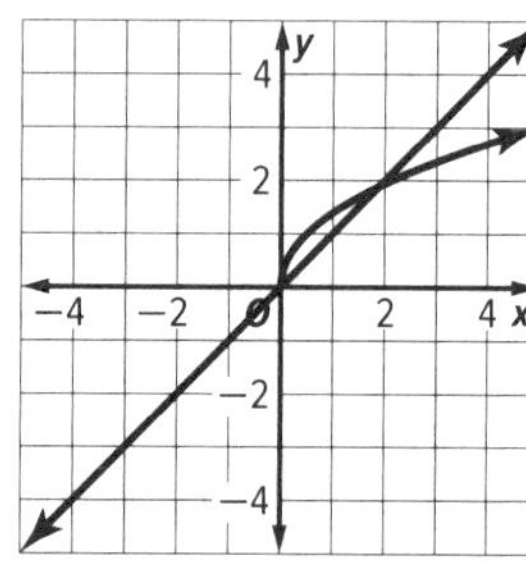

EXAMPLE 2 Create Logarithmic Equations CCSS A.CED.1

a. PLAN A SOLUTION Solve the equation $\log_2(x + 3) + \log_2 a = 3$ for a. If $a = x - 4$, find the values of x that are solutions of the equation. If a solution is extraneous, explain why. CCSS SMP 1

b. PLAN A SOLUTION What is the value of a in the equation $2\log_{10} a - \log_{10} (x - 3) = 2$ when $x = 7$? Explain your answer, and determine whether any solutions are extraneous. CCSS SMP 1

c. PLAN A SOLUTION Solve the equation $\log_6(x - 1) + \log_6 a = 2$ if $a = x - 1$. Explain your answer, and determine whether any of the solutions are extraneous. Does a graph of each side of the equation support your answer? CCSS SMP 1

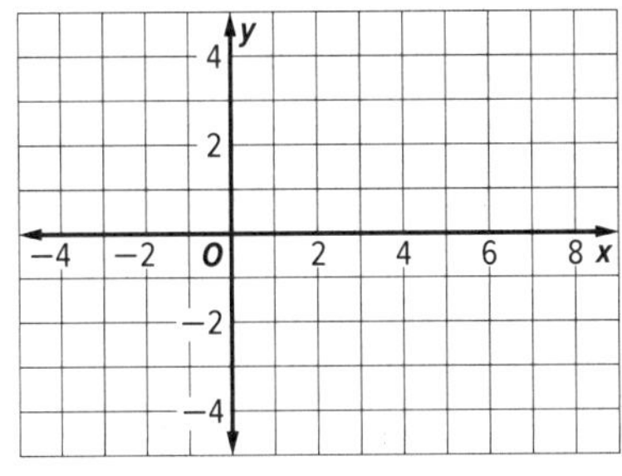

PRACTICE

1. **CRITIQUE REASONING** Consider the function $f(x) = \log_2 x$. Kayo believes that horizontally compressing the graph by a factor of 8 is exactly the same as translating the graph 3 units up. Is she correct? Explain your answer. CCSS A.CED.1, SMP 3

2. **USE STRUCTURE** If $\log_{10} 2 \approx 0.301$ and $\log_{10} 3 \approx 0.477$, explain how to determine the approximate solutions to $\log_{10} a = 0.903$ and $\log_{10} b = 2.477$, without using a calculator. CCSS A.CED.1, SMP 7

3. **PLAN A SOLUTION** Create a logarithmic equation in base 3 that has a true solution of $x = 3$ and an extraneous solution of $x = -9$. Explain your reasoning, and show how to confirm your answer graphically. CCSS A.CED.1, SMP 1

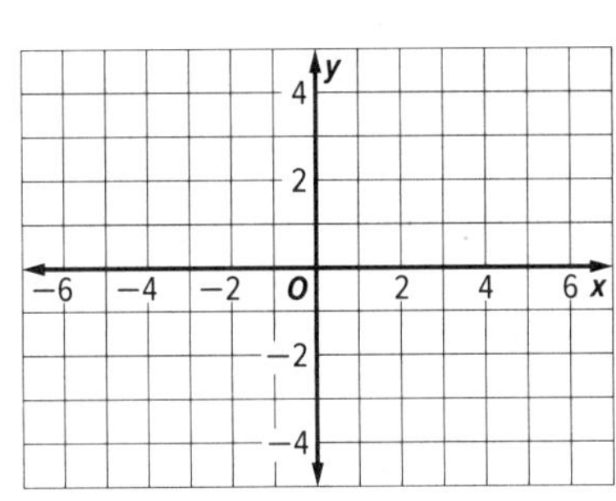

4. **USE STRUCTURE** If $\log_{10} 5 \approx 0.70$, create five logarithmic equations than can be solved with this information and without using a calculator. Give the solutions for each equation you create. *Hint: One possible equation is obtained by multiplying both sides of the equation $\log_{10} 5 = 0.70$ by 10 to get $\log_{10}(5^{10}) = 7$. Then replace 5^{10} with x.* CCSS A.CED.1, SMP 7

5. **USE TOOLS** Does $\log_5 x^2 = (\log_5 x)(\log_5 x)$ for all values of x? How can you prove or disprove this statement using a calculator? How can you prove or disprove this statement algebraically? CCSS A.CED.1, SMP 5

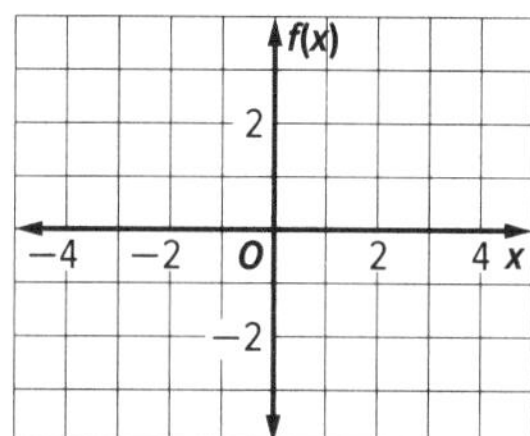

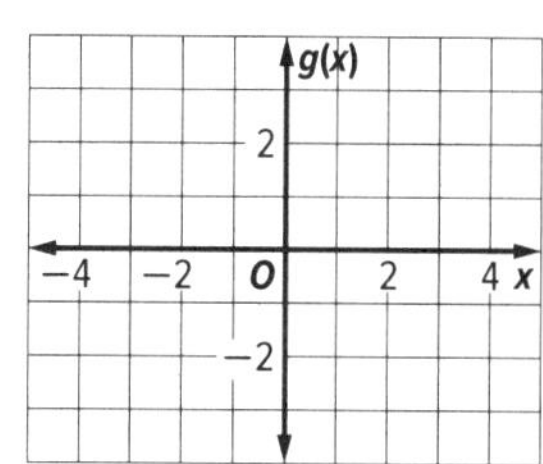

6. **PLAN A SOLUTION** If the equation $\log_4(x + 6) + \log_4 2a = 3$ has a solution of $x = 2$, what could a equal in terms of x? Explain your answer, and determine any other solutions. If a solution is extraneous, explain why. CCSS SMP 1

7. **PLAN A SOLUTION** If the equation $\log_8 a + \log_8 (x - 2) = 1$ has an extraneous solution of $x = 0$, what could a equal in terms of x? Explain your answer, and determine any valid solutions. CCSS SMP 1

8. **USE A MODEL** A population of insects is decreasing at a rate of 2.5% per year. Currently, the population is 250,000. Write a logarithmic function for time in years based upon the population of insects. Explain how to use the function to determine the number of years it will be until the population reaches 100,000. Round to the nearest year. CCSS A.CED.1, SMP 4

9. USE STRUCTURE Explain how to determine the solutions to $\frac{8}{\log_2 x} = 6 - \log_2 x$ algebraically. Then confirm your answer graphically. CCSS A.CED.1, SMP 7

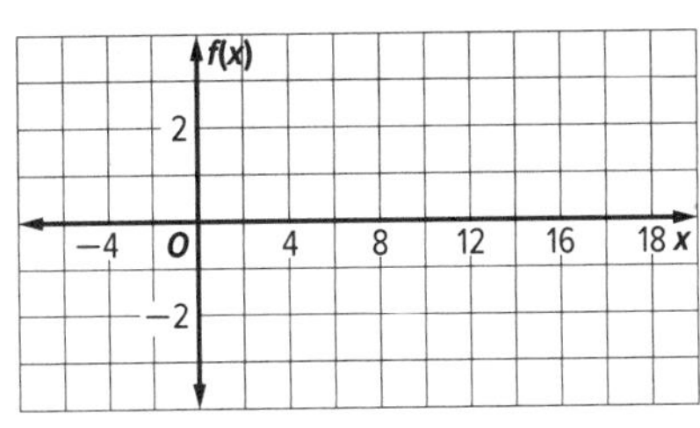

10. USE A MODEL A population of fruit flies is increasing at a rate of 5.3% per day. Currently, the population is 690. Write a logarithmic function for time in days based upon the population of fruit flies. Use the function to determine the approximate number of days that have passed if the population is 10,000. CCSS A.CED.1, SMP 4

11. PLAN A SOLUTION Shown is the graph of $f(x) = 2\log_{10}(2x)$ and $g(x) = 2 - \log_{10}(3x)$ as well as their point of intersection. Explain how to find the point of intersection algebraically. CCSS A.CED.1, SMP 1

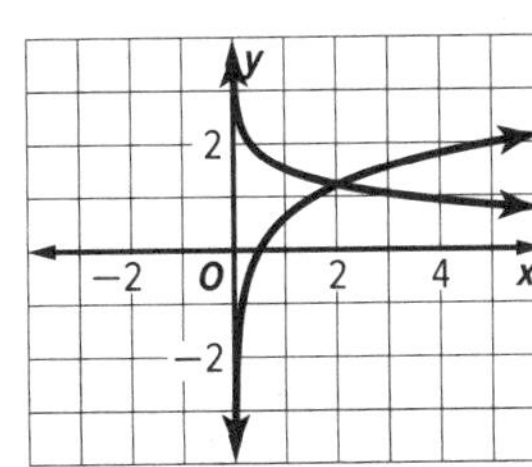

12. USE STRUCTURE Answer the questions about amounts of money invested into accounts compounded annually. CCSS A.CED.1, SMP 7

a. If \$5000 is initially deposited into an account that earns 4% compounded annually, write a logarithmic function $T(x)$ for the number of years it takes for the account to reach x dollars.

b. Suppose another account earns 2% interest compounded annually and that \$10,000 was initially deposited into the account. Write a logarithmic function $P(x)$ for the number of years it takes for the account to reach x dollars.

c. Is there an amount of money for which both accounts take the same amount of time to reach? Explain your reasoning.

7.6 Common Logarithms

Objectives

- Write and solve logarithmic equations and inequalities.
- Use the change of base formula to help simplify logarithmic expressions.

Content: A.SSE.2, A.CED.1
Practices: 1, 2, 3, 6, 7, 8
Use with Lesson 7-6

Base 10 logarithms are called **common logarithms** because they arise in a wide variety of applications. When dealing with such logarithms, the subscript is typically omitted. You can write equivalent logarithmic expressions with different bases using the change of base formula.

KEY CONCEPT Change of Base Formula

For positive real numbers *a*, *b*, and *n*, none of which equals 1,

$$\log_a n = \frac{\log_b n}{\log_b a}$$

EXAMPLE 1 Depreciation CCSS A.SSE.2, A.CED.1

EXPLORE Electronic devices depreciate in value over time because newer, improved models are manufactured regularly. Suppose that a certain device has a life expectancy of *E* years, has an initial cost of *D* dollars, and it depreciates to a value of *V* dollars after *n* years. These quantities are related by the equation $n = \frac{\log V - \log D}{\log\left(1 - \frac{2}{E}\right)}$.

a. Solve the equation for *E*. Show your work. CCSS SMP 2

b. **REASON ABSTRACTLY** Show that the number of years *n* it takes for the object to depreciate to the value *V* is given by the following formula, where *p* represents the percent of the initial cost *D*. Show your work. CCSS SMP 2

$$n = \frac{\log p - 2}{\log\left(1 - \frac{2}{E}\right)}.$$

c. **INTERPRET PROBLEMS** Suppose that $E = 14$ in **part b**. Express n as a function of p of the form $n = A \log p + B$, for appropriate real numbers A and B. Graph this function and label the axes appropriately. CCSS SMP 1

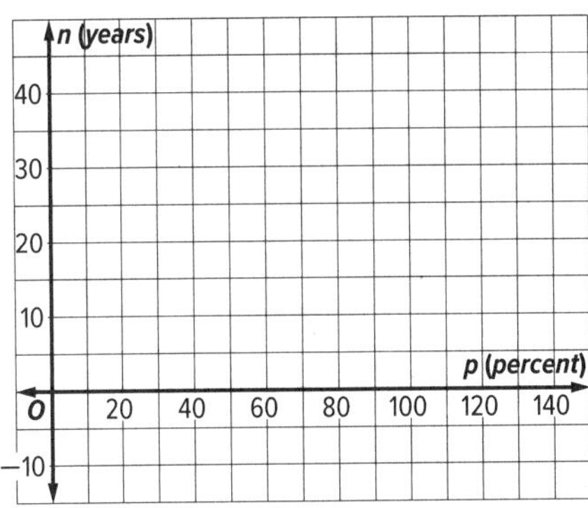

d. **INTERPRET PROBLEMS** What is the p-intercept of the graph? Why does this make sense in this context? CCSS SMP 1

e. **REASON QUANTITATIVELY** What is the domain of this function in this context? Discuss how realistic the model is. CCSS F.IF.5, SMP 2

EXAMPLE 2 Exponential Inequalities Involving a Parameter CCSS A.SSE.2

a. **REASON QUANTITATIVELY** Consider the inequality $5^{x+a} \leq 2^x$, where a is a real number. Show that the solution set for this inequality can be expressed as the set $\left\{x \mid x \leq \frac{a \log 5}{\log\left(\frac{2}{5}\right)}\right\}$. Show your work. CCSS SMP 2

b. **CONSTRUCT ARGUMENTS** Consider the inequality $b \cdot 5^{x+a} \leq 2^x$, where a and b are real numbers. Does this inequality have a nonempty solution set for all values of a and b? Explain your reasoning. For those values of a and b for which the solution set is nonempty, if any, express the solution set using common logarithms, if possible. CCSS SMP 3

PRACTICE

1. Biologists have found that the energy required to transport a substance from outside to inside of a certain type of living cell can be modeled by $E = 2.08(\log Y_1 - \log Y_0)$, measured in kilocalories per gram molecule, where Y_1 is the concentration of the substance inside the cell and Y_0 is the concentration of the substance outside the cell. 

   a. **REASON ABSTRACTLY** Show that $E = \frac{2.08}{\log_2 10} \cdot \log_2\left(\frac{Y_1}{Y_0}\right)$. Show your work. CCSS SMP 2

   b. **CALCULATE ACCURATELY** If $Y_1 = 2Y_0$, show that $E = 2.08 \log 2$. Show your work. CCSS SMP 6

   c. **DESCRIBE A METHOD** If $Y_1 = nY_0$, where n is a positive integer, determine a simplified expression for E similar to the one in **part b**. Show your work. CCSS SMP 8

   d. **REASON ABSTRACTLY** What is the smallest integer n for which the energy level is at least 3.51 kilocalories per gram-molecule? Explain your reasoning. CCSS A.CED.1, SMP 2

2. **REASON ABSTRACTLY** Consider the inequality $5^{ax} \leq 2^{-bx}$, where a and b are positive real numbers. Show that the solution set for this inequality does not depend on the values of a or b. Explain your reasoning. CCSS A.SSE.2, SMP 2

3. Claudia wants to find the value of $\log_7 8$ to the nearest thousandth. She uses the change of base formula and a calculator to find the approximation. The screen at right shows what she entered, and the result. CCSS A.SSE.2

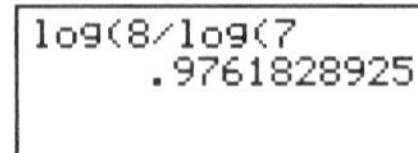

   a. **CRITIQUE REASONING** Did Claudia make a mistake? If so, explain how she can correct her mistake, referring to the calculator display. CCSS SMP 3

b. REASON QUANTITATIVELY Explain why the result of 0.976... should alert Claudia that she made some sort of mistake by considering the relationship between log 8 and log 7. CCSS SMP 2

4. INTERPRET PROBLEMS The equation $t = \frac{1}{\log(1 + r)}$ gives the number of years it takes \$1000 to increase to \$10,000 in a savings account earning annual interest rate r, compounded annually. If Jo wants the \$10,000 for retirement at age 62, when must she make the \$1000 deposit if she can earn 4.9% interest? Show your work or explain your reasoning. CCSS A.CED.1, SMP 1

5. USE STRUCTURE Rewrite the given equation using common logarithms and solve. CCSS A.SSE.2, SMP 7

a. $3\log_2(2x) - \log_4(3x) = 1$

b. $\log_6(x^2) + \log_3(x) = 3$

c. $2\log_2(3x) = -8\log_3(x)$

6. USE STRUCTURE Rewrite the given inequalities using common logarithms and solve. CCSS A.SSE.2, SMP 7

a. $\log_7(2x) < 2\log_6(x)$

b. $3\log_2(4x) \leq \log(8x)$

7.7 Base e and Natural Logarithms

Objectives

- Determine equivalent forms of base *e* and natural logarithm expressions.
- Use these equivalent forms to solve exponential and natural logarithm equations.

STANDARDS

Content: A.SSE.2
Practices: 1, 3, 4, 5, 6, 7
Use with Lesson 7-7

An exponential function is a function written in the form, $y = ab^x$, where $a \neq 0$ and $b > 0$ except $b \neq 1$. A **natural base exponential function** is an exponential function where b is the **natural base *e***, which is an irrational number approximately equal to 2.71828.... Natural base functions are used to show continuous exponential growth or decay, represented by $y = e^x$ or $y = e^{-x}$, respectively. The inverse of natural base e is the **natural logarithm** written as $\ln x$. One example of a natural base exponential function is the model for continuously compounded interest, $A(t) = Pe^{rt}$, where P is the principal or initial amount, r is the rate, t is time, and A is the amount at time t.

EXAMPLE 1 Using Natural Base Logarithms

EXPLORE Bradley invested in a retirement account that is compounded continuously. The table shows the amount of money he has accrued in the past five years, beginning at the end of the first full year of opening the account. CCSS A.SSE.2

# of Years since Opening Account	Amount of Money Accrued
1	\$3720.35
2	\$4226.26
3	\$4800.97
4	\$5453.83
5	\$6197.47

a. **PLAN A SOLUTION** Explain how to determine the amount of money Bradley initially deposited in the account and the interest rate, then find each value. CCSS SMP 1

b. **USE TOOLS** What interest rate would cause Bradley's deposit to double in five years? Round the percent to the nearest hundredth. CCSS SMP 5

c. **USE STRUCTURE** Explain why $\ln\left(\frac{4800.97}{3720.35}\right) + \ln\left(\frac{5453.83}{4800.97}\right) = \ln\left(\frac{5453.83}{3720.35}\right)$. CCSS SMP 7

EXAMPLE 2 Carbon Dating

Over time, the amount of carbon-14 present in non-living organic material decreases. Radiocarbon dating is a technique that measures the amount of carbon-14 that is present in these materials to estimate their age. The exponential function $P = P_0e^{kt}$, represents the relationship between the current amount P of carbon-14 in organic material and the initial amount P_0 of carbon-14 over time t with rate of decay k. CCSS A.SSE.2

a. **INTERPRET PROBLEMS** The amount of time it takes for the percent of carbon-14 present in dead organic material to decrease by half is 5730 years. This is called the half-life. Use this information to find the exponential rate of decay k associated with carbon-14. *Note: Don't simplify the natural logarithm in the answer until **part c**.* CCSS SMP 1

b. **USE TOOLS** A fossil initially contained 200 milligrams of carbon-14. Write an equation that represents the relationship between the age of the fossil and the amount of carbon-14 currently in the sample. Graph this relationship using a graphing calculator and sketch the graph. CCSS SMP 5

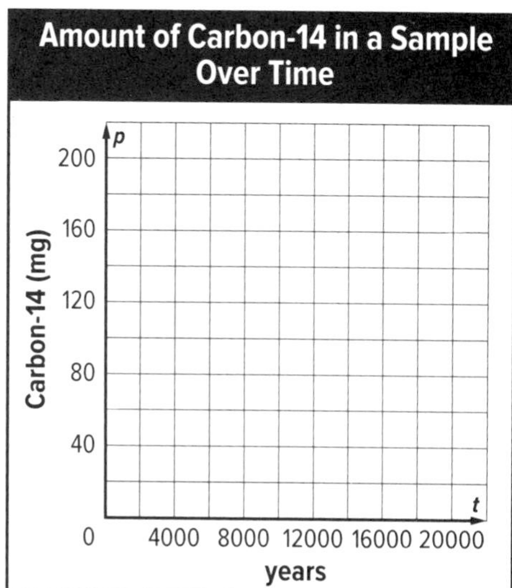

c. **USE A MODEL** Use the graph to estimate to the nearest year how long it will take the fossil to contain 40% or less of the carbon-14 it initially contained. Solve the exponential inequality. CCSS SMP 4

EXAMPLE 3 Exponential and Logarithmic Expressions and Equations CCSS A.SSE.2, SMP 6

CALCULATE ACCURATELY Write an equivalent exponential or logarithmic function.

a. $\ln x = e$

b. $e^{x+2} = 5$

c. $2 \ln 5 = x$

CALCULATE ACCURATELY Write each as a single logarithm.

d. $5 \ln \frac{2}{3} + 2 \ln x$

e. $2 \ln x^3 - 3 \ln 7$

f. $2 \ln (x + 1) + \frac{1}{3} \ln 8$

CALCULATE ACCURATELY Given $\ln 2 = 0.6931$ and $\ln 9 = 2.1972$, evaluate each expression without using a calculator to determine the natural log.

g. $\ln 36$

h. $\ln 2.25$

i. $\ln 1.5$

CALCULATE ACCURATELY Solve each equation or inequality. Round to the nearest ten-thousandth.

j. $4e^x + 2 < 18$

k. $\ln(2x+1)^5 \geq 14$

l. $9^x - 3^x = 6$

PRACTICE

1. A nuclear power plant is storing plutonium-238, a radioactive isotope with a half-life of 87.7 years, in the caves of a mountain. The plant has 2000 kg of the substance stored away.

a. **USE TOOLS** Write a function that represents the amount of plutonium-238 available over time. Use your graphing calculator to graph the equation and sketch the graph below. CCSS A.SSE.2, SMP 5

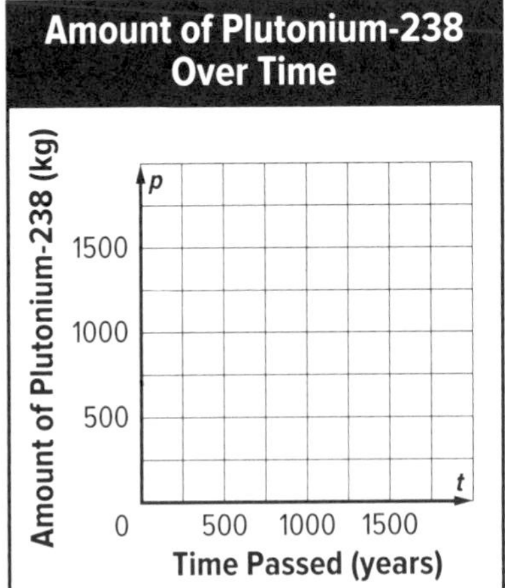

b. **USE A MODEL** Use the graph to estimate to the nearest year how long it will take for at least 90% of the original amount of plutonium-238 to decay. Then solve the inequality to find the answer. CCSS A.SSE.2, SMP 4

c. **COMMUNICATE PRECISELY** Another radioactive isotope, plutonium-239, has a half-life of 24,100 years. If the plant stored 20 kilograms of this isotope, after how many years will the amounts of the two types of plutonium be equal? *Note: Don't simplify the natural log until the final calculation.* CCSS A.SSE.2, SMP 6

2. **CALCULATE ACCURATELY** Write an equivalent exponential or logarithmic function. CCSS A.SSE.2, SMP 6

a. $e^2 = x^5$

b. $\ln(x-4) = 7$

c. $e^{3e} = x^3$

3. **CALCULATE ACCURATELY** Write each as a single logarithm. CCSS A.SSE.2, SMP 6

a. $3 \ln x - 2 \ln 2x$

b. $x \ln 25 + x \ln 5$

c. $8 \ln (x + 1) - 3 \ln (x + 1)$

4. **CALCULATE ACCURATELY** Given $\ln 5 = 1.6094$ and $\ln 8 = 2.0794$, evaluate each expression without using a calculator to determine the natural log. CCSS A.SSE.2, SMP 6

a. $\ln 200$

b. $\ln 3.125$

c. $\ln 10$

5. **CALCULATE ACCURATELY** Solve each equation or inequality. Round to the nearest ten-thousandth. CCSS A.SSE.2, SMP 6

a. $5e^{2x} - 4 > 11$

b. $-[\ln (x - 3)^{-2}] \leq 8$

c. $-5^{x+1} + 25^{x} = -6$

6. **CONSTRUCT ARGUMENTS** Gwen says that $e^{e^{x}} = x$. Do you agree or disagree? Justify your answer. CCSS A.SSE.2, SMP 3

7. **CALCULATE ACCURATELY** Suppose $10,000 was initially deposited into an account that is compounded continuously. If the amount after 5 years is $13,438, what is the interest rate of the account? Round to the nearest tenth of a percent. CCSS A.SSE.2, SMP 6

8. **USE STRUCTURE** If $5000 is invested into an account that earns 2% compounded continuously and at the same time $3000 is invested into an account that earns 4% compounded continuously, at what time will the two accounts contain the same amount of money? CCSS A.SSE.2, SMP 7

9. **USE STRUCTURE** Suppose $1000 is deposited in an account that earns 1.5% compounded continuously. Write a logarithmic equation for the time *t* in years it takes to reach the amount A. CCSS A.SSE.2, SMP 7

7.8 Modeling: Exponential and Logarithmic Functions

Objectives

- Use formulas to solve problems.
- Interpret expressions for exponential and logarithmic functions.
- Evaluate logarithms.

CCSS STANDARDS

Content: F.IF.4, F.IF.8b, F.LE.4, A.CED.1, A.CED.2, A.CED.3, A.SSE.1a, A.SSE.1b, A.SSE.2

Practices: 1, 2, 3, 4, 5, 6

Use with Lesson 7-8

Exponential Growth	Exponential Decay
Exponential growth can be modeled by: $f(x) = ae^{kt}$ where a is the initial value, t is the time in years, and k is a constant representing the rate of continuous growth.	Exponential decay can be modeled by: $f(x) = ae^{-kt}$ where a is the initial value, t is the time in years, and k is a constant representing the rate of decay.

EXAMPLE 1 Exponential Decay Model

EXPLORE A biology experiment starts with 1,000,000 cells and 30% of the cells are dying every minute. How long will it take to have less than 1000 cells?

a. **INTERPRET PROBLEMS** Complete the table of values and graph the points. Determine what kind of mathematical model best describes the points. CCSS F.IF.4, SMP 1

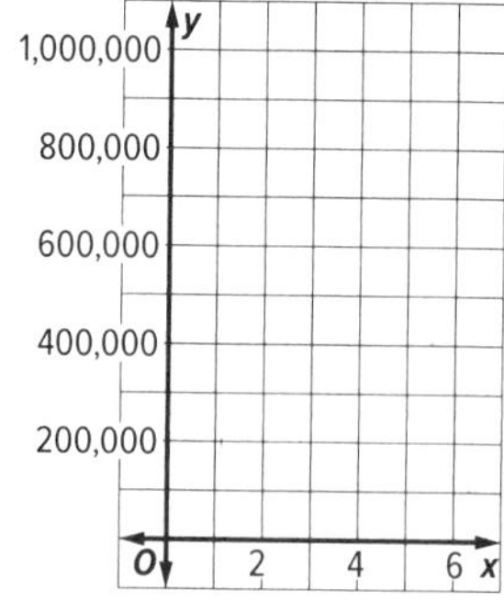

t (min)	30% dying, therefore 70% survive	$(t, f(t))$
0	Initial amount	(0, 1,000,000)
1	(0.70)(1,000,000) = 700,000 survive	(1, 700,000)
2	(0.70)(700,000) =	
3		

b. **COMMUNICATE PRECISELY** Using the exponential formula, define what each variable represents for this biology experiment. Rewrite the exponential formula with the information that is known. CCSS A.SSE.1a, A.SSE.2, SMP 6

c. **REASON QUANTITATIVELY** Find the value of the constant k, and tell whether it indicates growth or decay. From the table of values it is known that after 1 minute there are 700,000 cells left. Use this information to substitute into the function and solve for k to 6 decimal places. Write the model for this experiment. CCSS A.SSE.1b, SMP 2

d. **CALCULATE ACCURATELY** How long will it take to have less than 1000 cells? CCSS F.LE.4, SMP 6

EXAMPLE 2 Exponential Growth Model

The exponential growth model $f(t) = 250e^{0.01753t}$ describes the population of a city in the United States, in thousands, t years after 1992.

a. **USE A MODEL** Using the given exponential model, which variable represents population? What was the population of the city in 1992? CCSS A.SSE.1a, SMP 4

b. **USE A MODEL** By what percent does the population of the city increase each year in the given growth model? CCSS A.SSE.2, A.CED.1, F.IF.8b, SMP 4

c. **REASON QUANTITATIVELY** What was the population of the city in 2003? When will the city's population be 500,000? CCSS F.LE.4, SMP 2

d. **COMMUNICATE PRECISELY** Give the end behavior of $f(t)$. Based on the end behavior, is $f(t)$ an accurate model? Explain your reasoning. CCSS F.IF.4, SMP 6

Logistic Growth Function

Let a, b, and c be positive constants, where $b < 1$. The logistic growth function is represented by $f(t) = \frac{c}{1 + ae^{-bt}}$, where t represents time.

Exponential growth is unrestricted, meaning it will increase without bound. A logistic growth model represents growth that has a limiting factor.

The graphs of logistic growth functions have the following characteristics:

- The horizontal lines $y = 0$ and $y = c$ are asymptotes.
- The y-intercept is $\frac{c}{1 + a}$.
- The domain is all real numbers, and the range is $0 < y < c$.
- The graph is increasing from left to right.
- The point of maximum growth is $\left(\frac{\ln a}{b}, \frac{c}{2}\right)$.

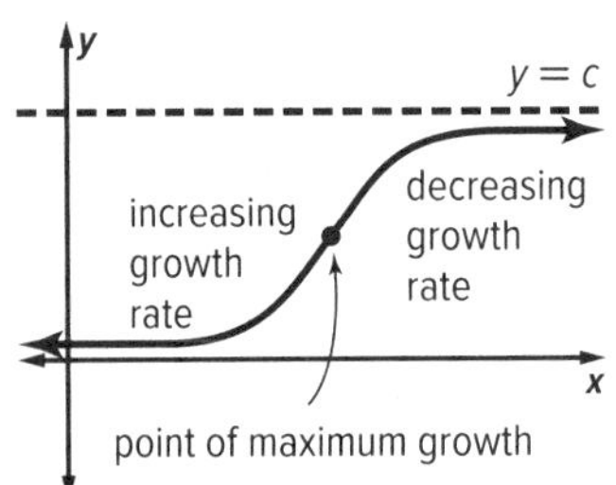

EXAMPLE 3 Logistic Growth Function CCSS A.CED.3

The logistic growth function $f(t) = \frac{400}{1 + 9e^{-0.22t}}$ describes the population of a species of grasshoppers introduced to a habitat, where t represents time in months.

a. COMMUNICATE PRECISELY What are the asymptotes of the function? What do they represent? CCSS A.SSE.1a, A.SSE.1b, SMP 6

b. USE A MODEL Find the y-intercept and the point of maximum growth. What do they represent? CCSS F.IF.4, SMP 4

c. USE A MODEL Sketch the graph of this function. Label asymptotes, y-intercepts, maximum growth point, where growth is increasing and growth is decreasing. CCSS F.IF.4, SMP 4

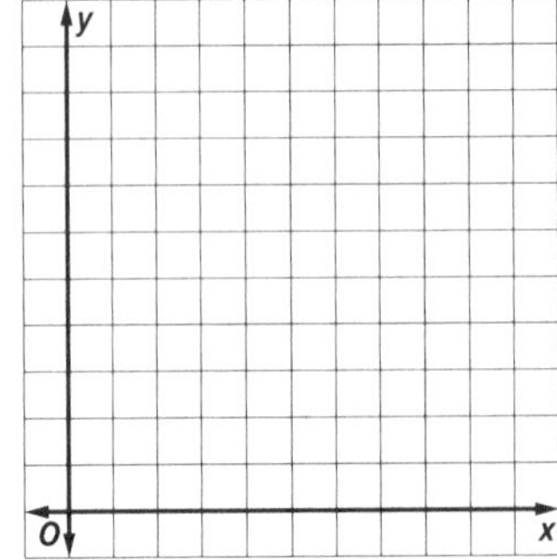

d. CALCULATE ACCURATELY How many grasshoppers are expected in the habitat in 6 months? CCSS SMP 6

PRACTICE

1. USE TOOLS Once an organism dies, Carbon-14 decays. Find the decay constant and the model that will represent this data. Verify your answer. CCSS A.CED.2, F.LE.4, SMP 5

t (years)	*f*(*t*)
0	1000
1	999.876
2	999.752
3	999.628

2. The logistic growth function $f(t) = \frac{320}{1 + 12.4e^{-.25t}}$ describes the population of a species of spider *t* months after they are introduced to a new zoo habitat.

a. USE A MODEL Sketch the graph of the function, labeling asymptotes, *y*-intercept, and maximum growth point. CCSS F.IF.4, SMP 4

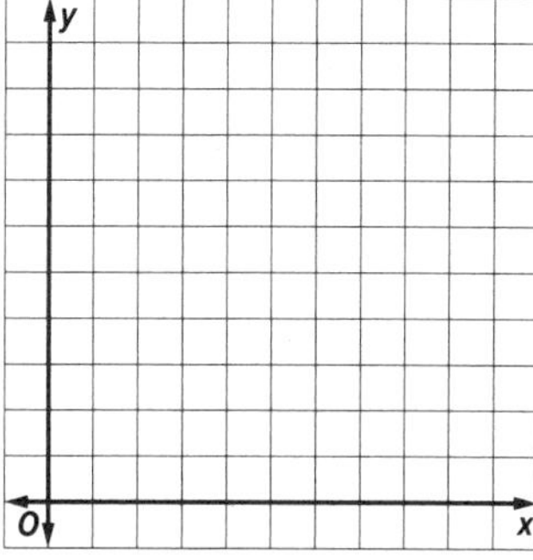

b. USE A MODEL How many spiders were initially introduced into the habitat, how do you know? CCSS F.IF.4, SMP 4

c. USE TOOLS Write an equation to solve for the number of months *t* it takes for the spider population to reach 200. Use a graphing calculator to solve the equation graphically. CCSS A.CED.1, SMP 5

3. **USE A MODEL** Cholera, an intestinal disease, is caused by cholera bacterium. Transmission occurs primarily by drinking water or eating food that has been contaminated. This bacteria multiplies exponentially by cell division as modeled by $f(t) = Ne^{1.386t}$, where $f(t)$ is the number of bacteria present after t hours and N is the number of bacteria present at $t = 0$. About 100 million bacteria must typically be ingested to cause cholera in a healthy adult. Fill in the table

t (hours)	$f(t)$
0	1
2	
4	
6	
8	
10	

Due to nearly universal advanced water treatment and sanitation practices, cholera is no longer a major health concern. According to the data, is this a fast multiplying disease? Why do you think it is important to understand how fast or slow a disease multiplies? CCSS F.IF.4, SMP 4

4. On the Richter scale, the magnitude R of an earthquake of intensity I is $R = \log \frac{I}{I_0}$, where $I_0 = 1$ is the minimum intensity used for comparison. Find the intensities per unit of area for the following earthquakes. Compare the intensities. CCSS A.SSE.2

   a. **MAKE A CONJECTURE** The 1960 earthquake in Chile where $R = 9.5$ CCSS SMP 3

   b. **MAKE A CONJECTURE** The 1923 earthquake in Japan where $R = 7.9$ CCSS SMP 3

   c. **REASON QUANTITATIVELY** There are 1,000,000 microns in one meter, or 1 meter $= 10^6$ microns. If the minimum intensity used for comparison is one micron, find the magnitude R of an earthquake with intensity 2 meters. CCSS SMP 2

   d. **REASON QUANTITATIVELY** If one earthquake is ten times the intensity of another earthquake, find the difference of their magnitudes on the Richter scale. CCSS SMP 2

5. The logistic growth function $g(t) = \frac{16000}{1 + 79e^{-0.57t}}$ describes the population of mice t months after they are introduced to a new habitat. CCSS F.IF.4

a. **COMMUNICATE PRECISELY** What are the asymptotes of the function? What do they represent? CCSS SMP 6

b. **USE A MODEL** Find the y-intercept and the point of maximum growth. What do they represent? CCSS SMP 4

c. **USE A MODEL** Sketch the graph of this function. Label asymptotes, y-intercepts, and the maximum growth point. CCSS SMP 4

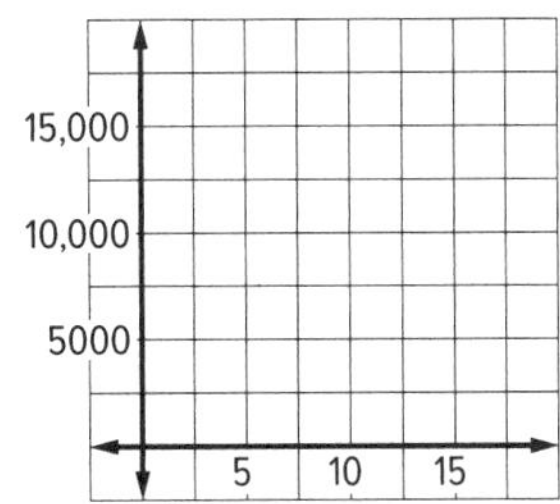

d. **USE A MODEL** What was the population of mice 10 months after they were introduced? Use the graph to estimate when the mice population will be 5000. CCSS SMP 4

6. **REASON QUANTITATIVELY** A population of birds can be modeled using a logistic growth function $f(t) = \frac{c}{1 + ae^{-bt}}$ where t represents time in years. CCSS A.CED.1, SMP 2

a. The initial population of birds is 20 and the point at which the population is at maximum growth is at 5 years when the population is 1000. Find the constants a, b, and c and give the logistic growth function $f(t)$.

b. As time goes on, what number will the population of birds approach? Explain your reasoning.

Performance Task

Modeling Bacterial Growth

Provide a clear solution to the problem. Be sure to show all of your work, include all relevant drawings, and justify your answers.

You are in charge of a team of scientists studying a new disease caused by Bacteria Q. You divided your team into two groups and asked each to model how a population of Bacteria Q grows with time. The teams derived the equations $Q_1(t) = 150e^{0.03t}$ and $Q_2(t) = 170e^{0.025t}$, where t represents time in hours and Q represents bacterial population.

Part A

Use a graphing calculator to help sketch the functions $Q_1(t)$ and $Q_2(t)$ on the grid below. Describe what the parameters in each equation represent. Which is the more important parameter? Explain.

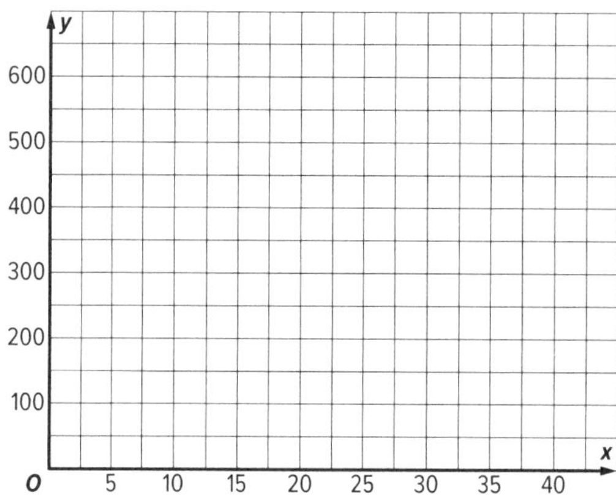

Part B

Given the two models, at what time do the teams predict their populations will be the same? What is that population? Find the exact answer and an approximation.

Part C

You conduct your own study, starting with 200 bacteria, and find that in 10 hours there are 257 bacteria. Which of the teams' models, if either, is supported by this outcome? Explain.

Performance Task

Logarithms On Shaky Ground

Provide a clear solution to the problem. Be sure to show all of your work, include all relevant drawings, and justify your answers.

The Richter Scale is used for measuring the magnitude of earthquakes. The magnitude of an earthquake is, by definition, measured 100 kilometers from its originating point, and is given by the equation $M = \log\frac{S}{10^{-4}}$, where S is the "strength" of the earthquake, as determined by the shockwaves it sends through the ground. Earthquake strengths vary greatly, from 0 in some cases to 800,000,000 or more in other cases, which is why a logarithmic scale is used to measure their magnitudes.

Part A

One of the strongest earthquakes ever recorded had a magnitude of 8.9 on the Richter Scale.

i) Determine the strength of this earthquake.

ii) What would be the magnitude of an earthquake four times weaker than this one?

iii) How many times stronger was this earthquake than the magnitude 2 earthquake that struck central New Jersey in 2012?

Part B

Suppose a given earthquake is twice as strong as another earthquake. Determine the difference in magnitude of these earthquakes.

Part C

The largest manmade explosion was the test detonation of the *Tsar Bomba* hydrogen bomb. It was equivalent to about 50 million tons of TNT. It was similar in intensity to a magnitude 8.5 earthquake. The strongest earthquake ever recorded was the 1960 Valdivia earthquake in Chile, with a magnitude of 9.5. How much stronger was this earthquake than the *Tsar Bomba*? How much TNT was the earthquake equivalent to? Show your work.

Standardized Test Practice

1. Which function is shown in the graph? CCSS F.IF.7e

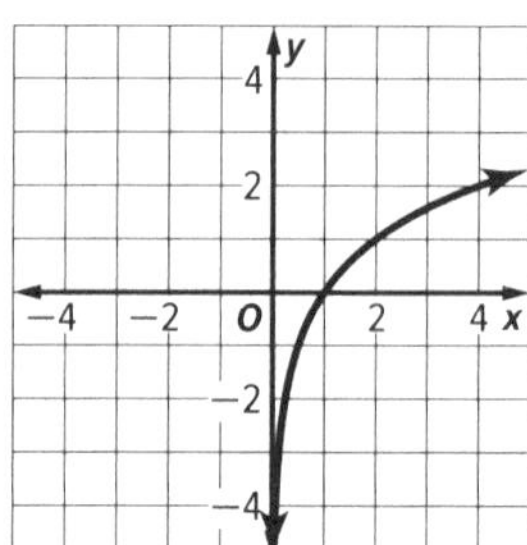

$f(x) = \log x$ $f(x) = \log_4 x$

$f(x) = \log_2 x$ $f(x) = \sqrt{x}$

2. Evaluate the following logarithms. CCSS A.SSE.2

$\log_{0.5} 0.125 = \square$

$\log_7 \left(\frac{1}{49}\right) = \square$

3. Solve for x. If necessary, round to the nearest thousandth. CCSS A.CED.1

$81^{x-1} = 27^{x+2}$ $\square$

$6.2^{x-3} = 8.9$ $\square$

$\left(\frac{1}{25}\right)^x = 125^{x+1}$ $\square$

4. The function $dB = 10\log\left(\frac{I}{I_0}\right)$ gives the loudness of a sound in decibels, where I is the intensity of the sound and I_0 is the intensity of threshold sound. The loudness of a dog's bark was found to be 72 decibels. How many times more intense is the dog's bark than threshold sound? CCSS A.CED.1

about $\square$ times as intense

5. Graph $f(x) = \left(\frac{1}{2}\right)^x$ and $g(x) = 3^x$. CCSS F.IF.1

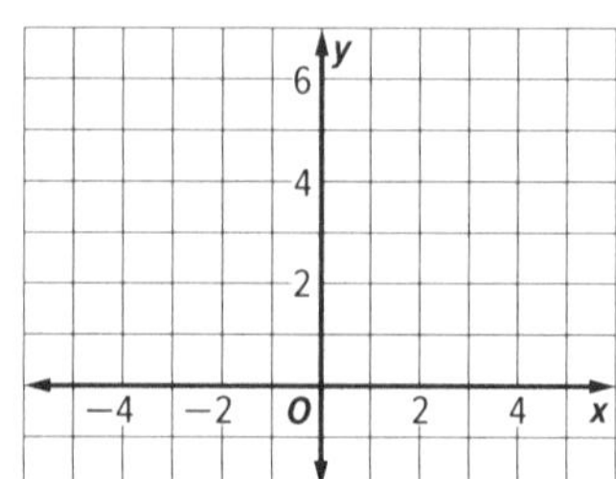

What is the point of intersection of these two functions? CCSS A.REI.11

$\square$

6. Solve for x. Give an exact value and a value rounded to the nearest hundredth. CCSS A.SSE.2

$\ln 2x - \ln (x - 3) = 1$

$x = \square$ $x \approx \square$

7. The graph of $y = 8e^{kx}$ passes through the point $(4, 20)$. What is the exact value of k? CCSS F.LE.4

$k =$

8. Write an equivalent expression for $\ln \frac{8}{9}$ in terms of $\ln 2$ and $\ln 3$. CCSS A.SSE.2

$\square$

9. Let $f(x) = 3 \log_4(x - 2)$. Which function results in a translation of $f(x)$ 2 units to the left and 1 unit down? CCSS F.BF.3

$h(x) = 3 \log_4(x - 4) - 1$ $d(x) = 3 \log_4(x - 3) + 2$

$g(x) = 3 \log_4(x) - 1$ $r(x) = 3 \log_4(x - 1) - 2$

10. Graph $f(x) = \log(x - 1)$ and identify the vertical asymptote. CCSS F.IF.7e

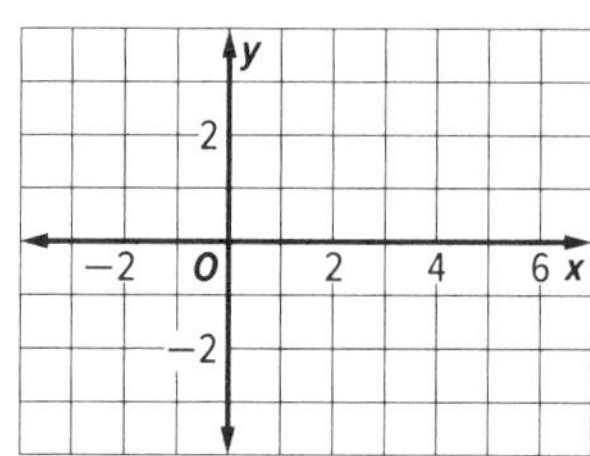

There is a vertical asymptote at $\square$.

11. Which function could be used to model the data shown in the table? CCSS A.CED.1

$A(t) = 2(4)^t$

$A(t) = 4\log_2(t)$

$A(t) = 4(t + 1)^2$

$A(t) = 4(2)^t$

Time	Amount
0	4
1	8
2	16
3	32

12. State the end behavior of $g(x) = -5(0.5)^x$. CCSS F.IF.7e

The end behavior is $g(x) \to$ ☐ as $x \to -\infty$ and $g(x) \to$ ☐ as $x \to \infty$.

13. Consider each equation or inequality in the table for $a > 0$ and $b > 0$, and determine whether it is true or false. Check True or False in each row. CCSS A.CED.3

Equation or Inequality	True	False
$\log_b a = \frac{1}{\log_a b}$		
If $a > b$ and $x \neq 0$, then $\log_a x > \log_b x$		
If $\log_b x < 0$, then $x < 1$		

14. The table shows the population of a town after it was incorporated. CCSS F.IF.4

Years Since Incorporation	2	5	8	10	15	18
Population	6329	7103	7289	7412	7609	7715

a. Does an exponential or logarithmic model better fit the data? Explain your reasoning.

b. What is an equation for the better model?

c. Use your model to predict the change in population between 20 and 25 years after incorporation.

15. The half-life of Cadmium-113 is 14.1 years. CCSS A.CED.1

a. What is the constant of decay for Cadmium-113? Show your work.

b. How long would it take for a 16-gram sample of Cadmium-113 to decay to 5 grams? Show your work.

c. Explain why you think your answer for **part b** is reasonable.

16. An amount P of money is invested in an account with an annual interest rate of r compounded continuously. CCSS A.CED.1

a. Write a function for the amount A of money in the account after t years if $P = \$10{,}000$ and $r = 2\%$.

b. Write and solve an equation for the time t it will take for the \$10,000 invested in **part a** to reach \$25,000.

c. If $P = \$5000$ is invested in the account and the amount grows to \$10,000 in 16 years, find the interest rate of the account.

8 Rational Functions

CHAPTER FOCUS Learn about some of the Common Core State Standards that you will explore in this chapter. Answer the preview questions. As you complete each lesson, return to these pages to check your work.

What You Will Learn	Preview Question
Lesson 8.1: Multiplying and Dividing Rational Expressions	
CCSS A.APR.7 Understand that rational expressions form a system analogous to the rational numbers, closed under addition, subtraction, multiplication, and division by a nonzero rational expression; add, subtract, multiply, and divide rational expressions.	CCSS SMP 6 Multiply these rational expressions. $\frac{x+3}{3x-1} \cdot \frac{x+1}{x+3} \cdot \frac{x+2}{6x+4}$
Lesson 8.2: Adding and Subtracting Rational Expressions	
CCSS A.APR.7 Understand that rational expressions form a system analogous to the rational numbers, closed under addition, subtraction, multiplication, and division by a nonzero rational expression; add, subtract, multiply, and divide rational expressions.	CCSS SMP 7 What is the perimeter of the rectangle? 
Lesson 8.3: Graphing Reciprocal Functions	
CCSS A.CED.2 Create equations in two or more variables to represent relationships between quantities; graph equations on coordinate axes with labels and scales. CCSS F.IF.5 Relate the domain of a function to its graph and, where applicable, to the quantitative relationship it describes. CCSS F.BF.3 Identify the effect on the graph of replacing $f(x)$ by $f(x) + k$, $k\,f(x)$, $f(kx)$, and $f(x + k)$ for specific values of k (both positive and negative); find the value of k given the graphs. Experiment with cases and illustrate an explanation of the effects on the graph using technology. Include recognizing even and odd functions from their graphs and algebraic expressions for them. **Also addresses:** F.IF.4, F.IF.6	CCSS SMP 7 The graph of $f(x) = \frac{1}{x}$ is shown below. Describe how the graph of $g(x) = \frac{2}{x+1} - 4$ is related to the graph of $f(x)$. Then graph $g(x)$. 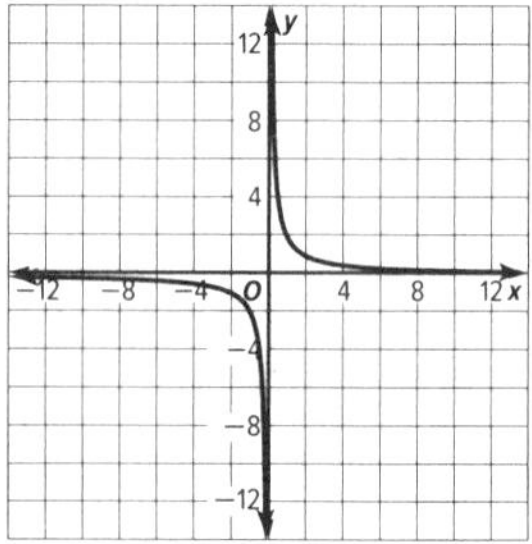

What You Will Learn	Preview Question
Lesson 8.4: Graphing Rational Functions	
CCSS F.IF.9 Compare properties of two functions each represented in a different way (algebraically, graphically, numerically in tables, or by verbal descriptions). For example, given a graph of one quadratic function and an algebraic expression for another, say which has the larger maximum. CCSS A.CED.2 Create equations in two or more variables to represent relationships between quantities; graph equations on coordinate axes with labels and scales. **Also addresses:** A.APR.7, F.IF.4, F.IF.5	CCSS SMP 6 Which function approaches $y = 0.5$ as x approaches infinity? Explain. A. $f(x) = \frac{2 - \frac{1}{x}}{4 - \frac{2}{x}}$ B. $f(x) = \frac{4 - \frac{1}{x}}{2 - \frac{2}{x}}$
Lesson 8.5: Modeling: Rational Functions	
CCSS A.SSE.1b Combine standard function types using arithmetic operations. For example, build a function that models the temperature of a cooling body by adding a constant function to a decaying exponential, and relate these functions to the model. CCSS F.IF.4 For a function that models a relationship between two quantities, interpret key features of graphs and tables in terms of the quantities, and sketch graphs showing key features given a verbal description of the relationship. **Also addresses:** A.SSE.1a, A.SSE.2, A.CED.4, F.IF.5, F.BF.1b, F.BF.4a	CCSS SMP 7 Use the diagram below to define a function for finding the dimensions of a rectangle of constant area. Show y as a function of x. 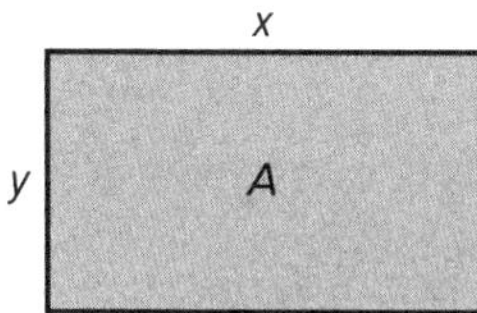
Lesson 8.6: Solving Rational Equations and Inequalities	
CCSS A.REI.2 Solve simple rational and radical equations in one variable, and give examples showing how extraneous solutions may arise. CCSS A.CED.1 Create equations and inequalities in one variable and use them to solve problems. CCSS A.REI.11 Explain why the x-coordinates of the points where the graphs of the equations $y = f(x)$ and $y = g(x)$ intersect are the solutions of the equation $f(x) = g(x)$; find the solutions approximately, e.g., using technology to graph the functions, make tables of values, or find successive approximations. Include cases where $f(x)$ and/or $g(x)$ are linear, polynomial, rational, absolute value, exponential, and logarithmic functions.	CCSS SMP 2 A region of area A is divided into two fractional amounts, where one fraction is half the other. Write the rational equation this represents.

8.1 Multiplying and Dividing Rational Expressions

Objectives

- Understand that rational expressions are closed under multiplication and division.
- Simplify rational expressions.
- Multiply and divide rational expressions

Content: A.APR.7
Practices: 1, 2, 3, 4, 6, 7
Use with Lesson 8-1

A **rational expression** is a ratio of two polynomial expressions. Operations with rational numbers and rational expressions are similar.

Multiplying and Dividing Rational Expressions

To **multiply** two rational expressions, multiply the numerators and multiply the denominators.
For all rational expressions, $\frac{a}{b}$ and $\frac{c}{d}$ with $b \neq 0$ and $d \neq 0$, $\frac{a}{b} \cdot \frac{c}{d} = \frac{ac}{bd}$.

To **divide** two rational expressions, multiply by the reciprocal of the divisor.
For all rational expressions, $\frac{a}{b}$ and $\frac{c}{d}$ with $b \neq 0$, $c \neq 0$, and $d \neq 0$, $\frac{a}{b} \div \frac{c}{d} = \frac{a}{b} \cdot \frac{d}{c} = \frac{ad}{bc}$.

EXAMPLE 1 Investigate Rational Expressions CCSS A.APR.7

EXPLORE In this exploration, you will explore how to multiply and divide rational expressions.

a. USE STRUCTURE Let $\frac{a}{b} = \frac{6}{x-2}$ and $\frac{c}{d} = \frac{3}{x+2}$. Find $\frac{a}{b} \cdot \frac{c}{d}$. Is the result a rational expression? State the restriction(s) that should be placed on x and why they exist. CCSS SMP 7

b. USE STRUCTURE Let $\frac{a}{b} = \frac{6}{x-2}$ and $\frac{c}{d} = \frac{3}{x+2}$. Find $\frac{a}{b} \div \frac{c}{d}$. Is the result a rational expression? State the restriction(s) that should be placed on x and why they exist. CCSS SMP 7

c. REASON ABSTRACTLY Look at the Key Concept box above for multiplying $\frac{a}{b} \cdot \frac{c}{d}$. Is the result a rational expression? What does this indicate about the closure of rational expressions under multiplication? Explain. CCSS SMP 2

d. REASON ABSTRACTLY Look at the Key Concept box above for $\frac{a}{b} \div \frac{c}{d}$. Is the result a rational expression? What does this indicate about the closure of rational expressions under division? Explain. CCSS SMP 2

e. **CONSTRUCT ARGUMENTS** Is the set of rational numbers closed under multiplication and division? Explain how you know. CCSS SMP 3

EXAMPLE 2 Simplify Rational Expressions CCSS A.APR.7

Simplify $\frac{5x^2 - 20x}{x^2 - x - 12}$.

a. **USE STRUCTURE** Write the expression in simplified form. Explain how multiplication of rational expressions is used to simplify the expression. CCSS SMP 7

b. **REASON QUANTITATIVELY** When is this rational expression undefined? Explain. How does this compare to how you know when a rational number is undefined? CCSS SMP 2

c. **INTERPRET PROBLEMS** Simplify $\frac{20x - 5x^2}{x^2 - x - 12}$. Explain how the original expression and this expression are related. CCSS SMP 1

EXAMPLE 3 Multiply Rational Expressions CCSS A.APR.7

Simplify $\frac{9x^2 - 25}{12x^3 + 21x^2 - 6x} \cdot \frac{x^3 - x^2 - 6x}{6x^2 - x - 15}$.

a. **CRITIQUE REASONING** Mr. Li and his daughter Julie both started to multiply these expressions. Their work is shown in the table below. Which method could lead to the correct answer? Explain why. CCSS SMP 3

Mr. Li's Method	Julie's Method
$\frac{9x^2 - 25}{12x^3 + 21x^2 - 6x} \cdot \frac{x^3 - x^2 - 6x}{6x^2 - x - 15} = \frac{9 - 25}{12x^3 + 21 - 6x} \cdot \frac{x^3}{-15}$	$\frac{9x^2 - 25}{12x^3 + 21x^2 - 6x} \cdot \frac{x^3 - x^2 - 6x}{6x^2 - x - 15} = \frac{(3x - 5)(3x - 5)}{3x(x + 2)(4x - 1)} \cdot \frac{x(x - 3)(x + 2)}{(2x + 3)(3x - 5)}$

b. CALCULATE ACCURATELY Simplify the expression. Explain why you would not want to multiply the numerators and multiply the denominators without factoring first. CCSS SMP 6

EXAMPLE 4 Divide Rational Expressions CCSS A.APR.7

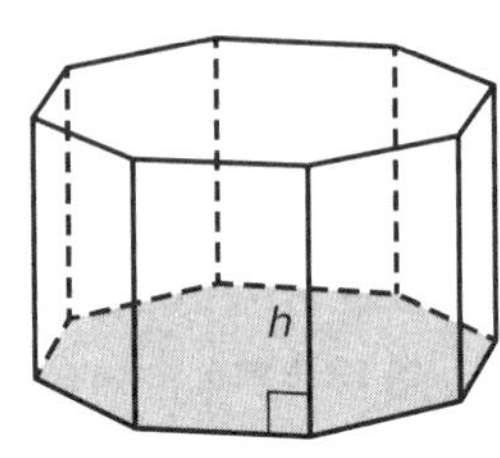

The volume of the prism shown is $\frac{x^4 - 1}{x^3 + 27}$ ft³. The area of the prism's base, B, is $\frac{x^3 + 3x^2 + x + 3}{2x^3 + 11x^2 + 12x - 9}$ ft², and its height is h. The formula for finding the volume of any prism is $V = Bh$.

a. PLAN A SOLUTION Write an expression that relates the volume and area of the base to the height of the prism. How does knowing how to solve for an unknown value in a formula involving rational numbers help you in this situation? CCSS SMP 1

b. CALCULATE ACCURATELY Find the height of the prism. How is the method you used similar to finding the height of a prism in which B and V are rational numbers? CCSS SMP 6

c. COMMUNICATE PRECISELY Can the value of x be -3? -2? Explain why or why not. CCSS SMP 6

A **complex fraction** is a rational expression that has a numerator and/or denominator that is also a rational expression. In other words, a complex fraction is the result of dividing two rational expressions. In order to simplify complex fractions, the method for dividing rational expressions can be utilized. Multiplying and dividing complex fractions is similar to multiplying or dividing rational expressions. However, simplifying the result may take several extra steps.

EXAMPLE 5 Simplify Complex Fractions CCSS A.APR.7

a. REASON QUANTITATIVELY For what values of a, b, c, and d is the complex fraction $\frac{\frac{a}{b}}{\frac{c}{d}}$ undefined? CCSS SMP 2

b. **USE STRUCTURE** Rewrite the complex fraction $\frac{\frac{w}{x}}{\frac{y}{z}}$ in the form $\frac{a}{b} \div \frac{c}{d}$ and simplify. CCSS SMP 7

c. **USE STRUCTURE** Simplify $\frac{\frac{x+3}{x-1}}{\frac{x-2}{x^2-1}}$. How can this be rewritten in the form $\frac{a}{b} \div \frac{c}{d}$? CCSS SMP 7

EXAMPLE 6 The Domain of a Simplified Rational Expression CCSS A.APR.7

Julie and Jared are looking at a math problem that begins with the expression $\frac{x+1}{x^2-1}$. Julie claims that this expression is equivalent to $\frac{1}{x-1}$.

a. **CRITIQUE REASONING** Why does Julie think the two expressions are equivalent? CCSS SMP 3

b. **CRITIQUE REASONING** Is Julie correct? Are the two expressions equal for all values of x? CCSS SMP 3

c. **CRITIQUE REASONING** Julie graphs $y = \frac{x+1}{x^2-1}$ and $y = \frac{1}{x-1}$ on her graphing calculator. She claims that the graphs are identical, and therefore, so are the expressions. How do you respond to Julie? CCSS SMP 3

PRACTICE

Use the following information to answer Exercises 1–2. CCSS A.APR.7

Quincy and Ariana both solve for x in $\frac{a^3-b^3}{4a^2-b^2}(x) = \frac{a^2+ab+b^2}{2a+b}$, where x represents the rational expression that will make this statement true. Quincy states you must multiply both sides of the equation by the reciprocal of $\frac{a^2+ab+b^2}{2a+b}$. Ariana states that you must multiply both sides by the reciprocal of $\frac{a^3-b^3}{4a^2-b^2}$.

1. **CRITIQUE REASONING** Without solving, determine whose method is correct: Quincy's or Ariana's? Explain how you know. CCSS SMP 3

2. **CALCULATE ACCURATELY** Find the value of x in the given equation. Check your answer. CCSS SMP 6

Use the following information to answer Exercises 3–4. CCSS A.APR.7

The time it takes Irfan to row 9 miles upstream is $\frac{9}{5-c}$. The time it takes him to row 9 miles downstream is $\frac{9}{5+c}$, where c is the speed of the current in miles per hour and $0 \leq c < 5$.

3. **INTERPRET PROBLEMS** Which way does Irfan travel faster, downstream or upstream? Give two reasons for your answer. CCSS SMP 1

4. **CALCULATE ACCURATELY** Using your answer to **Exercise 3**, determine how much faster Irfan travels in the faster direction than in the slower direction. Explain your method. CCSS SMP 6

5. **REASON ABSTRACTLY** During her first week of running, Sharia takes t minutes to run a mile. In miles per hour, her average speed is given by $\frac{60}{t}$. Sharia hopes to get her average time to $t - 2$ minutes to run a mile during her second week of running. On average, how many times faster does Sharia hope to run a mile during her second week than during her first week of running? Explain how you know. CCSS A.APR.7, SMP 2

6. **CRITIQUE REASONING** Shanna has graphed the function $y = \frac{5(x+1)(x+2)(x+3)}{(x+1)(x+2)(x+3)}$ on a graphing calculator and claims that the result is a horizontal line because of the following display. Is Shanna correct? Explain your reasoning. CCSS A.APR.7, SMP 3

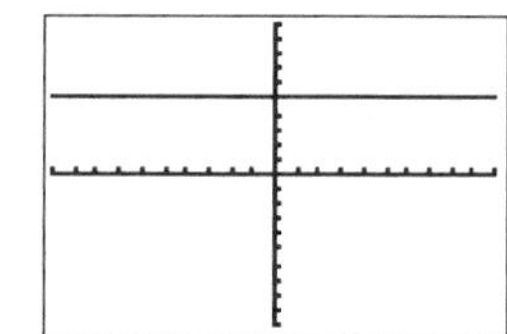

Use the following information to answer Exercises 7–8. CCSS A.APR.7

Ana's yard is being professionally landscaped. The final design will consists of a fountain x ft in diameter in square A surrounded by a 5-ft wide grassy area in square B, and a 3-ft wide gravel pathway in square C that borders the grassy area. The square areas will be centered on each other as shown in the diagram. Square A will have a side length of $2x$ feet.

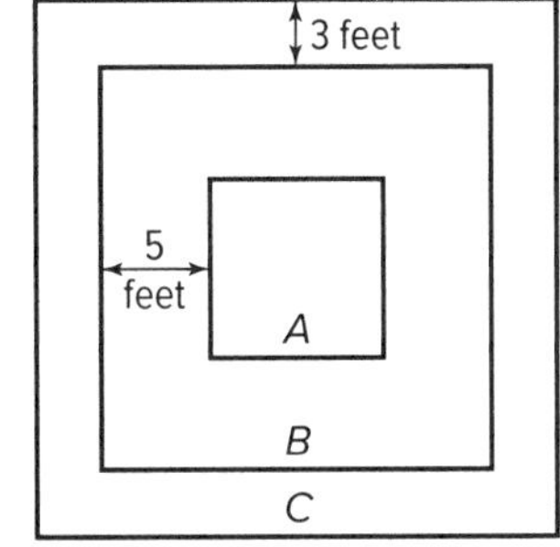

7. **USE A MODEL** Ana would like the lengths of the sides to be proportional. For what values of x will the ratio of the lengths of a side of square C to a side of square B equal the ratio of the lengths of a side of square B to a side of square A? Explain your reasoning. What diameter could the fountain have? CCSS SMP 4

8. **REASON QUANTITATIVELY** If the landscape architect changed the width of the gravel pathway to 4 ft and the width of the grassy area to 2 ft, is there a value for x that would make the ratios described in **Exercise 7** equal? Explain your reasoning. What diameter could the fountain have? CCSS SMP 2

9. **USE STRUCTURE** Simplify the complex fraction. CCSS A.APR.7, SMP 7

a. $\dfrac{\dfrac{(x^2-4)(x+1)}{x^2+x-12}}{\dfrac{(x-2)(x^2+1)}{(x+1)(x-3)}}$

b. $\dfrac{x^2-x-6}{\left[\dfrac{2(x-3)}{x^2-1}\right]}$

c. $\dfrac{\left[\dfrac{x^2+5x+4}{x-3}\right]}{(x+1)(x+2)}$

8.2 Adding and Subtracting Rational Expressions

Objectives

- Understand that rational expressions are closed under addition and subtraction.
- Add and subtract rational expressions.

Content: A.APR.7
Practices: 1, 2, 3, 4, 6, 7, 8
Use with Lesson 8-2

Adding and Subtracting Rational Expressions

To **add** rational expressions with unlike denominators, find the least common denominator (LCD). Rewrite each expression with the LCD. Then add.

For all $\frac{a}{b}$ and $\frac{c}{d}$ with $b \neq 0$ and $d \neq 0$, $\frac{a}{b} + \frac{c}{d} = \frac{ad}{bd} + \frac{bc}{bd} = \frac{ad + bc}{bd}$.

To **subtract** rational expressions with unlike denominators, find the least common denominator (LCD). Rewrite each expression with the LCD. Then subtract.

For all $\frac{a}{b}$ and $\frac{c}{d}$ with $b \neq 0$ and $d \neq 0$, $\frac{a}{b} - \frac{c}{d} = \frac{ad}{bd} - \frac{bc}{bd} = \frac{ad - bc}{bd}$.

EXAMPLE 1 Investigate the Addition and Subtraction of Rational Expressions CCSS A.APR.7

EXPLORE In this exploration, you will explore how to add and subtract rational numbers.

a. **USE STRUCTURE** Let $A = \frac{2x}{x-3}$ and $B = \frac{x+1}{4x-1}$. Find and simplify $A + B$. Is the result a rational expression? CCSS SMP 7

b. **USE STRUCTURE** Let $A = \frac{x+1}{x^2-2}$ and $B = \frac{3x}{x^2+3}$. Find and simplify $A - B$. Is the result a rational expression? CCSS SMP 7

c. **REASON ABSTRACTLY** Look at the Key Concept box above for adding the rational expressions $\frac{a}{b} + \frac{c}{d}$. Does the Closure Property of Addition apply to rational expressions? Explain. CCSS SMP 2

d. **REASON ABSTRACTLY** Look at the Key Concept box above for subtracting the rational expressions $\frac{a}{b} - \frac{c}{d}$. Does the Closure Property of Subtraction apply to rational expressions? Explain. CCSS SMP 2

e. CONSTRUCT ARGUMENTS Is the set of rational numbers closed under addition and subtraction? Explain how you know. CCSS SMP 3

EXAMPLE 2 Monomial Denominators CCSS A.APR.7

Simplify $\frac{7z^2}{3x^2y} + \frac{3x}{6y^3}$.

a. REASON QUANTITATIVELY What is the least common denominator of $\frac{7z^2}{3x^2y}$ and $\frac{3x}{6y^3}$? CCSS SMP 2

b. CALCULATE ACCURATELY Simplify the expression using the LCD. Explain how you know your answer is in lowest terms. CCSS SMP 6

EXAMPLE 3 Polynomial Denominators CCSS A.APR.7

Simplify $\frac{3x}{x^2 - 4} - \frac{3}{2x - 4}$.

a. CALCULATE ACCURATELY Find the LCD and simplify the expression. Why does the answer in lowest terms not have the LCD as its denominator? CCSS SMP 6

b. REASON QUANTITATIVELY How would the simplification be different for the expression $\frac{3x}{4 - x^2} - \frac{3}{4 - 2x}$? Find the LCD, and simplify this expression. How does it relate to the original expression? Why does this make sense? CCSS SMP 2

A complex fraction can have a sum or difference of rational expressions in the numerator and/or denominator. One method for simplifying a complex fraction of this form is to simplify the numerator and the denominator separately. Another method for simplifying a complex fraction of this form is to find the LCD of all of the denominators. Then, use that LCD to rewrite and simplify the complex fraction.

EXAMPLE 4 Two Methods for Simplifying Complex Fractions CCSS A.APR.7

Find the length of the unknown dimensions in the rectangular figure shown.

$A = \frac{3}{x^2} + \frac{1}{x}$ cm² $2 - \frac{4}{5y}$ cm

?

a. **USE A MODEL** Write a complex fraction that represents the length of the unknown dimension. Explain how you found the complex fraction. Are there any restrictions on the values for x and y? Explain. CCSS SMP 4

b. **USE STRUCTURE** Simplify the numerator and the denominator separately. Show your work. CCSS SMP 7

c. **CALCULATE ACCURATELY** What is the unknown dimension in simplified form? How does knowing how to multiply and divide rational expressions help you? CCSS SMP 6

d. **DESCRIBE A METHOD** Find the LCD for all of the denominators in the original complex fraction from **part a**. Use this LCD to simplify the equation. Explain what happens to the denominators. CCSS SMP 8

PRACTICE

1. **CALCULATE ACCURATELY** Find and simplify the sum or difference. Show your work. CCSS A.APR.7, SMP 6

   a. $\frac{2(x+5)}{2x^2-1}+\frac{4x}{x+1}$

   b. $\frac{x-1}{x^2+x-12}-2$

2. **USE A MODEL** Martha needs to buy fencing for her rectangular garden.

   a. Write an expression, in simplest form, that represents the number of feet of fencing Martha needs. Are there any restrictions on the variables? Explain. CCSS A.APR.7, SMP 4

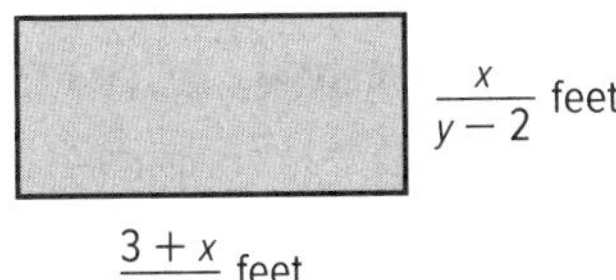

   b. Martha wants to remove a square corner from her garden. The square section removed will have sides the length of half the width of the original garden. What expression represents the perimeter of the new garden? Explain. CCSS A.APR.7, SMP 4

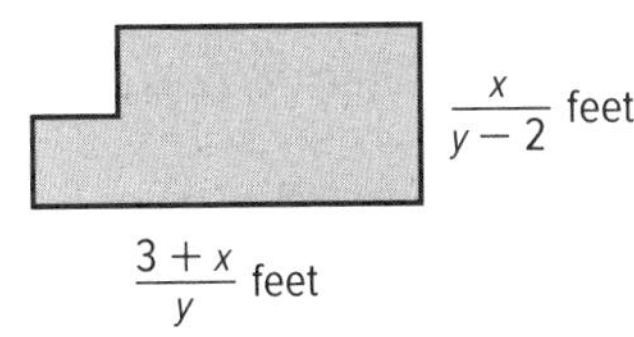

3. **COMMUNICATE PRECISELY** Simplify the rational expression, $\frac{\frac{1}{h}+\frac{3}{4}}{\frac{2}{h^2}-\frac{7}{h}}$. Describe the method you used. CCSS A.APR.7, SMP 6

4. **CALCULATE ACCURATELY** Determine the average of three rational numbers represented by these rational expressions: $\frac{1}{x}$, $\frac{1}{x-3}$, and $\frac{1}{2x}$, for $x \neq 3$, and $x \neq 0$. Explain how you found the average. CCSS A.APR.7, SMP 6

5. A resistor is an electrical component that reduces the flow of electrical current through a circuit. A resistor is connected in parallel when both of its terminals are connected to both terminals of an adjacent resistor. When three resistors are connected in parallel the total resistance, R_T, is given by $R_T = \frac{1}{\frac{1}{R_1} + \frac{1}{R_2} + \frac{1}{R_3}}$.

   a. **REASON ABSTRACTLY** Simplify the complex fraction. Explain how you know your result is simplified as much as possible. CCSS A.APR.7, SMP 2

   b. **EVALUATE REASONABLENESS** Timothy found this formula for total resistance, $\frac{1}{R_T} = \frac{1}{R_1} + \frac{1}{R_2} + \frac{1}{R_3}$. He said that this formula is equivalent to the original formula. Is Timothy correct? Explain. CCSS A.APR.7, SMP 8

6. **PLAN A SOLUTION** A local high school is landscaping a local park. John can plant a flower bed in 3 hours. Max can plant a flower bed of the same size in 4 hours. Using *t*, time in hours, write an expression representing how many flower beds they will complete in *t* hours working together. Explain your reasoning. CCSS A.APR.7, SMP 1

7. PLAN A SOLUTION Determine three real numbers that divide the real number line between $\frac{x}{6}$ and $\frac{x}{2}$ $(x \neq 0)$ into four equal parts. Explain your reasoning.

8. REASON QUANTITATIVELY Consider the expression $\frac{1}{1+\frac{1}{x}}$. CCSS A.APR.7, SMP 2

a. Simplify the expression.

b. Use your work from **part a** to simplify $\frac{1}{1+\frac{1}{1+\frac{1}{x}}}$.

c. Use your work from **part b** to simplify $\frac{1}{1+\frac{1}{1+\frac{1}{1+\frac{1}{x}}}}$.

d. Use your work from **part c** to simplify $\frac{1}{1+\frac{1}{1+\frac{1}{1+\frac{1}{1+\frac{1}{x}}}}}$.

e. For $x = 1$, find the values of each of the four expressions.

f. The Fibonacci numbers are a famous sequence starting with 1 and 1, where the next term is found by adding the previous two. So the Fibonacci sequence is 1, 1, 2, 3, 5, 8, 13, 21, 34, ... What does this sequence seem to have in common with your answers from **part e**? What would you conjecture the next value in the pattern from **part e** to be?

8.3 Graphing Reciprocal Functions

Objectives

- Create and graph a reciprocal function and interpret key features of the graph.
- Calculate and interpret the average rate of change of a function over an interval.

CCSS STANDARDS

Content: A.CED.2, F.IF.4, F.IF.5, F.IF.6, F.BF.3
Practices: 1, 2, 3, 4, 6, 7, 8
Use with Lesson 8-3

EXAMPLE 1 Create and Graph a Function

EXPLORE A *forest inventory* is a census of trees and their characteristics. Researchers often use a *fixed area plot* to analyze trees within a region of a given area. A current forest inventory is analyzing rectangular plots with an area of 100 m^2.

a. **USE A MODEL** One of the plots has a length of x meters and a width of y meters. Write an equation for y in terms of x to show how x and y are related to the area of the plot. CCSS A.CED.2, SMP 4

b. **INTERPRET PROBLEMS** Suppose another plot is to be made with length five meters less than that of the first plot. Write an equation for the width y_2 of the second plot in terms of x, the length of the first plot. CCSS A.CED.2, SMP 1

c. **USE A MODEL** Graph the equations for y and y_2 on the coordinate plane. CCSS A.CED.2, SMP 4

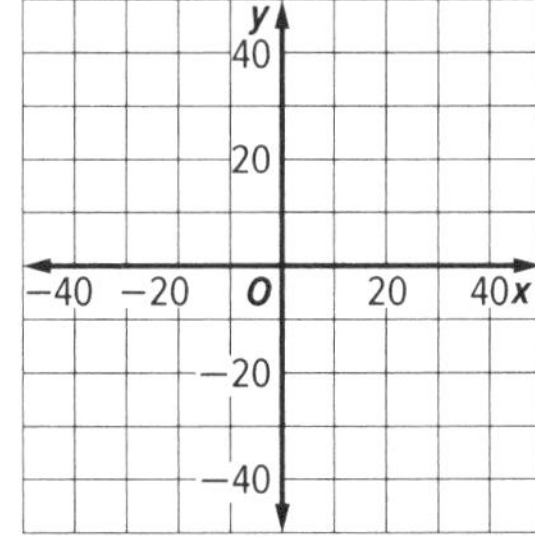

d. **REASON QUANTITATIVELY** What is an appropriate domain for the functions, considering the context in which the functions are used? Explain. CCSS F.IF.5, SMP 2

e. **COMMUNICATE PRECISELY** How does the graph of y_2 compare to the graph of y? Explain the change in graph based on the context. CCSS F.BF.3, SMP 6

f. **REASON QUANTITATIVELY** What happens to the graphs as x gets arbitrarily large ($x \to \infty$)? Why does this make sense? CCSS F.IF.4, SMP 2

g. INTERPRET PROBLEMS Consider the following functions, each a variation on the original function. Graph each and describe how the resulting graphs compare with the original. CCSS F.BF.3, SMP 1

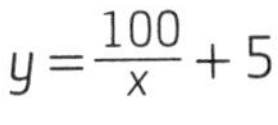

$y = \frac{100}{x} + 5$ $\qquad$ $y = 2\left(\frac{100}{x}\right)$ $\qquad$ $y = \frac{100}{2x}$

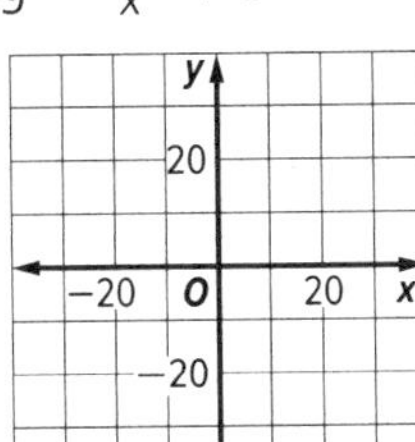

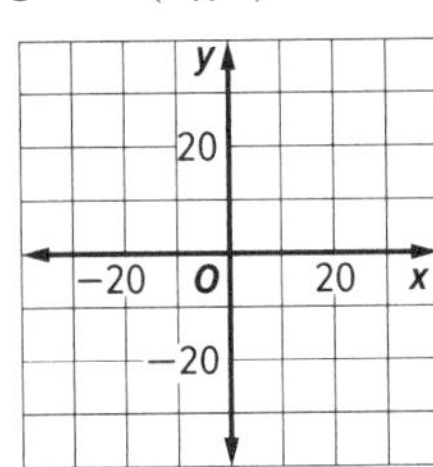

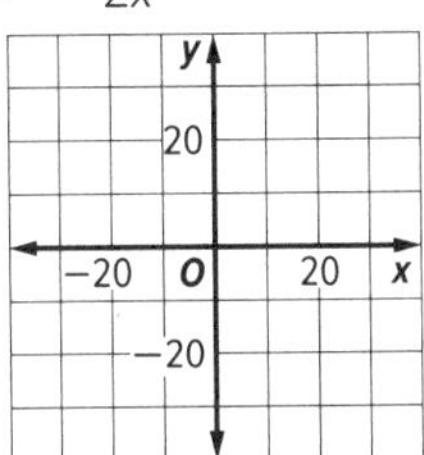

A **reciprocal function** has an equation of the form $f(x) = \frac{1}{a(x)}$, where $a(x)$ is a linear function and $a(x) \neq 0$. The parent function of reciprocal functions is $f(x) = \frac{1}{x}$, whose graph is a **hyperbola**.

KEY CONCEPT Transformations of Reciprocal Functions

Complete the table by describing the effect on the graph of $f(x) = \frac{1}{x}$ when $f(x)$ is replaced by each of the following.

$f(x) + k$ (Vertical Translation)		$f(x + k)$ (Horizontal Translation)	
$k > 0$	translation k units up	$k > 0$	translation k units left
$k < 0$	translation $\|k\|$ units down	$k < 0$	translation $\|k\|$ units right
$kf(x)$ (Vertical Stretch/Compression)		**$f(kx)$ (Horizontal Stretch/Compression)**	
$k < 0$	reflection in the x-axis	$k < 0$	reflection in the y-axis
$0 < \|k\| < 1$	vertical compression	$0 < \|k\| < 1$	horizontal stretch
$\|k\| > 1$	vertical stretch	$\|k\| > 1$	horizontal compression

EXAMPLE 2 Graph a Reciprocal Function CCSS F.BF.3

Follow these steps to analyze and graph $g(x) = \frac{2}{x-3} - 2$.

a. INTERPRET PROBLEMS Use transformations to explain how the graph of $g(x)$ is related to the graph of the parent reciprocal function $f(x) = \frac{1}{x}$. CCSS SMP 1

b. USE STRUCTURE Use your answer from **part a** to graph $g(x)$. CCSS SMP 7

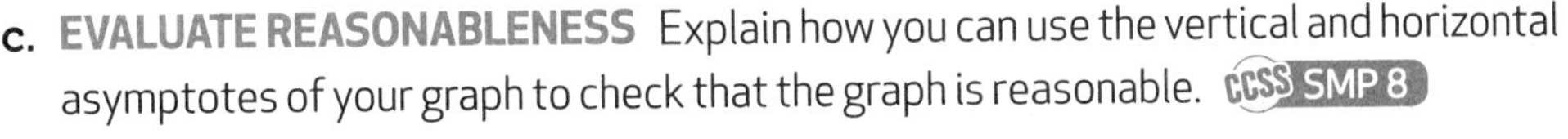

c. EVALUATE REASONABLENESS Explain how you can use the vertical and horizontal asymptotes of your graph to check that the graph is reasonable. CCSS SMP 8

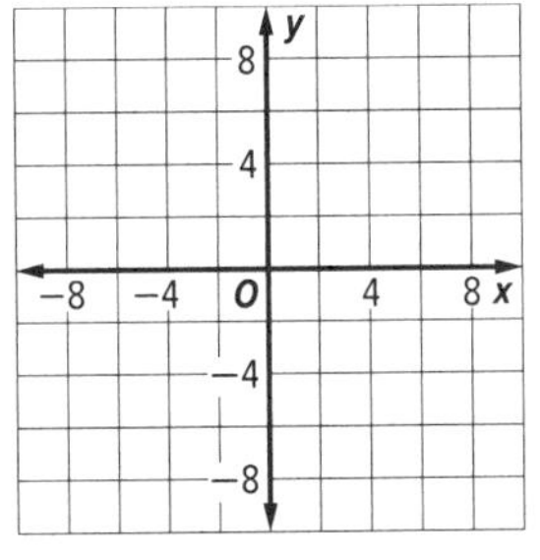

EXAMPLE 3 Analyze the Graph of a Reciprocal Function CCSS F.BF.3

The figure shows the graph of the parent reciprocal function, $f(x)$, and the graph of a reciprocal function, $h(x)$.

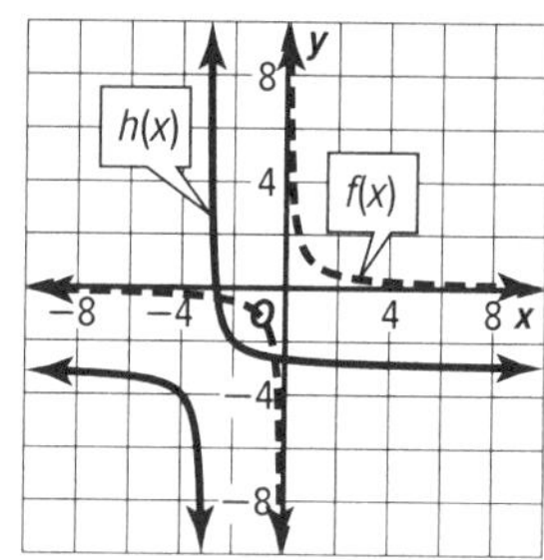

a. **INTERPRET PROBLEMS** How is the graph of $h(x)$ related to the graph of $f(x)$? CCSS SMP 1

b. **USE REASONING** What is the equation for the function $h(x)$? Explain how you know. CCSS SMP 3

EXAMPLE 4 Calculate a Rate of Change Over an Interval

Lakita is training for a triathlon, in which competitors run, bike, and swim. For today's session, she plans to bike for 10 miles and then run for 2 hours.

a. **USE A MODEL** Let r be the rate, in miles per hour, at which Lakita bikes. Write an equation that gives the total time t, in hours, for today's training session in terms of r. Explain how you wrote the equation. CCSS A.CED.2, SMP 4

b. **USE A MODEL** Use the coordinate plane at the right to graph the function you wrote in **part a**. CCSS A.CED.2, SMP 4

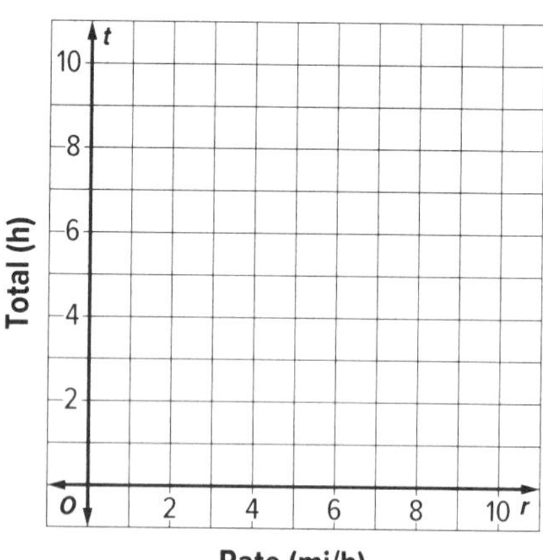

c. **CONSTRUCT ARGUMENTS** Lakita plans to calculate the average rate of change on the interval $r = 2$ to $r = 8$. Do you think this value will be positive or negative? Why? Why does this make sense? CCSS F.IF.6, SMP 3

d. **CALCULATE ACCURATELY** Find the average rate of change on the interval $r = 2$ to $r = 8$. Interpret the result. CCSS F.IF.6, SMP 6

PRACTICE

USE STRUCTURE **Graph each reciprocal function and identify any transformations from the graph of the parent function $f(x) = \frac{1}{x}$.** CCSS F.BF.3, SMP 7

1. $g(x) = \frac{-1}{x+2} + 1$

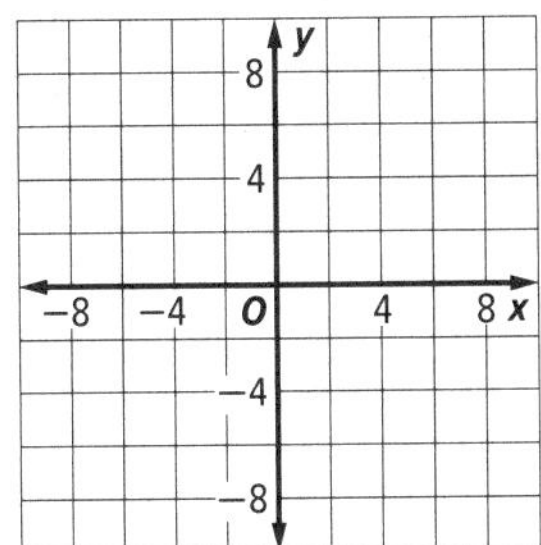

2. $h(x) = \frac{2}{x-4} - 5$

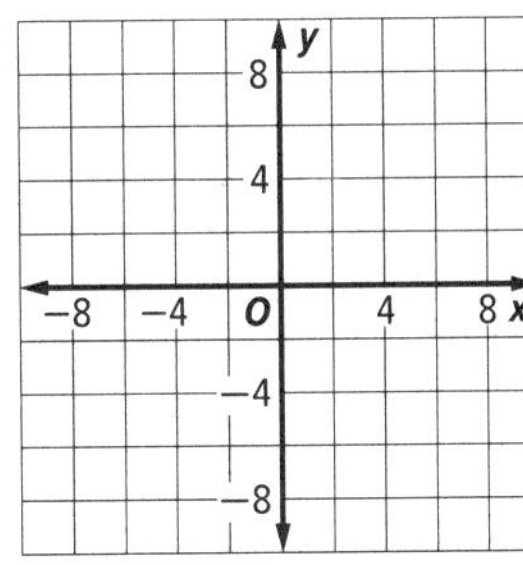

3. **INTERPRET PROBLEMS** The figure shows the graph of the parent reciprocal function, $f(x)$, and the graph of a reciprocal function, $k(x)$. Explain how to use transformations to write the equation for $k(x)$. CCSS F.BF.3, SMP 1

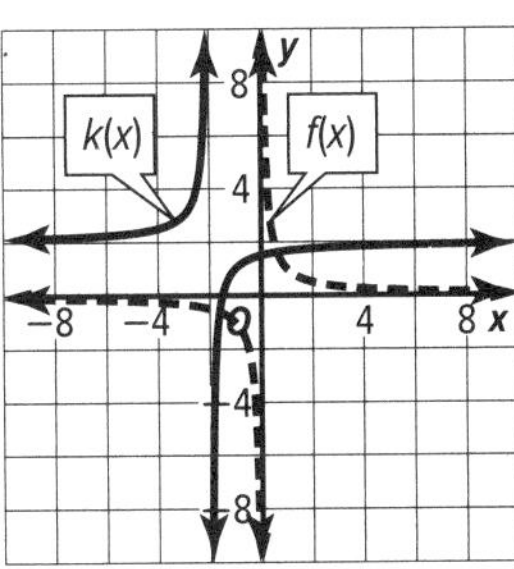

4. Some students are renting a bus for a trip to an aquarium. The bus costs \$200 and the aquarium tickets are \$25 per person.

a. **USE A MODEL** Write a function that gives the total cost per student y, assuming that there are x students and they share the cost of the bus equally. Then graph the function on the coordinate plane. CCSS A.CED.2, SMP 4

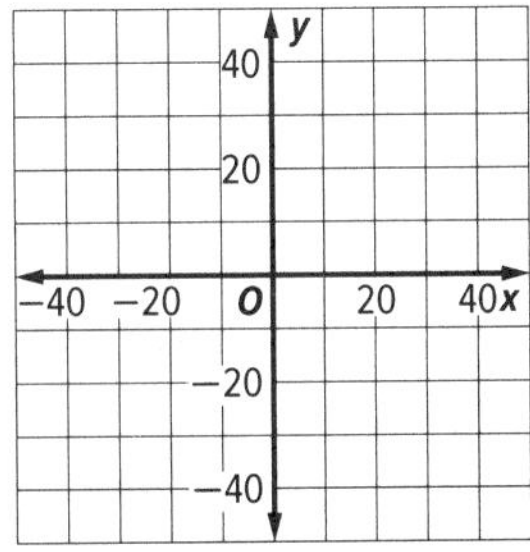

b. **REASON QUANTITATIVELY** What is an appropriate domain for the function? Explain. CCSS F.IF.5, SMP 2

c. **CALCULATE ACCURATELY** Find the average rate of change on the interval $x = 10$ to $x = 20$. Interpret the result. CCSS F.IF.6, SMP 6

8.4 Graphing Rational Functions

Objectives

- Graph rational functions.
- Compare the properties of rational functions, including zeros and asymptotes.
- Understand how restrictions on the domain of a rational function affect its graph.

CCSS STANDARDS

Content: F.IF.4, F.IF.5, F.IF.9, A.APR.7, A.CED.2

Practices: 1, 2, 3, 4, 5, 6, 7

Use with Lesson 8–4

A **rational function** has an equation of the form $f(x) = \frac{a(x)}{b(x)}$, where $a(x)$ and $b(x)$ are polynomial functions and $b(x) \neq 0$.

KEY CONCEPT Zeros and Vertical Asymptotes

Given a rational function $f(x) = \frac{a(x)}{b(x)}$ where $a(x)$ and $b(x)$ have no common factors other than 1:

$f(x)$ has a zero at all x values for which $a(x) = 0$,

so the graph of $f(x)$ has an x-intercept at all x values for which $a(x) = 0$.

$f(x)$ is undefined at all x values for which $b(x) = 0$,

so the graph of $f(x)$ has a **vertical asymptote** at all x values for which $b(x) = 0$.

EXAMPLE 1 Explore Domain, Zeros, and Vertical Asymptotes

a. **USE STRUCTURE** Consider the function $f(x) = \frac{x^3}{2x + 2}$. Describe any restrictions to its domain, and identify its zeros and vertical asymptotes. Explain your answers. CCSS F.IF.5, SMP 7

b. **USE STRUCTURE** Compare the function $g(x) = \left(\frac{x^3}{2x + 2}\right)\left(\frac{x}{x - 1}\right)$ to $f(x)$ from **part a**. How are their domains, zeros, vertical asymptotes, and end behaviors alike and different? *Recall that end behavior involves analyzing the function values as x approaches negative and positive infinity.* CCSS F.IF.5, SMP 7

c. COMMUNICATE PRECISELY Use the features of the graphs determined in **parts a** and **b** to graph $g(x)$ and $f(x)$ on the grids provided. Which features of the graphs could you predict from the equations, and which did you need the table of values to find? CCSS F.IF.4, SMP 6

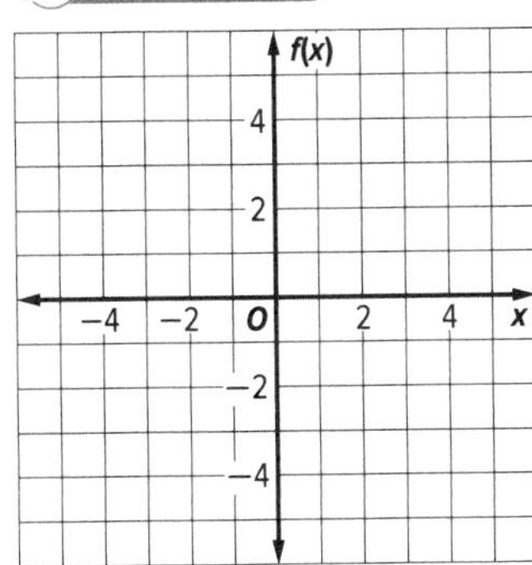

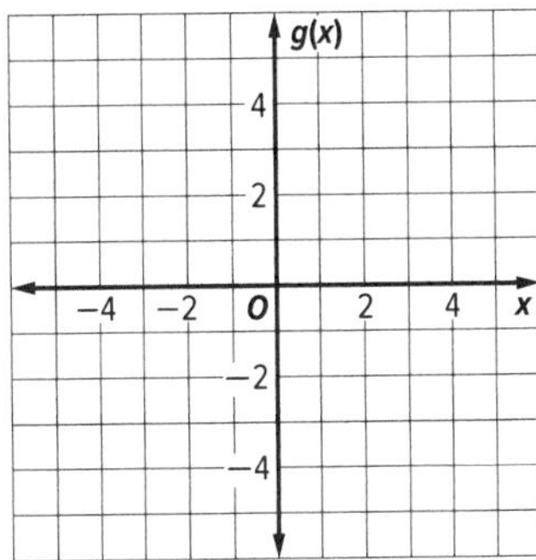

d. USE STRUCTURE The graph of $g(x)$ seems to be symmetric in the y-axis. Check this using the values $x = 2$, $x = -2$, $x = 3$, and $x = -3$. Why does it make sense from the equation that this symmetry exists? CCSS F.IF.4, SMP 7

In addition to vertical asymptotes, the graph of a rational function can have a horizontal asymptote of the form $y = c$, or an oblique asymptote which is neither vertical nor horizontal.

KEY CONCEPT Horizontal and Oblique Asymptotes

A rational function $f(x) = \frac{a(x)}{b(x)}$ can have at most one horizontal asymptote or one oblique asymptote. It cannot have both. The number and location of these asymptotes can be determined by comparing the degree of the numerator and denominator of the function.

Horizontal Asymptotes:

If the degree of $a(x)$ is less than the degree of $b(x)$, the horizontal asymptote is $y = 0$.

If the degree of $a(x)$ is equal to the degree of $b(x)$, the horizontal asymptote is $y = \frac{\text{leading coefficient of } a(x)}{\text{leading coefficient of } b(x)}$.

If the degree of $a(x)$ is greater than the degree of $b(x)$, there is no horizontal asymptote.

Oblique Asymptotes:

If the degree of $a(x)$ is one greater than the degree of $b(x)$, the oblique asymptote is $y = \frac{a(x)}{b(x)}$ without the remainder.

Otherwise, there is no oblique asymptote.

EXAMPLE 2 Compare Graphs of Rational Functions CCSS F.IF.9

a. **COMMUNICATE PRECISELY** Explain the similarities and differences between the graph of $g(x) = \frac{-3x^2 + 48}{x^2 - 9}$ and the graph of $f(x)$ shown at right. Then graph $g(x)$ to verify your answer. CCSS SMP 6

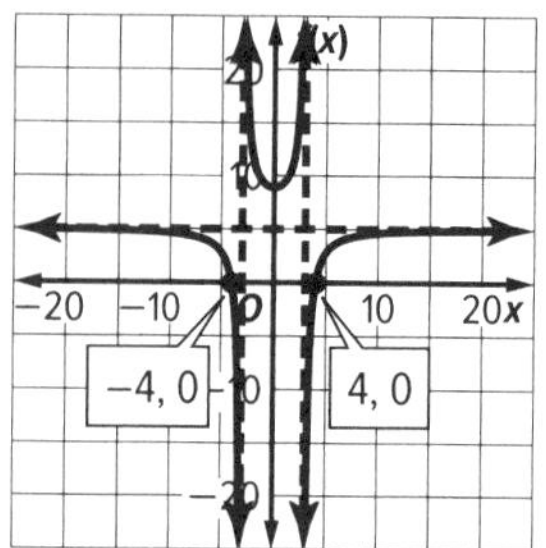

b. **USE TOOLS** Describe how the zeros and asymptotes of $h(x) = \frac{x^3 - 16x}{x^2 - 9}$ will be similar to and different from the features of the two functions in **part a**. Use a graphing calculator to help graph $h(x)$, then sketch the graph on the grid provided. CCSS SMP 5

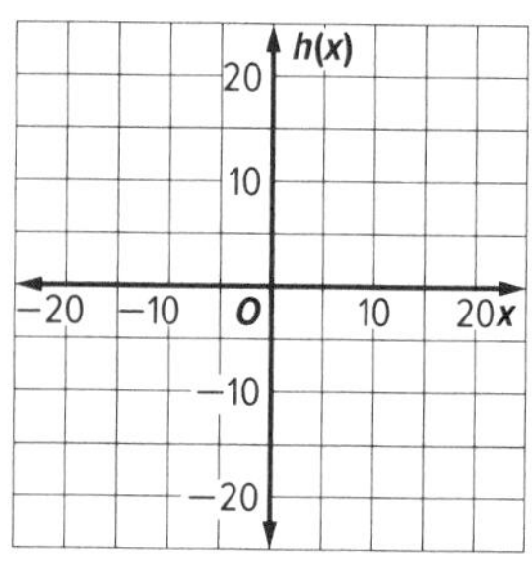

c. **USE STRUCTURE** Show that $\frac{x^3 - 16}{x^2 - 9} = x + \frac{9x - 16}{x^2 - 9}$. Use your calculator to find out what happens to the value of $\frac{9x - 16}{x^2 - 9}$ as x becomes very large. Use this information to explain why the equation of the oblique asymptote makes sense. CCSS SMP 7

KEY CONCEPT Point Discontinuity

If the numerator and denominator of a rational function have a common factor $(x - c)$, then the function is undefined when $x = c$.

This point is called a point discontinuity, and it looks like a hole in the graph of the function.

EXAMPLE 3 Graph Functions with Point Discontinuities

a. **USE STRUCTURE** What is the domain of $f(x) = \frac{x^3 + 2x^2 - 9x - 18}{x + 2}$? How is the graph of $f(x)$ related to the graph of $g(x) = x^2 - 9$? Graph the functions on the grids provided. CCSS F.IF.5, SMP 7

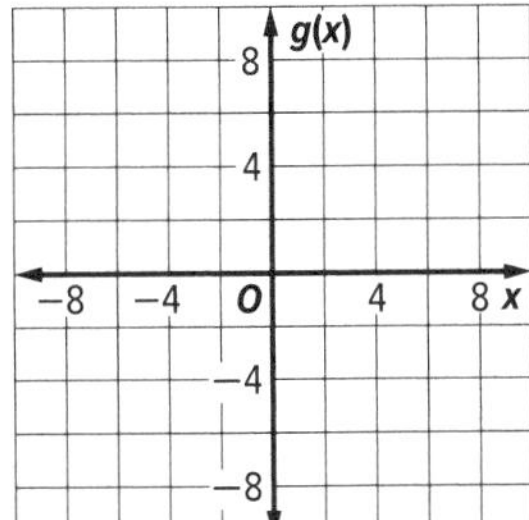

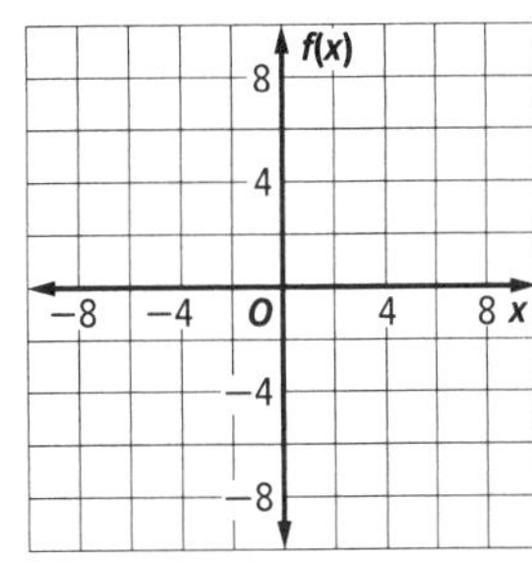

b. **USE STRUCTURE** What values of x are excluded from the domain of $f(x) = \frac{x^2 + x - 6}{2x^2 - 3x - 2}$? How are these excluded values shown on the graph of $f(x)$? Find the function that results if you remove a common factor from the numerator and denominator of $f(x)$, and identify the intercepts and asymptotes for the simplified function. Then use this information to graph $f(x)$. CCSS F.IF.5, SMP 7

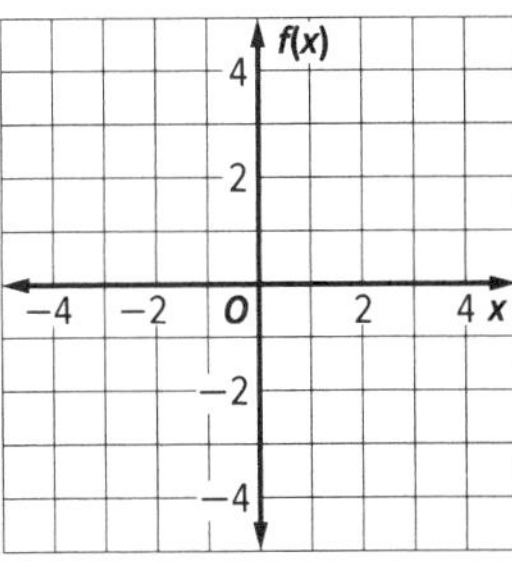

c. **USE STRUCTURE** Describe the features of the graph shown, and determine a function that is represented by the graph. CCSS A.CED.2, SMP 7

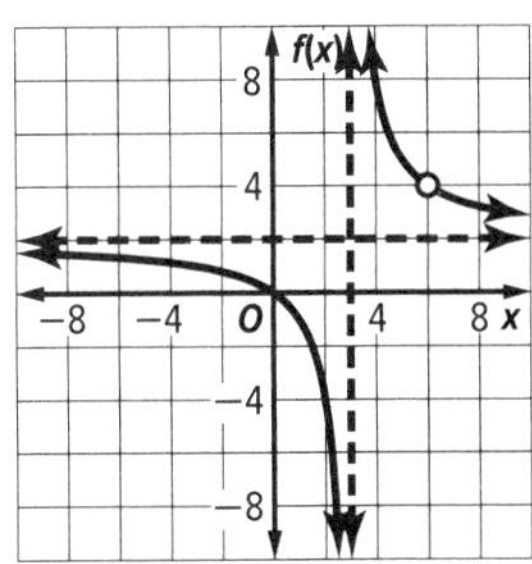

d. **CRITIQUE REASONING** Holden claims that the function $y = \frac{(2x)(x-6)^2}{(x-3)(x-6)^2}$ will also satisfy the graph. Is Holden correct? Is his function the same as yours? CCSS A.CED.2, SMP 3

PRACTICE

1. **USE A MODEL** A cable television company charges customers a \$60 installation fee plus \$30 per month for service. Write and graph a function representing the average monthly cost as a function of the number of months of service. Explain the meaning of the horizontal asymptote in the context of the problem. CCSS A.CED.2, F.IF.4, SMP 4

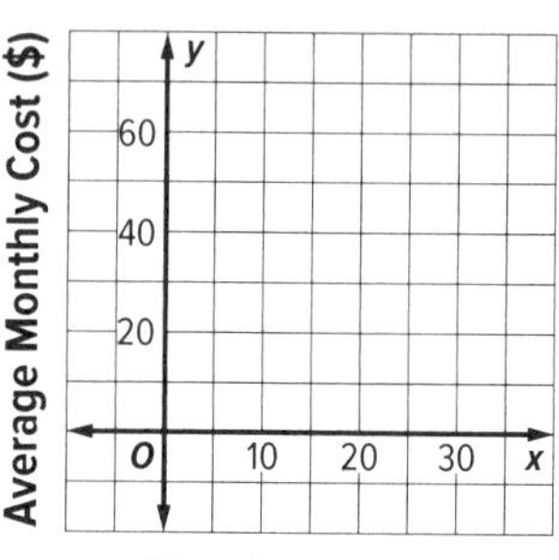

2. **CONSTRUCT ARGUMENTS** Petra is graphing $f(x) = \frac{x^2 - 6x + 8}{x^2 - 3x - 4}$. She factors out $(x - 4)$ from the numerator and denominator, then constructs the graph of $g(x) = \frac{x - 2}{x + 1}$ as her answer. Is she correct? Explain your answer in terms of the domain of $f(x)$. CCSS F.IF.5, SMP 3

3. On the drive to visit a nearby college, the Marshall family averages 40 miles per hour.

   a. **INTERPRET PROBLEMS** Determine and graph a function representing the average speed for the entire trip, in terms of the average speed for the drive home. (Hint: First write an expression for the average speed in terms of the distance and the times for the outgoing trip and the return trip. Then express time in terms of speed and distance.) CCSS A.CED.2, SMP 1

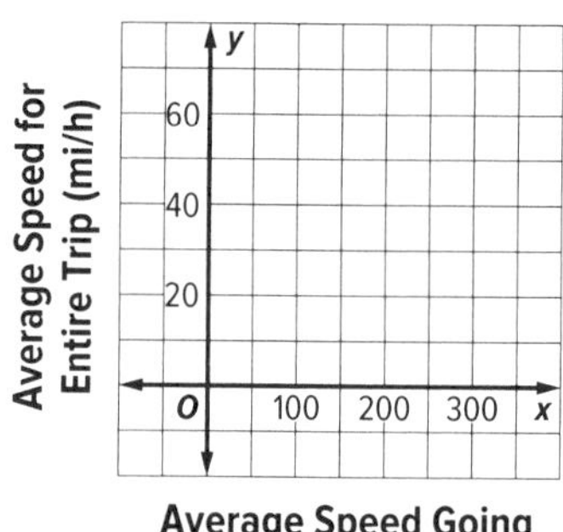

   b. **USE A MODEL** If the family averages 60 miles per hour driving home, is the average speed for the entire trip equal to 50 miles per hour? Why or why not? What is the horizontal asymptote, and what does it represent in the context of the problem? CCSS F.IF.4, SMP 4

4. **USE A MODEL** A music studio uses the function $f(x) = \frac{300x}{x^2 + 4}$ to estimate the number of downloads per hour (in thousands) in the hours after the release of a new song on the internet. Graph the function. Then restrict the domain of the function as required by the context, and graph the function with the restricted domain. Explain the shape of the graph in the context of the problem. CCSS F.IF.4, F.IF.5, SMP 4

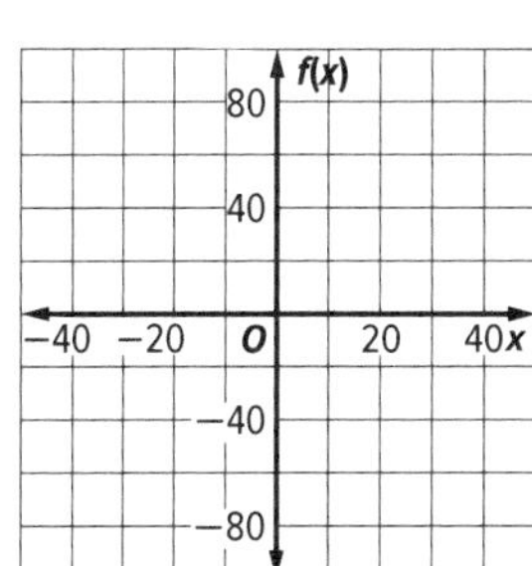

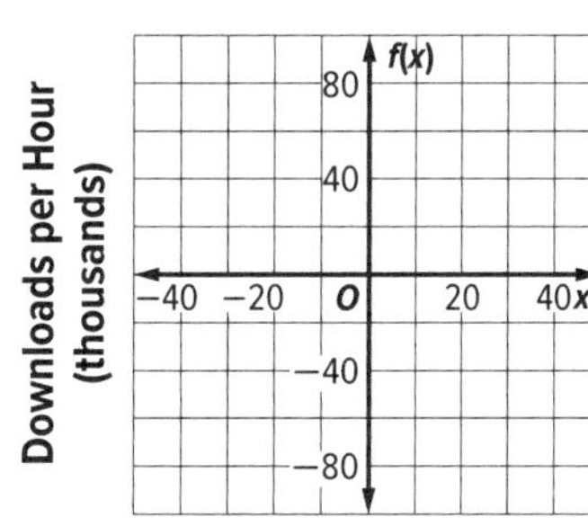

5. **USE STRUCTURE** Identify the domain, zeros, intercepts, and asymptotes of the graph, and determine a function that corresponds to the graph. CCSS F.IF.4, F.IF.5, SMP 7

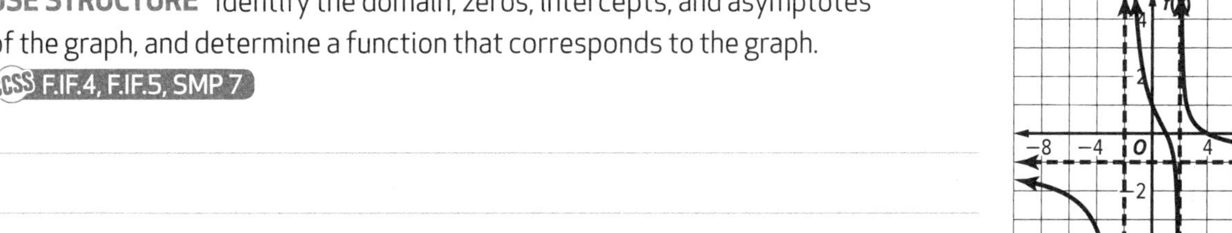

6. **USE STRUCTURE** Consider the functions $y = x + 1$, $y = \frac{(x + 1)(x - 1)}{x - 1}$, $y = \frac{(x + 1)(x - 1)^2}{x - 1}$, and $y = \frac{(x + 1)(x - 1)^2}{(x - 1)^2}$. Graph each of the functions. Which, if any, are equivalent? CCSS A.CED.2, SMP 7

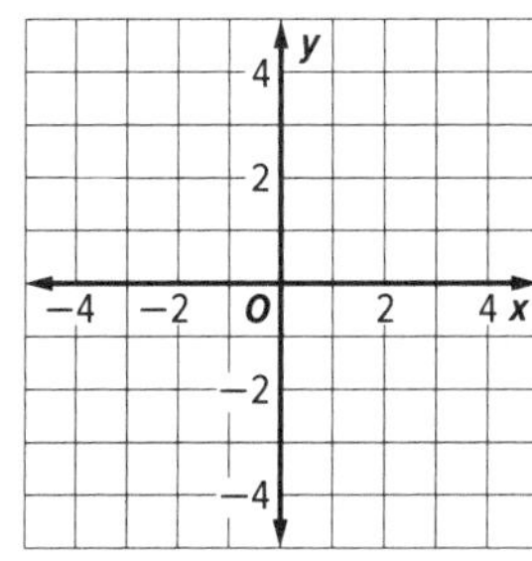

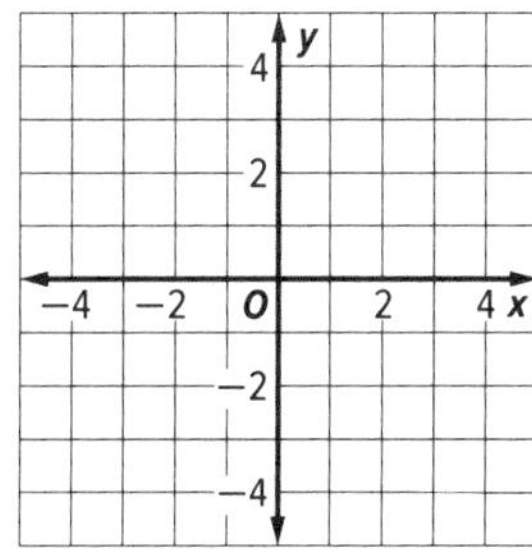

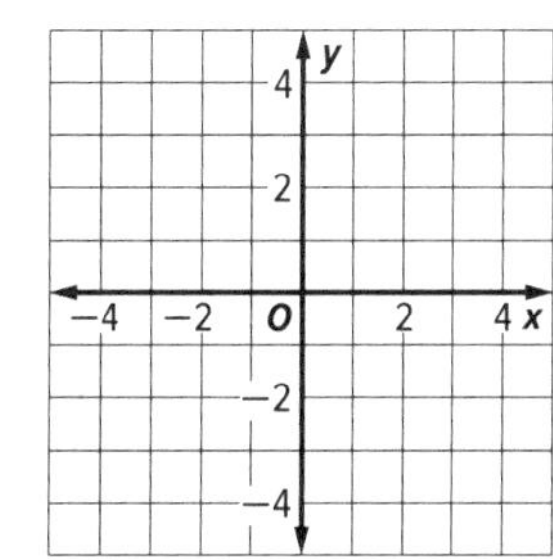

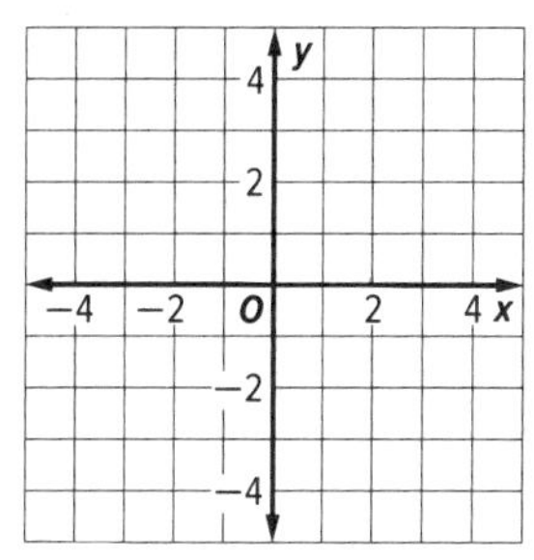

8.5 Modeling: Rational Functions

Objectives

- Write and interpret rational functions that model real-world situations.
- Investigate rational functions and their inverses.
- Manipulate rational functions algebraically in order to solve problems.

CCSS STANDARDS

Content: A.SSE.1a, A.SSE.1b, A.SSE.2, A.CED.4, F.IF.4, F.IF.5, F.BF.1b, F.BF.4a

Practices: 1, 2, 3, 4, 5, 7

Use with Lesson 8-4

Many real-world situations can be modeled with rational functions. Because rational functions represent fractions, any situation involving a ratio can be modeled with a rational function. This includes speed (miles per hour), fuel efficiency (miles per gallon), economic situations (cost per product), work productivity (products per hour), income (dollars per month), and pollution (volume per day). Graphing and modeling functions make it possible to visualize properties of the phenomenon being studied, and working with the equations makes it possible to come up with numeric answers to the question being asked.

EXAMPLE 1 Model Fuel Efficiency

EXPLORE The fuel efficiency of a vehicle, commonly reported in miles per gallon, is determined by dividing the distance traveled by the amount of gasoline consumed.

a. **REASON QUANTITATIVELY** Write a function $e(g)$ that gives the fuel efficiency of a vehicle that can drive 1500 miles on g gallons of gasoline. Then find the inverse of the function by solving the equation for g. What does the inverse function represent? CCSS A.CED.4, SMP 2

b. **REASON ABSTRACTLY** What are the restrictions on the domain of $e(g)$? Explain. CCSS F.IF.5, SMP 2

c. **INTERPRET PROBLEMS** Identify any asymptotes of $e(g)$. Explain the significance of any asymptotes based on the context. CCSS F.IF.4, SMP 1

EXAMPLE 2 Model Average Cost

As a fundraising project, the mathematics club is designing and selling t-shirts. The t-shirt printing company will charge a set-up fee for creating the design, plus an amount per t-shirt ordered. The costs are summarized in the table.

Number of Shirts	10	20	30
Total Cost	$280	$310	$340

a. **REASON QUANTITATIVELY** Write a function for total cost based on the number of shirts ordered. Then write a function for the average cost per shirt as a function of the number of shirts ordered. Explain what the constants in the functions represent. CCSS A.SSE.1a, F.BF.1b, SMP 2

b. **PLAN A SOLUTION** What is an appropriate domain for the average cost function? When graphing this function, which quadrants of the coordinate plane will you use? Explain. CCSS F.IF.5, SMP 1

c. **USE TOOLS** Graph the average cost functions using a graphing calculator. Sketch the graph on the grid provided, making sure to mark the scale and label the axes. Describe the end behavior of the function, and explain what it means in the real-world context. CCSS F.IF.4, SMP 5

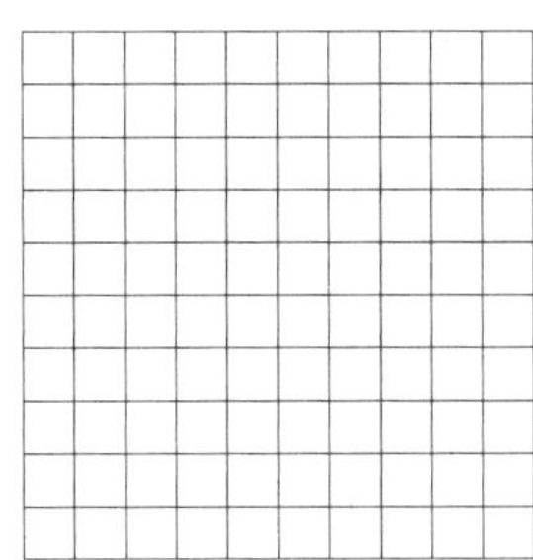

EXAMPLE 3 Model a Rate of Work

Larry is a mason who can lay the brick for a complete patio in 8 hours. He wants to hire a partner to help him finish the job faster.

a. **INTERPRET PROBLEMS** Write an expression that represents what fraction of the patio Larry and a partner, working together, can complete in t hours, if the partner could complete the job alone in x hours. Explain what each part of the expression represents. CCSS A.SSE.1a, A.SSE.1b, SMP 1

b. REASON ABSTRACTLY Use the expression from **part a** to write a function $t(x)$ that represents the time required for Larry and his partner to complete the patio. Simplify the rational expression in your equation. CCSS F.BF.1b, SMP 2

c. REASON QUANTITATIVELY Sketch the graph of $t(x)$ on the grid provided. What are the limitations on the domain and range? Interpret the end behavior of the graph in the context of the problem. CCSS F.IF.4, F.IF.5, SMP 2

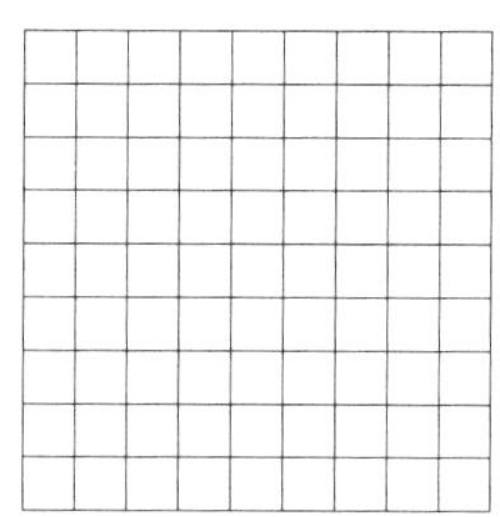

d. USE STRUCTURE Determine the inverse function for $t(x)$ and explain what it means in the context of this problem. CCSS A.SSE.2, F.BF.4a, SMP 7

PRACTICE

1. **REASON ABSTRACTLY** A gym charges an enrollment fee of \$120 plus \$50 per month for membership. Write a function that represents the average cost per month of belonging to the gym as a function of the number of months since joining. After how many months will the average monthly cost be \$60? CCSS F.BF.1b, SMP 2

2. **USE A MODEL** As CEO of a technology company, Patricia has found that the number of sales (in hundreds) of a new app on day x after its release is described by the model $f(x) = \frac{200x}{x^2 + 5}$. The company will release an upgraded app once daily sales fall below a quarter of the highest projected sales. According to the model, how long should Patricia wait before removing her app from the market and releasing the new one? Fully support your answer. CCSS F.IF.4, SMP 4

3. When a pollutant enters a fish tank and then is filtered out, the oxygen level in the water (as a fraction of its initial value) will change according to the model $O(t) = 1 - \frac{t}{t^2 + 1}$, where t is the number of days since the pollutant entered the water. CCSS F.IF.4

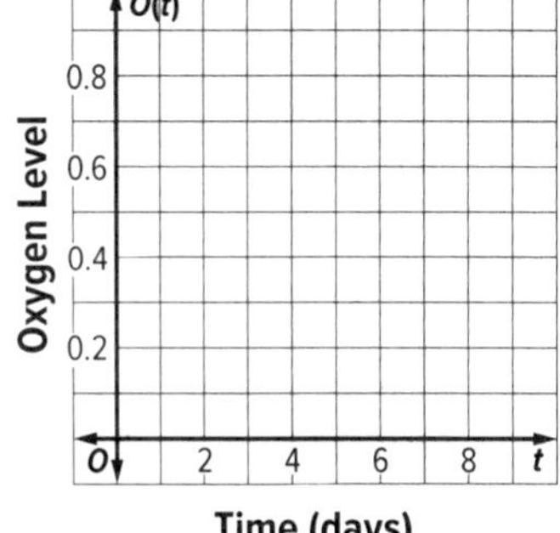

a. **REASON QUANTITATIVELY** What is an appropriate domain and range for this model? Graph the function, then describe how the oxygen level changes in the first week. CCSS SMP 2

b. **CRITIQUE REASONING** Marina argues that the oxygen level will never return to normal. Colin says it will. Who is correct? Fully support your position. CCSS F.IF.5, SMP 3

c. **REASON ABSTRACTLY** Does this function have an inverse function? If so, find it. If not, explain why. CCSS SMP 2

4. A marathon is a 26.2 mile race. A runner plans to compete in a marathon by running the first 20 miles at 7.5 miles per hour and then running the last 6.2 miles at a pace determined by how tired he is.

a. **INTERPRET PROBLEMS** Write a function $t(x)$ for the time in hours it takes the runner to complete the marathon if he runs at a pace of x miles per hour for the last 6.2 miles of the race. What time will he finish with if he runs the last 6.2 miles at 6.4 miles per hour? Round your answer to the nearest hundredth of an hour. CCSS F.BF.1b, SMP 1

b. **REASON QUANTITATIVELY** Does $t(x)$ have a horizontal asymptote? If so, what does it signify in the context of the problem? CCSS F.IF.4, SMP 2

c. **REASON ABSTRACTLY** Find the inverse of the function by solving the equation for x. What does the inverse function represent? CCSS A.CED.4, SMP 2

 Solving Rational Equations and Inequalities

Objectives

- Create rational equations and inequalities in one variable to solve problems.
- Solve rational equations and explain the significance of extraneous solutions.
- Connect solving rational equations and inequalities with solving systems of functions in the coordinate plane.

CCSS STANDARDS

Content: A.CED.1, A.REI.2, A.REI.11
Practices: 1, 2, 3, 4, 5, 6, 7
Use with Lesson 8-6

A **rational equation** contains one or more rational expressions. You can multiply both sides of a rational equation by the LCD, or Least Common Denominator, to simplify the rational expressions.

EXAMPLE 1 Write Equations and Inequalities CCSS A.CED.1

EXPLORE **During a basketball game, Madison scored 5 free throws out of 9 attempts.**

a. REASON ABSTRACTLY If she scores on each of her next x consecutive free throws, write and solve an equation that can be used to determine x, the number of consecutively scored free throws that will bring Madison's percentage up to 75%. Explain what your solution means. CCSS SMP 2

b. REASON ABSTRACTLY To make the travel team, Madison's free throw percentage needs to be at least 78%. Write and solve an inequality to determine how many consecutive free throws she needs to score in order to make the travel team. To what set of numbers should you restrict your solution and why? Explain what your result means in the context of the situation. CCSS SMP 2

Solving a rational equation using algebraic methods sometimes results in values of the solution that will make the denominator equal to zero. These are called **extraneous solutions** and must be excluded from the solution set. To see if a solution is extraneous, substitute the solution into the original function. If the result is undefined, the solution is extraneous.

EXAMPLE 2 Investigate Extraneous Solutions CCSS A.REI.2

The function $f(x)$ is defined by $f(x) = \frac{x^2 - 4x - 8}{x^2 - 8x + 12}$ and the function $g(x)$ is defined by $g(x) = \frac{1}{x - 6}$.

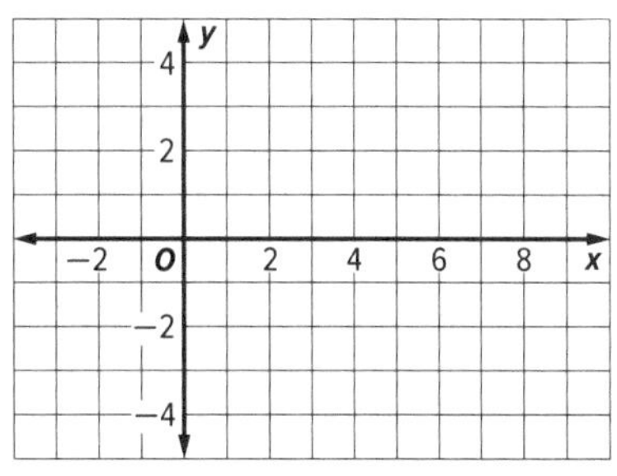

a. USE TOOLS Graph the functions $f(x)$ and $g(x)$ on your calculator and sketch the results on the coordinate grid provided. Based on the graphs, how many solutions do you expect the equation $f(x) = g(x)$ to have? Explain. CCSS A.REI.11, SMP 5

b. CALCULATE ACCURATELY Algebraically solve $f(x) = g(x)$ for all solutions. How can the graph from **part a** be used to help solve the equation? CCSS SMP 6

c. REASON QUANTITATIVELY Before solving $f(x) = g(x)$, is there any way to recognize potential extraneous solutions to the equation based on the equations of $f(x)$ and $g(x)$? Explain. CCSS SMP 2

When solving rational equations, special attention needs to be taken to avoid extraneous solutions. There are also challenges involved in solving rational inequalities. Recall that multiplying or dividing both sides of an inequality by a quantity does not necessarily yield an equivalent inequality, particularly when that quantity is negative.

EXAMPLE 3 Solving Rational Inequalities Algebraically CCSS A.REI.2

Let $f(x) = \frac{2}{x - 4}$ and $g(x) = x - 3$.

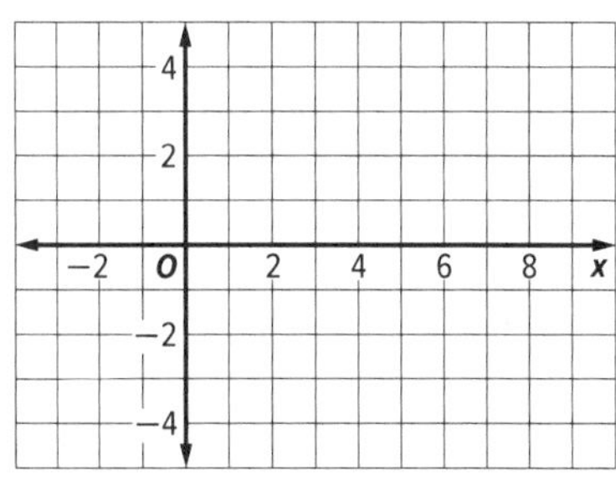

a. USE TOOLS Graph $f(x)$ and $g(x)$ with your graphing calculator and sketch the results on the coordinate grid provided. How can the inequality $x - 3 < \frac{2}{x - 4}$ be interpreted graphically? CCSS SMP 5

b. REASON ABSTRACTLY What can you determine about the x values at which the graph of $f(x)$ changes from below the graph of $g(x)$ to above the graph of $g(x)$ or vise versa? CCSS SMP 2

c. COMMUNICATE PRECISELY How can your answer from **part b** help determine the solution set to the inequality? Explain your reasoning. CCSS SMP 6

d. USE STRUCTURE Determine the x-values for which $f(x)$ or $g(x)$ is undefined and the solutions to $f(x) = g(x)$. Use these values and your sketch of the graphs of $f(x)$ and $g(x)$ from **part a** to write the solution set to the inequality $x - 3 < \frac{2}{x - 4}$. Graph the solution set on a number line. CCSS SMP 7

EXAMPLE 4 Connect Inequalities to the Coordinate Plane CCSS A.REI.11

Complete the following exploration to understand the relationship between a system of functions and solving a rational inequality graphically.

a. CALCULATE ACCURATELY Given $f(x) = \frac{x}{4} - \frac{5}{x + 2}$ and $g(x) = \frac{x + 1}{8}$, for what values of x is $f(x)$ undefined? $g(x)$? CCSS SMP 6

b. USE TOOLS Graph $f(x)$ and $g(x)$ with your graphing calculator. How might you identify the inequality $\frac{x}{4} - \frac{5}{x + 2} > \frac{x + 1}{8}$ on the graph? CCSS SMP 5

c. CRITIQUE REASONING Saraya sketched the graph shown below on the left to identify the solutions to the inequality in **part b**. Her classmate, Jacob, sketched the graph shown below on the right. Which student has properly identified the solution set for the inequality in **part b**? Justify your answer. CCSS SMP 3

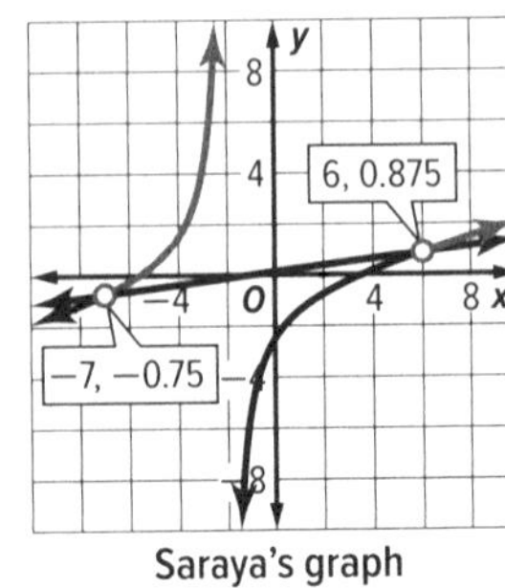

Saraya's graph

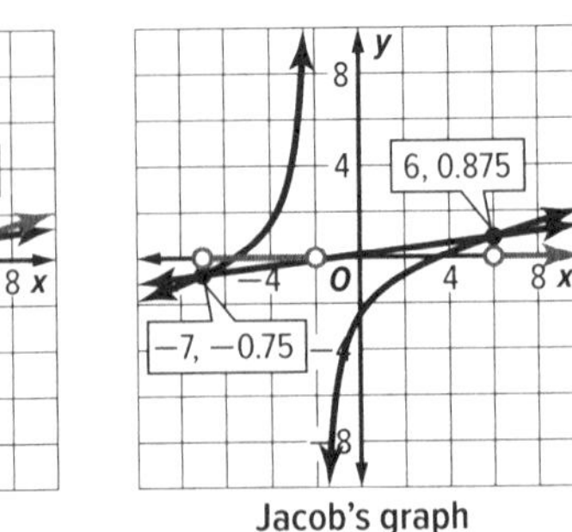

Jacob's graph

d. **CALCULATE ACCURATELY** Use the graph to help you write the intervals that correspond to the solutions for the inequality in **part b**. Explain why there is an open circle at $x = -2$. CCSS SMP 6

EXAMPLE 5 Higher Powers in Rational Inequalities CCSS A.CED.1

Solve the inequality $\frac{(x+1)(x-1)^2(x+2)}{(x-1)(x+3)^2(x-2)^3} < 0$.

a. **REASON ABSTRACTLY** Consider the factor $x + 1$. When is this factor 0? When is it positive? When is it negative? CCSS SMP 2

b. **REASON ABSTRACTLY** Consider the factor $(x + 3)^2$. When is this factor 0? When is it positive? When is it negative? CCSS SMP 2

c. **REASON ABSTRACTLY** What are all the points where any factor is zero? CCSS SMP 2

d. **REASON ABSTRACTLY** Consider all of the other factors, and fill in the table below with + or −. CCSS SMP 2

	$x < -3$	$-3 < x < -2$	$-2 < x < -1$	$-1 < x < 1$	$1 < x < 2$	$x > 2$
$x + 1$						
$(x-1)^2$						
$x + 2$						
$x - 1$						
$(x+3)^2$						
$(x-2)^3$						

e. **REASON ABSTRACTLY** Given that the function is a product and quotient of all of the factors above, complete the table below. CCSS SMP 2

	$x < -3$	$-3 < x < -2$	$-2 < x < -1$	$-1 < x < 1$	$1 < x < 2$	$x > 2$
$\frac{(x+1)(x-1)^2(x+2)}{(x-1)(x+3)^2(x-2)^3}$						

f. **REASON ABSTRACTLY** Using the table from **part e**, write down the solution for the original inequality $\frac{(x+1)(x-1)^2(x+2)}{(x-1)(x+3)^2(x-2)^3} < 0$. Explain your reasoning. CCSS SMP 2

1. **USE A MODEL** It takes one fuel line 3 hours to fill an oil tanker. How fast must a second fuel line be able to fill the oil tanker so that, when used together, the two lines will fill the tanker in 45 minutes? CCSS SMP 4

   a. Write an equation to solve for the number of hours, x, that it will take the second fuel line to fill a tanker. Explain your solution. CCSS A.CED.1

   b. Graph the function in the coordinate grid at right. Use your graph to approximate the solution. Explain your method. CCSS A.REI.11

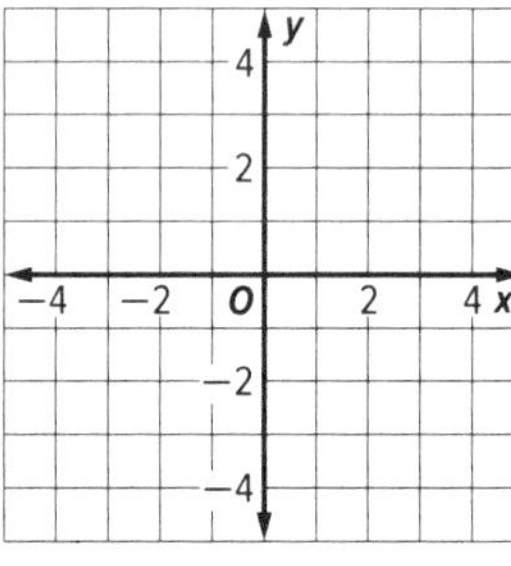

2. Given $\frac{5x}{x-2} = 7 + \frac{10}{x-2}$. CCSS A.REI.2

   a. **USE STRUCTURE** Solve the equation for x. Is your solution extraneous? Explain how you know. CCSS SMP 7

   b. **PLAN A SOLUTION** How can you use graphical methods to check your answer for **part a**? CCSS SMP 1

3. **USE TOOLS** Graph $f(x) = \frac{2}{x^2 - x}$ and $g(x) = \frac{1}{x-1}$ on your graphing calculator. Then use the graph to solve $\frac{2}{x^2 - x} = \frac{1}{x-1}$. Write your solution below and explain why this method works. CCSS A.REI.11, SMP 5

4. Let $f(x) = -x - 2$ and $g(x) = -\frac{3}{x}$. CCSS A.REI.2

   a. **CALCULATE ACCURATELY** Find any x values for which $f(x)$ or $g(x)$ is undefined and any solutions to $f(x) = g(x)$. CCSS SMP 6

   b. **USE STRUCTURE** Use your answer from **part a** and the aid of a graph to write the solution set to the inequality $f(x) \leq g(x)$. Graph your results on a number line. CCSS SMP 7

c. **COMMUNICATE PRECISELY** Explain why each value from **part a** should be included or excluded from the solution set to the inequality. CCSS SMP 6

5. The prom committee is spending \$1000 for the venue and DJ plus \$35 per catered meal. How many students must attend the prom to keep the ticket price below \$60?

a. **COMMUNICATE PRECISELY** Write an inequality that describes this situation. Explain each term in your inequality. CCSS A.CED.1, SMP 6

b. **USE TOOLS** A graph of the function that models the average ticket cost for prom is shown at right. Use it to approximate the answer to this problem. Explain your method. CCSS A.REI.11, SMP 5

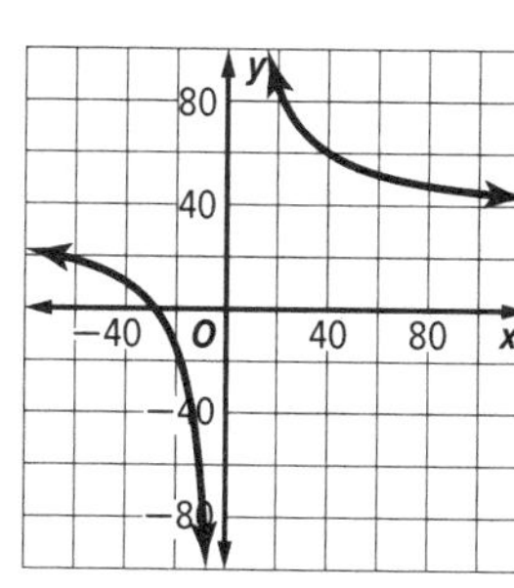

6. **CRITIQUE REASONING** Hideki used the rational function $f(x) = \dfrac{x^2 + x - 12}{x^2 - 2x - 3}$ to determine the solutions of $f(x) = 0$. He found that x could be 3 or −4. Check Hideki's work below and state whether you agree or disagree with his solutions and why. CCSS A.REI.2, SMP 3

Hideki's Work:

$\dfrac{(x^2 + x - 12)}{(x^2 - 2x - 3)} = 0$

$\dfrac{(x^2 + x - 12)}{(x^2 - 2x - 3)} \bullet (x^2 - 2x - 3) = 0 \bullet (x^2 - 2x - 3)$

$x^2 + x - 12 = 0$

$(x - 3)(x + 4) = 0$

$x = 3$ or $x = -4$

7. **CRITIQUE REASONING** A student is trying to solve the inequality $\frac{1}{x} < 2$ and finds the solution set to be $\frac{1}{2} < x$. Her work is shown below. State whether you agree or disagree with her solution set and why. What is one method this student could have used to check her work? CCSS A.REI.2, SMP 3

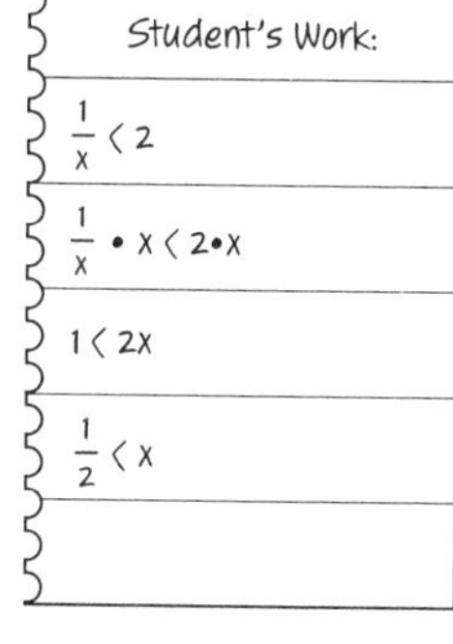
Student's Work:

$\frac{1}{x} < 2$

$\frac{1}{x} \bullet x < 2 \bullet x$

$1 < 2x$

$\frac{1}{2} < x$

Performance Task

Cost Over Time

Provide a clear solution to the problem. Be sure to show all of your work, include all relevant drawings, and justify your answers.

Ramon needs a new lamp. He has narrowed down his choices to the ones shown in the table.

Model	Initial Price	Energy Cost
A	50	$12/year
B	70	$15/year
C	80	$9/year
D	85	$12/year

Part A

Ramon really likes the looks of Model A, and the initial price is great. Assuming that the only costs associated with the lamp are the initial purchase and the energy costs, what is the annual cost of Model A as a function of years x? Explain your reasoning.

Part B

Sketch the graph of the function you found in **Part A.** Explain the domain and any zeros or asymptotes.

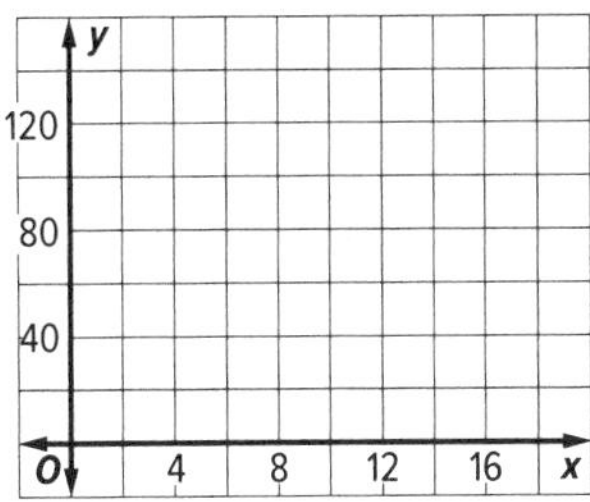

Part C

Ramon has decided he will either get Model A or Model C. Use the grid below to sketch a graph of both models. Then discuss which model you would prefer, using key features of the graph to justify your answer.

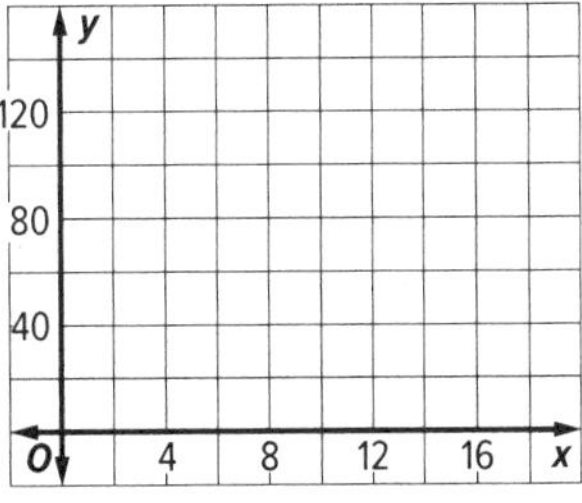

Performance Task

Flying Against The Wind

Provide a clear solution to the problem. Be sure to show all of your work, include all relevant drawings, and justify your answers.

When an airplane flies from one destination to another, the wind speed and direction have an effect on how long the trip takes. At high altitudes, such as those at which planes fly (30,000–40,000 feet), wind speeds and directions are fairly constant. This effect is known as the *jet stream*. For this task, we will assume that the wind speeds and directions are constant.

Part A

Suppose a plane has a constant speed of x on a trip for distance d. The wind speed is c. Write and simplify into a one-term rational function an expression for the round trip flight time if the plane flies against the wind traveling to its destination and with the wind on the return trip.

Part B

Airplanes typically travel at speeds around 470 mph, and the round trip from New York City to San Francisco and back takes about 12 hours. The flight path is the same on both trips, and is about 2800 miles one way. Assuming the wind speed and direction are constant, what is the wind speed for this trip?

Part C

Eric found data on the internet for the speed of the jet stream along the route that his flight would take, and used this data to calculate that the combined round-trip time for the 800-mile, one-way flight would be 50 hours if the speed of the airplane was a constant 500 miles per hour.

i) Determine the value Eric used for the wind speed.

ii) Do you think the data Eric used was valid? Explain your reasoning.

Standardized Test Practice

1. Multiply. Identify any values of x for which the expression is undefined. CCSS A.APR.7

$\frac{2x^2+7x+3}{x^2-2x-15} \cdot \frac{x^2-12x+32}{2x^2-7x-4}$

$=$ ☐

2. A rectangular playground is divided into a section for toddlers and a section for older children, as shown in the diagram. CCSS A.CED.2

Toddlers	Older Children
area = 104 yd^2	area = 195 yd^2

What expression gives the perimeter of the entire playground?

If the perimeter of the playground is 72 yards, what are the possible values of x?

3. Subtract. Identify any values of x for which the expression is undefined. CCSS A.APR.7

$\frac{2x-5}{x+3} - \frac{x+1}{x+5}$

$=$ ☐

4. What values of x satisfy the following inequality? CCSS A.REI.2

$\frac{8}{2x+1} \leq 3$

☐

5. Divide. Identify any values of x and y for which the expression is undefined. CCSS A.APR.7

$\frac{x-2}{x^2y} \div \frac{x^2-4}{x^2+2x}$

$=$ ☐

6. The sum of twice a number and three times its reciprocal is 8.3. What is the number? CCSS A.CED.1

7. The function shown in the graph below is of the form $y = \frac{ax+b}{x+c}$ and has a vertical asymptote at $x = -2$.

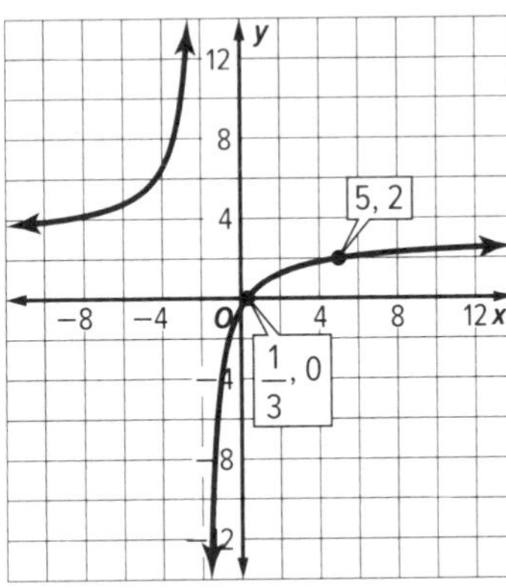

What are the values of a, b, and c? CCSS F.BF.3, A.CED.2

$a =$ ☐ $b =$ ☐ $c =$ ☐

8. The graphs of three functions are shown below. Which of these represents the graph of $y = \frac{2x^2-8}{x^2-x-6}$ CCSS F.IF.9

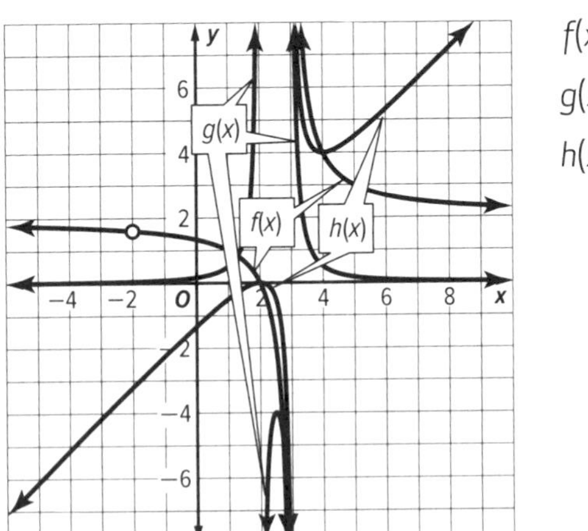

$f(x)$

$g(x)$

$h(x)$

9. Solve the equation $\frac{x+3}{x-1} = 2 + \frac{1}{x-1}$ and determine whether or not the solution is extraneous. CCSS A.REI.2

$x =$ ☐, the solution is ☐.

10. For each function, identify any zeros, vertical and horizontal asymptotes, and holes. CCSS A.CED.2, F.IF.4

Function	Zeros	Vertical Asymptotes	Horizontal Asymptotes	Holes
$y = \frac{1}{x+3} + 7$				
$y = \frac{x^2 + 2x - 15}{x + 5}$				
$y = \frac{9x^2 - 4}{4x^2 - 25}$				

11. Nina kayaks in still water at a speed of 6 miles per hour. She kayaks downstream, turns around, and returns upstream to her starting point. Her entire trip is 18 miles and takes 3.2 hours. What is the speed of the current? Show your work. CCSS A.CED.1, A.REI.2, A.CED.4

12. The track team wants to have water bottles made with their team logo. The table shows the set-up fee and the price per water bottle for two different companies.

Company	Set-up Fee	Price per Water Bottle
Sports Accessories	\$125	\$5
Outdoor Equipment	\$200	\$3

a. Write a function, $f(x)$, that gives the average cost per water bottle for an order of x water bottles from Sports Accessories. Write another function, $g(x)$, that gives the average cost per water bottle for an order of x water bottles from Outdoor Equipment. CCSS A.CED.2

b. Graph the two functions in the first quadrant. CCSS A.REI.11

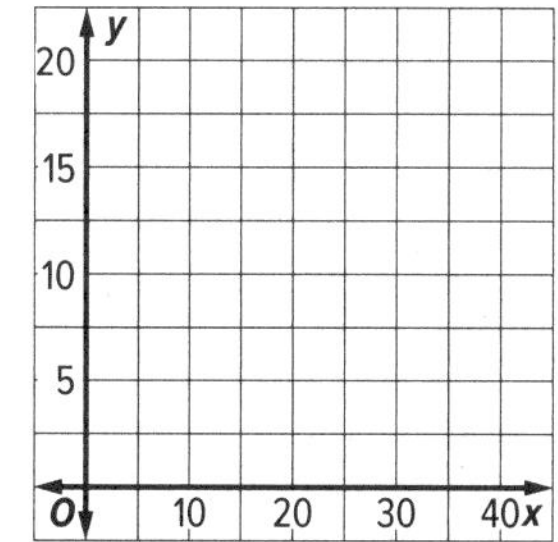

c. Explain what the graph tells you about the costs for the two companies. CCSS A.REI.11

9 Statistics and Probability

CHAPTER FOCUS Learn about some of the Common Core State Standards that you will explore in this chapter. Answer the preview questions. As you complete each lesson, return to these pages to check your work.

What You Will Learn	Preview Question
Lesson 9.1: Designing a Study	
CCSS S.IC.3 Recognize the purposes of and differences among sample surveys, experiments, and observational studies; explain how randomization relates to each. CCSS S.IC.5 Use data from a randomized experiment to compare two treatments; use simulations to decide if differences between parameters are significant.	CCSS SMP 1 Andrea wants to know how much students at her school spend on music, on average, each week. Describe the method she could use to collect the data. How could Andrea be sure that the data is representative of the students at her school?
Lesson 9.2: Simulations and Margins of Error	
CCSS S.IC.4 Use data from a sample survey to estimate a population mean or proportion; develop a margin of error through the use of simulation models for random sampling. CCSS S.IC.2 Decide if a specified model is consistent with results from a given data-generating process, e.g., using simulation. **Also addresses:** S.IC.6	CCSS SMP 1 Suppose you wanted to estimate the average age of a group of 5000 people. Which would give more accurate results, surveying 50 people or 500? Explain.
Lesson 9.3: Distributions of Data	
CCSS S.IC.1 Understand statistics as a process for making inferences about population parameters based on a random sample from that population.	CCSS SMP 4 Suppose you wanted a breakdown of the types of vehicles parked at a large football stadium. How could you estimate the number of cars, trucks, and SUVs?

What You Will Learn	Preview Question
Lesson 9.4: Probability Distributions	
CCSS S.MD.7 Analyze decisions and strategies using probability concepts (e.g., product testing, medical testing, pulling a hockey goalie at the end of a game).	CCSS SMP 5 If you were evaluating a new baseball player, what sorts of statistics would you look at? How would they influence you?
Lesson 9.5: The Binomial Theorem	
CCSS A.APR.5 Know and apply the Binomial Theorem for the expansion of $(x + y)^n$ in powers of x and y for a positive integer n, where x and y are any numbers, with coefficients determined for example by Pascal's Triangle.	CCSS SMP 2 Complete the binomial expansion of $(x + y)^4$. $\square x^4 + \square x^3y + \square x^2y^2 + \square xy^3 + \square y^4$
Lesson 9.6: The Binomial Distribution	
CCSS S.MD.6 Use probabilities to make fair decisions (e.g., drawing by lots, using a random number generator). CCSS S.MD.7 Analyze decisions and strategies using probability concepts (e.g., product testing, medical testing, pulling a hockey goalie at the end of a game). **Also addresses:** A.APR.5	CCSS SMP 6 Suppose you toss a coin five times and find repeatedly the probability of getting five heads is 1%. What can you conclude about the coin?
Lesson 9.7: The Normal Distribution	
CCSS S.ID.4 Use the mean and standard deviation of a data set to fit it to a normal distribution and to estimate population percentages. Recognize that there are data sets for which such a procedure is not appropriate. Use calculators, spreadsheets, and tables to estimate areas under the normal curve.	CCSS SMP 1 Which fraction best represents the amount of a normal distribution within one standard deviation from the mean?
Lesson 9.8: Confidence Intervals and Hypothesis Testing	
CCSS S.MD.7 Analyze decisions and strategies using probability concepts (e.g., product testing, medical testing, pulling a hockey goalie at the end of a game). CCSS S.IC.4 Use data from a sample survey to estimate a population mean or proportion; develop a margin of error through the use of simulation models for random sampling. **Also addresses:** S.IC.1	CCSS SMP 6 What is a standardized test? What does the average score on a standardized test mean?

9.1 Designing a Study

Objectives

- Identify the characteristics of different study types.
- Collect data from a study to make comparisons.

Content: S.IC.3, S.IC.5
Practices: 2, 3, 5, 6, 8
Use with Lesson 11-1

A statistical study involves the collection and analysis of data to answer questions about a population characteristic.

Samples of the population are most often used for studies where it is impractical or too expensive to collect data from a large population. A random sample, one where each member of the population has an equal chance of being selected, will provide data with the least chance of bias. The type of study used depends on the population and the information that will be collected.

KEY CONCEPT Types of Studies

Type	Definition	Sample Study
experiment	a study that compares characteristics of an experimental group to those of a control group	a study compares the growth of plants that are fertilized with those that are not fertilized
observational study	a study in which the sample population is measured or surveyed as is without any external influence	scientists observe the feeding habits of birds that live in the vicinity of two different lakes
survey	a study which collects data from responses given by members of a population regarding their characteristics, behaviors, or opinions	a teacher asks a random sample of students about the number of text messages they send each day

EXAMPLE 1 Analyze a Study on Exercise Habits CCSS S.IC.3

EXPLORE **Students in a health class are conducting a study on exercise habits. They want to determine whether the time spent exercising outside of school hours by those taking a physical education class is generally greater than, less than, or the same as the time spent exercising by those not enrolled in a physical education class.**

a. **DESCRIBE A METHOD** Which study type should the class use to collect the data? Compare and contrast the study type you choose with the other types in the context of the problem to explain your choice. CCSS SMP 8

b. CRITIQUE REASONING Lucy suggests questioning students on the track team when they stay after school for yearbook photos because they will all be in one place. She states that these students could be used to represent a random sample of those who do not take a physical education class. Describe two errors in Lucy's reasoning. CCSS SMP 3

c. DESCRIBE A METHOD Suggest a method for selecting an unbiased random sample of students who are not taking a physical education class. How would the random selection process be different if a different type of study were being used? CCSS SMP 8

d. COMMUNICATE PRECISELY The graph shows the results of the study. Group *A* is the students taking physical education, where group *B* is the students not taking physical education. The vertical axis is the average number of minutes spent exercising per day. Is there a difference between the two groups? Explain. CCSS S.IC.5, SMP 6

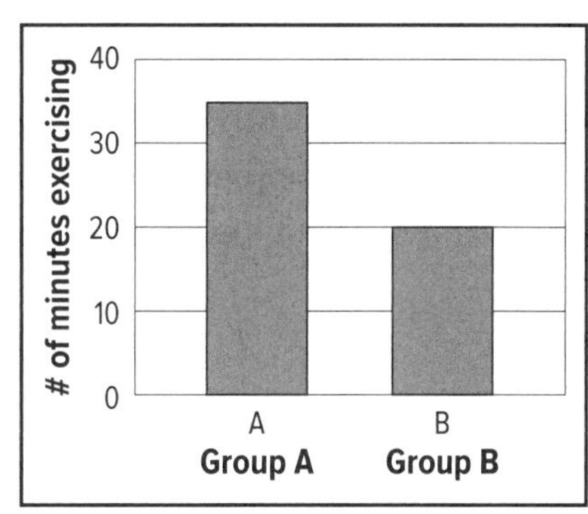

EXAMPLE 2 Compare Two Treatments CCSS S.IC.5

A Web site offers a new technique that the developers claim improves one's memory. To test this hypothesis, researchers set up an experiment in which 20 adults were randomly divided into two groups. An experimental group of 10 adults were asked to visit the Web site and learn the memory technique. Then members of both groups were given a memory test. Scores by members of the two groups are shown in the table.

Memory Test Scores					
Control Group	242	248	246	246	245
	247	244	246	247	244
Experimental Group	250	249	251	247	248
	252	249	251	250	248

a. CALCULATE ACCURATELY Determine the mean score, $\overline{X}_c$, for the control group and the mean score, $\overline{X}_e$, for the experimental group. What is the difference in the means, $\overline{X}_e - \overline{X}_c$? CCSS SMP 6

b. **REASON ABSTRACTLY** The researchers want to know if the difference in the mean scores is significant (that is, due to the memory technique) or if it is simply due to chance. One way to determine this is to "rerandomize" the two groups. That is, randomly divide the 20 participants into two groups many times and calculate the difference in the means for each new grouping. What could you conclude if the difference in the means that you found in **part a** occurs frequently? What could you conclude if it does not occur frequently? CCSS SMP 2

c. **USE TOOLS** Number the participants in the table from 1 to 20. Then use your calculator's random number generator to simulate choosing a random set of 10 participants for the control group and a random set of 10 participants for the experimental group. Calculate the mean score for each group and the difference of the means. Repeat the process three more times. Record your results below. CCSS SMP 5

Simulation	$\overline{X}_c$	$\overline{X}_e$	$\overline{X}_e - \overline{X}_c$
1			
2			
3			
4			

d. **CONSTRUCT ARGUMENTS** Compare your results with those of other students. Does a difference between the means of 4 or more appear frequently? State a conclusion about the results of the experiment and justify your answer. CCSS SMP 3

PRACTICE

DESCRIBE A METHOD **For Exercises 1–2, explain whether an experiment, an observational study, or a survey is best for achieving the indicated goal. If applicable, indicate under what conditions another type of study might be used.** CCSS S.IC.3, SMP 8

1. The buyer of a new car wants to determine whether it will be more cost efficient to run the car she chose using E85 flex-fuel or regular gasoline.

2. A pizza shop owner wants to determine how satisfied his customers are with a new special that he has on the menu.

3. **DESCRIBE A METHOD** A playground designer wants to complete a table like the one below for 30 children at a playground in order to determine the equipment and safety features to use in the construction of a new neighborhood playground. Explain which type of study the designer will use to complete the table. Explain whether or not random sampling is likely to be part of the process. CCSS S.IC.3, SMP 8

Subject Number	Approximate Age and Weight	Structures Played On and Approximate Length of Time	Incidents
1	6; 45 lb	Swing; 5 min.	None

4. **COMMUNICATE PRECISELY** Jamar wants to know the average time his schoolmates spend traveling to school in the morning. Describe the method he should use to collect the data and explain how randomization relates to your choice. CCSS S.IC.3, SMP 6

5. **CONSTRUCT ARGUMENTS** A cat rescue group wants to determine whether feeding formerly stray cats a new grain-free food helps them gain weight more quickly. They perform an experiment by giving a control group of 10 cats regular food and an experimental group of 10 cats the new food. They record the cats' weight gains after two weeks, as shown in the table. Is the mean weight gain with the new food greater than that with the regular food? If so, do you think this is significant (that is, due to the new food) or due to chance? Explain. CCSS S.IC.5, SMP 3

Weight Gain (oz)			
Control Group		Experimental Group	
6	5	4	4
3	6	7	8
8	4	5	3
5	9	7	6
7	4	9	5

9.2 Simulations and Margins of Error

Objectives

- Decide if a model will simulate a desired set of outcomes.
- Develop a margin of error.
- Evaluate reports based on data.

CCSS STANDARDS

Content: S.IC.2, S.IC.4, S.IC.6
Practices: 1, 2, 3, 4, 5, 6, 7
Use with Extend 11-1

When creating a survey or poll it is not always possible to survey the whole population. Often, a **random sample** of the population is taken instead. To test the results of the sample, more samples can be taken. This can often be more time consuming and costly than is possible. A simulation or model is often used to imitate a sampling situation.

To design a simulation:

- Identify the event to be simulated.
- Identify a model for the event (coin flip, spinner, random numbers) so that the outcomes with the model have the same probability as the outcomes for the event.
- Define what each outcome in the simulation will represent.
- Define what each trial will consist of and determine the number of trials needed to simulate the event. Then conduct the simulation.
- Summarize the outcomes of the simulation in terms of the event that was modeled.

EXAMPLE 1 Using a Simulation to Test Data CCSS S.IC.2

EXPLORE **The principal of a high school randomly chooses 1 of the four classes each week to be first in the lunch line. Three weeks in a row he has chosen the junior class. The senior class president states that there is something wrong with the principal's selection process.**

a. **USE STRUCTURE** Create a model for obtaining a sample of picking 1 of the four classes for 3 weeks. Describe how to use your model to test the principal's selection process. CCSS SMP 7

b. USE A MODEL Use the model to conduct 50 trials and record each set of 3 numbers spun in the table below. Based on your results, what is the experimental probability of getting 3-3-3? CCSS SMP 4

c. MAKE A CONJECTURE Find the theoretical probability of choosing the junior class three weeks in a row. CCSS SMP 3

d. CRITIQUE REASONING Do the outcomes of the simulation support the process used by the principal or the statement made by the senior class president? CCSS SMP 3

Random samples are used to estimate the mean of a parameter for an entire population. A simulation can then be used to model the results of the sample.

EXAMPLE 2 Use Data from a Random Sample to Find a Population Mean CCSS S.IC.4, S.IC.6

Ms. Johnson's class wanted to find out the average number of hours that ninth graders were spending doing homework during a week. The class randomly surveyed ten students, and the survey produced the following data set.

Student	A	B	C	D	E	F	G	H	I	J
Hours of Homework in a Week	5	4.5	5	3	3.5	6	4	4.5	3.5	5.5

a. PLAN A SOLUTION Identify the sample and the population of this study. CCSS SMP 1

b. REASON QUANTITATIVELY Use the sample mean to estimate the population mean. Why is this only an estimate? What factor would contribute to the accuracy of the estimate? CCSS SMP 2

c. CRITIQUE REASONING JoAnna from Ms. Johnson's class claims that the original sample is inadequate because of the small size. She has collected a sample that is much bigger. JoAnna surveys all 30 students in Ms. Johnson's class and find that the average number of hours spent on homework per week among this sample of size 30 is 7.2 hours a week. JoAnna claims that this is a much better estimate. Is she correct? Explain. CCSS SMP 3

d. COMMUNICATE PRECISELY Describe why it would be irresponsible to use this sample to generalize to the entire population of freshmen in the United States. What would need to be done in order to make this generalization? CCSS SMP 6

Conclusions based upon random samples are subject to the laws of probability that are introduced when the sample is selected. The **margin of error** is a range of values above and below the random sample proportion that indicate the likeliness of the mean for the entire population. The greater the margin of error, the less likely the sample is a true representation of the population. If the probability of a particular outcome in a random sample is $P(X)$ with a margin of error $\pm E$, then the probability of that outcome in the population is most likely within the interval $(P(X) - E, P(X) + E)$. The margin of error is related inversely to the sample size n. As the sample size increases, the margin of error decreases. The margin of error is approximated by the following formula.

Margin of Error Percent

$$\text{Margin of error} = \pm\frac{1}{\sqrt{n}}(100)$$

EXAMPLE 3 Margin of Error

As part of a class project, students are asked to survey a random sample of the entire school population and determine which sport they most like to watch on TV.

a. **REASON QUANTITATIVELY** Cindy made handouts of her survey to be given to 250 randomly chosen students. The results of her survey indicated that 32% said they most liked to watch soccer. Find the margin of error for her sample. What is the likely interval that contains the percent of the population who claim that soccer is their favorite sport to watch on TV? CCSS S.IC.4, SMP 2

b. **REASON QUANTITATIVELY** Jarod conducted a survey of 20 people, of which 70% said that football is their favorite sport to watch on TV. What is the margin of error and interval for this survey? What is the meaning of the interval in the context of the problem? CCSS S.IC.4, SMP 2

c. **COMMUNICATE PRECISELY** Why is the margin of error so different for Cindy and Jarod? CCSS S.IC.4, SMP 6

d. **REASON QUANTITATIVELY** Bambi surveyed the 35 members of the softball team and found that 95% stated that baseball was their favorite sport to watch on TV. What is the margin of error in her sample? Of the 3 students, whose study is the most reliable and why? CCSS S.IC.6, SMP 2

PRACTICE

1. **USE A MODEL** Forty-two students participated in a lottery with three prizes. Fifteen of the participants were members of the basketball team. All three winners were members of the basketball team. The other students claim the lottery was unfair. CCSS S.IC.2, S.IC.6, SMP 4

   a. **COMMUNICATE PRECISELY** Describe a model or simulation you could use to test the results of the lottery. CCSS SMP 6

b. **USE TOOLS** Run your simulation and record your results in the table. Then analyze the data. Is the data from your simulation consistent with the model? CCSS SMP 5

c. **CRITIQUE REASONING** One of the students who is not on the team says that your simulation proves that the lottery was not fair. Do you agree? Explain your reasoning. CCSS SMP 3

2. **REASON QUANTITATIVELY** A local television station reports that a sample poll indicated that 46% to 58% of the registered voters in a city plan to vote for Candidate A. CCSS S.IC.4, SMP 2

a. What is the sample mean of the population? What does this mean?

b. What is the margin of error? What does this mean?

c. How many registered voters were in the sample?

d. How can the television station reduce the margin of error to be more confident in applying the sample mean to the population of registered voters?

3. **CRITIQUE REASONING** Two students are campaigning for Student Council President. A survey of 25 students indicates that 62% of students plan to vote for Candidate A for Student Council President. Another survey of 50 students shows that 25% of students plan to vote for Candidate B. Karla says that they cannot both be right. Do you agree? Assume that all students vote for one of the candidates. Explain your reasoning. CCSS S.IC.6, SMP 3

4. **REASON QUANTITATIVELY** It is hypothesized that teens aged 16–18 tend to have more traffic violations than the rest of drivers. A random study of 10,000 drivers between the ages of 16 and 18 found that 18% of the drivers received a traffic citation in the last twelve months. CCSS S.IC.4, SMP 2

   a. What is the margin of error?

   b. Using the sample mean and the margin of error, what can you conclude? Be sure to correctly identify the population.

   c. A different group of researchers then randomly selected a group of only 100 drivers of all ages and asked the same question. Ten drivers of the 100 indicated that they had received a traffic citation in the last twelve months. What is the margin of error, the sample mean, and what can you conclude? Again, be sure to correctly identify the population.

   d. Is there anything that can be reasonably concluded from these two studies about the traffic citations of teenaged drivers compared to all drivers? What would be needed to make this study better?

9.3 Distributions of Data

Objectives

- Determine the shape of the distribution for a set of data.
- Use the proper measures to make inferences and justify conclusions from random samples.

Content: S.IC.1
Practices: 3, 4, 5, 6
Use with Lesson 11-2

Analyzing the shape of a distribution for a random sample of data can help you decide which measure of center or spread best describes a set of data. Histograms and box-and-whisker plots can be used to better see the shape of the distribution.

Symmetric and Skewed Distributions

Symmetric Distribution

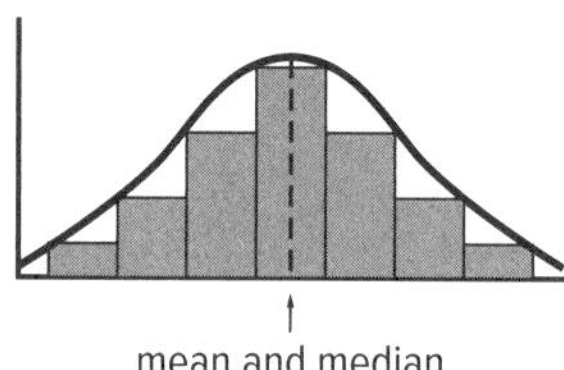

The mean and the median are approximately equal, and the data are evenly distributed on both sides of the mean. Use the mean and standard Deviation to describe the center and spread of data with a symmetric distribution.

Negatively Skewed Distribution

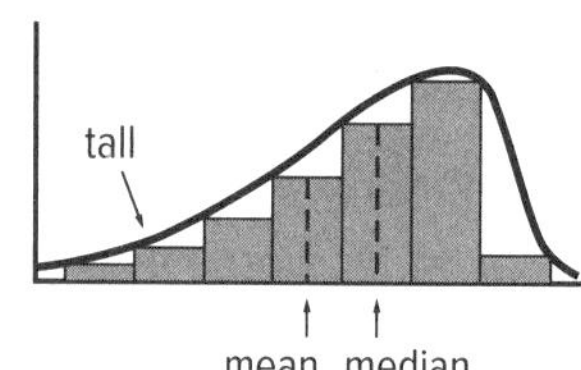

The mean is less than the median, and the majority of the data are on the right of the mean. Use the five-number summary to describe the center and spread of data with a skewed distribution.

Positively Skewed Distribution

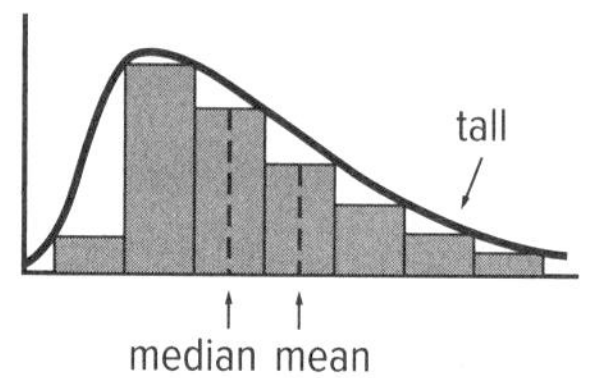

The mean is greater than the median, and the majority of the data are on the left of the mean. Use the five-number summary to describe the center and spread of data with a skewed distribution.

EXAMPLE 1 Using a Histogram CCSS S.IC.1

EXPLORE **A scientist is collecting data on the wingspan of a certain type of insect in a field near his laboratory. He captured and released 60 of these insects and measured their wingspan, rounding to the nearest millimeter. He recorded the number of times each wingspan occurred in the following chart.**

Wingspan (mm)	10	11	12	13	14	15	16	17	18	19	20
Frequency	4	5	7	14	8	6	5	3	4	3	1

a. **USE A MODEL** Draw a histogram to represent the wingspan data. CCSS SMP 4

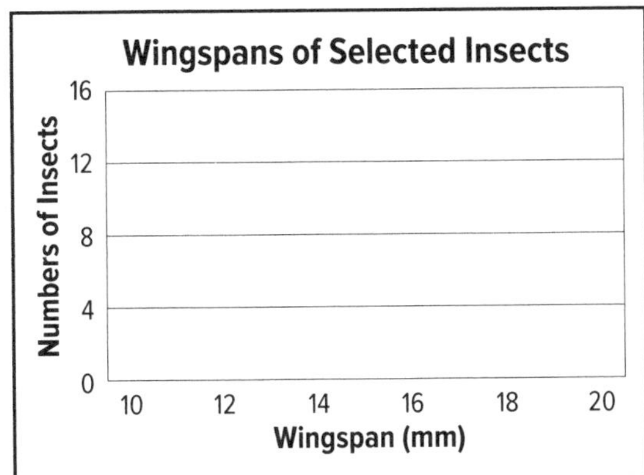

b. **COMMUNICATE PRECISELY** What is the shape of the data? How is it skewed? What does this mean? CCSS SMP 6

c. **CALCULATE ACCURATELY** What is the mean and median of the data? Describe the spread of the data given your description of skew in **part b**. CCSS SMP 6

d. **COMMUNICATE PRECISELY** What does the histogram tell you about the wingspan of these types of insects? CCSS SMP 6

EXAMPLE 2 Using a Box-and-Whisker Plot CCSS S.IC.1

The table shows the results of a survey that asked people how many minutes they spent driving to and from work per week.

Minutes Spent Driving to and from Work			
240	148	300	40
192	182	95	212
200	56	124	210
174	264	276	50

a. **USE A MODEL** Enter the data on a graphing calculator and create a box-and-whisker plot for the data. CCSS SMP 4

b. **COMMUNICATE PRECISELY** Describe the shape of the distribution. Explain your choice. What measures should be used to describe the center and spread of the data? CCSS SMP 6

c. **USE TOOLS** Use a graphing calculator to find the measures you chose in **part b**. CCSS SMP 5

d. **COMMUNICATE PRECISELY** What is true about the data that lies within the box for the plot in **part a**? Make sure to reference actual values in your answer. CCSS SMP 6

When making comparisons of two different random samples of data for a similar parameter, first find the shape of each distribution. Compare the center and spread of the two sets using the mean and standard deviation if the two sets have a relatively symmetric distribution. Use the five-number summaries if either, or both, of the sets has a skewed distribution.

EXAMPLE 3 Comparing Box-and-Whisker Plots CCSS S.IC.1, SMP 3

CONSTRUCT ARGUMENTS The results from a random sample of test scores on a Math Placement exam by students from the year 1990 and from the year 2000 are illustrated in the box-and-whisker plots.

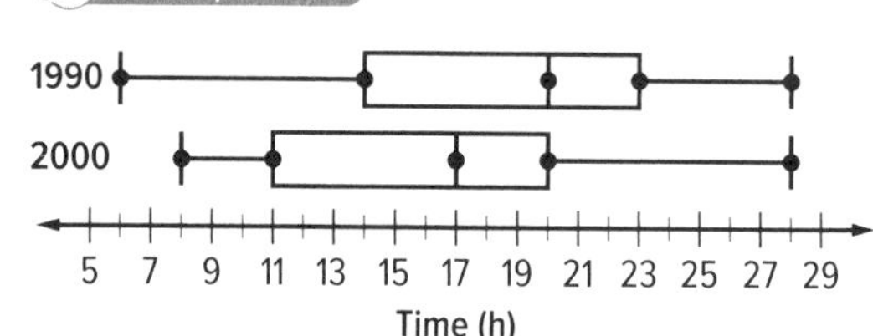

a. What is the shape of each distribution?

b. Describe the center and spread of each distribution.

c. What do these results tell us about how students performed on the 29 question test for the two years?

PRACTICE

1. John and Debra decide to do a study on the weight of rocks that occur in the field outside of their house. They collect 26 different rocks and record their weights in a frequency table. The results are shown below. CCSS S.IC.1

Weight (lbs)	2	3	4	5	6	7	8	9	10	11	12
Frequency	1	2	3	4	5	6	5	4	3	2	1

a. **USE A MODEL** Create a histogram to model the data shown in the frequency table. CCSS SMP 4

b. **USE TOOLS** Describe the histogram. Is it equally likely to get a rock weight greater than 7 lbs. as it is to get a rock weight less than or equal to 7 lbs.? CCSS SMP 5

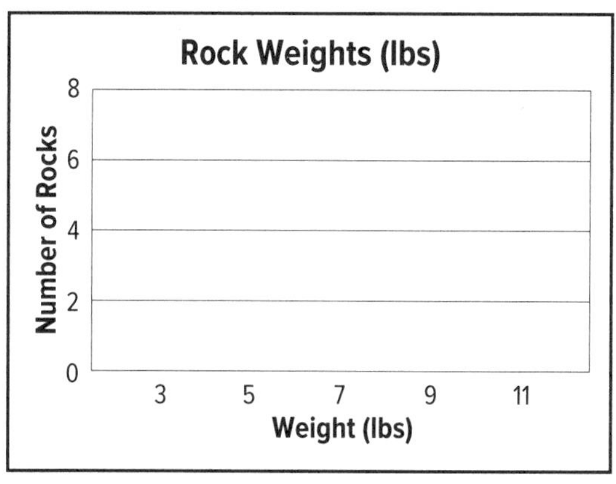

2. **COMMUNICATE PRECISELY** The following box-and-whisker plots show the results of a random sample of scores from two tests. CCSS S.IC.1, SMP 6

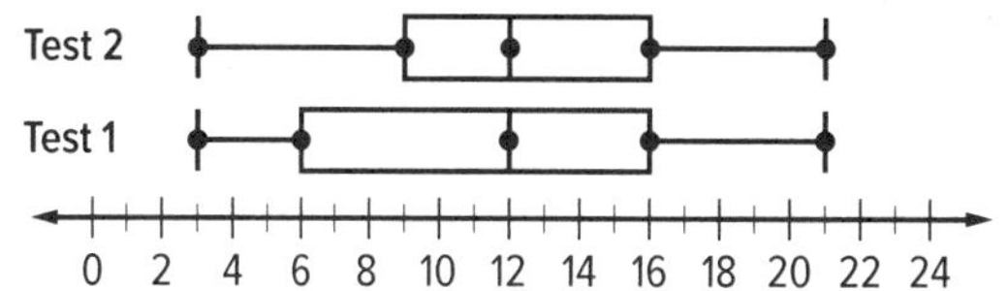

a. Describe the shape and spread for each of the box-and-whisker plots.

b. What do these box-and-whisker plots show about how the random sample of scores from Test 2 compared to Test 1?

3. The Table shows the results of math SAT scores in Mr. Schmetz's first period class. CCSS S.IC.1

710	600	490	500
490	410	520	580
590	610	570	570
520	600	420	490

a. **USE A MODEL** Enter the data on a graphing calculator. Find the quartile values and create a box-and-whisker plot for the data on the graph shown. CCSS SMP 4

410 470 530 590 650 710

b. **COMMUNICATE PRECISELY** Describe the shape of the distribution. What measures should be used to describe the center and spread of the data? Find these measures. CCSS SMP 6

9.4 Probability Distributions

Objectives

- Construct and analyze a probability distribution.
- Calculate the expected value of a random variable of a discrete random variable of a probability distribution.
- Use the statistics of a probability distribution to analyze a decision or strategy.

Content: S.MD.7
Practices: 1, 2, 3, 4, 6, 7, 8
Use with Lesson 11-3

The set of all possible outcomes for a distribution of data is its **sample space**. A **random variable** is defined as the numerical value of a random event. For instance, the roll of a fair die has the sample space {1, 2, 3, 4, 5, 6} and the random variable X is the result of any roll of the die.

A **discrete random variable** has a sample space that is a countable set of values such as the sum of the numbers of two cards drawn from a set of 10 cards numbered 1 through 10. A **continuous random variable** has a sample space consisting of a range of values such as the weight of fish caught by a contestant in a tournament, which might be any number from 0 up to 10 pounds.

EXAMPLE 1 Identify and Classify Random Variables

USE A MODEL **Identify the random variable for each distribution and describe the sample set. Is the random variable discrete or continuous?** CCSS S.MD.7, SMP 4

a. the height in real number inches of all female students at a high school

b. the number of foul shots made successfully out of 10 chances

c. CONSTRUCT ARGUMENTS The graph at right shows the sum of two dice that have been randomly rolled. If Janet rolls a sum of 7, she will not have to take the math test. If she rolls any other sum, she will take the test. Janet claims she has a better chance of not taking the test than taking it. Do you agree with her claim? Why or why not? CCSS SMP 3

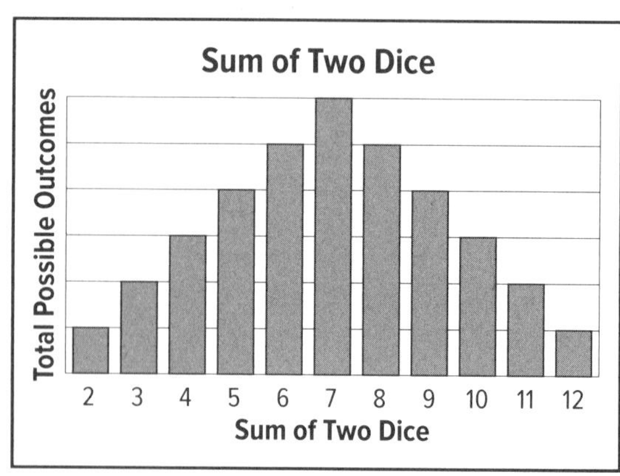

The **probability distribution** for a random variable is a function that pairs the theoretical or experimental probability of each outcome in the sample space with the entire sample space. Experimental probabilities are based on actual outcomes from an experiment while theoretical probabilities are based on the outcomes that are expected. The probability of a random variable is denoted by $P(X)$ and the sum of all the probabilities of a sample set must equal 1. That is, $\Sigma P(X) = 1$. A relative frequency table is a useful tool for constructing a probability distribution.

EXAMPLE 2 Construct and Interpret a Probability Distribution 

Two number cubes are rolled and the product of the resulting numbers is formed.

a. **INTERPRET PROBLEMS** Complete the table for all the possible products, the expected frequencies, and the relative frequencies. Interpret the meaning of the relative frequencies. CCSS SMP 1

Product	1	2	3															
Frequency	1	2	2															
Relative Frequency																		

b. **FIND A PATTERN** Graph the probability distribution for the table in **part a**. Describe the shape of the distribution. Does this represent a theoretical or experimental probability distribution? CCSS SMP 7

c. **EVALUATE REASONABLENESS** Tomas says that he has a higher probability of rolling 12 or less than a product greater than 12. Do you agree? Explain why or why not. CCSS SMP 8

d. **INTERPRET PROBLEMS** Salim, a member of the high school basketball team, is shooting hoops from the three-point line. He records how many hoops he makes in a row before a miss during a practice session. Complete the table by calculating the relative frequencies for how many shots in a row he made and how many times he achieved that number. What pattern do you notice in the relative frequencies? CCSS SMP 1

Number of Hoops	1	2	3	4	5	6	7	12
Frequency	9	4	2	5	1	2	1	1
Relative Frequency	36%							

e. **FIND A PATTERN** Graph the probability distribution for the table in **part d**. Describe the shape of the probability distribution. Is this a theoretical or experimental probability distribution? CCSS SMP 7

The **expected value** of a discrete random variable is the weighted average of the values of the variable, given by $E(X) = \sum[X \cdot P(X)]$, where X is the value of each outcome and $P(X)$ is the probability for that outcome. For example, if you have ten cards with the values 1 through 10 printed on them, the expected value of random draws from these cards would approach 5.5 after an increasing number of draws.

Expected value, though, does not always provide ample information about the data. Consider, for example, two rooms of people, Room A and Room B, which have the same expected value for height. Based on this information alone, the heights in the two rooms appear to be similar; however, while Room A is full of people that are all within 1 inch of the expected value, Room B is full of people that range from below and above 10 inches away from the expected value. This aspect of the data can be captured by the standard deviation, which measures the "spread" of the data from the mean.

The **standard deviation** for a probability distribution is defined as the square root of the variance, which is the sum of the products of the probability of each value of the variable and the square of the difference between each value and the expected value, or mean, of the probability distribution. That is, $\sigma^2 = \sum([X - E(X)]^2 \cdot P(X))$ and $\sigma = \sqrt{\sigma^2}$, where σ^2 is the variance and σ is the standard deviation.

EXAMPLE 3 Compare Distributions Using Mean and Standard Deviation

New products often suffer from a high failure rate during the manufacturing process. Two different companies are producing a ten-pack of lightbulbs using a brand new low-energy technology. Both companies are concerned about quality control, so they randomly pull packs and determine how many faulty bulbs are in each of the ten. The number of faulty lightbulbs in each batch is given in the following lists. CCSS S.MD.7

Company A	0, 0, 0, 1, 1, 1, 1, 1, 1, 1, 1, 1, 1, 2, 3
Company B	0, 0, 0, 0, 0, 0, 0, 0, 0, 1, 2, 3, 3, 3, 3

a. **CALCULATE ACCURATELY** Construct experimental probability distributions for each companies failure rate in ten-packs. CCSS SMP 6

Number of Faulty Bulbs in a Ten-Pack	0	1	2	3
Relative Frequency (Company A)				
Relative Frequency (Company B)				

b. **CALCULATE ACCURATELY** Calculate the mean and standard deviation of each distribution. CCSS SMP 6

c. **REASON QUANTITATIVELY** Based on the results of the two distributions, which company is in a better place with respect to quality control? CCSS SMP 2

PRACTICE

1. a. **MAKE A CONJECTURE** Without making any calculations, predict the general shape of the probability distribution for the *sum* of two number cubes. Explain your reasoning. CCSS S.MD.7, SMP 3

b. **CALCULATE ACCURATELY** Check your prediction by graphing the theoretical probability distribution for the sum of two number cubes. CCSS SMP 6

c. **USE STRUCTURE** Use the structure of the distribution to predict its expected value or mean, explaining your reasoning. Then, check your prediction. CCSS SMP 7

2. A basketball player is training for a three-point competition. The competition will consist of shooting 20 shots. So to practice, the player shoots 20 three-point shots every morning before heading to school. The following table shows the number of shots out of 20 that the player made over the course of two weeks. CCSS S.MD.7

Monday	Tuesday	Wednesday	Thursday	Friday
17	16	16	17	18
15	17	16	18	16

a. **CALCULATE ACCURATELY** Find the probability distribution for this situation, and then find the mean. CCSS SMP 6

Number Made (x)	15	16	17	18	19	20
$P(X = x)$						

b. **REASON QUANTITATIVELY** Given this distribution, how many shots can the player expect to make during the competition? Why might this data be limited for predicting his success? CCSS SMP 2

3. Eggs are sold in packs of 12, but occasionally there are cracked eggs in a pack. The following is a relative frequency table of the number of cracked eggs that show up in two different egg companies' products. CCSS S.MD.7

Number of Cracked Eggs in a Package	0	1	2	3	4	5	6	7	8	9	10	11	12
Relative Frequency for Company A	76%	13%	8%	3%	0%	0%	0%	0%	0%	0%	0%	0%	0%
Relative Frequency for Company B	71%	19%	7%	2%	1%	0%	0%	0%	0%	0%	0%	0%	0%

a. **CALCULATE ACCURATELY** For each distribution, find the mean and standard deviation. CCSS SMP 6

b. **COMMUNICATE PRECISELY** Interpret the mean for each distribution in the context of cracked eggs. Which company seems to be doing a better job of quality control? CCSS SMP 6

4. **a.** **CRITIQUE REASONING** Kara presents the following probability distribution as an answer to an in-class exercise. Maia says right away that Kara cannot possibly be correct even though she doesn't know the exact context of the problem. How does Maia know this? CCSS S.MD.7, SMP 3

Value (x)	1	2	3
$P(X = x)$	$\frac{2}{5}$	$\frac{1}{5}$	$\frac{3}{5}$

b. **REASON ABSTRACTLY** Fix one of the values in the table so that it represents a valid probability distribution, and then find the mean. CCSS S.MD.7, SMP 2

5. The archery target on the right has a radius of 40 cm. The inner circle has a radius of 10 cm., the next ring, 20 cm., the next, 30 cm., and the outer ring, 40 cm. The table below shows the probability of a dart that is thrown randomly hitting each of the rings. CCSS S.MD.7

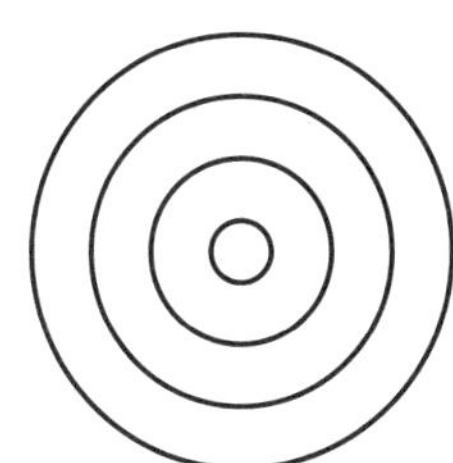

Ring 1 (innermost ring)	Ring 2	Ring 3	Ring 4 (outermost ring)
0.0625	0.1875	0.3125	0.4375

a. **REASON QUANTITATIVELY** Describe why the probability of hitting Ring 3 is 0.3125. CCSS SMP 2

b. **CALCULATE ACCURATELY** If you are awarded 10 points for hitting the innermost ring, 8 points for Ring 2, 5 points for Ring 3, and 1 point for Ring 4, find the expected value of a random shot at the target. CCSS SMP 6

c. **COMMUNICATE PRECISELY** Describe two reasons why this expected value would not work for someone who is actually throwing a dart at the target. CCSS SMP 6

9.5 The Binomial Theorem

Objectives

- Expand powers of binomials using Pascal's Triangle and the Binomial Theorem.
- Apply the Binomial Theorem to solve problems.

Content: A.APR.5
Practices: 1, 2, 3, 6, 7
Use with Lesson 10-6

KEY CONCEPT Pascal's Triangle

Pascal's Triangle shows a pattern for the powers of a binomial:	coefficients
$(a+b)^0$	1
$(a+b)^1$	1 1
$(a+b)^2$	1 2 1
$(a+b)^3$	1 3 3 1
$(a+b)^4$	1 4 6 4 1

The numbers in Pascal's Triangle are generated by adding together the two numbers directly above. For example, the 4 in the bottom row is generated by adding together the 1 and the 3 that are above it. The numbers in each row are the coefficients of the expansion. This pattern was discovered by the Chinese, and Pascal discovered that the numbers in the triangle were the coefficients in the expansion of the binomial $(a+b)^n$. Pascal's Triangle has many important and interesting mathematical properties.

EXAMPLE 1 Flipping a Coin CCSS A.APR.5

EXPLORE **At the county fair, there is a booth that offers a coin flipping game. A participant chooses "heads" or "tails," and wins a prize if at least 3 out of 4 flips show the chosen result.**

a. REASON ABSTRACTLY Expand $(H+T)^4$ using Pascal's Triangle. Explain the connection between the expansion and the probabilities involved in flipping a coin. CCSS SMP 2

b. REASON ABSTRACTLY Using the expression from **part a**, how many different ways are there of getting "heads" 3 times in the 4 flips? CCSS SMP 2

c. REASON ABSTRACTLY How many ways are there of getting *at least* 3 "heads" in 4 flips? CCSS SMP 2

d. REASON ABSTRACTLY Add up the coefficients of the expansion from **part a**. What does this number represent, and why does it make sense? CCSS SMP 2

Pascal's Triangle gives the coefficients in the expansion of $(a + b)^n$. However, it turns out that these coefficients are also combinations. When $(a + b)^n$ is expanded, each $(a + b)$ factor "donates" either an a or a b. If we are looking for the coefficient of $a^{n-k}b^k$, we can count the number of ways of choosing k of the $(a + b)$ factors to "donate" a b. There are n factors total, so the coefficient of $a^{n-k}b^k$ is ${}_nC_k$.

KEY CONCEPT The Binomial Theorem

The expansion of $(a + b)^n$, where n is a whole number, is given by:

$$(a+b)^n = \sum_{k=0}^{n} {}_nC_k (a)^{n-k}(b)^k$$

EXAMPLE 2 Using the Binomial Theorem CCSS A.APR.5

a. USE STRUCTURE What is the fourth term in the expansion of $(2x - 5y)^5$? CCSS SMP 7

b. PLAN A SOLUTION What is the expansion of $(2m + \frac{1}{2}n)^6$? CCSS SMP 1

c. USE STRUCTURE What is the coefficient of the term containing x^2 in the expansion of $(3x - 4)^4$? Show your work. CCSS SMP 7

d. PLAN A SOLUTION The coefficients in the binomial theorem are not necessarily the coefficients in the final answer to an expansion, as seen in **parts a–c**. Consider expanding $(6x - 2y)^4$. Complete the following table to accomplish this expansion. One column has been filled in for you. CCSS SMP 1

Coefficient from the Binomial Theorem		4			
$(6)^{4-k}(-2)^k$		$(6)^{4-1}(-2)^1 = -432$			
Term		$-1728x^3y$			

e. USE STRUCTURE Find the coefficient of x^{20} in $(x - 2)^{100}$. Do not simplify. CCSS SMP 7

EXAMPLE 3 Applying the Binomial Theorem CCSS A.APR.5

a. **REASON ABSTRACTLY** Jasmine is going to enter a free-throw contest. The contestants are to take five free throws as rapidly as possible. Use the Binomial Theorem to expand $(B + M)^5$. If "B" stands for "basket" and "M" stands for "miss," interpret the coefficients of the expansion in the context of the free-throw contest. Use the expansion to find the number of ways Jasmine can make *at least* 3 of the 5 shots. CCSS SMP 2

b. **CONSTRUCT ARGUMENTS** Martin completed part a but misunderstood the question. He instead answered, "How many ways can Jasmine miss at least 3 of the 5 shots." He claims that he got the "right" answer even though he did the problem incorrectly. Is this true? Explain. CCSS SMP 3

PRACTICE

1. **REASON ABSTRACTLY** A test consists of 10 questions, with five answer choices for each question. Matthew forgets to study and must guess on every question. How many ways can he get 8 or more correct answers on the test? Show your work using Pascal's Triangle. CCSS A.APR.5, SMP 2

2. **CALCULATE ACCURATELY** Use Pascal's Triangle to find the fourth term in the expansion of $(2x + 7)^6$. Why is it the same as the fourth term in the expansion of $(7 + 2x)^6$? CCSS A.APR.5, SMP 6

3. **USE STRUCTURE** Find the term containing x^3y^5 in the expansion of $(2x + 5y)^8$. Show your work and explain your reasoning. CCSS A.APR.5, SMP 7

4. **REASON ABSTRACTLY** A company that makes circuit boards uses a robotic welder in the creation process. Some of these are produced accurately, and others are not. Use the Binomial Theorem to find out how many ways exactly 5 of 7 circuit boards would be produced accurately. Show your work. CCSS A.APR.5, SMP 2

5. **REASON ABSTRACTLY** A manufacturing process is known to produce a defect in 1 out of 200 chairs. If a sample of 20 chairs is selected, how many different ways can *no more than* 2 of the chairs be defective. Show your work. CCSS A.APR.5, SMP 2

6. a. **CONSTRUCT ARGUMENTS** Write out the expansion of $(a + b)^4$. What do you notice about the sum of the exponents in each term? CCSS A.APR.5, SMP 3

   b. **CONSTRUCT ARGUMENTS** Find the term in $(a + b)^{12}$ where the exponent of a is 5. What is the sum of the exponents? CCSS A.APR.5, SMP 3

   c. **CONSTRUCT ARGUMENTS** What is the sum of the exponents in any term of the expansion for $(a + b)^n$? Why do you think this is true? CCSS A.APR.5, SMP 3

7. **REASON ABSTRACTLY** Shanna's dog had 4 puppies. How many ways are there of having 3 of one gender and 1 of another? Use Pascal's Triangle to help solve this problem. CCSS A.APR.5, SMP 2

8. **REASON ABSTRACTLY** If we classify each day in a particular city as "sunny" or "cloudy," how many ways are there in a 7-day period of having more sunny days than cloudy days? Use the Binomial Theorem to answer this question. CCSS A.APR.5, SMP 2

9.6 The Binomial Distribution

Objectives

- Construct a binomial distribution from a binomial experiment.
- Use probabilities from a binomial distribution to analyze decisions and strategy.

CCSS STANDARDS

Content: A.APR.5, S.MD.6, S.MD.7
Practices: 1, 2, 4, 5, 6, 7, 8
Use with Lesson 11-4

A **binomial experiment** is a probability experiment with a fixed number, n, of independent identical trials. Each trial can have only two outcomes, success or failure. The probability of success p and probability of failure q will always total 1, so $q = 1 - p$. The associated random variable X is the number of successes in n trials.
The probability distribution of each value of the random variable X that represents the number of successes in a binomial experiment is called a **binomial distribution**. If the probability of success is p and the probability of failure is q, then the probability of X successes in n trials is given by $P(X) = {}_nC_x p^x q^{n-x}$.

EXAMPLE 1 Construct a Binomial Distribution for an Experiment CCSS S.MD.6

EXPLORE A probability experiment consists of rolling two number cubes 100 times and recording the number of "doubles"—both cubes showing the same number.

a. **USE A MODEL** Explain why this is a binomial experiment. CCSS SMP 4

b. **USE TOOLS** Design a simulation using appropriate technology to perform the experiment 20 times. Record the number of doubles out of 100 trials in each simulation and display the results as relative frequencies using a bar graph. CCSS SMP 5

c. **FIND A PATTERN** Describe any patterns in the simulation result. Use your graph from **part b** to draw an approximation of the probability distribution. CCSS SMP 7

d. **USE STRUCTURE** Describe the shape of the graph, and determine the expected number of rolls of doubles. How does this expected number relate to the probability distribution? CCSS SMP 7

The binomial distribution can be found by expanding $(p + q)^n$ using Pascal's Triangle or the Binomial Theorem. Binomial distributions are useful when a range of successes is desired.

EXAMPLE 2 Construct a Binomial Distribution Using the Binomial Theorem

You can model binomial success-failure sequences and their theoretical probabilities using the expression $(p + q)^n$, with $q = 1 - p$. CCSS A.APR.5, S.MD.7

a. **USE STRUCTURE** The probability of drawing a heart from a standard deck of cards is 25%. Consider the process of repeating this experiment 5 times. Expand $(0.25 + 0.75)^5$ using the Binomial Theorem, but do not simplify the terms. What does each term in the expansion mean? CCSS SMP 7

b. **REASON QUANTITATIVELY** Use the expansion to find the probability of pulling at least three hearts in five trials. If you were asked to guess whether it was likely that at least three hearts were to be pulled, how would you respond? CCSS SMP 2

c. **REASON QUANTITATIVELY** Simplify the entire expression from **part a**. What is the meaning of your result? CCSS SMP 2

For a binomially distributed random variable X with parameters n and p, the probability of k successes is $P(X) = {}_nC_x p^k q^{n-k}$. A *cumulative* binomial function, available on most graphing calculators and spreadsheets, is used to determine the probability of *at most* k successes, $P(X \leq k)$.

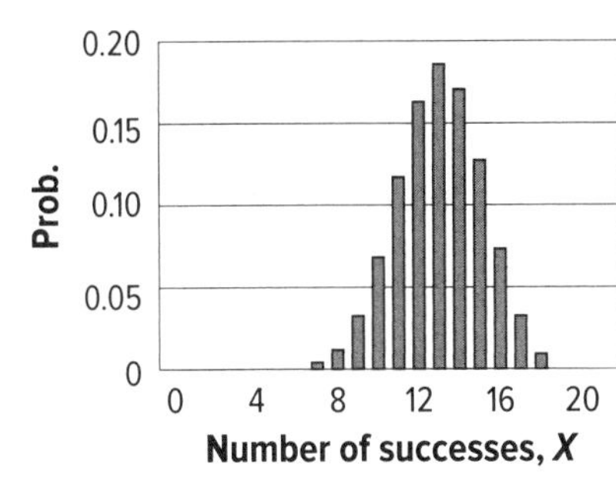

The mean μ of a binomial distribution with random variable X with parameters n and p is given by $\mu = np$. The expected value $E(x)$ is the mean rounded to the nearest whole number. The standard deviation of the binomial distribution is given by $\sigma = \sqrt{npq}$, where npq is the variance. A random variable is considered unusual if it is more than 2 standard deviations from the mean of the distribution.

EXAMPLE 3 Use the Mean and Standard Deviation to Describe a Binomial Distribution

Mr. Jameson has found that 40% of the students who enroll in his high school Calculus class are females. This year his class will have 20 students. CCSS A.APR.5, S.MD.7

a. **CALCULATE ACCURATELY** What is the mean of the binomial distribution for the number of females in his class? CCSS SMP 6

b. **CALCULATE ACCURATELY** What is the standard deviation of the binomial distribution for the number of females in his class? CCSS SMP 6

c. **REASON QUANTITATIVELY** Should Mr. Jameson be surprised if the class has only 4 females? Why or why not? Is it unlikely that the class will have 15 or more females? Explain. CCSS SMP 2

PRACTICE

1. **INTERPRET PROBLEMS** Which of these are binomial experiments? Justify your answer in each case. CCSS S.MD.7, SMP 1

   a. A coin is tossed 100 times and the number of heads is recorded.

   b. A magazine is checked for the total number of advertisements it contains.

   c. A coin is tossed 100 times and the number of runs of two or more heads is recorded.

   d. 52 consecutive issues of the same magazine are checked to see how many issues contain more than 60 advertisements.

2. In NASA rocket satellite launches, it is desirable for the probability of avoiding a launch failure of any kind to be 0.9999. In practice, however, rocket launches have not achieved better than 0.99 as a probability for the rocket to successfully launch. CCSS S.MD.7

 a. **CALCULATE ACCURATELY** Compare the probabilities of not losing a rocket in 100 launches for $p = 0.99$ and $p = 0.9999$. CCSS SMP 6

 b. **EVALUATE REASONABLENESS** We often exaggerate the chances of failure in our common language. Jeremy claims that the chances of NASA failing a rocket mission is "1 in a million." If Jeremy is correct, what is the probability of getting 1 rocket failure in 100 launches? Is this reasonable knowing that NASA has indeed failed several rocket launches in its history? CCSS SMP 8

3. **CALCULATE ACCURATELY** The probability of a child being left-handed is 11%. A family has six children. What is the probability that at least one of the children will be left-handed? CCSS S.MD.7, SMP 6

4. **CALCULATE ACCURATELY** A carnival game offers two options for play. Four cards are dealt, and you can either choose to win based on "Aces" or "Spades." If you choose "Aces," you win if at least one ace shows up in the four cards. If you choose "Spades," you win if at least two spades shows up in the four cards. CCSS S.MD.7, SMP 6

 a. What is the probability of winning the "Aces" game? Show your work.

 b. What is the probability of winning the "Spades" game? Which game should you choose to play?

5. **CALCULATE ACCURATELY** A factory produces light bulbs. After a quality control experiment, it was determined that 2% of the bulbs are defective. CCSS S.MD.7, SMP 6

 a. The company sells these bulbs in packs of 4. What is the probability that the package contains at least one defective bulb?

 b. The company also sells the bulbs in packs of 6. What is the probability that the package contains at least one defective bulb?

9.7 The Normal Distribution

Objectives

- Model data sets by fitting them to normal distributions.
- Use normal models to find probabilities and estimate percentages of populations.

CCSS STANDARDS

Content: S.ID.4
Practices: 1, 2, 3, 4, 5, 6, 8
Use with Lesson 11-5

A **normal distribution** is a continuous probability distribution with a symmetric "bell" shape that is the most commonly occurring continuous probability distribution. The exact shape of a normal distribution is defined by its mean μ and standard deviation σ. Its characteristics are:

- the graph of its curve is continuous (smooth), bell-shaped, and symmetric about the line $x = \mu$;
- the mean, median, and mode are equal;
- its curve approaches, but never touches, the x-axis;
- the area under its curve equals 1 or 100%.

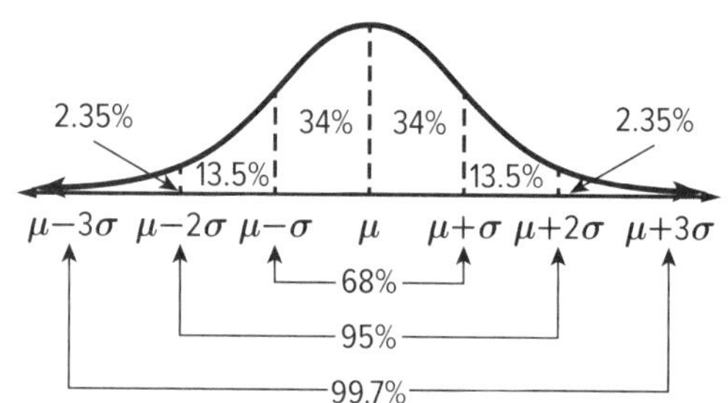

The **Empirical Rule** determines the area under the normal curve for various intervals, which is equivalent to the amount of data within those intervals.

EXAMPLE 1 Use the Empirical Rule to Analyze a Distribution

EXPLORE Scores on a national standardized mathematics test are believed to be a good fit to a normal distribution with mean 51 and standard deviation 12. CCSS S.ID.4

a. **USE A MODEL** Draw the normal curve replacing the labels on the number line using the mean and standard deviation. Shade the region under the curve that represents a score of at least 39. CCSS SMP 4

b. **CALCULATE ACCURATELY** Use the Empirical Rule to find the probability of a randomly chosen student scoring at least 39. CCSS SMP 6

c. **REASON QUANTITATIVELY** Predict an interval that should contain 68% of scores in any given year. Explain your answer. CCSS SMP 2

d. **INTERPRET PROBLEMS** Suppose approximately 430,000 students take the math test across the country one year. According to the normal model, approximately how many students would have mathematical ability more than three standard deviations above the mean? CCSS SMP 1

e. **CRITIQUE REASONING** JoAnna is in a mathematics class of 25 students that participated in this test. JoAnna knows that she scored an 80 on the test and reasons that, since only 2.5% of the students score above a 75 and there are only 25 students in her class, that it is highly likely that she received the top score of all the students in her class. Is this good reasoning? CCSS SMP 3

The **z-value, given by** $z = \frac{X - \mu}{\sigma}$, of a data value X from a distribution with mean μ and standard deviation σ is the number of standard deviations the value lies above or below the mean. This number is useful because it allows any data value to be located relative to the normal distribution to which the data are being fitted.

EXAMPLE 2 Use z-Values to Interpret Normally Distributed Data

The heights of 17-year-old males in the United States are believed to be normally distributed with mean 175 cm and standard deviation 7.3 cm. CCSS S.ID.4

a. **REASON QUANTITATIVELY** What is the percent of the males that are taller than 160.4 cm? CCSS SMP 2

b. **COMMUNICATE PRECISELY** Find the z-score of a height of 183 cm, to two decimals, and explain its meaning. Make a sketch of a normal distribution to illustrate your answer. CCSS SMP 6

c. **REASON QUANTITATIVELY** Based on your sketch and the Empirical Rule, estimate the probability of a 17-year-old male student being at least 183 cm tall. Justify your estimate. CCSS SMP 2

When you convert a set of data to z-values, you transform its probability distribution t into the **standard normal distribution**—which has a mean of $\mu = 0$ and a standard deviation of $\sigma = 1$. This allows for the comparison of similar sets of normally distributed data that have different means and standard deviations. All that is needed is to know how the position of each z-value is related to the area under the curve. Tables were originally created showing z-values rounded to one or two decimal places and the area under the curve either to the left or right of that value. The partial table below uses rows and columns to find z-values rounded to the nearest hundredth and the associated area under the curve to the left of that number of standard deviations. The highlighted cell shows that 0.0505 or 5.05% of the data is to the left of a z-value of -1.64.

It is more common today to use the normalcdf or ShadeNorm functions under the DISTR menu on a graphing calculator to find these values rather than look them up in a table. To find the probability of a data value being within an interval, type the endpoints of the interval followed by the mean and standard deviation or enter only the z-values of the endpoints after either function.

EXAMPLE 3 Find Probabilities for a Normal Distribution

Prior to calculators, standard normal distribution tables based on z-scores were used to find probabilities for normal models. The table below shows $P(Z \leq z)$, where Z is a standard normal random variable and z is a z-score between -1.69 and -1.5. CCSS S.ID.4

z	0.00	0.01	0.02	0.03	0.04	0.05	0.06	0.07	0.08	0.09
−1.6	0.0548	0.0537	0.0526	0.0516	0.0505	0.0495	0.0485	0.0475	0.0465	0.0455
−1.5	0.0668	0.0655	0.0643	0.0630	0.0618	0.0606	0.0594	0.0582	0.0571	0.0559

a. **USE TOOLS** Use the table to find $P(Z \geq -1.58)$. CCSS SMP 5

b. **USE TOOLS** Outline and use a method to find $P(Z < 1.67)$. (Hint: Use the symmetry of the standard normal distribution.) CCSS SMP 5

c. **USE TOOLS** Graphing calculators have a function that allow you to work with normal distributions. The **normalcdf(a, b, μ, σ)** function finds the probability of a data value lying between a and b for a normal distribution having mean μ and standard deviation σ. Use the **normalcdf** function on your calculator to find $P(32 \leq X \leq 49)$, where X is normal with $\mu = 42$ and $\sigma = 7$. CCSS SMP 5

d. **USE TOOLS** You can also use the **normalcdf** function with only two z-scores as arguments. Convert 32 and 49 to z-scores and use the **normalcdf** function to find the same probability. Confirm that your answer is the same as **part c**. CCSS SMP 5

PRACTICE

1. Wingspans of a certain species of insect are a good fit to a normal distribution with mean 64 mm and standard deviation of 4 mm. CCSS S.ID.4

   a. **USE A MODEL** Draw the normal curve, replacing the labels on the number line using the mean and standard deviation. Shade the region under the curve that represents a wing length no greater than 56 mm. CCSS SMP 4

   b. **CALCULATE ACCURATELY** Use the Empirical Rule to find the probability of a randomly chosen insect having a wingspan of at most 56 mm. CCSS SMP 6

2. a. **USE TOOLS** Using the table from **Example 3**, find $P(-1.66 \leq Z \leq -1.5)$. Describe your method. CCSS S.ID.4, SMP 5

   b. **USE TOOLS** Confirm your answer with a graphing calculator and the **normalcdf** function. Show your work. CCSS S.ID.4, SMP 5

3. **COMMUNICATE PRECISELY** 250 students in a high school took the ACT test. The mean score was 26.2 with a standard deviation of 4.1. Can we use the Empirical Rule to determine the probability that a student scored less than 22? CCSS S.ID.4, SMP 6

4. a. **DESCRIBE A METHOD** Describe and use a method that uses normalcdf to find the third-quartile score on the math test from Example 1 (75% of the population scores will likely be this much or less). CCSS S.ID.4, SMP 8

   b. **USE TOOLS** Graphing calculators generally have an invNorm function which, given a probability p, returns the z-score such that $P(Z \leq z) = p$. Use this function to complete the table. CCSS S.ID.4, SMP 5

Percent Scoring $\leq x$	5	10	15	20	25	30	40	50	60	70	75	80	85	90	95
Equivalent z-Score for x	−1.64														
Math SAT Score x															

9.8 Confidence Intervals and Hypothesis Testing

Objectives

- Use the maximum error of estimate to find a confidence interval to estimate a population parameter from a random sample.
- Use a hypothesis test to analyze the results of an experimental study or survey.

CCSS STANDARDS

Content: S.MD.7, S.IC.1, S.IC.4
Practices: 1, 2, 3, 4, 6, 7, 8
Use with Lesson 11-6

The statistics from a random sample are used to estimate population parameters. For example, if a random sample of 100 cereal boxes gives a sample mean of 312.3 grams of cereal per box, this figure can be used as an estimate of the population mean for this cereal brand. The value of an estimate of a population parameter depends on how well the random sample represents the population and the confidence interval within which it is most likely located. The sample of cereal boxes must all be similar sizes of the same brand to make the stated estimate.

A **confidence interval** is a stated range for an estimate of a population parameter with a given confidence level. The most commonly used levels are 90%, 95%, and 99%. For instance, you might be 95% confident that the population mean for the cereal boxes is in a range of 312.3 ± 2.5 grams. The confidence interval CI is found using the sample mean $\bar{x}$ and the maximum error of estimate E in the formula $CI = \bar{x} \pm E$.

To calculate the **maximum error of estimate** use $E = z \cdot \frac{s}{\sqrt{n}}$, where z is the z-value corresponding to the desired confidence level, s is the standard deviation of the sample, and n is the size of a sample of at least 30. See the table at right for the z-values that correspond to the most commonly used confidence levels.

Confidence Level	z-value
90%	1.64
95%	1.96
99%	2.58

EXAMPLE 1 Find and Interpret a Confidence Interval for a Population Mean

In a science experiment 35 tomato plants are grown using a new fertilizer. After 6 weeks of growth, the heights have a sample mean $\bar{x}$ of 95.4 centimeters and a standard deviation of 11.3 cm.

a. INTERPRET PROBLEMS Use the sample size and standard deviation to find the maximum errors of estimate for confidence levels of 90% and 99%. Interpret the results in the context of the population mean μ. CCSS S.IC.4, SMP 1

b. COMMUNICATE PRECISELY Write 90% and 99% confidence intervals for the population mean height μ after 6 weeks of tomato plants grown in these conditions. CCSS S.IC.4, SMP 6

c. **EVALUATE REASONABLENESS** A study by the fertilizer company states that the mean height of tomato plants after 6 weeks will be 89.8 centimeters. Is this possible? Explain your answer. CCSS S.IC.4, S.MD.7, SMP 8

If our sample estimate is given as a **population proportion** for a parameter p of a population, then the confidence interval is given by $CI = \hat{p} \pm ME$, where $\hat{p}$ ("p-hat") is the sample proportion being used to estimate the population proportion p, and the maximum error of estimation is given by $ME = z\sqrt{\frac{\hat{p}\hat{q}}{n}}$, where z and n are defined as before, and $\hat{q} = 1 - \hat{p}$.

EXAMPLE 2 Find and Interpret a Confidence Interval for a Population Proportion

In an opinion poll 5 days ahead of a Senate election, 1065 prospective voters were asked their voting intentions. 51.7% said that they planned to vote for the incumbent (the current Senator).

a. **INTERPRET PROBLEMS** Construct a 95% confidence interval for the proportion of all prospective voters p intending to vote for the incumbent, based on the data. CCSS S.IC.4, SMP 1

b. **USE A MODEL** The incumbent's campaign is considering stepping up their final push to secure their candidate's re-election. Would you advise them to do this? Explain. CCSS S.IC.4, S.MD.7, SMP 4

c. **COMMUNICATE PRECISELY** Opinion polls are often quoted in the media in the format "In a recent poll, accurate 19 times out of 20, the incumbent's support was at 51.7% plus or minus 3%." Explain the meaning of the phrases "19 times out of 20," "support was at 51.7%," and "plus or minus 3%" in terms of the confidence level, margin of error, and population proportion. CCSS S.IC.4, SMP 6

A **hypothesis test** is used to assess a specific claim about a population parameter. Usually, the claim will compare the population mean μ to a critical value k. The test is comprised of two hypotheses. The null hypothesis H_0 is a statement of equality which is the one tested by collecting and analyzing a sample. The alternative hypothesis H_a is a statement of inequality that is the complement of the null hypothesis. Forms of the null and alternative hypotheses are shown in the table at right with the actual claim being either the null or alternative hypothesis. The claim is tested by either rejecting or not rejecting the null hypothesis.

Null H_0	Alt. H_a
$\mu \leq k$	$\mu > k$
$\mu = k$	$\mu \neq k$
$\mu \geq k$	$\mu < k$

EXAMPLE 3 Identify Null and Alternative Hypotheses

Based on the data from Example 1, Jessica believes that the mean height of a 6-week-old tomato plant is less than a meter. CCSS S.IC.1

a. **INTERPRET PROBLEMS** Express Jessica's claim as an equality or inequality involving the population mean. Then, write the null and alternative hypotheses for a test of this claim. CCSS SMP 1

b. **CONSTRUCT ARGUMENTS** Explain, in terms of the null and alternative hypotheses, how a hypothesis test would determine whether or not the claim could be accepted. CCSS SMP 3

Based on the sample data, either H_0 is rejected or H_0 is not rejected. This depends on whether the sample mean falls in the **critical region**, which is the range of values suggesting a significant enough difference from H_0 to reject it. The critical region is determined by the inequality sign of H_a and the z-value determined by significance level α, usually 1%, 5%, or 10%. The three types of critical regions are shown below.

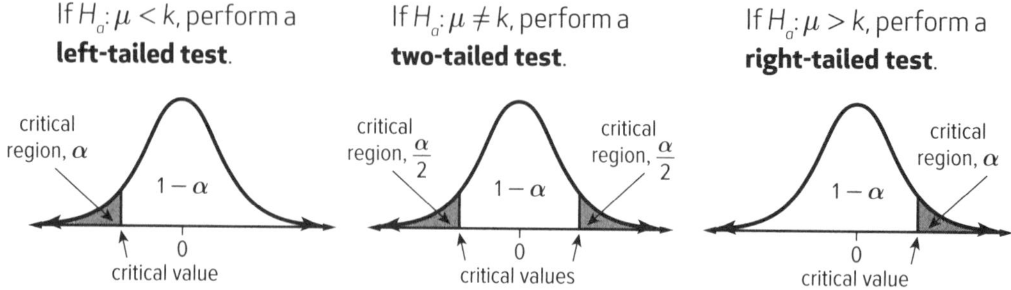

One of the tests that can be used is known as a z-test. For each type of critical region, a *z-statistic* is calculated using the sample mean $\bar{x}$, the standard deviation of the sample s, the sample size n, and the population mean from the null hypothesis μ in the formula. $z = \frac{\bar{x} - \mu}{\frac{s}{\sqrt{n}}}$. If the z-statistic falls in the critical region, H_0 is rejected.

EXAMPLE 4 Conduct a One-Sided Hypothesis Test

Danny, an engineer with a private space contractor, is looking at burn times for the first stage of the contractor's new launch vehicle. The stage burned for 72 s on three of its first five tests, with 71 s and 74 s on the remaining tests. Danny believes the average burn time is more than 71 s. CCSS S.IC.1

a. **INTERPRET PROBLEMS** State the hypotheses for a test of Danny's claim. Also determine the critical value and critical region at a significance level of 1%. (Hint: Use invNorm to find the critical value.) CCSS SMP 1

b. **REASON QUANTITATIVELY** Calculate the *z*-statistic for this test, and use it to reject, or fail to reject, the null hypothesis. Interpret the result of the test. CCSS SMP 2

c. **CONSTRUCT AN ARGUMENT** Is the test of Danny's claim reliable? Support your answer. CCSS SMP 3

EXAMPLE 5 Conduct a Two-Sided Hypothesis Test

Jessica is reviewing the data from the experiment in Example 1 and decides that she would like to test, at a level of significance of 5%, the claim that the mean height of a 6-week-old tomato plant is 95 centimeters. CCSS S.IC.1

a. **COMMUNICATE PRECISELY** State the hypotheses for this test. CCSS SMP 6

b. **USE STRUCTURE** For a two-tailed test, you need to use $\frac{\alpha}{2}$ to construct the critical region. Explain why this is so, and determine the *z*-values and the critical regions for this test. CCSS SMP 7

c. **INTERPRET PROBLEMS** Calculate the *z*-statistic, and use it to reject or not reject the null hypothesis. Then, discuss the results of the test. CCSS SMP 1

PRACTICE

1. Enrique is assessing his morning commute times for the first two weeks of his new job. For these 10 commutes, Enrique has calculated $\bar{x} = 30.8$,minutes and $s \approx 4.83$ minutes.

   a. **CALCULATE ACCURATELY** Find 90%, 95%, and 99% confidence intervals for Enrique's mean commute time. Express these as ranges (from x min to y min). CCSS S.IC.4, SMP 6

   b. **EVALUATE REASONABLENESS** If Enrique is looking for a "sense" of his average journey time, which is the most appropriate confidence interval to use, and why? CCSS S.IC.4, SMP 8

   c. **INTERPRET PROBLEMS** Enrique collects another two weeks' data, which gives him $\bar{x} = 31.1$ min and $s \approx 6.40$ min for the 20-workday period. During this time, Enrique is slightly late for work twice. Which confidence level should Enrique pick if he wants to avoid being late, and why? Calculate this confidence interval for μ, and explain how Enrique should use it with his new estimate of σ to ensure he is on time. CCSS S.IC.4, S.MD.7, SMP 1

2. **COMMUNICATE PRECISELY** Karen, an artisan potter, is concerned that her kiln is not heating evenly. She finds it acceptable to throw away no more than 10% of the pieces she fires in the kiln, but recently has had to discard 32 pieces out of a sample of 205. Find a 95% confidence interval for the population proportion of discards. Based on your finding, write a mathematically precise statement to help Karen decide whether to buy a new kiln. CCSS S.IC.4, S.MD.7, SMP 6

3. A pharmaceutical company is testing a new drug that is supposed to shorten the number of days a patient suffers from influenza symptoms. It was tested on 45 patients, and the average number of days until symptoms cleared was 5.3 with a standard deviation of 1.1 days.

   a. **INTERPRET PROBLEMS** Use the sample size and standard deviation to find maximum error of estimate for a confidence level of 99%. Interpret the results in the context of the population mean. CCSS S.IC.4, SMP 1

   b. **COMMUNICATE PRECISELY** Write the 99% confidence interval for the population mean number of days until recovery using this new drug. CCSS S.IC.4, SMP 6

   c. **EVALUATE REASONABLENESS** The drug company claims that the average number of days until recovery using their new drug can be as low as four days. Is this reasonable? CCSS S.IC.4, SMP 8

4. At a particular hospital, 50 newborns' records were randomly pulled for a study on birth weight. The weights have a sample mean of 7.8 pounds and a standard deviation of 1.25 pounds. CCSS S.IC.1

   a. **USE STRUCTURE** We want to test, at a level of significance of 5%, the claim that the mean weight of newborns in this hospital is 8 pounds. State the hypothesis, and determine the z-values and critical regions for this test. CCSS SMP 7

   b. **INTERPRET PROBLEMS** Calculate the z-statistic, and use it to reject or not reject the null hypothesis. Then discuss the results of the test. CCSS SMP 1

Performance Task

Exploring the Binomial Theorem

Provide a clear solution to the problem. Be sure to show all of your work, include all relevant drawings, and justify your answers.

In this performance task, you will prove the Binomial Theorem by *induction*. Induction is a method of proving statements involving natural numbers. To prove that a statement is true for all natural numbers n, start by proving that the statement is true for $n = 1$. Then, assume that the statement is true for $n = k$, where k is some natural number. Finally, prove that if the statement is true for $n = 1$ and $n = k$, then it is true for $n = k + 1$. This step completes the proof by induction.

Part A

Prove the Binomial Theorem for the case $(x + y)^1$ and write out the general Binomial Theorem in summation notation for the case $(x + y)^{n-1}$.

Part B

Derive a summation expression for the case $(x + y)^n$ using *only* known algebra and your work for the case $(x + y)^{n-1}$. Your summation indices must be different for different summations.

Part C

An open question in mathematics is whether or not $0^0 = 1$. Mathematicians do not all agree. Using the fact that ${}_nC_k = \frac{n!}{k!(n-k)!}$ and the definition $0! = 1$, substitute the expression $(x - x)^0$ into the Binomial Theorem. Do you think this is a valid proof? Explain. What do you think 0^0 is equal to?

Performance Task

Record Weights

Provide a clear solution to the problem. Be sure to show all of your work, include all relevant drawings, and justify your answers.

From the 1940s until the 1980s, long-playing vinyl records were the preferred format for storing and distributing music albums. Today, some of these records are considered extremely valuable and collectible. One of the things collectors look for is the weight of the record. A sample collection is shown below. The weight of each record, the company that released each record, and the release year are given.

Record Weights and Release Years					
Record Company A		Record Company B		Record Company C	
Weight (grams)	Year	Weight (grams)	Year	Weight (grams)	Year
195	1954	188	1957	196	1971
122	1974	129	1971	166	1956
180	1966	121	1968	167	1955
181	1959	190	1956	186	1955
201	1958	187	1961	156	1971
166	1969	190	1965	176	1970
144	1971	157	1961	149	1971
114	1979	145	1961	112	1968
188	1958	168	1968	213	1951
161	1957	171	1959	177	1968
131	1971	174	1953	198	1971
125	1969	166	1955	176	1965

Part A

What is the average weight of a record in this collection? Can you draw any conclusions about average record weight in general? Is that average biased? Is the sample truly random?

Part B

Find the average and margin of error (using the $\frac{1}{\sqrt{n}}$ method) for the entire sample, 1950s record, 1960s record, 1970s records, Company A records, Company B records, and Company C records.

Part C

An oil embargo in the 1970's made vinyl more expensive, which some collectors say caused a decrease in average vinyl weight from 1970 onward. Does the data support this claim? Why or why not? Be sure to discuss averages, margins of error, and anything else that is relevant in your answer.

Standardized Test Practice

1. Scores on a test taken by 186 students are normally distributed with a mean of 93 and a standard deviation of 9. Marcos had a score of 111. About how many students scored better than Marcos? CCSS S.ID.4

about ☐ students

2. Expand using the Binomial Theorem. CCSS A.APR.5

$(3a + 2b)^4 =$

☐

3. For a Statistics Class game, letter tiles corresponding to the letters in the word "statistics" are placed in a bag. For each turn a player can draw one letter or pass. CCSS S.MD.7

Complete the probability distribution table.

a	c	i	s	t

If a vowel is drawn from the bag, the player loses all current points. If a "c" or "s" is drawn, the player's current points are doubled. Joseph thinks he has a better chance of drawing a "c" or "s" than a vowel, so he decides to draw. Find each of the probabilities and state if Joseph is correct.

P(vowel) = ☐ P("c" or "s") = ☐

Is Joseph correct? Yes No

4. The table shows the ages of a random sample of 20 participants in a marathon. CCSS S.IC.1

67	17	48	31	27
23	35	21	22	52
46	27	37	18	33
39	63	28	59	40

Find the mean and median of the sample. Round to the nearest tenth. Is the data symmetric or skewed?

Mean: ☐ Median: ☐

Shape: symmetric skewed

5. Dominik is taking a multiple-choice test. Each question has four answer choices. If he guesses on three questions, what is the probability that he will guess correctly on at least two of the questions? Round to the nearest tenth of a percent. CCSS A.APR.5

☐

6. Laura is playing a game that she has a 20% chance of winning. She designed a simulation to determine the experimental probability of winning 2 or more times out of 6 tries. The table shows the data she collected, where 1 and 2 represent a win and 3 through 10 represent a loss. She repeated the simulation 5 times. CCSS S.IC.2

	A	B	C	D	E	F
Sim 1	4	7	6	1	7	6
Sim 2	4	6	10	5	4	5
Sim 3	4	2	5	3	7	3
Sim 4	5	2	8	2	3	8
Sim 5	9	10	2	5	8	1

Based on the simulation, what is the experimental probability of winning 2 or more games out of 6 games? ☐

7. The heights of a population of students are normally distributed with mean 61 inches and standard deviation 6 inches. Round to the nearest percent. CCSS S.ID.4

What percent of students do you expect to have heights in the range

49 inches to 73 inches? ☐

55 inches to 67 inches? ☐

8. Find the real numbers a and b that make the equation true. CCSS A.APR.5

$(x - y)^5 = x^5 - 5x^4y + ax^3y^2 - 10x^2y^3 + bxy^4 - y^5$

$a =$ ☐ $b =$ ☐

9. A box-and-whisker plot illustrates the distribution of a random sample of data. Which values should be used to describe the center and spread of the data? CCSS S.ID.4

five-number summary   mean and standard deviation

10. The school cafeteria director is deciding what new items to add to the school lunch menu. To determine what items students would like, he plans to give students a survey as they pay for their school lunch. Is this a good method? Why or why not? CCSS S.IC.3

11. The table below shows the number of students in each grade who are members of the school band.

9th Grade	10th Grade	11th Grade	12th Grade
14	15	12	9

The band director selects 5 band members at random to help him select music for the next concert. CCSS A.APR.5

a. What is the probability that exactly 2 of them will be 12th graders? Show your work.

b. What is the probability that at most 3 of them will be 10th graders? Show your work.

12. The table below shows the hourly wages of 20 recent college graduates. CCSS S.IC.1, S.IC.4

$15.12	$35.00	$12.95	$18.20	$52.65	$31.75	$25.00	$23.85	$14.60	$29.30
$17.84	$27.80	$33.05	$22.10	$25.40	$20.65	$19.80	$26.00	$24.10	$34.62

a. What are the mean and standard deviation of the data?

b. What is the margin of error for a 95% confidence level? Show your work. Explain.

c. What is the confidence interval for the mean? Explain.

13. A farmer harvests a field of watermelons that she has grown and weighs each of them. She finds that the data is normally distributed with mean 21.7 pounds and standard deviation 2.6 pounds. CCSS S.ID.4

a. The gardener only sells watermelons that weigh more than 16.5 pounds. What percent of the watermelons can the gardener expect to sell? Round to the nearest tenth.

b. According to the gardener, watermelons in the range 19.1 lbs. to 24.3 lbs. have the best taste. What percent of watermelons are expected to be in this range? Round to the nearest percent.

Trigonometric Functions

CHAPTER FOCUS Learn about some of the Common Core State Standards that you will explore in this chapter. Answer the preview questions. As you complete each lesson, return to these pages to check your work.

What You Will Learn	Preview Question
Lesson 10.1: Angles and Angle Measure	
CCSS F.TF.1 Understand radian measure of an angle as the length of the arc on the unit circle subtended by the angle.	CCSS SMP 4 Over a 20-minute period, a satellite sweeps out a 25°-angle. What is its speed in radians per hour?
Lesson 10.2: Circular and Periodic Functions	
CCSS F.TF.2 Explain how the unit circle in the coordinate plane enables the extension of trigonometric functions to all real numbers, interpreted as radian measures of angles traversed counterclockwise around the unit circle. **Also addresses:** F.TF.1, F.IF.4	CCSS SMP 6 Use this diagram to find the value of x.
Lesson 10.3: Graphing Trigonometric Functions	
CCSS F.TF.5 Choose trigonometric functions to model periodic phenomena with specified amplitude, frequency, and midline. CCSS F.IF.7e Graph exponential and logarithmic functions, showing intercepts and end behavior, and trigonometric functions, showing period, midline, and amplitude. CCSS F.IF.4 For a function that models a relationship between two quantities, interpret key features of graphs and tables in terms of the quantities, and sketch graphs showing key features given a verbal description of the relationship. **Also addresses:** F.IF.5, A.CED.2, A.SSE.2	CCSS SMP 6 A pendulum has repeated, periodic movement that can be modeled by the periodic function shown. How would A be represented in the diagram? 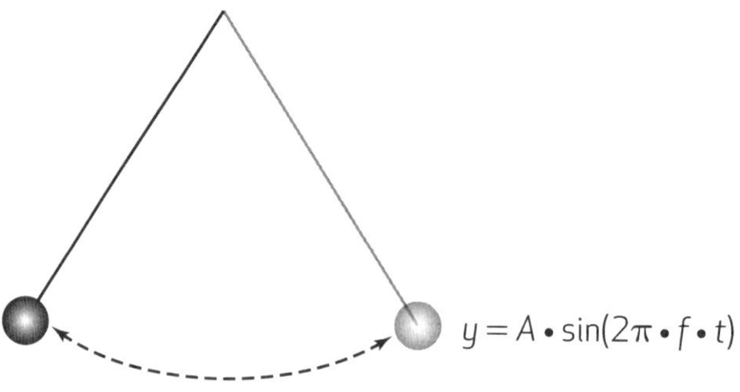

What You Will Learn	Preview Question
Lesson 10.4: Translating Trigonometric Functions	
CCSS F.TF.5 Choose trigonometric functions to model periodic phenomena with specified amplitude, frequency, and midline. CCSS F.IF.7e Graph exponential and logarithmic functions, showing intercepts and end behavior, and trigonometric functions, showing period, midline, and amplitude. **Also addresses:** F.IF.4, F.BF.1b, F.BF.3, A.CED.2	CCSS SMP 7 Use this diagram to show that cosine is an even function and sine is an odd function. 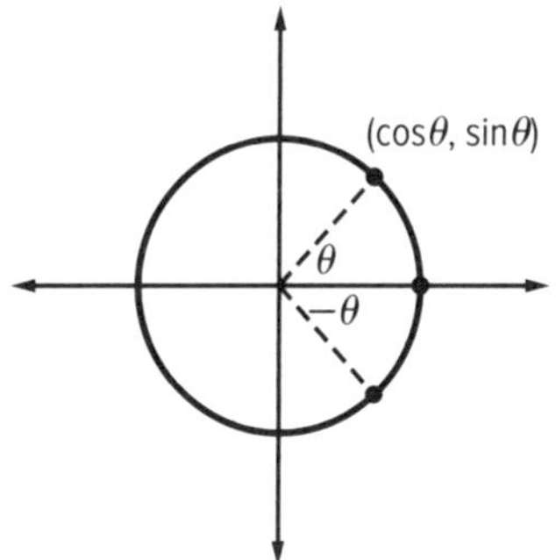
Lesson 10.5: Modeling: Trigonometric Functions	
CCSS A.SSE.1a Interpret parts of an expression, such as terms, factors, and coefficients. CCSS F.TF.5 Choose trigonometric functions to model periodic phenomena with specified amplitude, frequency, and midline. **Also addresses:** A.SSE.1b, F.IF.4, F.IF.5, F.BF.1b	CCSS SMP 6 A sound wave can be modeled by the function $y = A\sin 2ft\pi$, where f is the frequency of the sound. What is the relationship between frequency and wavelength?
Lesson 10.6: Inverse Trigonometric Functions	
CCSS A.CED.2 Create equations in two or more variables to represent relationships between quantities; graph equations on coordinate axes with labels and scales.	CCSS SMP 1 Suppose $f(x)$ is a trigonometric function. What can you conclude about the graph of $f^{-1}(x)$?
Lesson 10.7: Verifying the Pythagorean Identity	
CCSS F.TF.8 Prove the Pythagorean identity $\sin^2(\theta) + \cos^2(\theta) = 1$ and use it to find $\sin(\theta)$, $\cos(\theta)$, or $\tan(\theta)$ given $\sin(\theta)$, $\cos(\theta)$, or $\tan(\theta)$ and the quadrant of the angle.	CCSS SMP 2 Write a version of the Pythagorean Theorem in which all sides are divided by the length of the hypotenuse. Rewrite in trigonometric form.
Lesson 10.8: Angle Sum and Difference Formulas	
CCSS F.TF.9 Prove the addition and subtraction formulas for sine, cosine, and tangent and use them to solve problems.	CCSS SMP 6 Find the sine and cosine of angles α and β. 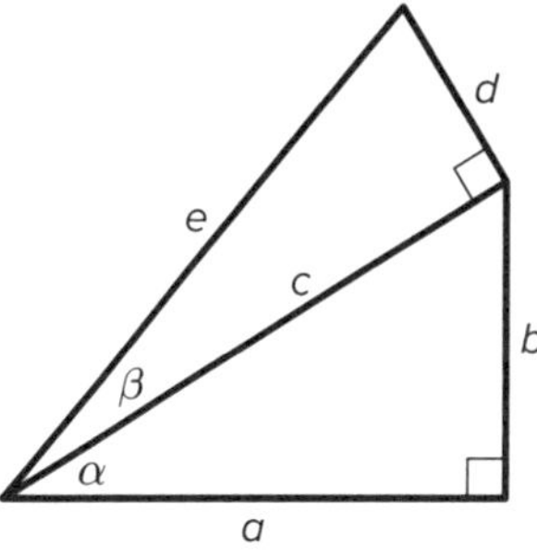

10.1 Angles and Angle Measure

Objectives

- Understand radian measure of an angle as the length of the arc on the unit circle subtended by the angle.
- Convert between radian measure and degree measure.
- Use radian measure to calculate arc length.

CCSS STANDARDS

Content: F.TF.1
Practices: 1, 2, 3, 4, 6, 7, 8
Use with Lesson 12-2

A **unit circle** is a circle of radius 1 whose center is at the origin of a coordinate plane. A **central angle** of a circle is an angle whose vertex is at the center of the circle. In the unit circle at the right, $\angle AOB$ is a central angle that *subtends* $\overset{\frown}{AB}$.

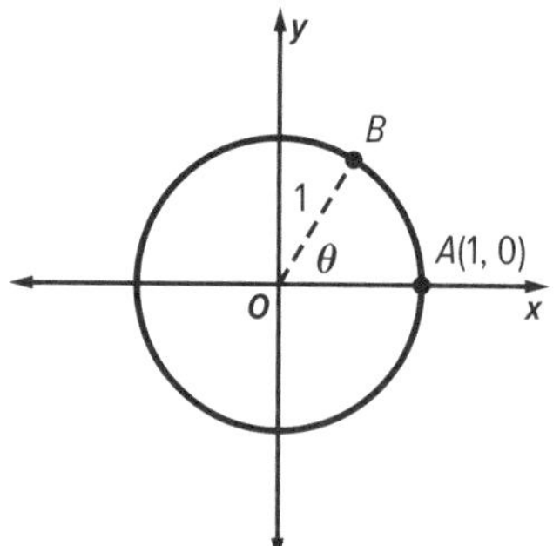

EXAMPLE 1 **Analyze Angles and Arc Lengths** CCSS F.TF.1

EXPLORE **Investigate the connection between the measure of a central angle of a unit circle and the length of the arc subtended by the angle.**

a. **CALCULATE ACCURATELY** Suppose $m\angle AOB = 60$ in a unit circle. Explain how to find the length of $\overset{\frown}{AB}$. Express the length in terms of π. CCSS SMP 6

b. **FIND A PATTERN** Complete the table. Express arc lengths in terms of π. CCSS SMP 7

$\angle AOB$	10°	20°	30°	43°	87°
Length of $\overset{\frown}{AB}$					

c. **CALCULATE ACCURATELY** Suppose that you know the length of $\overset{\frown}{AB}$ is $\frac{\pi}{2}$. Explain how to find $m\angle AOB$. CCSS SMP 6

d. **FIND A PATTERN** Complete the table. Leave answers in terms of π when necessary. CCSS SMP 7

Length of $\overset{\frown}{AB}$	$\frac{\pi}{20}$	$\frac{\pi}{10}$	$\frac{\pi}{5}$	$\frac{\pi}{4}$	1
$\angle AOB$					

e. **DESCRIBE A METHOD** Describe a quick way to find the length of $\overset{\frown}{AB}$ if you know $m\angle AOB$. CCSS SMP 8

f. **DESCRIBE A METHOD** Describe a quick way to find $m\angle AOB$ if you know the length of $\overset{\frown}{AB}$. CCSS SMP 8

In the exploration, you may have noticed a connection between the degree measure of a central angle and the length of the arc it subtends on the unit circle. You can use this relationship to define radian measure. The **radian measure** of an angle is the length of the arc on the unit circle subtended by the angle.

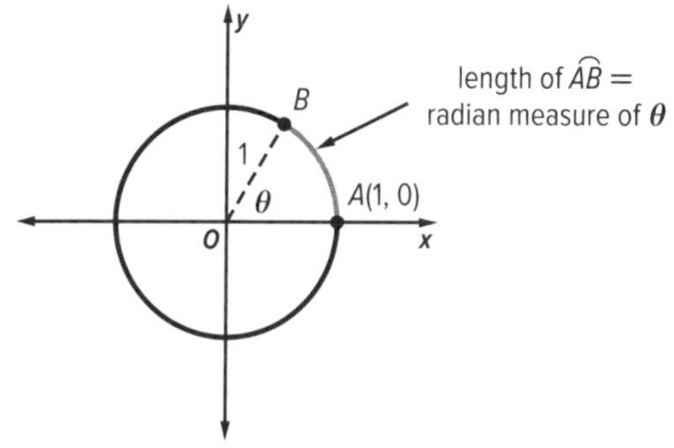

KEY CONCEPT Convert Between Degrees and Radians

Use your results from the previous exploration to help you complete the following.

Degrees to Radians	Radians to Degrees
To convert from degrees to radians, multiply the angle measure in degrees by ________.	To convert from radians to degrees, multiply the angle measure in radians by ________.

EXAMPLE 2 Convert Between Degrees and Radians CCSS F.TF.1

It is convenient to know the measures of some specific angles in both degree and radian measures. Use the conversion factors and/or proportional reasoning to complete the following.

a. **REASON ABSTRACTLY** What degree measure corresponds to a radian measure of 2π? Why does this make sense? CCSS SMP 2

b. **CALCULATE ACCURATELY** Complete the figure at the right by writing the radian measure for each angle. CCSS SMP 6

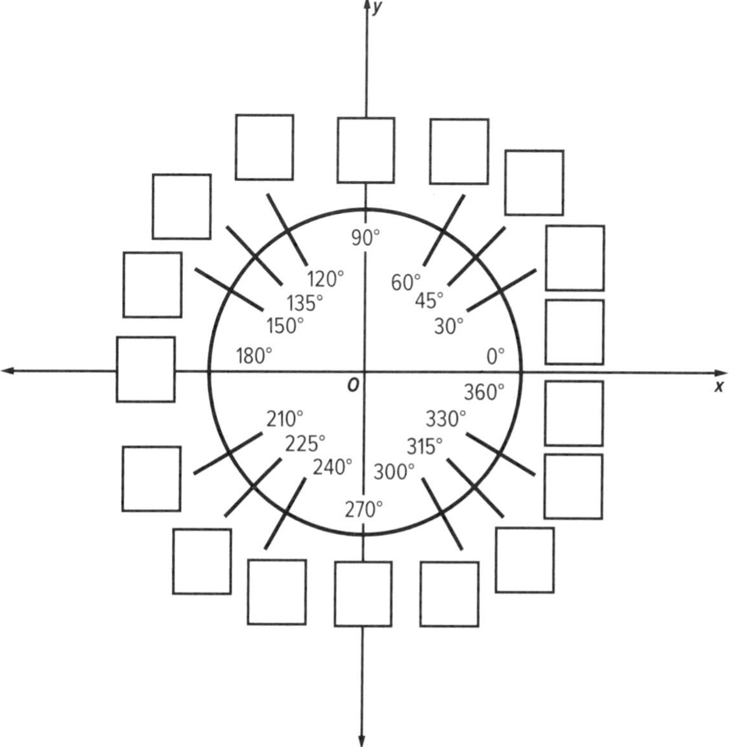

c. **FIND A PATTERN** Describe at least one pattern or symmetry that you notice in the completed figure. CCSS SMP 7

EXAMPLE 3 Develop an Arc Length Formula CCSS F.TF.1

Consider the circle shown at the right.

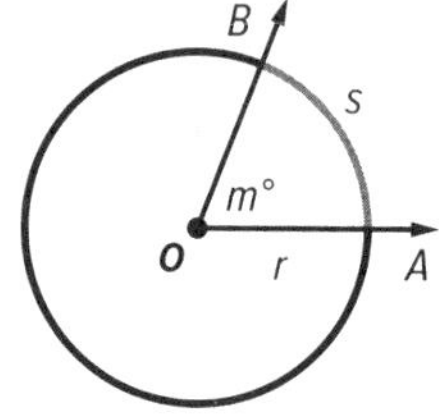

a. **REASON ABSTRACTLY** Write a formula for s in terms of m and r. CCSS SMP 2

b. **USE STRUCTURE** If you write the formula from **part a** in the form $s = [\text{coefficient}] \cdot r$, what is the coefficient of r? What do you notice about this coefficient? CCSS SMP 7

c. **DESCRIBE A METHOD** Let θ be the radian measure of the central angle. Describe a method for finding the arc length by rewriting the formula from **part a** in terms of r and θ. CCSS SMP 8

KEY CONCEPT Arc Length

Use your results from Example 3 to complete the following.

For a circle with radius r and central angle θ where the measure of θ is given in radians, the length s of the arc subtended by the central angle equals ________.

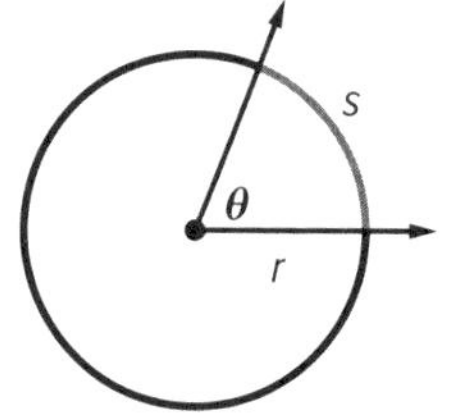

EXAMPLE 4 Apply Radian Measure CCSS F.TF.1

The London Eye is one of the world's largest Ferris wheels. The diameter of the wheel is 135 meters and it makes a complete rotation in 30 minutes. A passenger gets on the ride and travels for 5 minutes before the ride stops. The passenger wants to know how far she traveled during this time.

a. **INTERPRET PROBLEMS** During the 5-minute interval, what is the measure of the central angle of the wheel's rotation in radians? Explain. CCSS SMP 1

b. **CALCULATE ACCURATELY** Show how to find the distance the passenger traveled to the nearest meter. CCSS SMP 6

c. **EVALUATE REASONABLENESS** Explain how you know your answer is reasonable. CCSS SMP 8

PRACTICE

CALCULATE ACCURATELY Convert each degree measure to the equivalent measure in radians. Convert each radian measure to the equivalent measure in degrees. CCSS F.TF.1, SMP 6

1. $25°$
2. $200°$
3. $-12°$
4. $-117°$
5. $\frac{\pi}{40}$
6. $-\frac{8\pi}{9}$
7. 2
8. -3π

9. **CALCULATE ACCURATELY** A circular pizza with a diameter of 18 inches is cut into 8 congruent slices. What is the radian measure of the central angle of each slice? Explain. CCSS F.TF.1, SMP 6

10. **REASON ABSTRACTLY** In the figure, $s = r$. What is $m\angle PQR$ in radians? What is the measure of the angle to the nearest tenth of a degree? CCSS F.TF.1, SMP 2

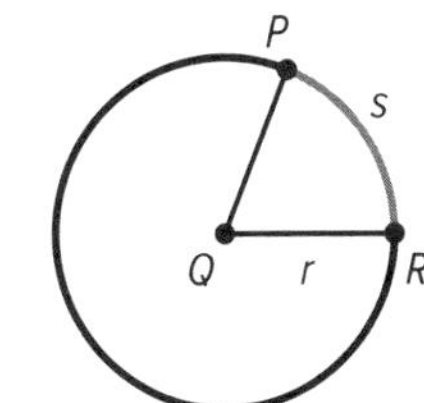

11. **USE A MODEL** Alex wraps a measuring tape that is 30-inches long around the circumference of an oil drum, but he finds that it does not go all the way around. What is the radian measure of the arc that is determined by the measuring tape? CCSS F.TF.1, SMP 4

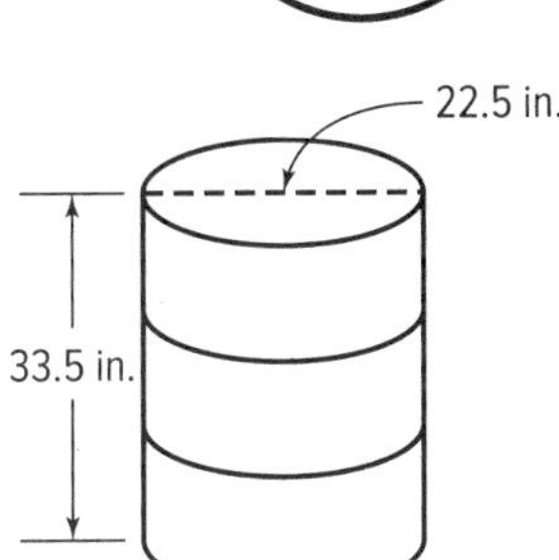

12. **CRITIQUE REASONING** A lawn sprinkler produces a stream of water that reaches 15 feet from the sprinkler head. The sprinkler rotates to sweep out part of a circle. The area of the lawn that gets watered as the sprinkler moves back and forth is 75π square feet. Melinda says it is possible to determine the radian measure of the angle that is swept out by the sprinkler. Caitlyn says there is not enough information to determine this. Who is correct? Justify your answer. CCSS F.TF.1, SMP 3

13. **COMMUNICATE PRECISELY** Circle C has a radius of 5 centimeters. The function $f(x)$ gives the length in centimeters of an arc of circle C that is subtended by a central angle of x radians. Describe the graph of $f(x)$ as precisely as possible and justify your answer. CCSS F.TF.1, SMP 6

14. **REASON ABSTRACTLY** In the figure, $2s = r$. What is $m\angle PQR$ in radians? What is the measure of the angle to the nearest tenth of a degree? CCSS F.TF.1, SMP 2

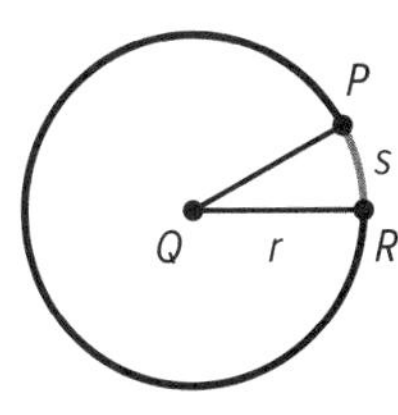

10.2 Circular and Periodic Functions

Objectives

- Explain how a unit circle allows trigonometric functions to be extended to all real numbers.
- Interpret key features of the graphs of trigonometric functions.

CCSS STANDARDS

Content: F.TF.1, F.TF.2, F.IF.4
Practices: 2, 4, 6, 7
Use with Lesson 12-6

EXAMPLE 1 Analyze Trigonometric Functions on the Unit Circle

EXPLORE The figure shows the unit circle. Let $P(x, y)$ be a point on the unit circle in the first quadrant. Let θ be an angle in standard position in which the terminal side intersects the unit circle at point P.

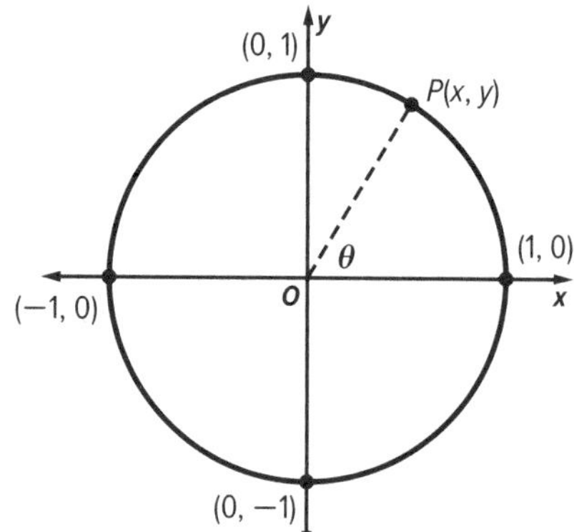

a. USE STRUCTURE Draw a perpendicular segment from P to the x-axis. This forms a right triangle. Label the length of each side of the triangle in terms of x and y. CCSS F.TF.2, SMP 7

b. COMMUNICATE PRECISELY Explain how to write and simplify expressions for $\sin\theta$ and $\cos\theta$ in terms of x and y. CCSS F.TF.2, SMP 6

c. REASON ABSTRACTLY You can define the sine and cosine of any angle θ, where θ is measured in radians, by drawing the angle in standard position, determining the point P at which the angle intersects the unit circle, and using the expressions you wrote in **part b**. Show how to use the unit circle at the right to plot the point P when $\theta = \frac{2\pi}{3}$. Then explain how to find the coordinates of point P. CCSS F.TF.1, F.TF.2, SMP 2

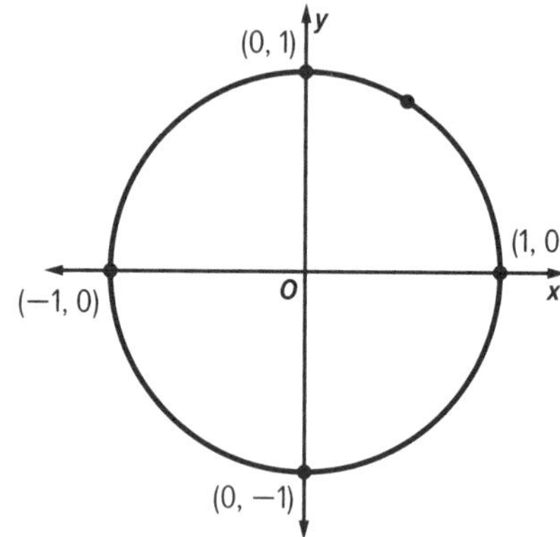

d. REASON ABTRACTLY Explain how to use the coordinates for P that you wrote in **part c** to find $\sin\frac{2\pi}{3}$ and $\cos\frac{2\pi}{3}$. CCSS F.TF.2, SMP 2

e. USE STRUCTURE What are the greatest and least possible values for the sine of an angle? What are the greatest and least possible values for the cosine of an angle? Explain. CCSS F.TF.2, SMP 7

You can extend the definition of the sine and cosine to all real numbers, interpreted as the radian measure of angles in standard position around the unit circle. In other words, the sine and cosine ratios can be considered functions in which the domain is all real numbers.

KEY CONCEPT The Sine and Cosine Functions

Complete the following.

If the terminal side of an angle θ in standard position intersects the unit circle at $P(x, y)$, then $\cos\theta =$ ________ and $\sin\theta =$ ________.

P(x, y)
y
θ
O
x

EXAMPLE 2 Use the Unit Circle to Determine Values of Sine and Cosine CCSS F.TF.2

The figure below shows the radian measure of angles as you move counterclockwise around the unit circle.

a. FIND A PATTERN Complete the figure by writing the coordinates of each given point around the unit circle. CCSS SMP 7

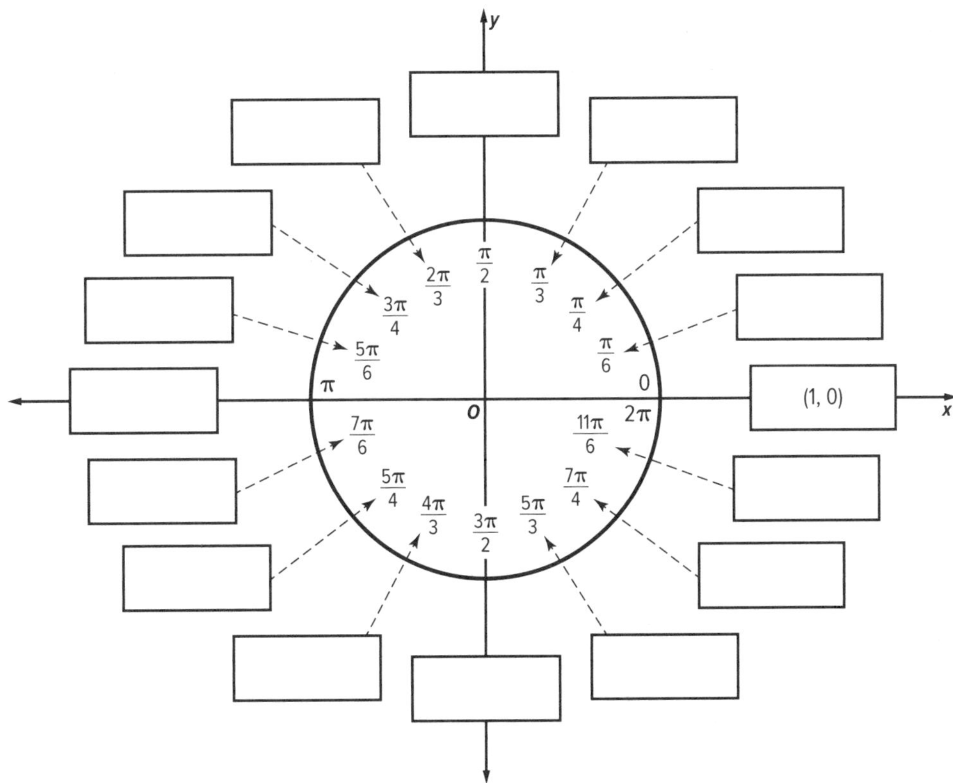

b. COMMUNICATE PRECISELY Explain how to use your work from **part a** to find $\cos\frac{4\pi}{3}$. CCSS SMP 6

c. **USE STRUCTURE** What is the value of $\sin \frac{13\pi}{6}$? Explain. CCSS SMP 7

d. **FIND A PATTERN** What can you conclude about the sine and cosine of an angle that is greater than π radians but less than $\frac{3\pi}{2}$ radians? Explain. CCSS SMP 7

EXAMPLE 3 Graph a Trigonometric Function

The wheel at a water park has a radius of 1 meter. As the water flows, the wheel turns counterclockwise, as shown. A point P on the edge of the wheel begins at the surface of the water. The function $f(x) = \sin x$ represents the height of P above or below the surface of the water as the wheel rotates through an angle of x radians.

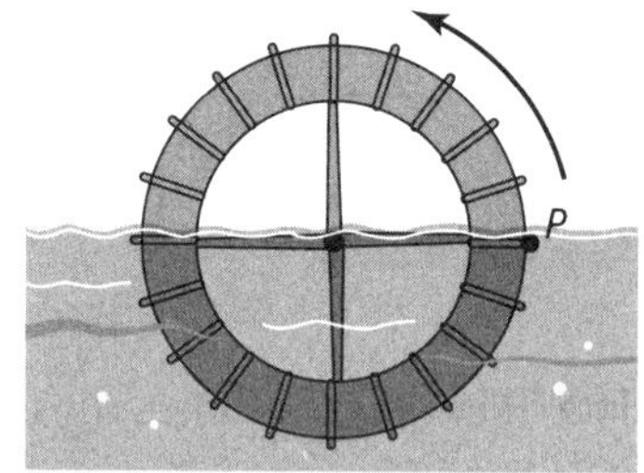

a. **REASON QUANTITATIVELY** How far does point P travel as the wheel rotates through an angle of $\frac{3\pi}{4}$ radians? Explain. CCSS F.TF.1, SMP 2

b. **CALCULATE ACCURATELY** Graph $f(x) = \sin x$ on the coordinate plane at the right. CCSS F.IF.4, SMP 6

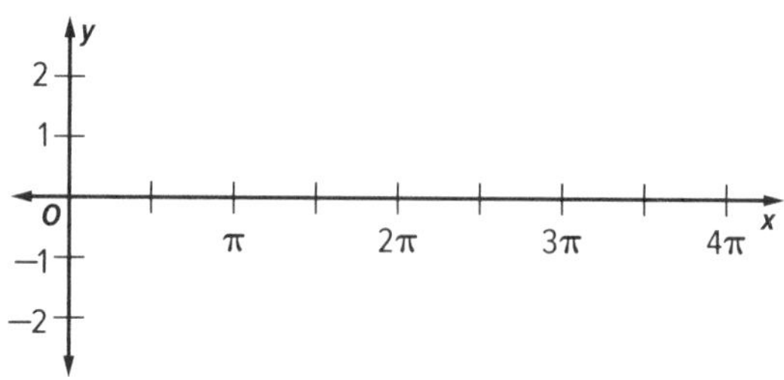

c. **USE A MODEL** What is the period of the function? Explain how you know and explain how the period is shown in the graph. What does the period tell you about point P? CCSS F.IF.4, SMP 4

d. **USE A MODEL** What are the x-intercepts? What do these represent? CCSS F.IF.4, SMP 4

e. **USE A MODEL** Identify an interval where the function is decreasing. What does this represent? CCSS F.IF.4, SMP 4

PRACTICE

REASON ABSTRACTLY Determine whether each statement is *always, sometimes,* or *never true*. Explain. CCSS F.TF.2, SMP 2

1. If k is a real number, then there is a value of θ such that $\cos \theta = k$.

2. $\sin \theta = \sin(\theta + 2\pi)$

3. If $\theta = n\pi$, where n is a whole number, then $\cos \theta = 1$.

4. If θ is an angle in standard position in which the terminal side lies in Quadrant IV, then $\sin \theta$ is positive.

5. A machine in a factory has a gear with a radius of 1 foot. A point P on the edge of the gear begins at the furthest point from a wall and then the gear begins to rotate counterclockwise. The function $f(x) = \cos x + 2$ represents the distance of P from the wall as the gear rotates through an angle of x radians.

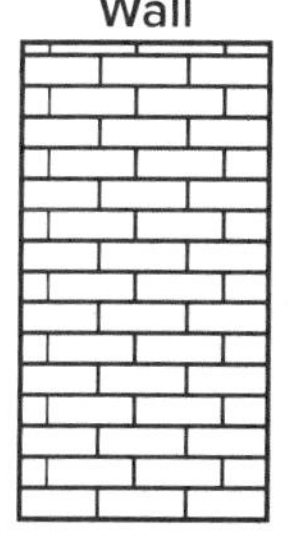

a. **USE A MODEL** What is $f\left(\frac{\pi}{2}\right)$? What does it represent? CCSS F.TF.2, SMP 4

b. **CALCULATE ACCURATELY** Graph $f(x)$ on the coordinate plane at the right. CCSS F.IF.4, SMP 6

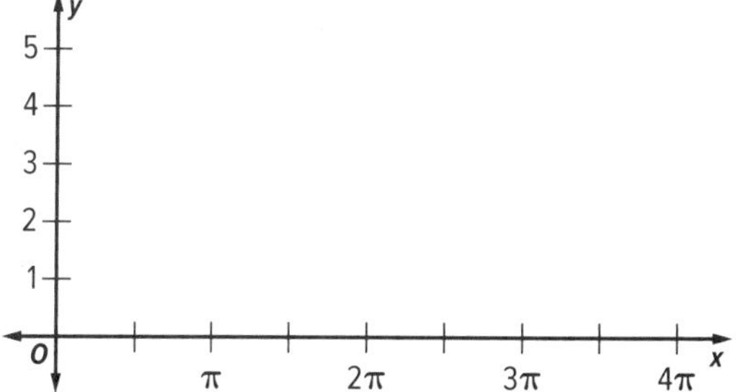

c. **USE A MODEL** What is the period of the function? What does this tell you about P? CCSS F.IF.4, SMP 4

d. **USE A MODEL** What are the maximum and minimum values of the function? What do these tell you about P? CCSS F.IF.4, SMP 4

6. **REASON ABSTRACTLY** Point P lies on the unit circle and on the line $y = x$. If θ is an angle in standard position in which the terminal side contains P, what can you conclude about $\sin \theta$ and $\cos \theta$? Explain. CCSS F.TF.2, SMP 2

10.3 Graphing Trigonometric Functions

Objectives

- Graph trigonometric functions.
- Use trigonometric functions to model periodic phenomena.

CCSS STANDARDS

Content: F.IF.7e, F.TF.5, F.IF.4, F.IF.5, A.CED.2, A.SSE.2
Practices: 1, 2, 3, 4, 5, 6, 7, 8
Use with Lesson 12-7

Graphs of trigonometric functions have repeating cycles. Recall that the horizontal length of one cycle is the *period* of the function.
The **amplitude** of the graph of a sine or cosine function is half the difference between the maximum and minimum values of the function. In the graph of the trigonometric function at the right, the period of the function is 2π and the amplitude is 4.

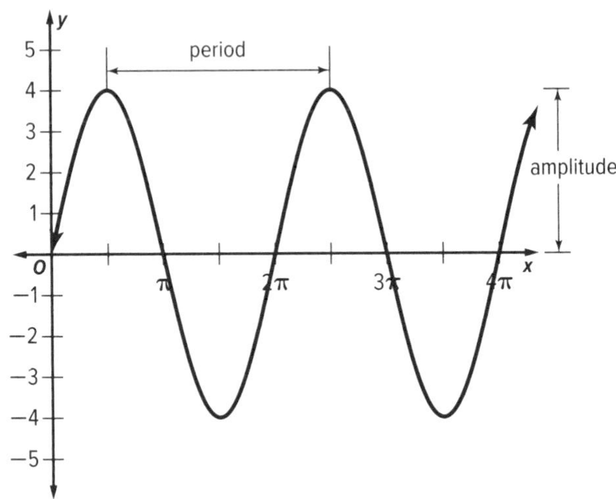

EXAMPLE 1 Analyze Characteristics of Trigonometric Functions

CCSS A.CED.2

EXPLORE Use a graphing calculator to help you sketch the graphs in this exploration.

a. **USE TOOLS** Sketch and label the graphs of $f(\theta) = \sin(-\theta)$, $g(\theta) = \sin 2\theta$ and $h(\theta) = \sin 0.5\theta$ on the coordinate plane at the right. CCSS SMP 5

b. **USE STRUCTURE** Complete the table. CCSS SMP 7

Function	Period
$f(\theta) = \sin(-\theta)$	
$g(\theta) = \sin 2\theta$	
$h(\theta) = \sin 0.5\theta$	

c. **MAKE A CONJECTURE** What is the period of the function $y = \sin(b\theta)$? Are there any restrictions on the value of b? CCSS SMP 3

d. **USE TOOLS** Sketch and label the graphs of $f(\theta) = 2\cos\theta$, $g(\theta) = -\cos\theta$, and $h(\theta) = 0.5\cos\theta$ on the coordinate plane at the right. CCSS SMP 5

e. **USE STRUCTURE** Complete the table. CCSS SMP 7

Function	Amplitude
$f(\theta) = 2\cos\theta$	
$g(\theta) = -\cos\theta$	
$h(\theta) = 0.5\cos\theta$	

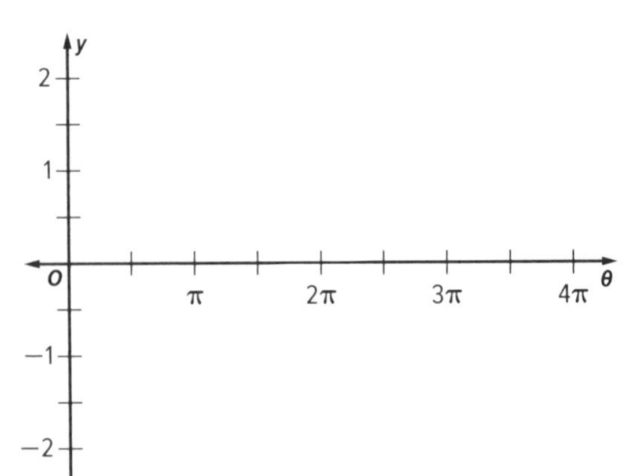

f. **MAKE A CONJECTURE** What is the amplitude of the function $y = a \cos \theta$? Are there any restrictions on the value of a? CCSS SMP 3

KEY CONCEPT Characteristics of Trigonometric Functions

Use your findings from the previous exploration to help you complete the following.

Function	$y = a \sin b\theta$	$y = a \cos b\theta$
Domain		
Range		
Period		
Amplitude		
Example	$y = 2 \sin 2\theta$	$y = 1.5 \cos 0.5\theta$

EXAMPLE 2 Graph a Sine Function CCSS A.CED.2

Graph $f(\theta) = -\frac{3}{4} \sin \frac{\pi}{2}\theta$.

a. **REASON ABSTRACTLY** Show how to determine the amplitude and period of the function. CCSS SMP 2

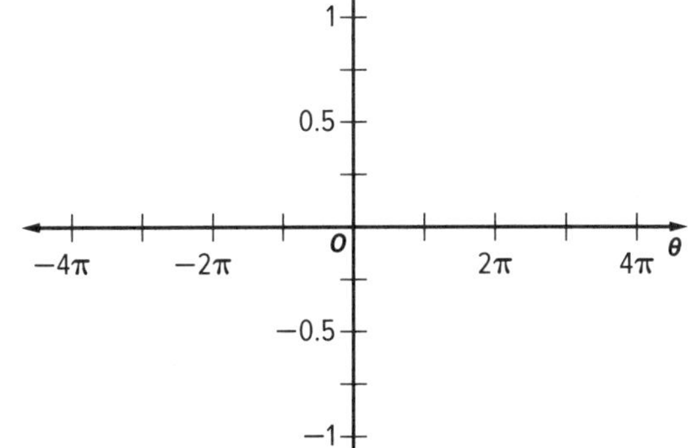

b. **USE STRUCTURE** Graph the function on the coordinate plane at the right. CCSS SMP 7

c. **EVALUATE REASONABLENESS** Explain how you know your graph is reasonable. CCSS SMP 8

d. **USE STRUCTURE** Explain how the graph of $g(\theta) = \frac{3}{4} \sin \frac{\pi}{2}\theta$ would compare to the graph of $f(\theta)$. CCSS SMP 7

You can use trigonometric functions to model a wide range of periodic relationships. In many cases, the given information that you are provided will include the frequency. The **frequency** is the number of cycles in a given unit of time. It is the reciprocal of the period of the function.

EXAMPLE 3 Model a Periodic Relationship

A Ferris wheel at a state fair has a diameter of 65 feet and makes 4 complete revolutions each minute. Dustin boards a car of the Ferris wheel at the car's lowest point and he rides for 2 minutes. He wants to develop a model for his height above or below the axle of the Ferris wheel θ seconds after the ride starts.

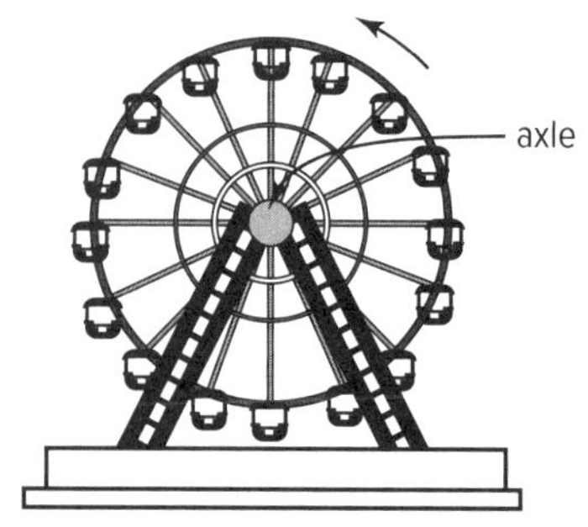

a. **INTERPRET PROBLEMS** Suppose $f(\theta)$ is a sine or cosine function that models Dustin's height above or below the axle. What will be the amplitude of the function? Explain. CCSS F.TF.5, SMP 1

b. **REASON ABSTRACTLY** Explain how to find the period of the function. CCSS F.TF.5, SMP 2

c. **USE STRUCTURE** What should be the value of $f(0)$? Why? Explain how you can use this to decide whether the function should have the form $f(\theta) = a \sin b\theta$ or $f(\theta) = a \cos b\theta$. CCSS F.TF.5, SMP 7

d. **USE A MODEL** Write the function $f(\theta)$. CCSS F.TF.5, SMP 4

e. **USE STRUCTURE** Graph $f(\theta)$ on the coordinate plane at the right. CCSS F.IF.7e, SMP 7

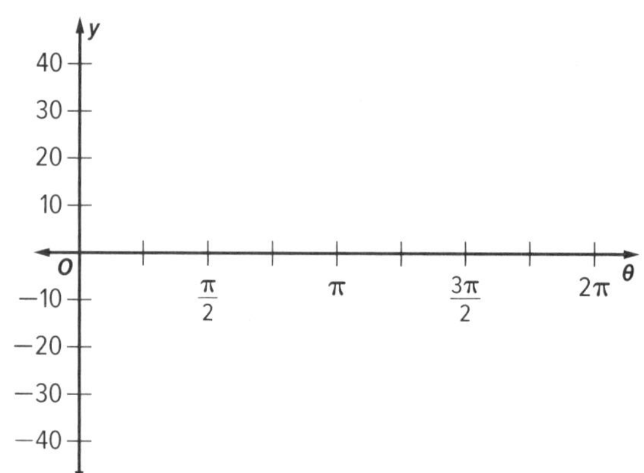

f. **USE A MODEL** What is the domain of the function? How is this related to the real-world situation? CCSS F.IF.5, SMP 4

g. **USE A MODEL** How many θ-intercepts does your graph have? What does this tell you about Dustin's ride? CCSS F.IF.4, SMP 4

h. **USE A MODEL** How is the period of the function shown in the graph? What does this tell you about Dustin's ride? CCSS F.TF.4, SMP 4

EXAMPLE 4 Graph a Tangent Function CCSS A.CED.2

Graph $f(\theta) = \tan \theta$.

a. **USE STRUCTURE** For what values of θ is $f(\theta)$ undefined? What does this tell you about the domain of the function? CCSS SMP 7

b. **USE STRUCTURE** Where does the graph of $f(\theta)$ have vertical asymptotes? CCSS SMP 7

c. **CRITIQUE REASONING** A student said that the amplitude of $f(\theta)$ is 1 because the function has the form $y = a \tan b\theta$, where $a = 1$. Do you agree? Explain. CCSS SMP 3

d. **USE STRUCTURE** Graph $f(\theta)$ on the coordinate plane at the right. CCSS SMP 7

e. **COMMUNICATE PRECISELY** Explain how you can use the graph to determine the period of $f(\theta)$. How does the period compare to that of $y = \sin \theta$ and $y = \cos \theta$? CCSS SMP 6

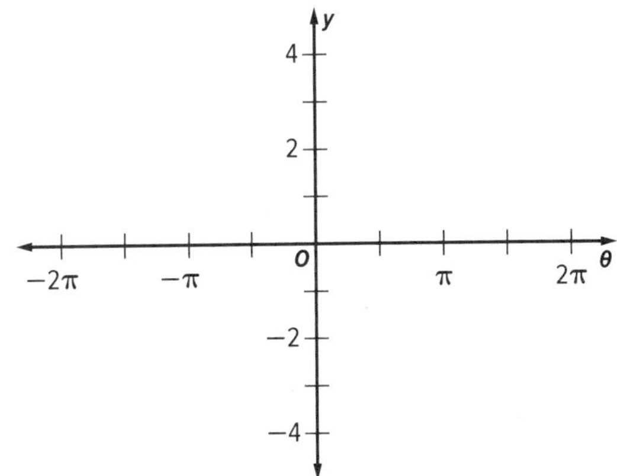

In order to graph a secant, cosecant, or cotangent function, you can use structure to write the function as a cosine, sine, or tangent function.

EXAMPLE 5 Graph a Secant Function

Graph $f(\theta) = \sec 4\theta$.

a. **USE STRUCTURE** Explain how to write $f(\theta)$ as a cosine function. CCSS A.SSE.2, SMP 7

b. **USE STRUCTURE** Graph $g(\theta) = \cos 4\theta$ on the coordinate plane at the right. CCSS A.CED.2, SMP 7

c. **COMMUNICATE PRECISELY** Describe how you can use the graph of $g(\theta)$ to determine where the graph of $f(\theta)$ will have asymptotes. CCSS A.CED.2, SMP 6

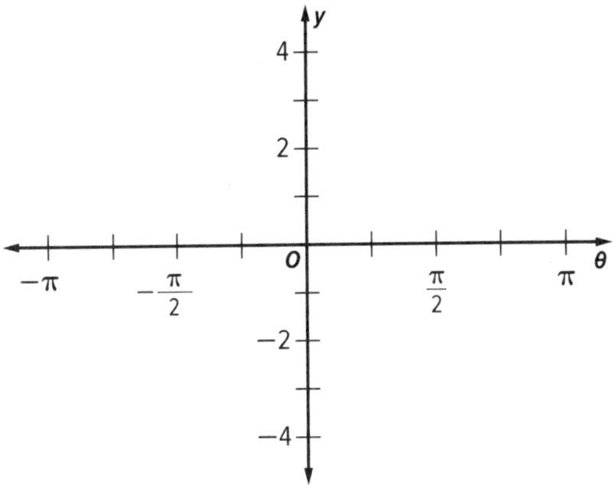

d. **USE STRUCTURE** Use your answer to **part c** and the graph of $g(\theta)$ to help you graph $f(\theta)$ on the coordinate plane at the right. CCSS A.CED.2, SMP 7

PRACTICE

USE STRUCTURE Graph each function. CCSS A.CED.2, SMP 7

1. $f(\theta) = -4 \sin \pi\theta$

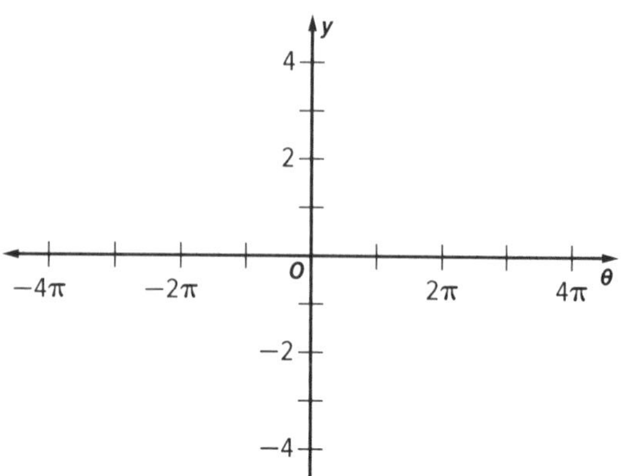

2. $g(\theta) = 3.5 \cos 0.4\theta$

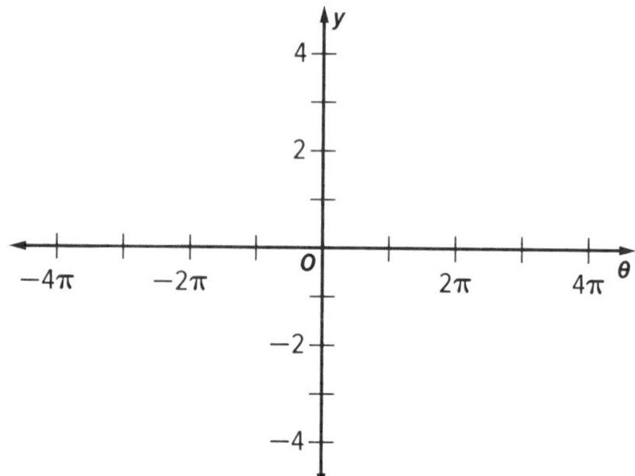

3. $h(\theta) = -\tan 2\theta$

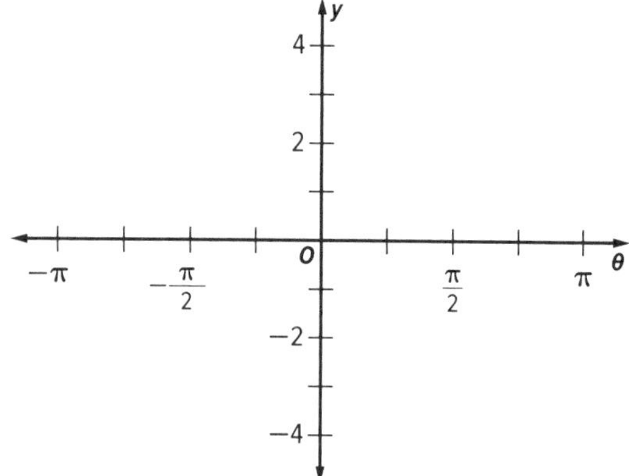

4. $f(\theta) = 0.4 \sec 2\pi\theta$

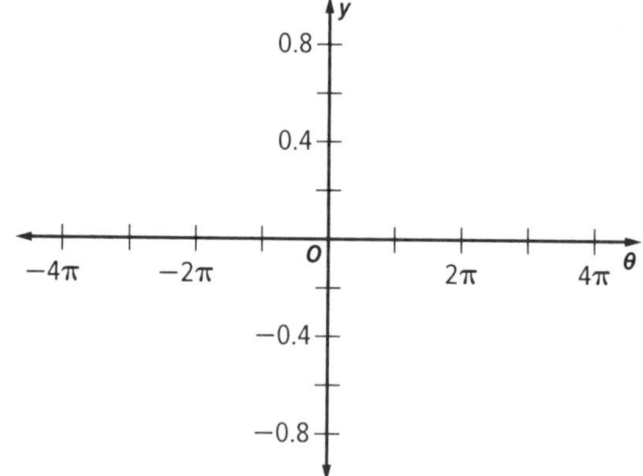

5. $g(\theta) = 2 \cot \frac{\pi}{2}\theta$

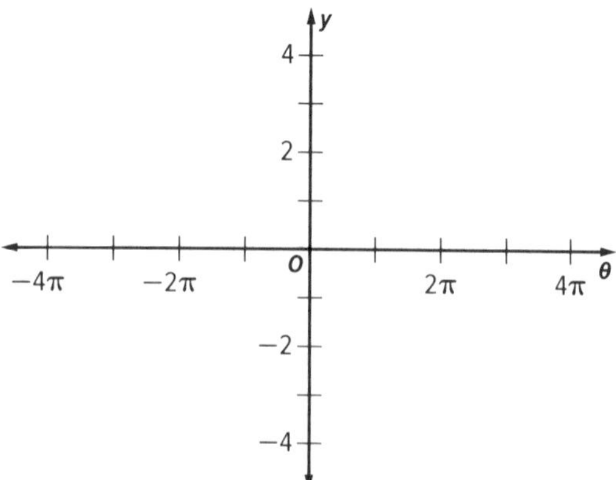

6. $h(\theta) = \frac{1}{2}\csc 0.8\theta$

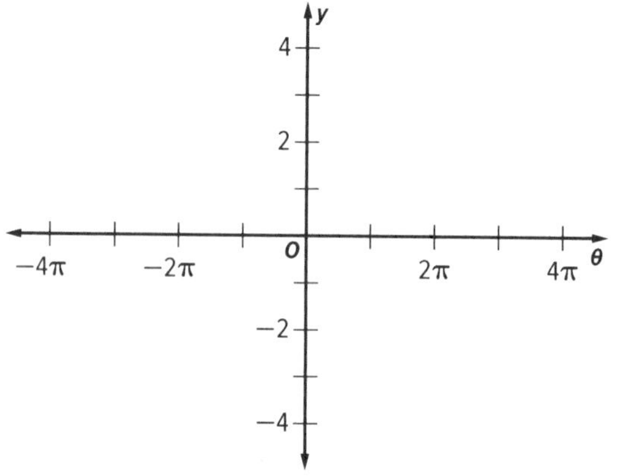

7. **REASON ABSTRACTLY** Write a trigonometric function in which the amplitude is 2 and its graph has 3 complete cycles on the interval $0 \leq \theta \leq \pi$. Justify your answer. CCSS A.CED.2, SMP 2

8. A boat that is tied to a dock moves vertically up and down with the waves. Katie watches the boat for 30 seconds and notes that the boat moves up and down a total of 6 times. The difference between the boat's highest point and lowest point is 3 feet. Katie wants to write a trigonometric function that models the boat's vertical position x seconds after she began watching. Assume that when Katie began watching the boat, it was at its highest point and that its average vertical position was 0 feet.

a. INTERPRET PROBLEMS What is the amplitude of the function? Explain. CCSS F.TF.5, SMP 1

b. REASON ABSTRACTLY What is the period of the function? Explain. CCSS F.TF.5, SMP 2

c. USE A MODEL Explain how to write the function $h(x)$. CCSS F.TF.5, SMP 4

d. USE STRUCTURE Graph $h(x)$ on the coordinate plane at the right. CCSS F.IF.7e, SMP 7

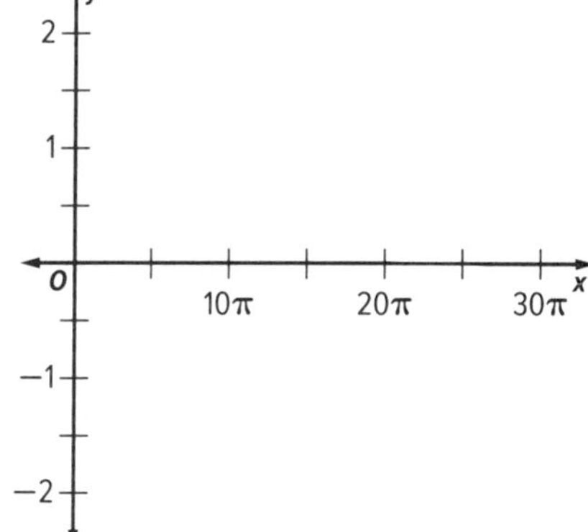

e. USE A MODEL What is the domain of the function? How is this related to the real-world situation? CCSS F.IF.5, SMP 4

f. USE A MODEL What is the minimum value of the function? How many times does the function attain this value in your graph? What does this tell you about the motion of the boat? CCSS F.IF.4, SMP 4

9. COMMUNICATE PRECISELY Write a sine or cosine function for the graph shown at the right. Explain your method. CCSS A.CED.2, SMP 6

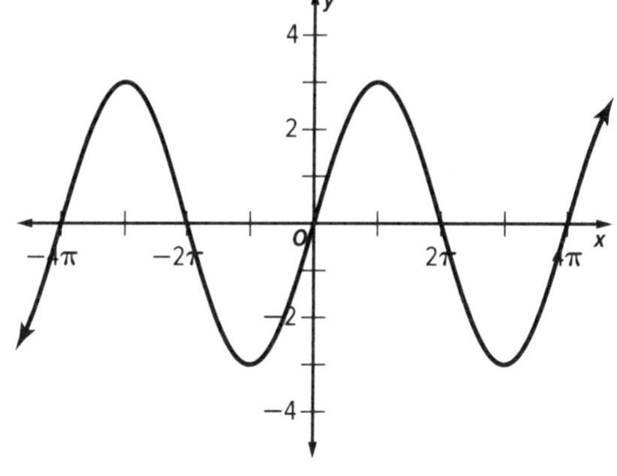

10.4 Translating Trigonometric Functions

Objectives

- Translate the graphs of trigonometric functions.
- Choose trigonometric functions to model periodic phenomena.

CCSS STANDARDS

Content: F.TF.5, F.IF.4, F.IF.7e, F.BF.1b, F.BF.3, A.CED.2
Practices: 1, 2, 3, 4, 5, 6, 7
Use with Lesson 12-8

EXAMPLE 1 Analyze Translations of Trigonometric Graphs

EXPLORE Use a graphing calculator to help you sketch the required graphs in this exploration.

a. **USE TOOLS** Sketch and label the graphs of $f(\theta) = \sin \theta$, $g(\theta) = \sin\left(\theta - \frac{\pi}{2}\right)$, and $h(\theta) = \sin(\theta + \pi)$ on the coordinate plane at the right. CCSS A.CED.2, SMP 5

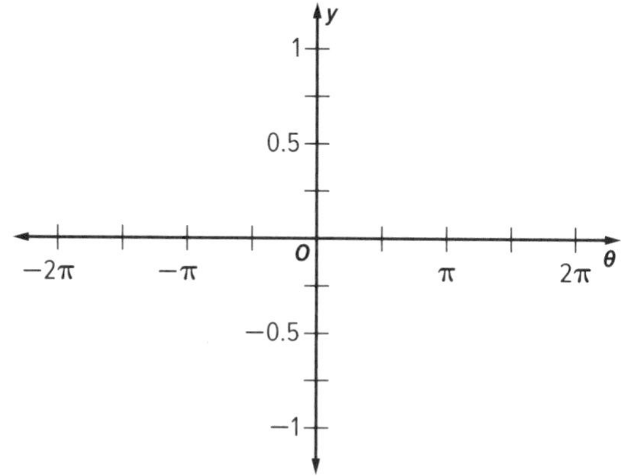

b. **MAKE A CONJECTURE** Describe how the graph of $y = \sin(\theta - h)$ compares to graph of the parent function $y = \sin \theta$. What characteristics do the graphs have in common? What characteristics are different? CCSS F.BF.3, SMP 3

c. **USE TOOLS** Sketch and label the graphs of $f(\theta) = \cos \theta$, $g(\theta) = \cos \theta + 2$ and $h(\theta) = \cos \theta - 3$ on the coordinate plane at the right. CCSS A.CED.2, SMP 5

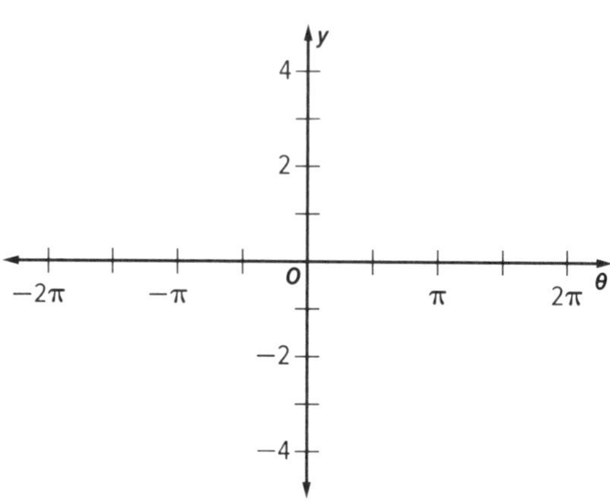

d. **MAKE A CONJECTURE** Describe how the graph of $y = \cos \theta + k$ compares to graph of the parent function $y = \cos \theta$. What characteristics do the graphs have in common? What characteristics are different? CCSS F.BF.3, SMP 3

e. **USE STRUCTURE** The graph of the function $y = \cos \theta$ oscillates about the horizontal line $y = 0$ (the θ-axis). Use your results from **parts c** and **d** to write the equation of the horizontal line about which the graph of the function $y = \cos \theta + k$ oscillates. CCSS F.BF.3, SMP 7

f. **USE STRUCTURE** Write the equation of a sine function with period 2π and amplitude 1 that oscillates about the line $y = 3.3$. Is there more than one possibility? Explain. CCSS F.BF.3, SMP 7

A horizontal translation of a trigonometric function is called a **phase shift**. A vertical translation of a trigonometric function is called a **vertical shift**. The horizontal line about which a trigonometric function oscillates is called the **midline**.

KEY CONCEPT Translations of Trigonometric Graphs

Use your findings from the previous example to help you complete the following.

	Phase Shift	Vertical Shift
Description	The phase shift of the function $y = a \sin b(\theta - h)$, where $b > 0$, is h.	The vertical shift of the function $y = a \sin b\theta + k$ is k.
Direction	If ________, the shift is h units to the right. If ________, the shift is $\lvert h \rvert$ units to the left.	If ________, the shift is k units up. If ________, the shift is $\lvert k \rvert$ units down.
Example	$y = \sin\left(\theta - \frac{\pi}{4}\right)$	$y = \sin \theta - 0.5$
The same rules can be used to graph cosine, tangent, secant, cosecant, and cotangent functions.		

EXAMPLE 2 Graph a Trigonometric Function CCSS F.BF.3, A.CED.2

Graph $f(\theta) = \frac{3}{2}\cos 2\left(\theta - \frac{\pi}{2}\right) + 1$.

a. REASON ABSTRACTLY Explain how to determine the amplitude, period, phase shift, vertical shift, and midline of the function. CCSS SMP 2

b. USE STRUCTURE Graph the function on the coordinate plane at the right, including a graph of the midline. CCSS SMP 7

c. COMMUNICATE PRECISELY Describe how the graph of $f(\theta)$ compares to the graph of $g(\theta) = \frac{3}{2}\cos 2\left(\theta + \frac{\pi}{2}\right) + 1$. CCSS SMP 6

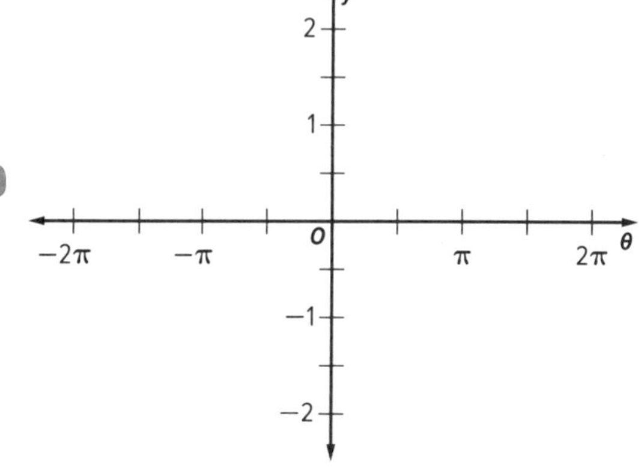

EXAMPLE 3 Model a Periodic Relationship

As the tide moves in and out of a bay, the depth of the water in the bay varies from a low of 15 meters to a high of 18 meters. It takes 6.2 hours for the tide to come in and 6.2 hours for the tide to go out. A marine biologist starts recording the depth of the water in the bay at high tide. She would like to develop a model that gives the depth of the water in the bay at any time *t* hours after she starts recording the data.

a. **INTERPRET PROBLEMS** Suppose $f(t)$ is a sine or cosine function that models the depth of the water in the bay. What will be the amplitude of the function? What will be the midline of the function? Explain. CCSS F.TF.5, SMP 1

b. **REASON ABSTRACTLY** Explain how to find the period of the function. CCSS F.TF.5, SMP 2

c. **USE STRUCTURE** Is it more convenient to use a sine function or a cosine function to model the depth of the water? Justify your choice. CCSS F.TF.5, SMP 7

d. **USE A MODEL** Write a function $f(t)$ that models the depth of the water. CCSS F.TF.5, SMP 4

e. **USE STRUCTURE** Graph $f(t)$ on the coordinate plane at the right. CCSS F.IF.7e, SMP 7

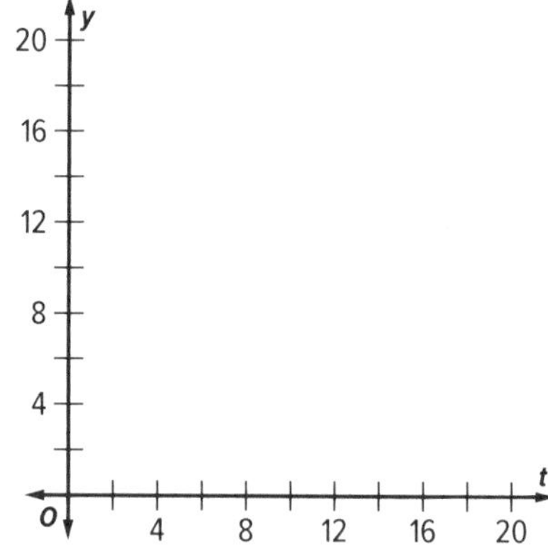

f. **USE STRUCTURE** Explain how you can think of the function you wrote as a sum of two functions, $g(t)$ and $h(t)$. What do the functions g and h represent? CCSS F.BF.1b, SMP 7

g. **USE A MODEL** Identify an interval where the graph is increasing. What does this represent in the real-world situation? CCSS F.IF.4, SMP 4

h. **USE A MODEL** Explain how you can write the model using a different trigonometric function, and explain how you know this new model is also correct. CCSS F.TF.5, SMP 4

PRACTICE

USE STRUCTURE **Graph each function on the coordinate plane provided.** 

1. $f(\theta) = -\tan\frac{1}{2}\left(\theta + \frac{\pi}{2}\right) - 2$

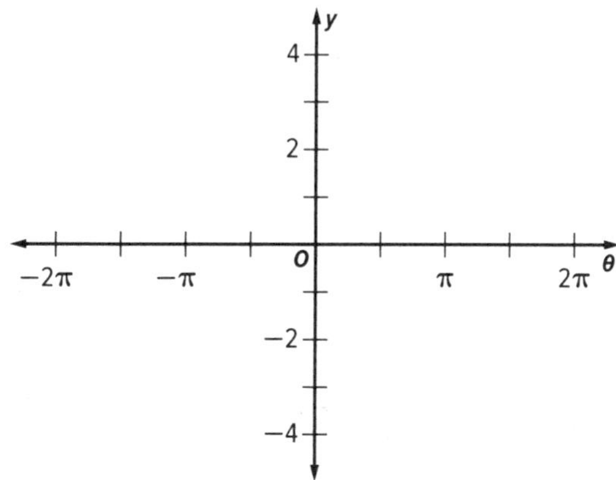

2. $g(\theta) = 2\sec 2(\theta - \pi) + 0.5$

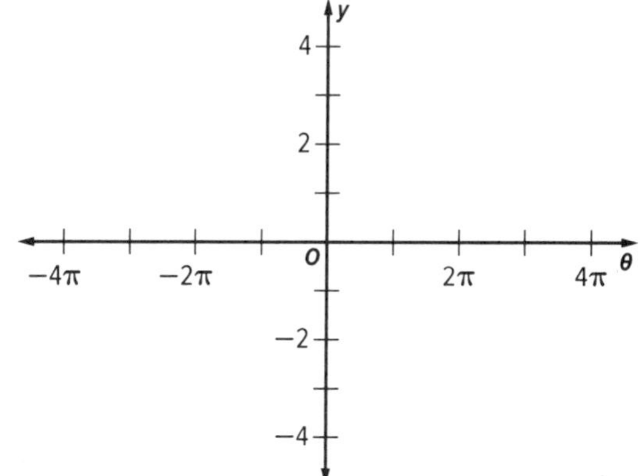

3. An office building has a large clock on its exterior. The center of the clock is located 79 feet above ground level and the minute hand of the clock is 3 feet long. The function $h(t)$ gives the height of the tip of the minute hand above ground level in feet at any time t minutes after 12:00 noon.

   a. **INTERPRET PROBLEMS** What is the amplitude of $h(t)$? What is the midline? What is the period? CCSS F.TF.5, SMP 1

   b. **USE A MODEL** Write the function $h(t)$ as a cosine function. CCSS F.TF.5, SMP 4

   c. **USE STRUCTURE** Graph $h(t)$ on the coordinate plane at the right. CCSS F.IF.7e, SMP 7

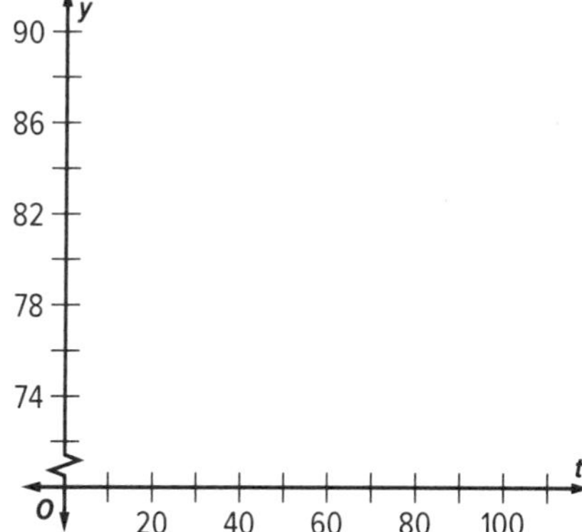

   d. **USE A MODEL** On the interval $t = 0$ to $t = 120$, how many times does the function attain its minimum value? What does this represent in the real world-situation? CCSS F.IF.4, SMP 4

   e. **USE A MODEL** Write $h(t)$ using a sine function. CCSS F.TF.5, SMP 4

4. **USE STRUCTURE** Let $f(\theta) = 2.7\cos 3(\theta - \pi) + 1$. Write a sine function $g(\theta)$ that has the same graph as $f(\theta)$. CCSS A.CED.2, SMP 7

5. **USE STRUCTURE** Write a cosine function $h(\theta)$ with midline $y = -2$ and period π, that has no θ-intercepts. CCSS A.CED.2, SMP 7

10.5 Modeling: Trigonometric Functions

Objectives

- Choose trigonometric functions to model periodic phenomena.
- Interpret features of graphs and parts of expressions that model periodic phenomena.

CCSS STANDARDS

Content: A.SSE.1a, A.SSE.1b, F.TF.5, F.IF.4, F.IF.5, F.BF.1b
Practices: 1, 2, 4, 5, 6, 7, 8
Use with Lesson 12-8

EXAMPLE 1 Modeling Motion CCSS F.TF.5

EXPLORE **Marisol sets up an experiment with a spring at a physics lab. She hangs the spring from a hook and attaches a weight to the bottom of the spring. She records the length of the spring when is it fully compressed and fully extended, as shown. When she releases the spring from the fully-compressed position, she finds that it takes 2 seconds to come back to this position. Marisol wants to write a function $f(t)$ that models the length of the spring, in centimeters, t seconds after it has been released.**

Fully Compressed
3 cm
Fully Extended
12 cm

a. **INTERPRET PROBLEMS** What is the amplitude of $f(t)$? What is the period? What is the midline? CCSS SMP 1

b. **USE A MODEL** Write a function $f(t)$ that models the length of the spring. CCSS SMP 4

c. **USE STRUCTURE** Graph $f(t)$ on the coordinate plane at the right. CCSS SMP 7

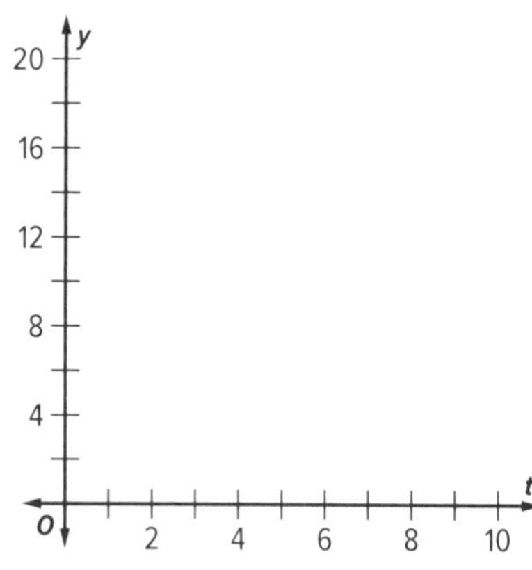

d. **EVALUATE REASONABLENESS** Explain how you know that your model is reasonable. CCSS SMP 8

e. **USE A MODEL** What is the frequency of the function? Explain how you found the frequency and explain what it tells you about the spring. CCSS SMP 4

f. **USE TOOLS** Marisol's teacher asks her to find the approximate times during the first 2 seconds when the length of the spring is 5 centimeters. Find these times to the nearest tenth of a second and explain your method. CCSS SMP 5

EXAMPLE 2 Use Sinusoidal Regression

Aaron is planning a softball tournament in Chicago, Illinois. He would like the tournament to take place when the mean daily temperature is likely to be 65° F. To help him decide when to hold the tournament, he collects the data shown in the table.

Mean Daily Temperatures in Chicago						
Date	Jan 15	Mar 15	Jun 15	Aug 15	Nov 15	Dec 15
Day of Year	15	74	166	227	319	349
Mean Temperature (°F)	25	39	71	74	43	30

a. **USE TOOLS** Enter the data for the day of the year and the corresponding temperature into two lists in your calculator. Then use the calculator's sinusoidal regression tool to find a sine function, $f(x)$, that models the mean daily temperature, in degrees Fahrenheit, on day x of the calendar year. Write the function, rounding each parameter to the nearest thousandth. CCSS F.TF.5, SMP 5

b. **USE A MODEL** If you write $f(x)$ in the form $f(x) = a \sin b(x - h) + k$, what are the values of a and k? What do these tell you about the mean daily temperatures in Chicago? CCSS A.SSE.1a, SMP 4

c. **USE A MODEL** Explain how to view $f(x)$ as a sum of two functions, $g(x)$ and $h(x)$. Then explain what $g(x)$ and $h(x)$ represent. CCSS F.BF.1b, A.SSE.1b, SMP 4

d. **USE STRUCTURE** Graph $f(x)$ on the coordinate plane at the right. Based on the graph, identify the approximate period of the function. Then explain what this represents and explain why it makes sense. CCSS F.IF.4, SMP 7

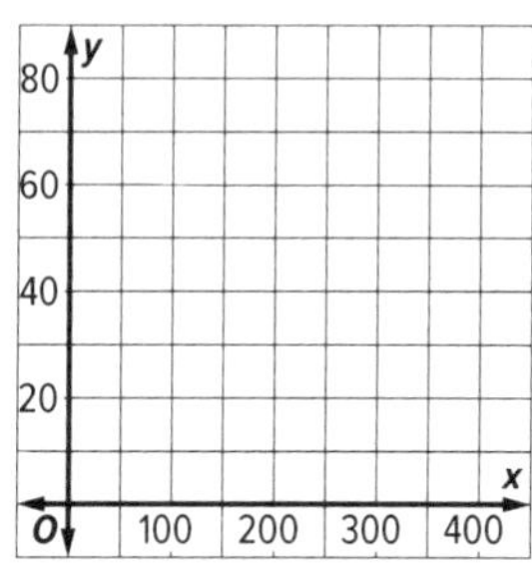

e. **USE TOOLS** Use your calculator to determine the day or days of the year when Aaron should consider holding the tournament. Explain your answer. CCSS F.TF.5, SMP 5

It is often convenient to model periodic phenomena with a sine or cosine function. However, the other trigonometric functions can also be used to develop a model.

EXAMPLE 3 Write a Model Using Other Trigonometric Functions

An ambulance stops outside a hospital as shown in the figure. The ambulance is 10 feet from the hospital's outer wall. The rotating light at the top of the ambulance projects a beam of light $\overline{AB}$ that shines on the wall. In the figure, *c* represents the length of the beam and θ represents the measure of the angle in radians that the beam makes with the perpendicular segment to the wall. The light makes one complete rotation every second.

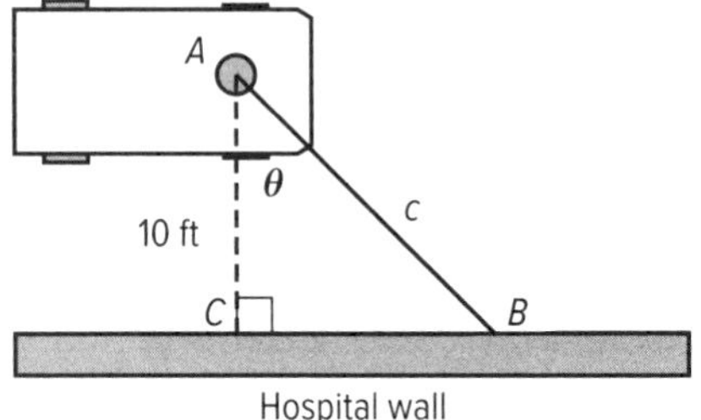

a. **INTERPRET PROBLEMS** Use the figure to write a trigonometric ratio involving θ. CCSS F.TF.5, SMP 1

b. **USE A MODEL** Use your answer to **part a** to help you write a function $c(x)$ that gives the length c of the beam of light, in feet, as a function of the time x in seconds. Assume the beam starts by shining directly on the wall at $x = 0$ seconds. CCSS F.TF.5, SMP 4

c. **USE STRUCTURE** Graph $f(x)$ on the coordinate plane at the right. CCSS F.TF.5, SMP 7

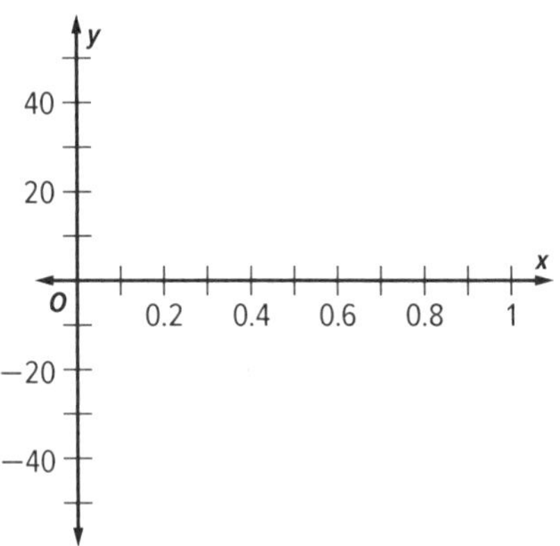

d. **USE STRUCTURE** What is the domain of the function? What does this tell you about the graph? CCSS F.IF.5, SMP 7

e. **COMMUNICATE PRECISELY** What do the asymptotes of the graph represent in the real-world situation? CCSS F.IF.4, SMP 6

f. **USE A MODEL** Does the graph have x-intercepts? Explain why your answer makes sense in the context of the real-world situation. CCSS F.IF.4, SMP 4

g. **USE A MODEL** For what values of x on your graph is $c(x) < 0$? What does this represent in the real-world situation? CCSS F.IF.4, SMP 4

PRACTICE

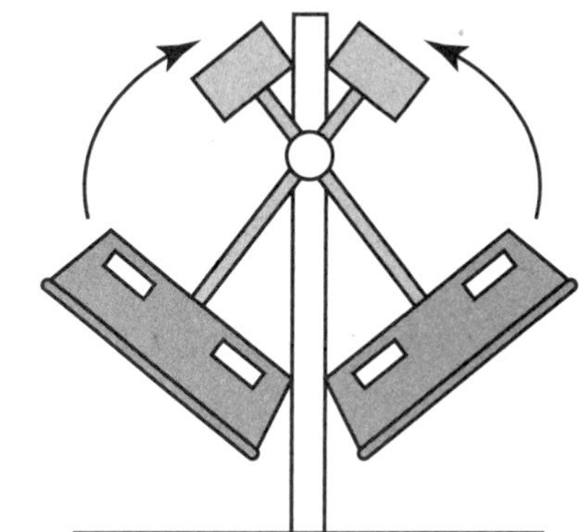

1. An amusement park ride consists of two 16-foot arms that each have a car on one end of the arm and a counterweight on the other end. The arms rotate in opposite directions. Each car has a minimum height of 3 feet above the ground. It takes each car 3 seconds to make a complete rotation. The function $f(x)$ models the height of one car above the ground x seconds after the ride starts. (Assume the cars make complete rotations as soon as the ride begins.)

   a. **USE A MODEL** Write a function $f(x)$ that models the motion of a car. CCSS F.TF.5, SMP 4

   b. **USE STRUCTURE** Graph $f(x)$ on the coordinate plane. CCSS F.IF.4, SMP 7

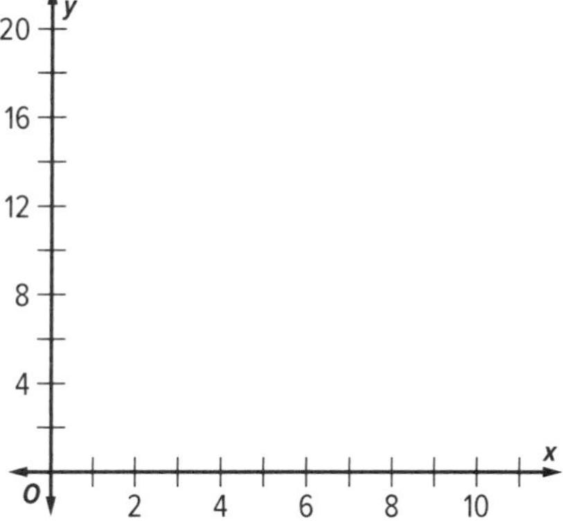

   c. **USE A MODEL** What is the first interval of your graph in which the function is increasing? What does it represent? CCSS F.IF.4, SMP 4

   d. **USE A MODEL** Explain how your graph shows the period of the ride. CCSS F.IF.4, SMP 4

   e. **USE TOOLS** To the nearest tenth of a second, when is the first time that the height of the car is exactly 12 feet? Explain your method. CCSS F.TF.5, SMP 5

   f. **REASON ABSTRACTLY** The cars rotate in opposite directions. Does it matter which car you use when you develop the function $f(x)$? Explain why or why not. CCSS F.TF.5, SMP 2

2. **COMMUNICATE PRECISELY** The function $h(x) = 6 \sin 30\pi x + 10$ models the height above ground in inches of a point P at the tip of a blade of a floor fan x seconds after the fan is turned on. What is the speed of the fan in rotations per minute (rpm)? Explain. CCSS A.SSE.1a, A.SSE.1b, SMP 6

3. The table shows the number of hours of daylight on various days of the year in Seattle, Washington.

Number of Hours of Daylight in Seattle						
Date	Jan 1	Feb 11	May 3	Aug 22	Oct 15	Nov 29
Day of Year	1	42	123	234	288	333
Hours of Daylight	8.5	10.1	14.6	13.9	10.9	8.8

a. **USE TOOLS** Use your calculator to write a function $f(x)$ that models the number of hours of daylight in Seattle on day x of the calendar year. Round each parameter to the nearest thousandth. CCSS F.TF.5, SMP 5

b. **USE STRUCTURE** Graph $f(x)$ on the coordinate plane. CCSS F.IF.4, SMP 7

c. **USE A MODEL** What is the maximum value of $f(x)$? What does this tell you? CCSS F.IF.4, SMP 4

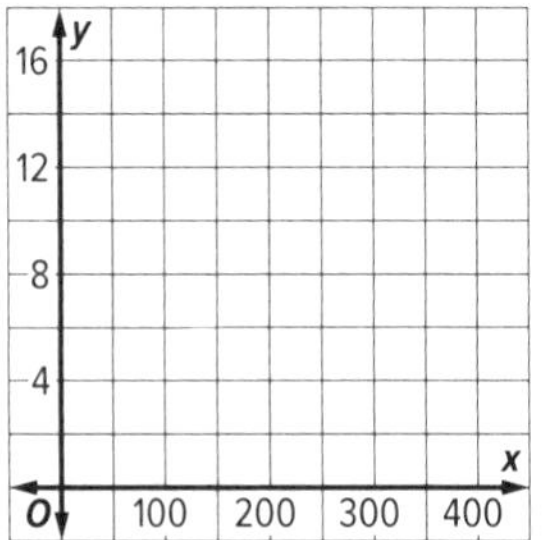

d. **USE TOOLS** During what period of the year does Seattle get more than 14 hours of daylight per day? Explain. CCSS F.IF.4, SMP 5

4. Carissa has a small business installing and repairing air conditioners. She has kept track of her monthly profit since she started the business. The table shows data for several months.

Monthly Profit						
Months Since Start of Business	8	12	16	21	23	31
Monthly Profit ($)	3550	480	1820	2480	950	4030

a. **USE TOOLS** Write a function $P(x)$ that models the monthly profit in dollars, x months since the start of the business. Round each parameter to the nearest thousandth. CCSS F.TF.5, SMP 5

b. **REASON QUANTITATIVELY** Determine the approximate period of the function. What does this tell you about Carissa's business? CCSS A.SSE.1a, SMP 2

c. **USE A MODEL** Next month will mark the 6th anniversary since Carissa started the business. What profit should she expect to make next month? Explain. CCSS F.TF.5, SMP 4

5. Jamal's house has a thermometer on an outside wall. One day last June he recorded temperatures from the thermometer at several times during the day, as shown.

Outside Temperature Readings						
Hours Since Midnight	1	4	8.5	14	19.5	22
Temperature (° F)	59	50	65	84	72	60

a. **USE TOOLS** Write a function $T(x)$ that models the temperature in degrees Fahrenheit, x hours after midnight. Round each parameter to the nearest thousandth. CCSS F.TF.5, SMP 5

b. **USE A MODEL** What was the approximate high temperature on the day that Jamal recorded the data? At approximately what time did this occur? Explain. CCSS F.TF.5, SMP 4

6. A garden sprinkler is attached to a spigot by a straight hose that is 3 meters long. The sprinkler makes 6 complete rotations every minute. As it rotates, it sprays a stream of water that hits a wall, as shown. Let d be the length of the stream of water from the sprinkler to the wall, and assume the stream of water hits the spigot when the sprinkler is first turned on.

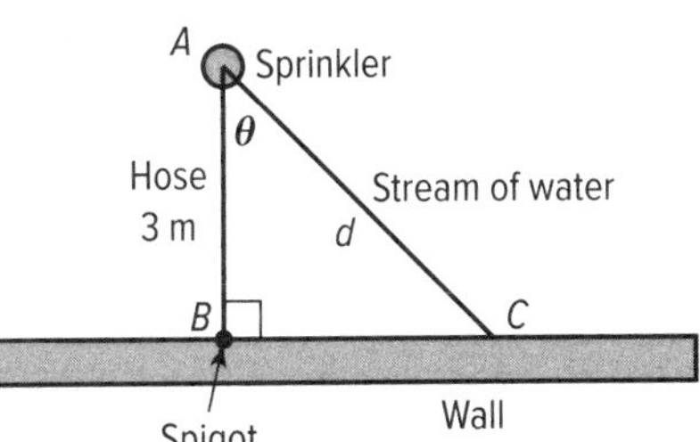

a. **USE A MODEL** Write a function $d(x)$ that models the length of the stream in meters as a function of the time x in seconds. CCSS F.TF.5, SMP 4

b. **USE STRUCTURE** Graph $d(x)$ on the coordinate plane. CCSS F.IF.4, SMP 7

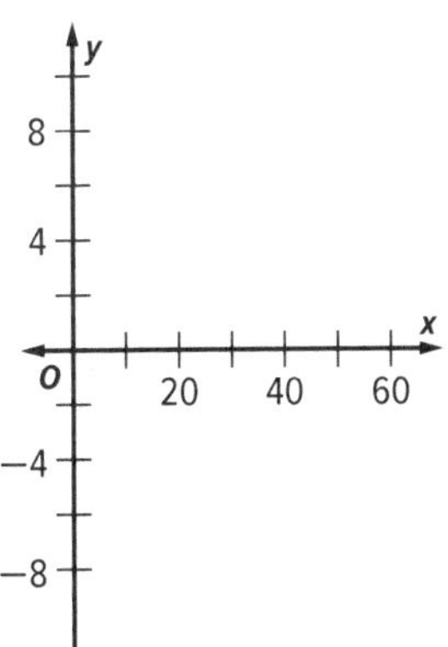

c. **USE STRUCTURE** What is the domain of the function? Explain how this is shown in the graph and explain what it tells you about the sprinkler. CCSS F.IF.5, SMP 7

d. **USE TOOLS** After the sprinkler is turned on, what is the first time when the stream of water is 5 meters long? Explain. CCSS F.TF.5, SMP 5

7. **USE A MODEL** A cyclist pedals at a rate of 6 rotations every 5 seconds. The motion of the pedals is circular with a radius of 7 inches. The closest the pedals get to the ground is four inches away. Write a function $h(t)$ that models the height of a pedal in inches as a function of the time t in seconds. Assume the pedal starts at its highest point. CCSS F.TF.5, SMP 4

10.6 Inverse Trigonometric Functions

Objectives

- Graph inverse trigonometric functions.
- Write equations involving inverse trigonometric functions.

Content: A.CED.2
Practices: 1, 2, 4, 6, 7, 8
Use with Lesson 12-9

Recall that an **inverse function** is the function in which domain and range of the initial function are reversed.

EXAMPLE 1 Analyze the Inverse of the Sine Function CCSS A.CED.2

EXPLORE **The inverse of the function $y = \sin \theta$ is $\theta = \sin y$. You will investigate this inverse in the following exploration.**

a. **COMMUNICATE PRECISELY** Describe how the graph of $\theta = \sin y$ is related to the graph of $y = \sin \theta$. CCSS SMP 6

b. **USE STRUCTURE** Use your answer to **part a** to graph $\theta = \sin y$ on the coordinate plane at the right. CCSS SMP 7

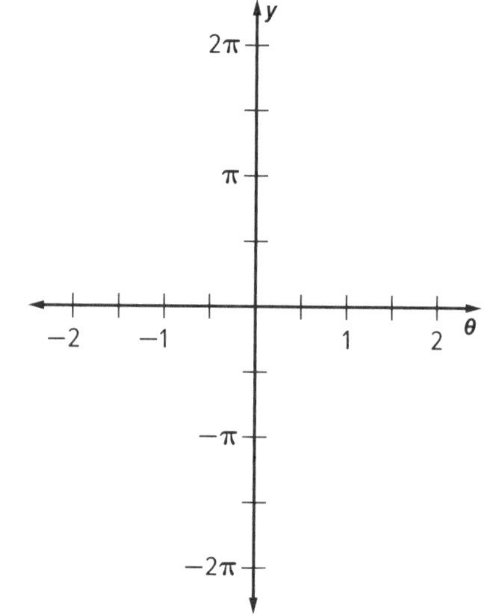

c. **REASON ABSTRACTLY** Suppose $\theta = \frac{1}{2}$. What is the value of y? Is there more than one possible y-value? Explain. CCSS SMP 2

d. **COMMUNICATE PRECISELY** Is the relation $\theta = \sin y$ a function? Why or why not? CCSS SMP 6

e. **USE STRUCTURE** Consider a new function, $y = \text{Sin}\,\theta$, which is the sine function with its domain restricted to $\left\{-\frac{\pi}{2} \leq \theta \leq \frac{\pi}{2}\right\}$. Graph $\theta = \text{Sin}\,y$ on the coordinate plane provided. CCSS SMP 7

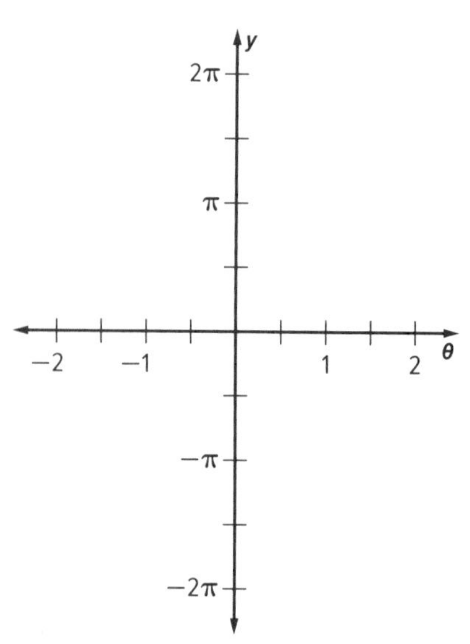

f. **COMMUNICATE PRECISELY** Is the relation $\theta = \text{Sin}\,y$ a function? Why or why not? CCSS SMP 6

g. **REASON ABSTRACTLY** For $\theta = \text{Sin}\,y$, suppose you know that $\theta = \frac{1}{2}$. What is the value of y? Is there more than one possible y-value? Explain. CCSS SMP 7

In order to define inverse trigonometric functions, it is necessary to restrict the domains of the trigonometric functions. Values in the restricted domains are called **principle values**. The trigonometric functions with restricted domains are as follows.

- $y = \text{Sin}\,\theta$ is $y = \sin\theta$, where $D = \left\{-\frac{\pi}{2} \leq \theta \leq \frac{\pi}{2}\right\}$.
- $y = \text{Cos}\,\theta$ is $y = \cos\theta$, where $D = \{0 \leq \theta \leq \pi\}$.
- $y = \text{Tan}\,\theta$ is $y = \tan\theta$, where $D = \left\{-\frac{\pi}{2} \leq \theta \leq \frac{\pi}{2}\right\}$.

KEY CONCEPT Inverse Trigonometric Functions

Use your findings about $y = \text{Sin}\,\theta$ from the previous Example 1 to help you complete the following.

Inverse Function	Symbols	Domain	Range	Sample Graph
Arcsine	$y = \text{Arcsin}\,\theta$ $y = \text{Sin}^{-1}\,\theta$			$y = \text{Sin}^{-1}\theta$ y, $\frac{\pi}{2}$, −1, 1, θ, $\frac{-\pi}{2}$
Arccosine	$y = \text{Arccos}\,\theta$ $y = \text{Cos}^{-1}\,\theta$	$\{-1 \leq \theta \leq 1\}$	$\{0 \leq y \leq \pi\}$	
Arctangent	$y = \text{Arctan}\,\theta$ $y = \text{Tan}^{-1}\,\theta$	{all real numbers}	$\left\{-\frac{\pi}{2} < y < \frac{\pi}{2}\right\}$	

EXAMPLE 2 Graph an Inverse Trigonometric Function CCSS A.CED.2

Graph $g(\theta) = \text{Cos}^{-1}\left(\theta - \frac{1}{2}\right) + \frac{\pi}{2}$.

a. COMMUNICATE PRECISELY Describe how the graph of $g(\theta)$ is related to the graph of the parent function $f(\theta) = \text{Cos}^{-1}\,\theta$. CCSS SMP 6

b. USE STRUCTURE Graph the parent function $f(\theta) = \text{Cos}^{-1}\,\theta$ on the coordinate plane below using a dashed line. CCSS SMP 7

c. USE STRUCTURE Graph $g(\theta)$ on the coordinate plane using a solid line. CCSS SMP 7

d. EVALUATE REASONABLENESS Explain how you can check that your graph of $g(\theta)$ is reasonable. CCSS SMP 8

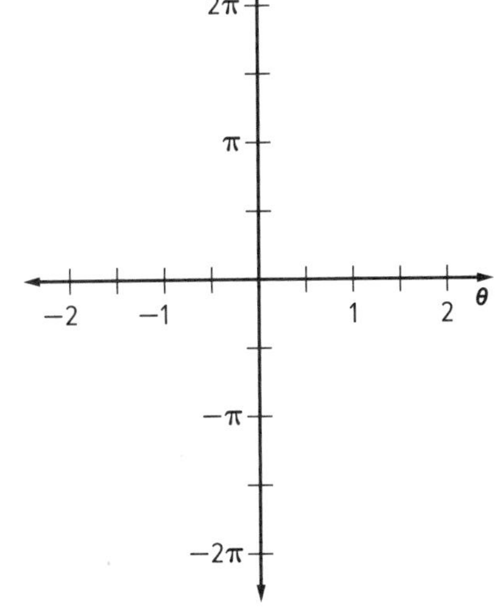

EXAMPLE 3 Develop a Model with an Inverse Trigonometric Function CCSS A.CED.2

A movie crew is preparing to film a train as it passes through a station on straight tracks. The camera is 20 feet from the tracks and the train will be traveling at 60 miles per hour. The crew members need a function that tells them the angle θ, in radians, through which the camera should be turned so that it points at the front of the train x seconds after it passes in front of the camera.

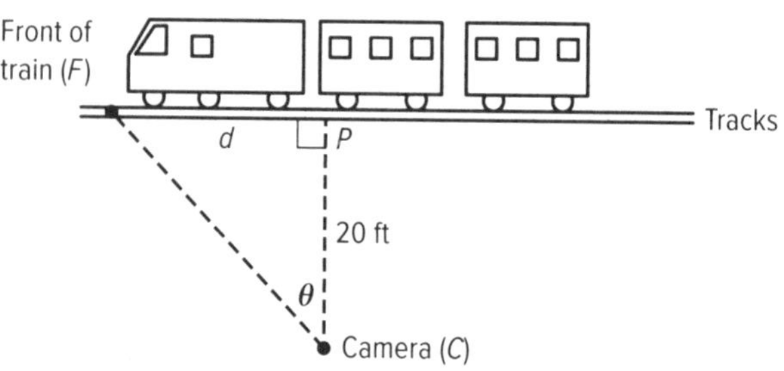

a. **INTERPRET PROBLEMS** Let d be the distance in feet that the train has traveled since it passed in front of the camera. Explain how to write an expression that gives d in terms of x. CCSS SMP 1

b. **USE A MODEL** Explain how to write a function $f(x)$ that gives the measure of θ at any time x seconds after the train has passed in front of the camera. CCSS SMP 4

c. **USE STRUCTURE** Graph $f(x)$ on the coordinate plane at the right. CCSS SMP 7

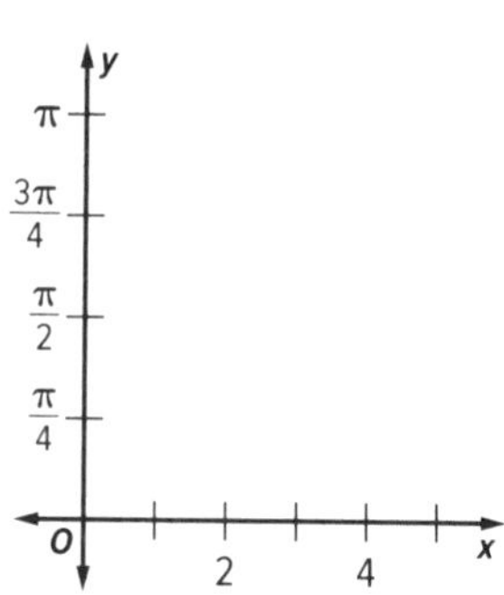

d. **USE STRUCTURE** Does the graph have any asymptotes? If so, give their equations and explain what the asymptotes represent in the real-world situation. CCSS SMP 7

PRACTICE

1. **COMMUNICATE PRECISELY** Write an inverse trigonometric function with domain $\{-2 \leq \theta \leq 2\}$ and range $\{1 \leq y \leq \pi + 1\}$. Justify your answer and graph the function at the right. CCSS A.CED.2, SMP 6

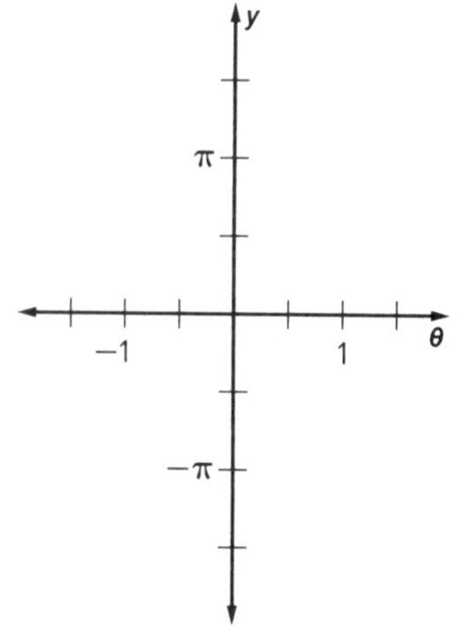

USE STRUCTURE **Graph each function on the coordinate plane provided.** CCSS A.CED.2, SMP 7

2. $f(\theta) = -\text{Tan}^{-1}\,\theta - \frac{\pi}{2}$

3. $g(\theta) = \frac{1}{2}\text{Sin}^{-1}\left(\theta + \frac{1}{2}\right)$

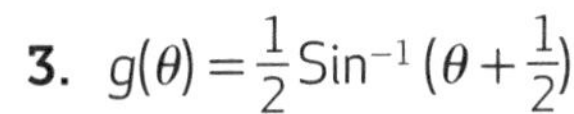

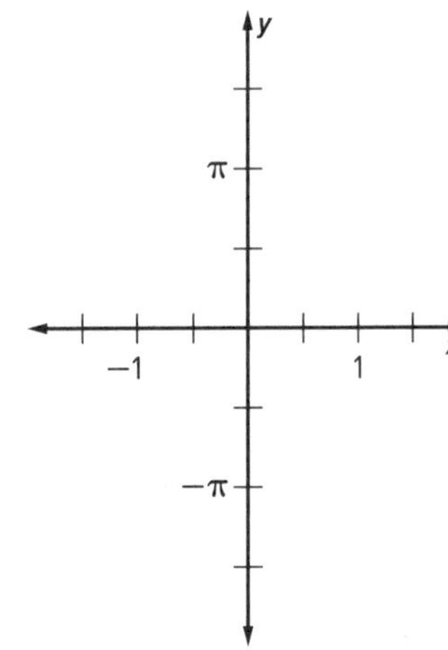

4. Keisha places a 12-foot ladder against a wall so that the base of the ladder is 4 feet from the bottom of the wall. She pulls the base of the ladder away from the wall at a rate of 2 feet per second. CCSS A.CED.2

a. **USE A MODEL** Let θ be the angle, measured in radians, that the ladder makes with the floor. Write a function $f(x)$ that gives the measure of θ after x seconds. CCSS SMP 4

b. **USE STRUCTURE** Graph $f(x)$ on the coordinate plane at the right. CCSS SMP 4

c. **USE A MODEL** What is the domain of $f(x)$? Why? CCSS SMP 7

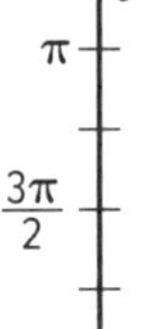

d. **USE A MODEL** What is the y-intercept of the graph? Explain what this represents CCSS SMP 4

5. **USE STRUCTURE** It is possible to define an inverse function for the secant function. CCSS A.CED.2, SMP 7

a. Explain how you can restrict the domain of the function $y = \sec\theta$ so that the inverse, $y = \text{Sec}^{-1}\,\theta$, is a function.

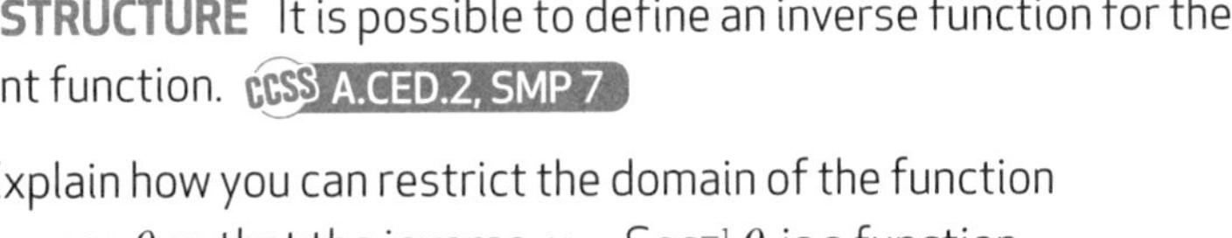

b. Graph $y = \text{Sec}^{-1}\,\theta$ on the coordinate plane at the right.

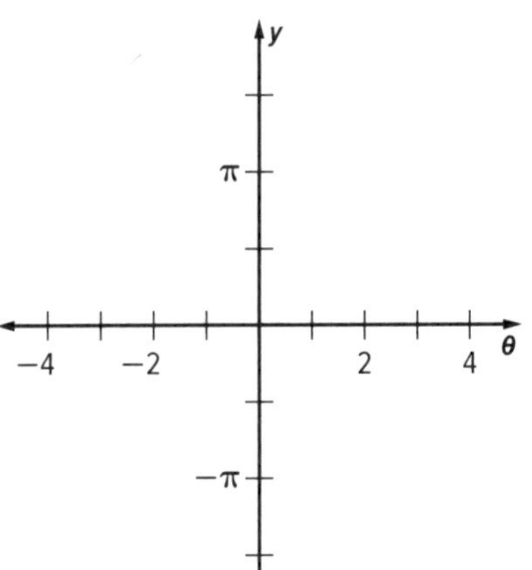

10.7 Verifying the Pythagorean Identity

Objectives

- Verify the Pythagorean Identity.
- Use the Pythagorean Identity to solve problems.

Content: F.TF.8
Practices: 1, 2, 3, 5, 6, 7
Use with Lessons 13-1, 13-2

Recall that an identity is an equation that is true for all values of the variable. A **trigonometric identity** is an equation involving trigonometric functions that is true for all values of the variable for which every expression in the equation is defined.

KEY CONCEPT The Pythagorean Identity

Complete the following identity.

For any angle θ, $\sin^2\theta + \cos^2\theta =$ ____________.

EXAMPLE 1 Verifying the Pythagorean Identity CCSS F.TF.8

EXPLORE **Consider the unit circle and a point $P(x, y)$ that is located on the unit circle.**

a. **USE STRUCTURE** Construct a right triangle with one side on the x-axis and a vertex at P, and express x and y in terms of a reference angle θ. CCSS SMP 7

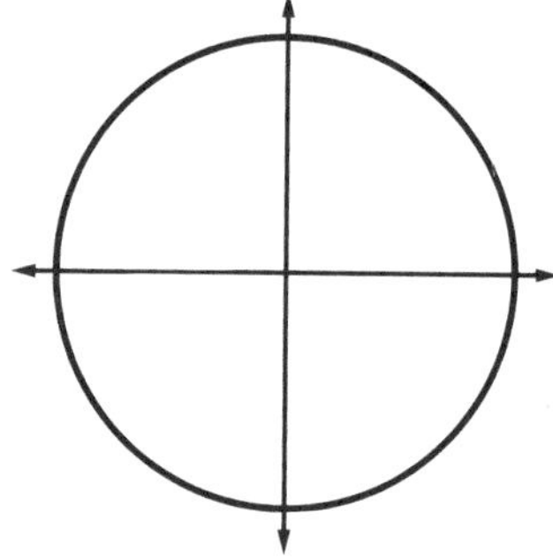

b. **REASON QUANTITATIVELY** Explain how to use the result from **part a** to verify the Pythagorean Identity. CCSS SMP 2

c. **USE STRUCTURE** Show that the Pythagorean Identity is true for circles with a radius not equal to 1. Use a diagram to help explain your reasoning. CCSS SMP 7

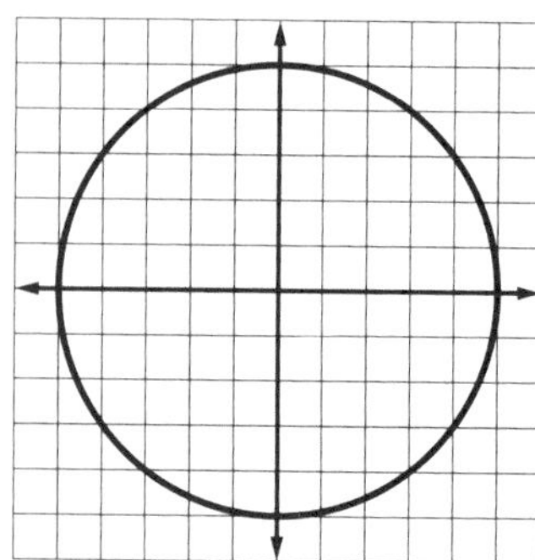

EXAMPLE 2 Use the Pythagorean Identity to Solve Problems 

If you know the value of one trigonometric function for an angle, you can sometimes use the Pythagorean Identity to find the values of other trigonometric functions for the angle.

a. **PLAN A SOLUTION** If $\sin \theta = \frac{\sqrt{7}}{4}$, can you determine the value of $\cos \theta$? Explain your reasoning. CCSS SMP 1

b. **PLAN A SOLUTION** If $\cos \theta = -\frac{1}{a}$ with $a > 0$, and $\tan \theta > 0$, find the value of $\tan \theta$. Show your work. CCSS SMP 1

c. **PLAN A SOLUTION** If $\sin \theta = \frac{x+3}{x+4}$, $\cos \theta = \frac{x-4}{x+4}$, and $\cos \theta < 0$, show how to find the possible value(s) of x. CCSS SMP 1

EXAMPLE 3 Other Pythagorean Identities CCSS F.TF.8

You can use the Pythagorean Identity to derive other useful identities.

a. **USE STRUCTURE** Explain how to derive another version of the Pythagorean Identity that involves only $\tan \theta$ and $\sec \theta$. CCSS SMP 7

b. **PLAN A SOLUTION** If $\sec \theta + \tan \theta = 3$, is it possible to determine the value of $\sec \theta - \tan \theta$? Explain your reasoning. CCSS SMP 1

c. **USE TOOLS** Another version of the Pythagorean Identity is $\csc^2 \theta - \cot^2 \theta = 1$. How can you verify this identity using your calculator? Explain your answer and show the graph you used. CCSS SMP 5

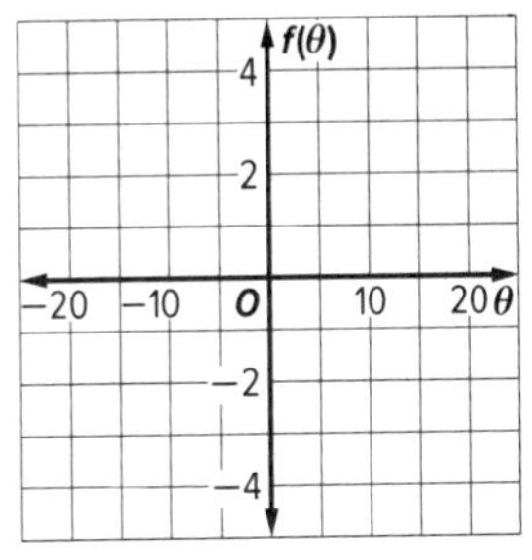

PRACTICE

1. **PLAN A SOLUTION** If $\sin\theta = \frac{2}{a}$ and $\tan\theta = \frac{2}{b}$, express a in terms of b. Show your work. CCSS F.TF.8, SMP 1

2. **PLAN A SOLUTION** If $\sec\theta = \frac{x+10}{x-8}$ and $\tan\theta = \frac{x+9}{x-8}$, show how to find the possible value(s) of x. Show your work. CCSS F.TF.8, SMP 1

3. **CRITIQUE REASONING** Diego decides that if $\sin^2 A + \cos^2 B = 1$, and A and B both have measures between 0° and 180°, then $A = B$. Is he correct? Explain your reasoning. CCSS F.TF.8, SMP 3

4. **USE STRUCTURE** Verify the identity $1 = (\sec^2\theta - 1)(\csc^2\theta - 1)$ algebraically. CCSS F.TF.8, SMP 7

5. **PLAN A SOLUTION** If $\cos\theta = -\frac{\sqrt{10}}{5}$ and $\tan\theta > 0$, explain how to find the value of $\sin\theta$. Does the answer change if $\tan\theta < 0$? CCSS F.TF.8, SMP 1

6. **USE TOOLS** How can you use your calculator to show that the identity $\sin^2\theta + \csc^2\theta = 1$ is false? Explain your answer and show the graph you used. CCSS F.TF.8, SMP 5

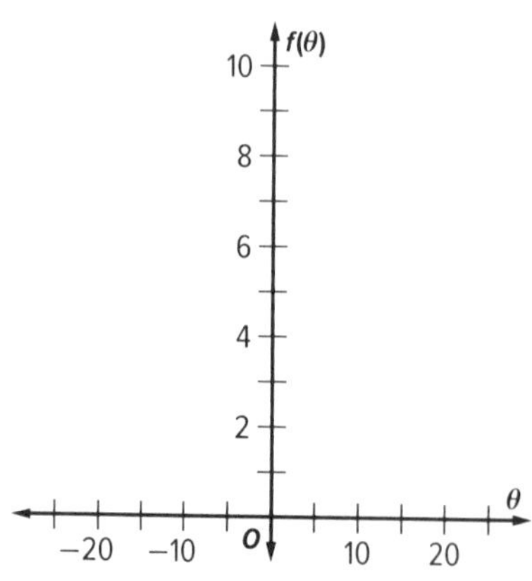

7. **CALCULATE ACCURATELY** Use the Pythagorean Identity to answer the questions. Round your answers to the nearest ten-thousandth. CCSS F.TF.8, SMP 6

a. Find $\cos \theta$ if $\sin \theta = 0.6321$ and θ is in the second quadrant.

b. Find $\sin \theta$ if $\cos \theta = -0.2863$ and θ is in the third quadrant.

c. Find $\cos \theta$ if $\sin \theta = -0.9376$ and θ is in the fourth quadrant.

d. Find $\sin \theta$ if $\cos \theta = 0.4284$ and θ is in the fourth quadrant.

8. **CALCULATE ACCURATELY** Use the Pythagorean Identity $\tan^2 \theta + 1 = \sec^2 \theta$ to answer the questions. Round your answers to the nearest ten-thousandth. CCSS F.TF.8, SMP 6

a. Find $\tan \theta$ if $\sec \theta = 4.6$ and θ is in the first quadrant.

b. Find $\sec \theta$ if $\tan \theta = -37.1$ and θ is in the second quadrant.

c. Find $\sec \theta$ if $\tan \theta = 5.6$ and θ is in the third quadrant.

d. Find $\tan \theta$ if $\sec \theta = -12.3$ and θ is in the second quadrant.

9. **INTERPRET PROBLEMS** Let $f(\theta) = 3 - \sin \theta$ and $g(\theta) = 3 + \sin \theta$. CCSS F.TF.8, SMP 1

a. Find and simplify an expression for $f(\theta)g(\theta)$ in terms of $\cos \theta$.

b. If $f(\theta) - g(\theta) = 1.54$ and θ is in the third quadrant, find $\cos(\theta)$.

10. **PLAN A SOLUTION** If $\sin(\theta) = \frac{2x-1}{x-5}$ and $\cos(\theta) = \frac{6+5x}{x-5}$, find the possible values of x. CCSS F.TF.8, SMP 1

10.8 Angle Sum and Difference Formulas

Objectives

- Prove the addition and subtraction formulas for sine, cosine, and tangent.
- Use the addition and subtraction formulas to solve problems.

CCSS STANDARDS

Content: F.TF.9
Practices: 1, 2, 3, 4, 6, 7, 8

In general, if $f(x)$ is a trigonometric function, then $f(A + B) \neq f(A) + f(B)$. For instance, $\sin (A + B) \neq \sin A + \sin B$. The table summarizes the addition and subtraction formulas for the sine, cosine, and tangent functions.

KEY CONCEPT Angle Sum and Difference Formulas

Sum Formulas	Difference Formulas
• $\sin (A + B) = \sin A \cos B + \cos A \sin B$	• $\sin (A - B) = \sin A \cos B - \cos A \sin B$
• $\cos (A + B) = \cos A \cos B - \sin A \sin B$	• $\cos (A - B) = \cos A \cos B + \sin A \sin B$
• $\tan (A + B) = \frac{\tan A + \tan B}{1 - \tan A \tan B}$	• $\tan (A - B) = \frac{\tan A - \tan B}{1 + \tan A \tan B}$

EXAMPLE 1 Prove the Addition and Subtraction Formulas for Sine CCSS F.TF.9

EXPLORE **Prove that $\sin (A \pm B) = \sin A \cos B \pm \cos A \sin B$.**

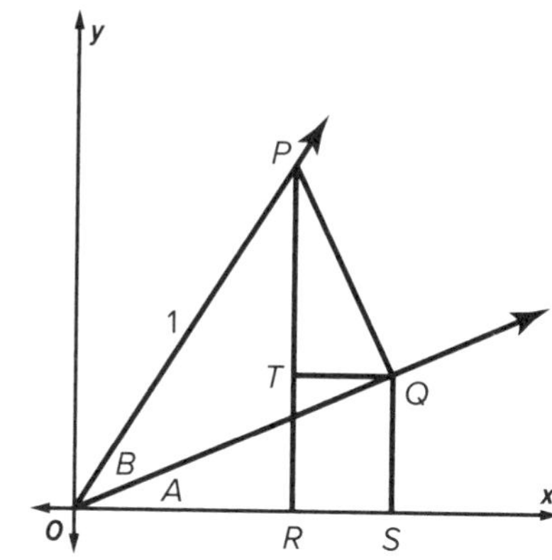

a. **INTERPRET PROBLEMS** Draw an angle with measure A in standard position and draw an angle with measure B so that it shares a side with the first angle, as shown. Let P be a point on the side of the angle with measure $A + B$ so that $OP = 1$. Draw a perpendicular, $\overline{PQ}$, from P to the terminal side of the angle with measure A. Draw a perpendicular, $\overline{QS}$, from Q to the x-axis and a perpendicular, $\overline{PR}$, from P to the x-axis. Finally, draw a perpendicular, $\overline{QT}$, from Q to $\overline{PR}$. Mark the right angles in the figure. CCSS SMP 1

b. **USE STRUCTURE** Determine the measure of $\angle TPQ$. Mark this angle measure on the figure and explain how you found your answer. CCSS SMP 7

c. **REASON ABSTRACTLY** Explain how to write the lengths PQ and OQ in terms of $\sin B$ and/or $\cos B$. (*Hint*: Focus on $\triangle POQ$.) CCSS SMP 2

d. **REASON ABSTRACTLY** Use the side lengths of $\triangle TPQ$ to write a ratio for cos *A*. Then use this, and your work from **part c**, to write an expression for the length *PT* that involves sines and/or cosines of *A* and/or *B*. CCSS SMP 2

e. **REASON ABSTRACTLY** Use the side lengths of $\triangle QOS$ to write a ratio for sin *A*. Then use this, and your work from **part c**, to write an expression for the length *QS* that involves sines and/or cosines of *A* and/or *B*. CCSS SMP 2

f. **CONSTRUCT ARGUMENTS** Explain how you can prove the sum formula for sines by using your work from the previous steps and by using $\triangle POR$ to write an expression for $\sin(A + B)$. (*Hint:* Use the Segment Addition Postulate to write the length *PR* as a sum of lengths you have already expressed as products of sines and cosines.) CCSS SMP 3

g. **CONSTRUCT ARGUMENTS** Explain how to use the sum formula for sines to prove the difference formula for sines. Use the fact that $\sin(A - B) = \sin(A + (-B))$. CCSS SMP 3

EXAMPLE 2 Find an Exact Trigonometric Value CCSS F.TF.9

Follow these steps to find the exact value of $\cos\frac{7\pi}{12}$.

a. **PLAN A SOLUTION** Complete the table. Then explain how you can use the table to help you find the exact value of $\cos\frac{7\pi}{12}$. CCSS SMP 1

	$\frac{\pi}{6}$	$\frac{\pi}{4}$	$\frac{\pi}{3}$	$\frac{3\pi}{4}$
sine				
cosine				

b. **CALCULATE ACCURATELY** Show how to find the exact value of $\cos\frac{7\pi}{12}$. CCSS SMP 6

c. **EVALUATE REASONABLENESS** Is there a different way to find the exact value of $\cos\frac{7\pi}{12}$? Does it give the same result? Explain. CCSS SMP 8

EXAMPLE 3 Use an Addition or Subtraction Formula to Solve a Problem CCSS F.TF.9

A radio tower is anchored by a cable that runs from the top of the tower to the ground. The cable is 160 feet long and makes an angle of 60° with the tower. A crew is installing a new cable that will run from the top of the tower to the ground and make an angle of 45° with the original cable. The crew wants to know the exact distance *d* from the base of the tower at which the new cable should be attached the ground.

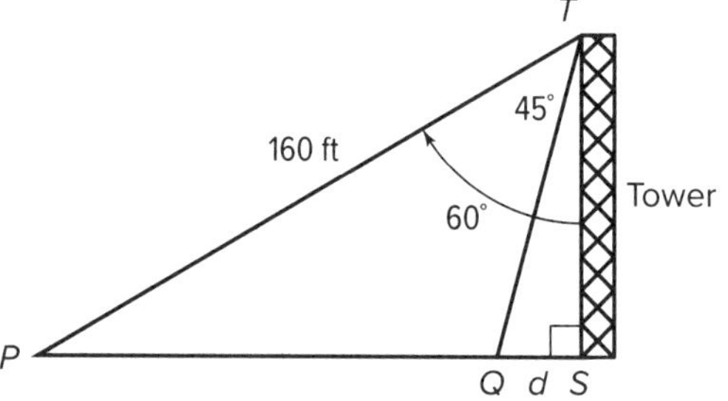

a. **USE A MODEL** Explain how to find the height of the tower. CCSS SMP 4

b. **USE STRUCTURE** Write a trigonometric ratio that involves *d* and the height of the tower. Explain your answer. CCSS SMP 7

c. **CALCULATE ACCURATELY** Show how to use your answer to **part b** and a sum or difference formula to find the exact value of *d*. CCSS SMP 6

PRACTICE

1. Use the figure from **Example 1** to prove that $\cos(A + B) = \cos A \cos B - \sin A \sin B$. The proof begins in the same way as **parts a–c** of **Example 1**. CCSS F.TF.9

   a. **REASON ABSTRACTLY** Use the side lengths of $\triangle QOS$ to write a ratio for $\cos A$. Then use this, and the expressions for *PQ* and *OQ* from **part c** of **Example 1**, to write an expression for the length *OS* that involves sines and/or cosines of *A* and/or *B*. CCSS SMP 2

   b. **REASON ABSTRACTLY** Use the side lengths of $\triangle TPQ$ to write a ratio for $\sin A$. Then use this, and the expressions for *PQ* and *OQ* from **part c** of **Example 1**, to write an expression for the length *QT* that involves sines and/or cosines of *A* and/or *B*. CCSS SMP 2

   c. **CONSTRUCT ARGUMENTS** Explain how to prove the sum formula for cosines by using your work in the previous steps. CCSS SMP 3

2. **CONSTRUCT ARGUMENTS** Explain how to use the sum formula for cosines to prove the difference formula for cosines. CCSS F.TF.9, SMP 3

3. **CONSTRUCT ARGUMENTS** You can use the sum and difference formulas for sines and cosines to prove the sum and difference formulas for tangents. CCSS F.TF.9, SMP 3

   a. Prove the sum formula for tangents.

   b. Prove the difference formula for tangents.

4. The figure shows the graphs of $y = \sin \theta$ and $y = \cos \theta$.

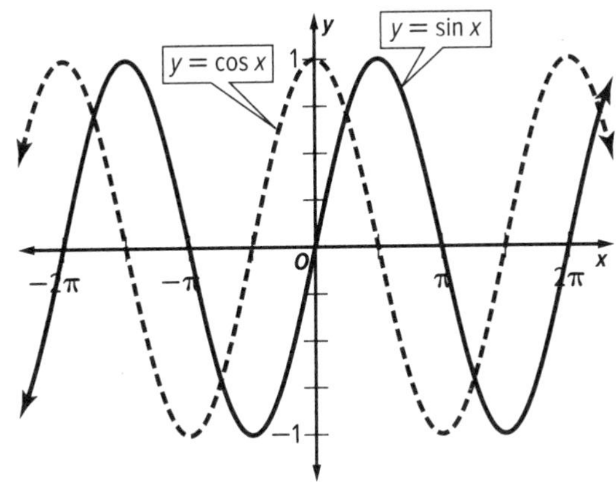

   a. **USE STRUCTURE** Explain how to use the graphs to find the value of h in the equation $\sin(\theta - h) = \cos \theta$. CCSS F.TF.9, SMP 7

   b. **CONSTRUCT ARGUMENTS** Use one or more sum or difference formulas to prove that you found the correct value of h. CCSS F.TF.9, SMP 3

5. **CALCULATE ACCURATELY** Show how to find the exact value of $\sin \frac{13\pi}{12}$. CCSS F.TF.9, SMP 6

6. **USE A MODEL** Demetri stands 6 feet from the base of a flagpole and sights the top of the pole with an angle of elevation of 75°. His eyes are 5 feet above the ground. Find the exact height of the flagpole. Then find the height to the nearest tenth of a foot. CCSS F.TF.9, SMP 4

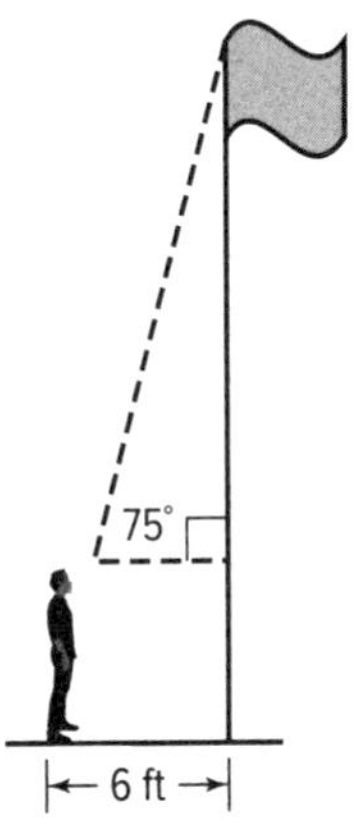

7. **USE STRUCTURE** Solve $\cos(\theta + \pi) = \sin(\theta - \pi)$ algebraically, given that $0 \leq \theta \leq \pi$. Explain your steps. CCSS F.TF.9, SMP 7

Performance Task

A Complicated Greeting

Provide a clear solution to the problem. Be sure to show all of your work, include all relevant drawings, and justify your answers.

At an amusement park, you and a friend decide to ride separate rides and attempt to wave at each other. You get on the Bungee Drop, and your friend gets on the Ferris wheel.

Part A

The Bungee Drop starts halfway up a column and alternately raises and lowers with constant frequency, each time changing the height of the rise and drop. Its vertical position in feet t seconds after starting is given by $p(t) = \left[\frac{-152}{65}|t - 65| + 152\right]\sin\left(\frac{9\pi}{130}t\right) + 152$. Use technology to sketch a graph of this function for $t = [0, 130]$. Label several key points, and describe its behavior. Determine when the ride reaches its maximum height. What is that maximum height?

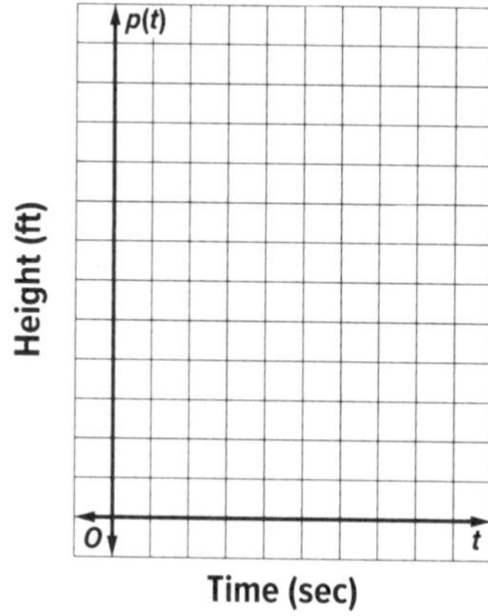

Part B

The Ferris wheel has a radius of 150 feet, and its lowest point is located 4 feet from the ground. It makes a complete rotation every 40 seconds. Write a function for the vertical height *h* of a point on the wheel at a certain time *t* such that it starts at a height of 4 feet.

Part C

If the Ferris wheel starts at the same time that the Bungee Drop starts, and assuming the rides are continuous, how many times will you and your friend be at the same height during 3 minutes of riding? At what time can you first wave at each other from the same height?

Performance Task

Constructive and Destructive Interference

Provide a clear solution to the problem. Be sure to show all of your work, include all relevant drawings, and justify your answers.

When two waves meet such that their crests coincide, they create a larger wave. This is called *constructive interference*. When they meet such that the crest of one wave coincides with the trough of the other wave they tend to create a smaller wave, or cancel each other. This is called *destructive interference*. This is the principle behind noise-canceling headphones. Interference can be explored by adding two waves mathematically and analyzing the resulting function. Examples are given below.

Part A

What conclusions can you draw about the phase shifts of interfering waves from the given illustration?

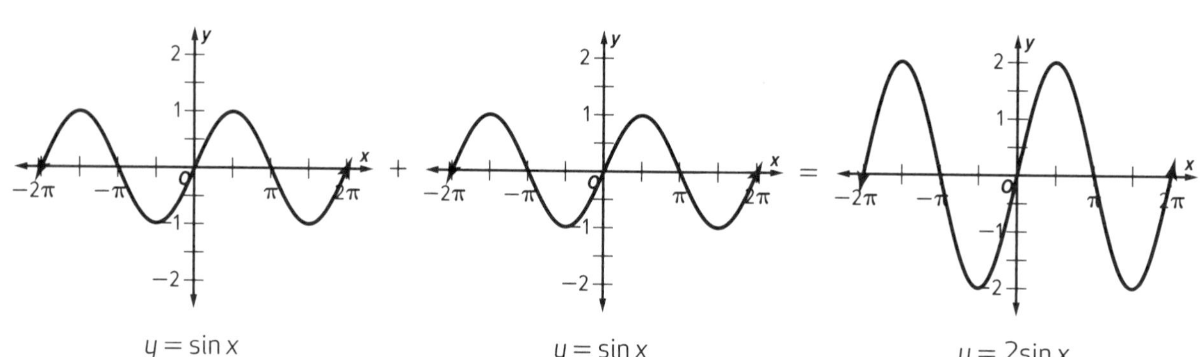

$y = \sin x$ + $y = \sin x$ = $y = 2\sin x$

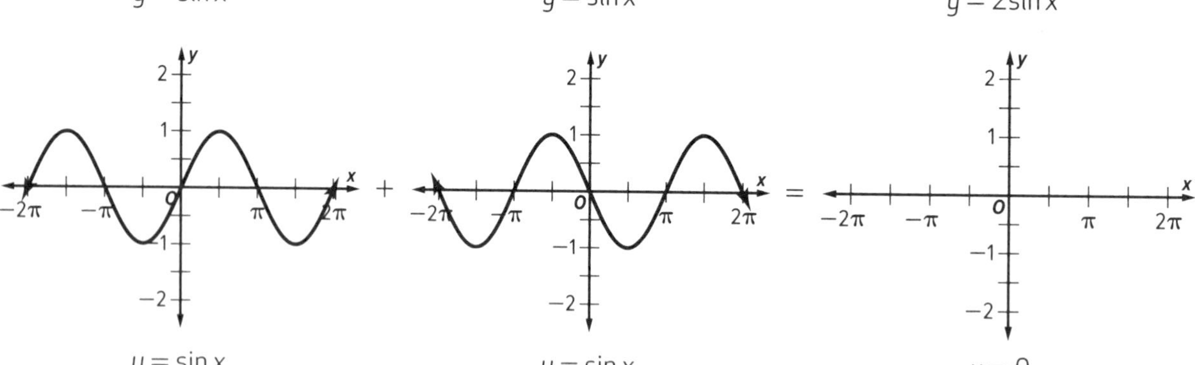

$y = \sin x$ + $y = \sin x$ = $y = 0$

Part B

Consider two sine waves. The first oscillates 5 times every second, has a midline of $y = 0$, and an amplitude of 3. It includes the point $\left(\frac{1}{20}, 3\right)$. The second also oscillates 5 times every second with a midline of $y = 0$. Its amplitude is 1 and it includes the point $\left(\frac{1}{20}, -1\right)$. Determine the difference in phase shifts of these waves, sketch a graph of their interference, name the interference, and draw a conclusion about amplitudes of waves interfering in this way.

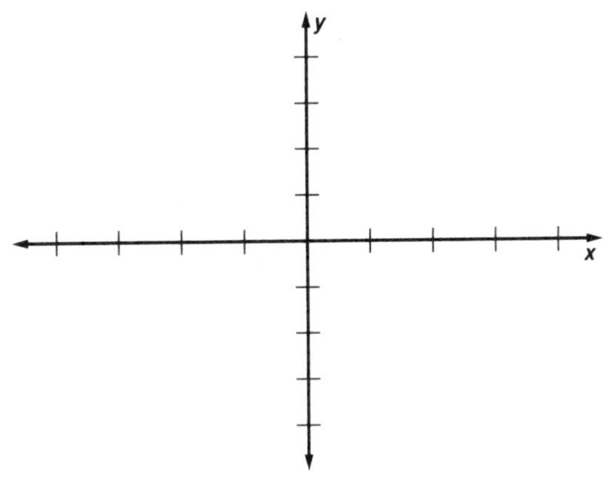

Part C

Consider the wave given by $y = \frac{2}{5}\sin\left(\frac{2\pi}{3}x - \frac{\pi}{3}\right)$. Derive a *cosine* wave that will cancel this wave by destructive interference.

Standardized Test Practice

1. In the gears shown below, the smaller gear drives the larger gear.

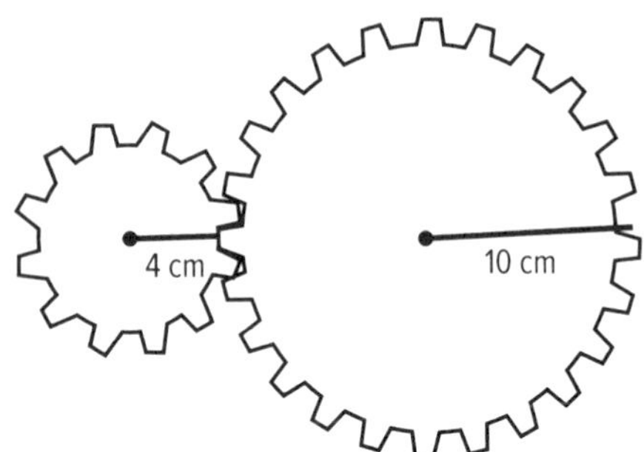

If the smaller gear rotates through an angle of $\frac{5\pi}{6}$ radians, how many radians will the larger gear rotate? CCSS F.TF.1

2. Use a sum or difference formula to write an expression that can be used to evaluate cos 75°. CCSS F.TF.9

What is the exact value of this expression?

3. Consider the graph of a function shown below.

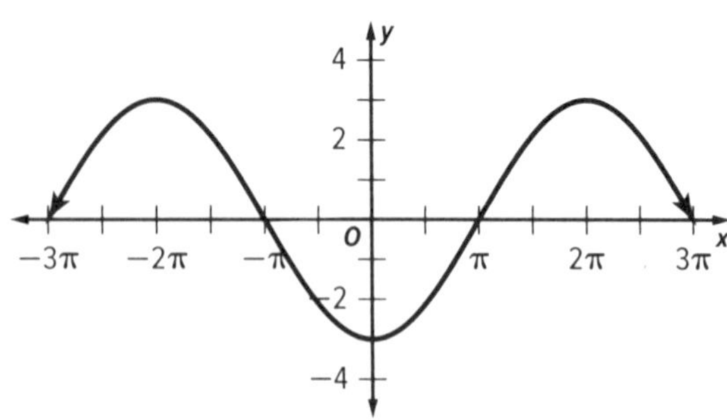

Find both a sine function and a cosine function that represent the function. CCSS F.IF.7e, F.BF.3

sine function:

cosine function:

4. If $\cos\theta = 0.7$ and $270° < \theta < 360°$, what is the value of θ to the nearest tenth of a degree? CCSS A.CED.2

5. The function $f(x) = \sec x$ is transformed so that it has a period of $\frac{\pi}{2}$. What is the transformed function? CCSS F.IF.4

$g(x) =$

What are the asymptotes of the graph of the transformed function? CCSS F.IF.4

Graph the transformation. CCSS F.IF.7e

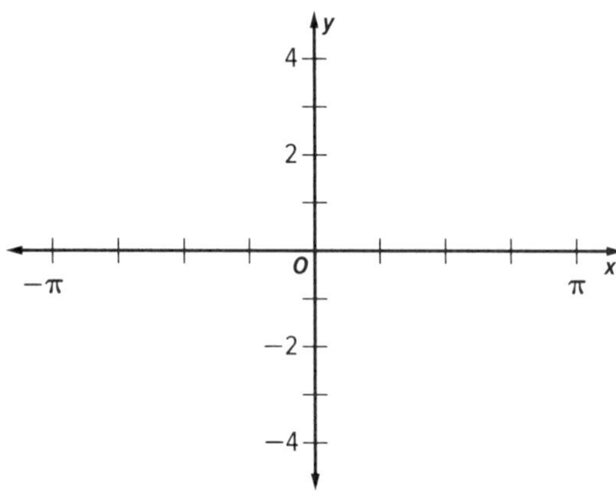

6. Find the principal value of the inverse trigonometric function. State your answer in radians and degrees. CCSS F.TF.2

$\tan^{-1}(-\sqrt{3})$

7. Write the expression $\csc\theta - \cot\theta\cos\theta$ in terms of $\sin\theta$. CCSS A.SSE.2

8. Use the Pythagorean Identity to determine the value of $\sin\theta$ if $\cos\theta = -0.7193$ and θ is in the third quadrant. Round your answer to the nearest ten-thousandth. CCSS F.TF.8

$\sin\theta =$

9. Consider the angle shown in the following graph. Complete each of the following. CCSS F.TF.2

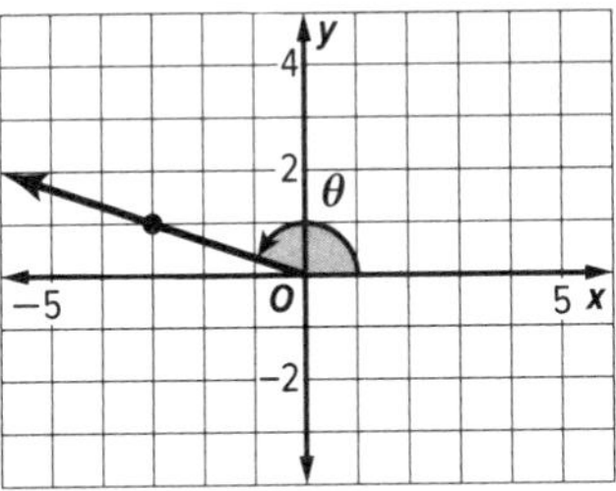

$\sin\theta =$ ☐ $\cos\theta =$ ☐

$\tan\theta =$ ☐ $\csc\theta =$ ☐

$\sec\theta =$ ☐ $\cot\theta =$ ☐

10. A programmable thermostat is set to a high temperature of 72°F at 2 P.M., and a low temperature of 60°F at 2 A.M. The temperature throughout the day can be modeled by a sine function.

a. Write a sine function that represents the temperature throughout the day, where x is the hours since 12:00 A.M. CCSS F.TF.5

b. What are the amplitude and period of the function?  CCSS A.SSE.1a

c. Graph the function. CCSS F.IF.7e

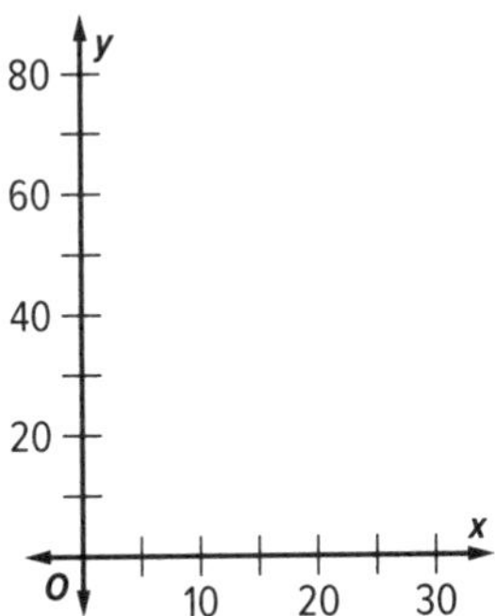

11. If $\cos\theta = -\frac{5}{7}$ and is in Quadrant II, what is the value of $\sin\theta$? Show your work. CCSS F.TF.8

12. Verify the identity $\tan^2\theta + \cos^2\theta - 1 = \tan^2\theta\sin^2\theta$. Show your work. CCSS F.TF.8

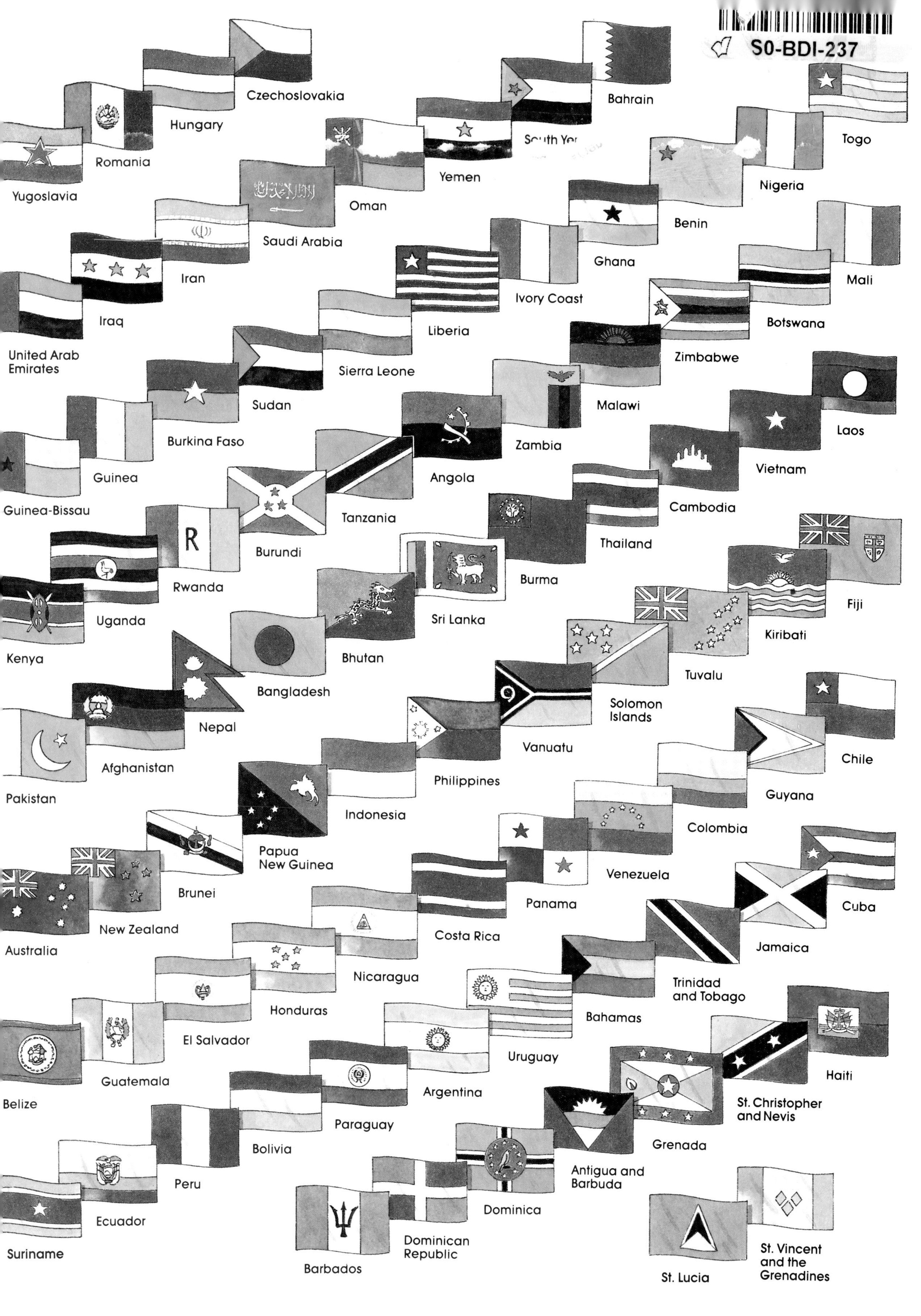

S0-BDI-237
Czechoslovakia
Hungary
Romania
Yugoslavia
Bahrain
Togo
Yemen
Oman
Saudi Arabia
Iran
Iraq
United Arab Emirates
Nigeria
Benin
Ghana
Ivory Coast
Liberia
Sierra Leone
Sudan
Burkina Faso
Guinea
Guinea-Bissau
Mali
Botswana
Zimbabwe
Malawi
Zambia
Angola
Tanzania
Burundi
Rwanda
Uganda
Kenya
Laos
Vietnam
Cambodia
Thailand
Burma
Sri Lanka
Bhutan
Bangladesh
Nepal
Afghanistan
Pakistan
Fiji
Kiribati
Tuvalu
Solomon Islands
Vanuatu
Philippines
Indonesia
Papua New Guinea
Brunei
New Zealand
Australia
Chile
Guyana
Colombia
Venezuela
Panama
Costa Rica
Nicaragua
Honduras
El Salvador
Guatemala
Belize
Cuba
Jamaica
Trinidad and Tobago
Bahamas
Uruguay
Argentina
Paraguay
Bolivia
Peru
Ecuador
Suriname
Haiti
St. Christopher and Nevis
Grenada
Antigua and Barbuda
Dominica
Dominican Republic
Barbados
St. Lucia
St. Vincent and the Grenadines

MY FIRST ATLAS

By Kate Petty

Illustrated by Colin King

WARNER JUVENILE BOOKS

A Time Warner Inc. Company

New York

Contents

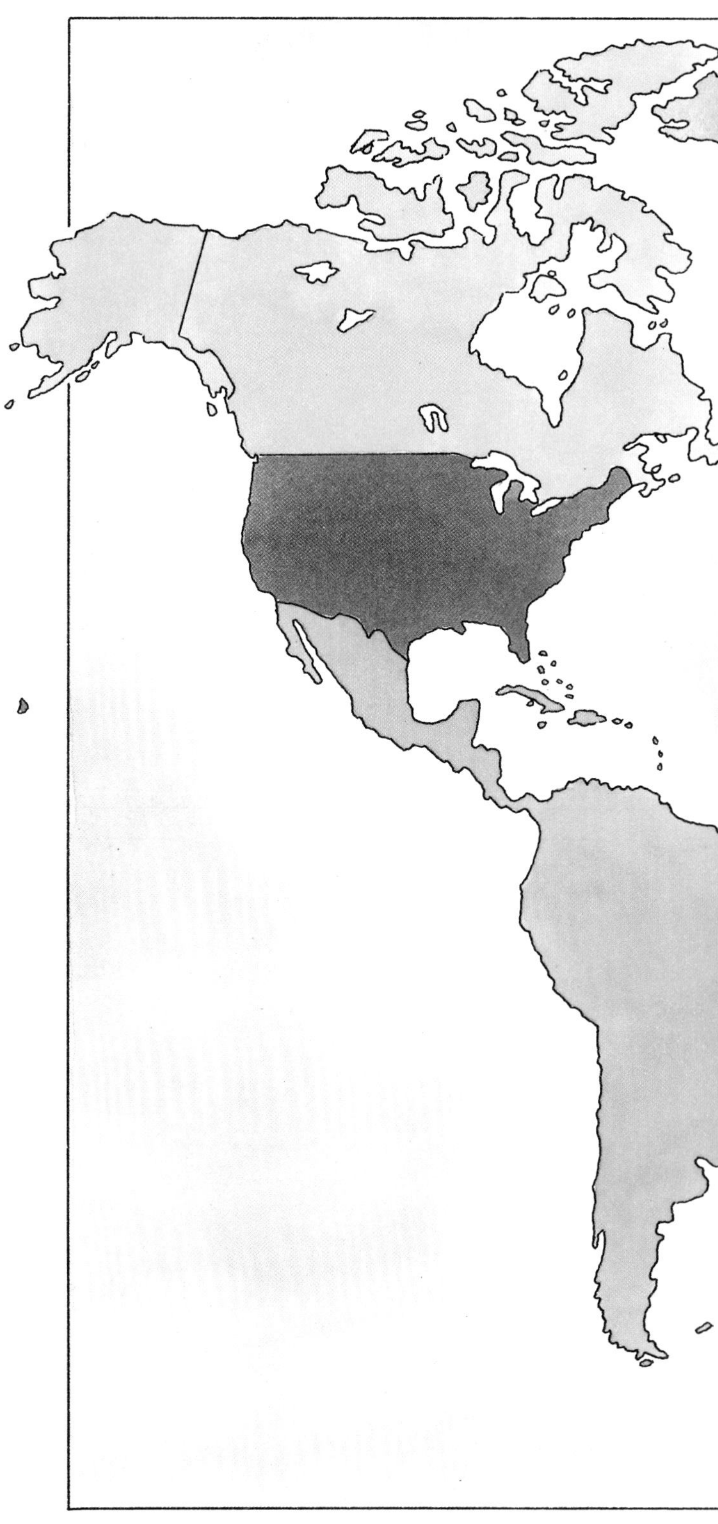

Warner Juvenile Books Edition
First American Edition

This Warner Juvenile Books edition is published by arrangement with Conran Octopus Limited, London.

Warner Books, Inc., 666 Fifth Avenue, New York, NY 10103

A Time Warner Inc. Company

Printed in Great Britain
First Warner Juvenile Books Printing: May 1991

10 9 8 7 6 5 4 3 2 1

Library of Congress Catalog Card Number: 90-37926
ISBN: 1-55782-361-8

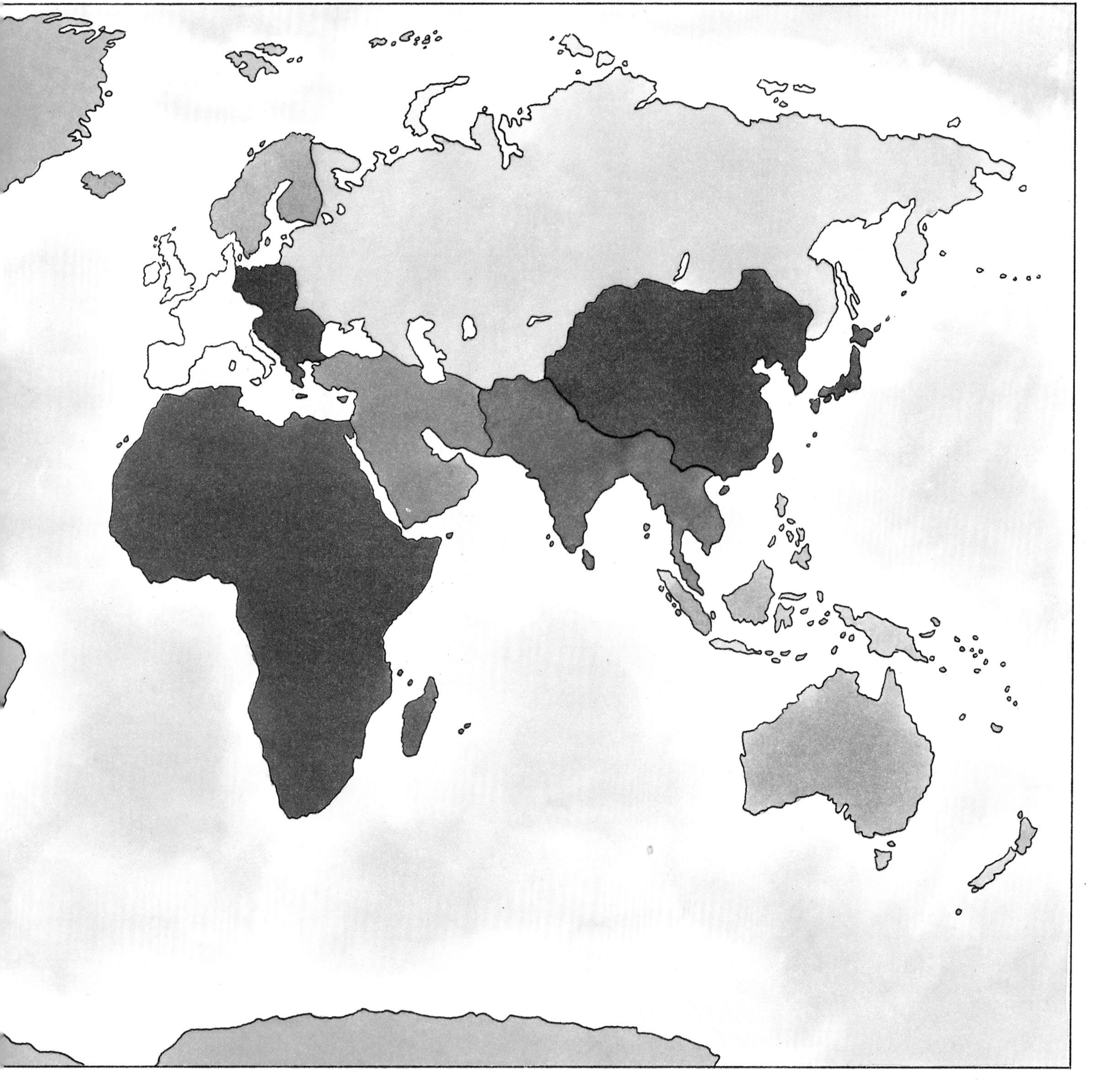

Let's Look at Maps

How we show flat pictures of a round world.

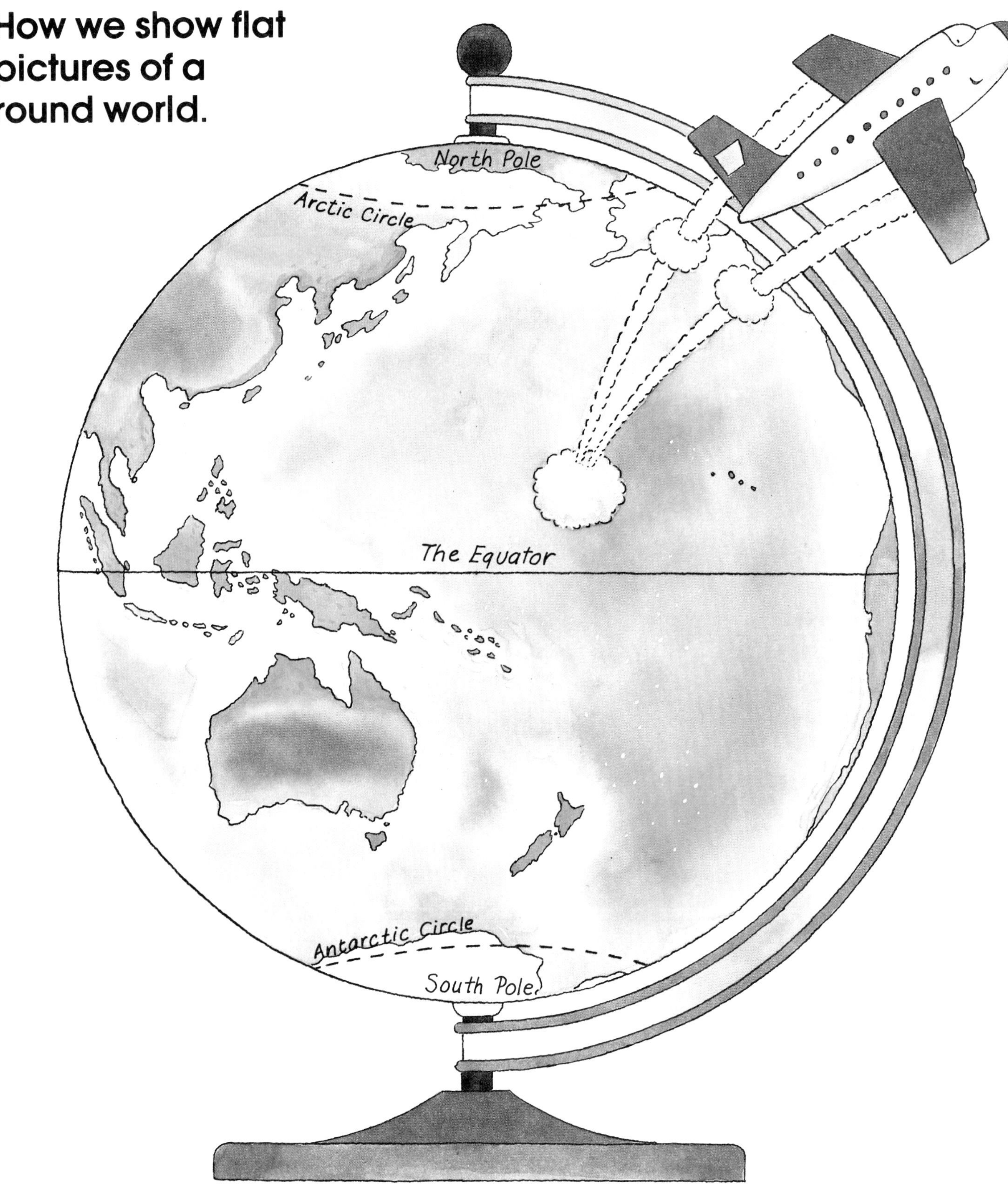

Our world is enormous. The only way you could see it all would be to fly out into space. Or you could look at a little model of it, like the globe shown here.

The North Pole is at the top and the South Pole is at the bottom. The "pretend" line around the middle is the Equator. Maps are just flat pictures of the globe.

Continents

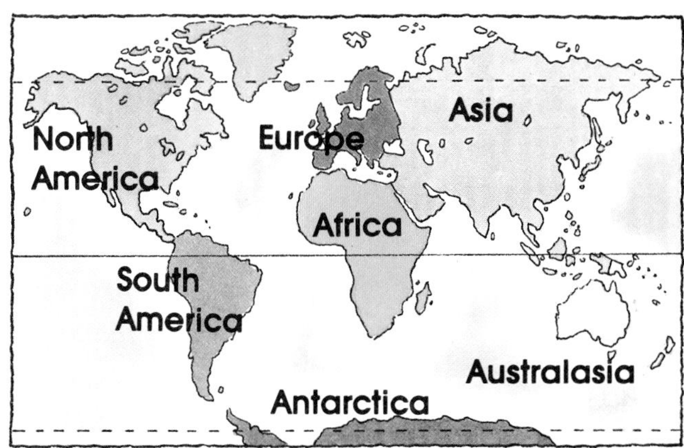

Your country is in one of the big continents shown here. Do you know which one it is?

People and what they do

Look for people making and growing things the world needs.

The countryside around

You can tell from the map if a place is mountainous or flat, desert or ice, wooded or bare.

Countries

There are almost 200 countries in the world. In many of them different languages are spoken.

Animals

Different animals live all over the world. Find them on the maps.

The compass

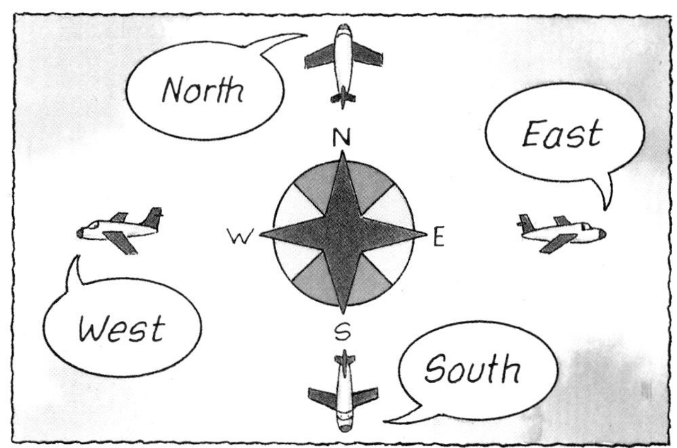

The points of the compass show you where to find North, South, East and West on a map.

The United States of America

This map shows 48 of the 50 United States of America. The other two are Alaska (shown on page 8) and Hawaii in the Pacific Ocean, shown in the location map, above. The first Europeans went to live in North America nearly 500 years ago. North American Indians have always lived there. The U.S.A. is one of the richest countries in the world. The president lives in the White House in Washington, D.C.

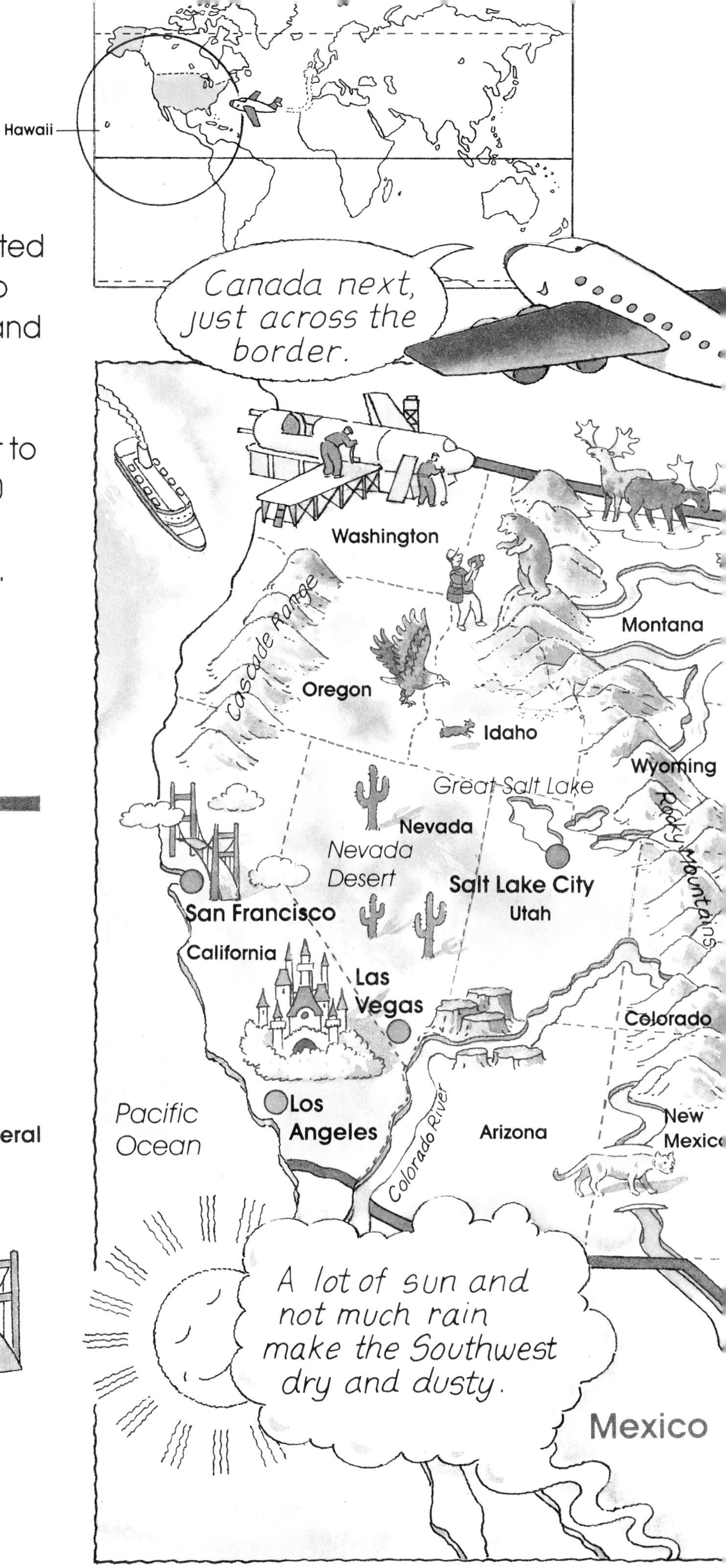

Can you find? . . .

Hawaii

Canada next, just across the border.

Washington
Montana
Cascade Range
Oregon
Idaho
Wyoming
Great Salt Lake
Rocky Mountains
Nevada
Nevada Desert
Salt Lake City
Utah
San Francisco
California
Las Vegas
Colorado
Pacific Ocean
Los Angeles
Colorado River
Arizona
New Mexico

A lot of sun and not much rain make the Southwest dry and dusty.

Mexico

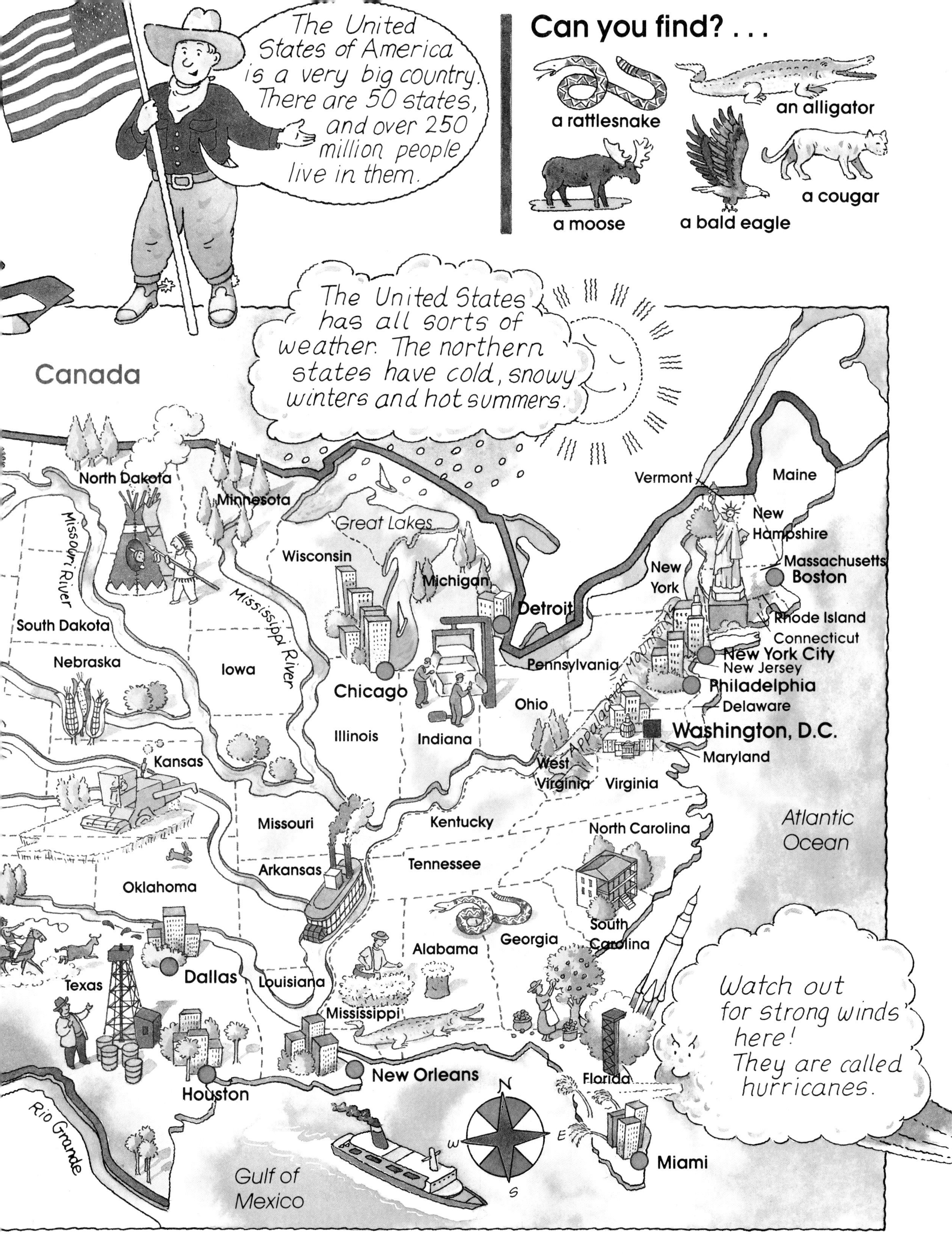
The United States of America is a very big country. There are 50 states, and over 250 million people live in them.
Can you find? . . .
a rattlesnake
an alligator
a moose
a bald eagle
a cougar
The United States has all sorts of weather. The northern states have cold, snowy winters and hot summers.
Canada
North Dakota
Minnesota
Missouri River
Mississippi River
Great Lakes
Wisconsin
Michigan
Detroit
Vermont
Maine
New Hampshire
Massachusetts
Boston
New York
Rhode Island
Connecticut
New York City
New Jersey
Philadelphia
Delaware
South Dakota
Nebraska
Iowa
Chicago
Pennsylvania
Appalachian Mountains
Ohio
Illinois
Indiana
Washington, D.C.
Maryland
Kansas
West Virginia
Virginia
Missouri
Kentucky
North Carolina
Atlantic Ocean
Arkansas
Tennessee
Oklahoma
South Carolina
Georgia
Alabama
Texas
Dallas
Louisiana
Mississippi
Watch out for strong winds here! They are called hurricanes.
New Orleans
Houston
Florida
Rio Grande
N
W
E
S
Miami
Gulf of Mexico

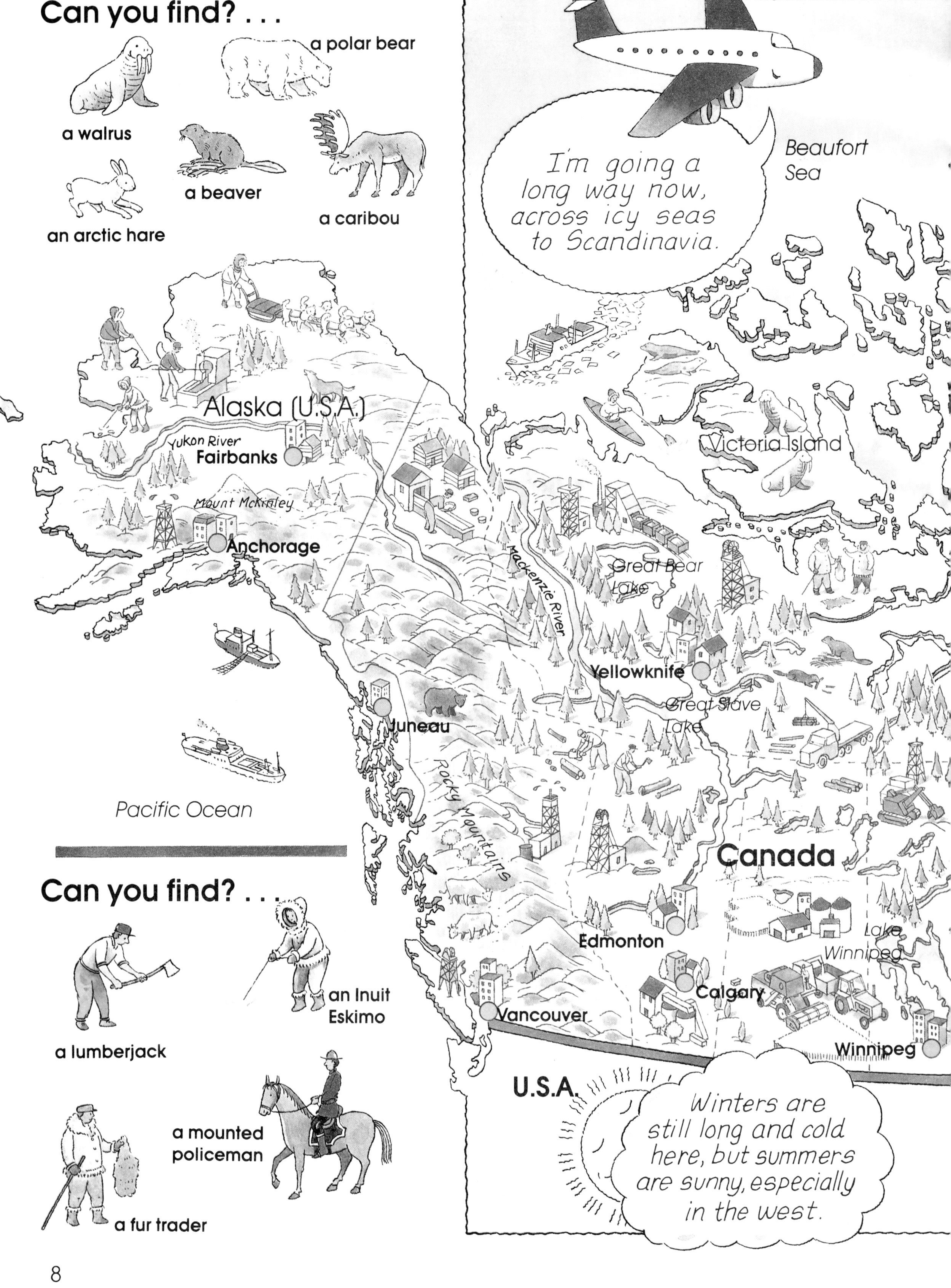
Can you find? . . .
a polar bear
a walrus
a beaver
a caribou
an arctic hare
I'm going a long way now, across icy seas to Scandinavia.
Beaufort Sea
Alaska (U.S.A.)
Yukon River
Fairbanks
Mount McKinley
Anchorage
Victoria Island
Great Bear Lake
Mackenzie River
Yellowknife
Great Slave Lake
Juneau
Rocky Mountains
Pacific Ocean
Canada
Can you find? . . .
an Inuit Eskimo
a lumberjack
Edmonton
Lake Winnipeg
Calgary
Vancouver
Winnipeg
U.S.A.
a mounted policeman
a fur trader
Winters are still long and cold here, but summers are sunny, especially in the west.

Canada and Alaska

Alaska, which is the biggest state in the U.S.A., and Canada cover huge areas of land, much of it frozen. The Inuit people (called Eskimos) know how to survive the snow and ice, but most Americans and Canadians live in the south. The Canadian cities of Montreal and Toronto are much bigger than the capital, Ottawa. French and English are both spoken in Montreal, so signs are posted in two languages.

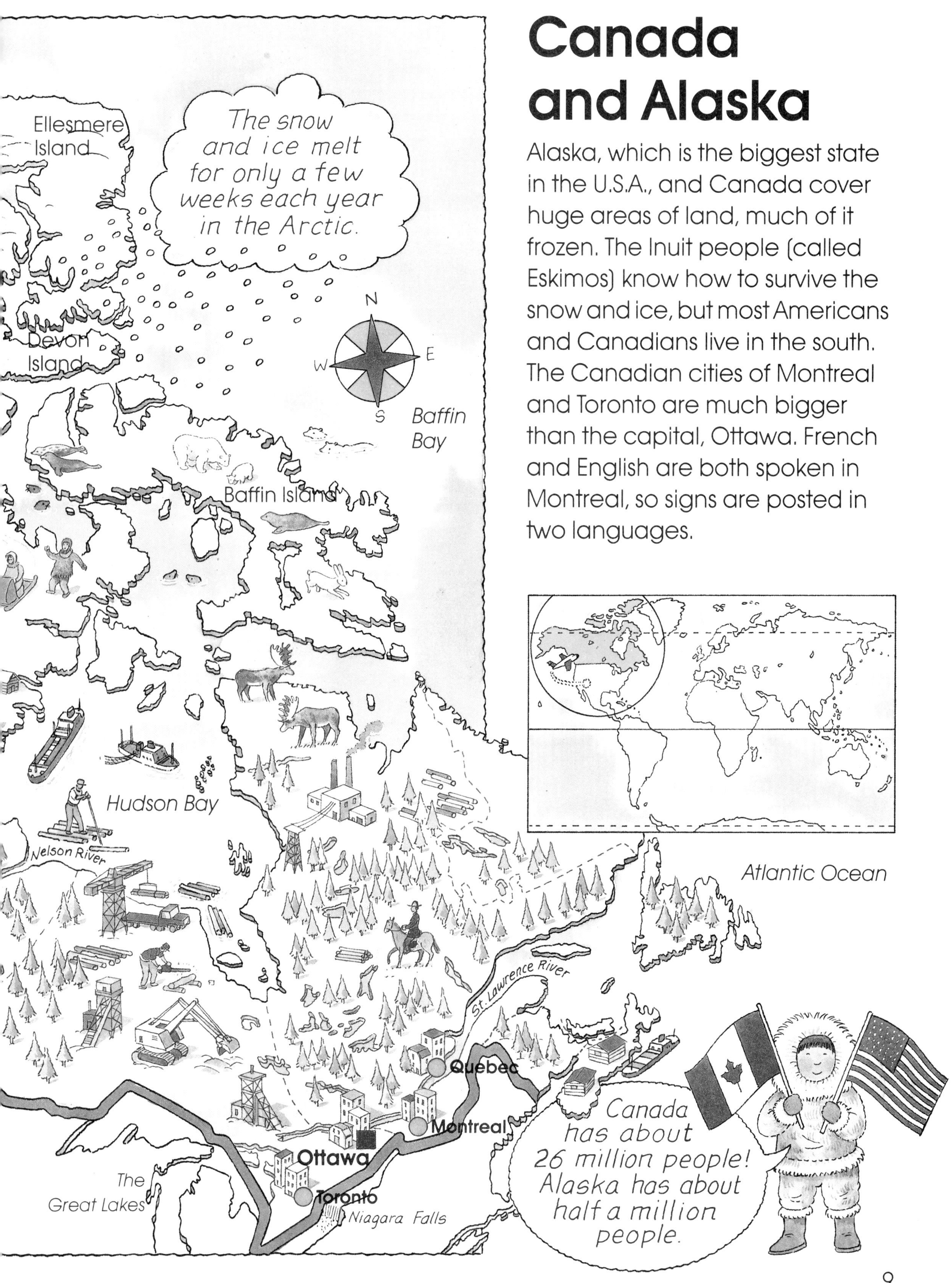

The correct name for these northern lights is aurora borealis.
Arctic Ocean
Iceland
Reykjavík
Mount Hekla
Vatnajökull Volcano
Sweden
Narvik
Norway
Trondheim fjord
North Sea
Trondheim
Iceland is more than 600 miles away from Norway across the sea.
Mount Glittertind
Sogne fjord
Bergen
Hardanger fjord
Oslo
Uppsala
Stavanger
Stockholm
Lake Vänern
Oslo fjord
Denmark is the warmest Scandinavian country.
Jönköping
Göteborg
Aarhus
Esbjerg
Malmö
Copenhagen
Denmark
Baltic Sea
Southwest now, across the North Sea, to the British Isles.

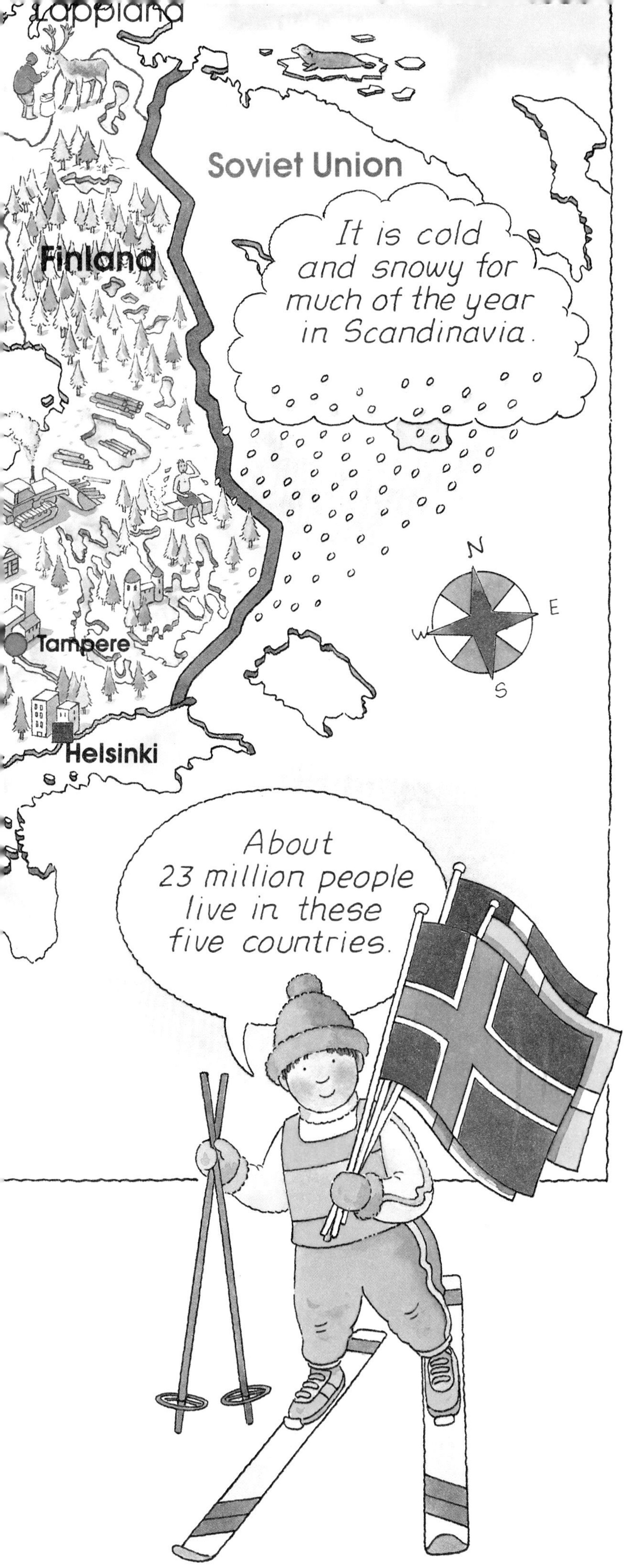

Lappland

Soviet Union

Finland

It is cold and snowy for much of the year in Scandinavia.

Tampere

Helsinki

About 23 million people live in these five countries.

Scandinavia

In Scandinavia the winter months are dark and cold, but in the middle of summer the days are very long. Most people have everything they need to live comfortably in the towns of the south. The Norwegian fjords are like deep rivers coming in from the sea. Finland is a country with 55,000 lakes. Iceland is more than 600 miles away from Norway across the sea.

Can you find?...

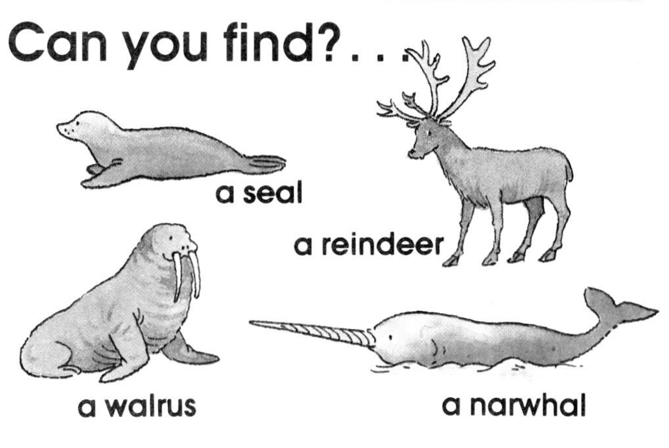

Can you find?...

The British Isles

Great Britain and Ireland are the two biggest islands of the British Isles. You can see that there are some smaller ones as well. One queen (or king) rules Northern Ireland, England, Scotland and Wales, which is why they are called the United Kingdom. The Republic of Ireland is a separate country.

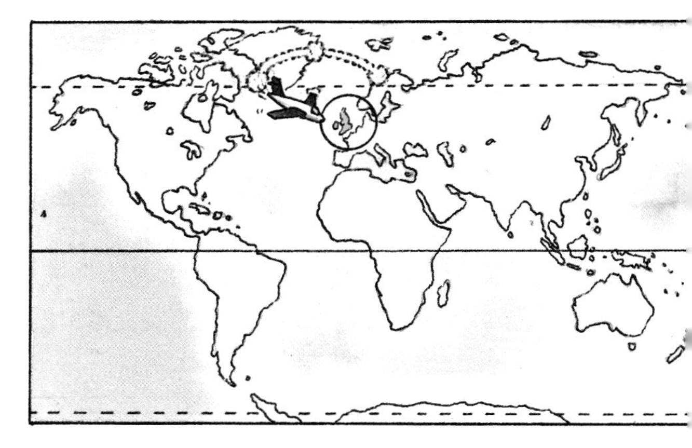

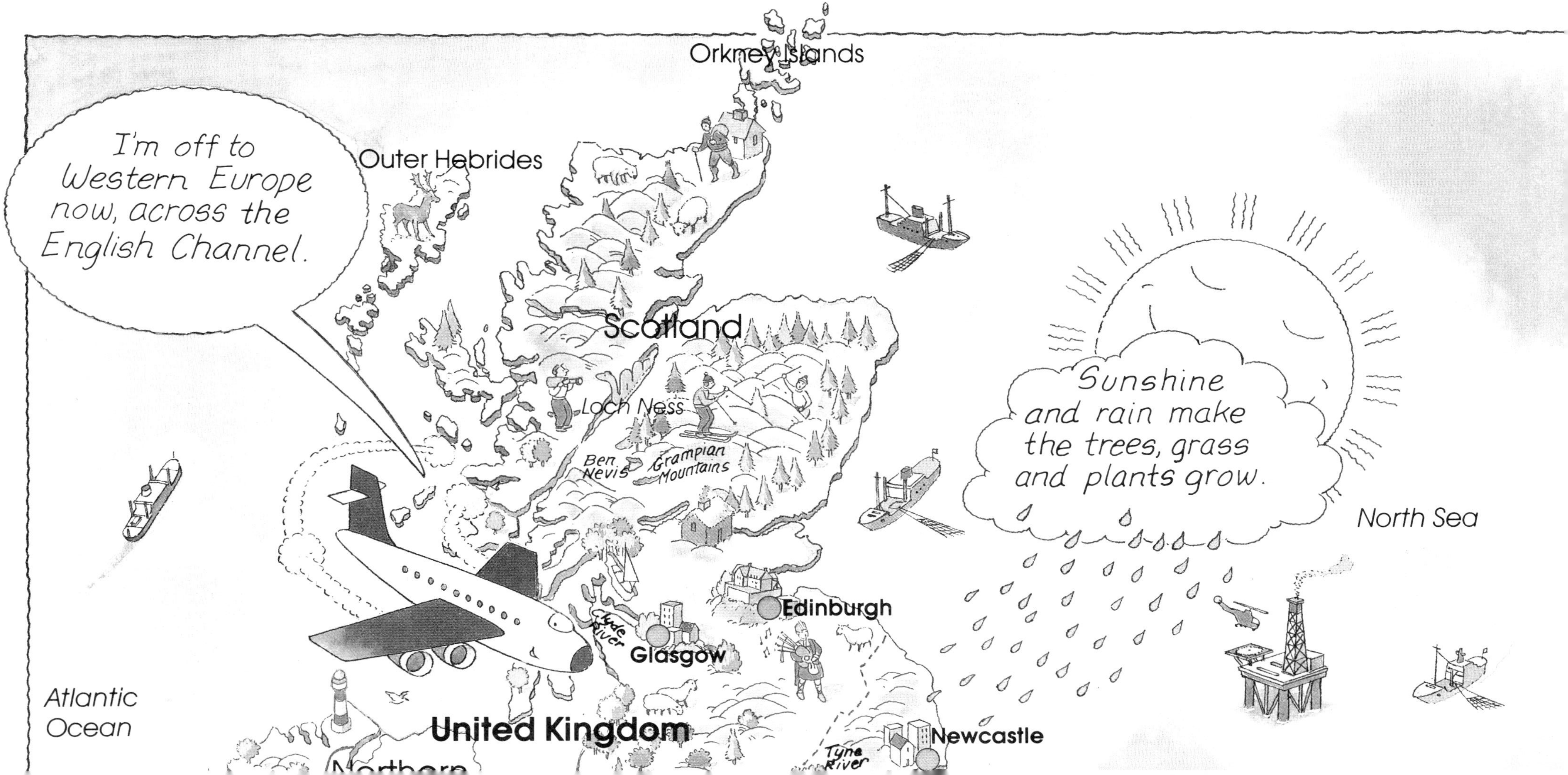

I'm off to Western Europe now, across the English Channel.

Sunshine and rain make the trees, grass and plants grow.

Orkney Islands

Outer Hebrides

Scotland

Loch Ness

Ben Nevis

Grampian Mountains

North Sea

Edinburgh

Clyde River

Glasgow

Atlantic Ocean

United Kingdom

Newcastle

Tyne River

56 million people live in the United Kingdom, but only 3½ million live in the Republic of Ireland.
Belfast
Mourne Mountains
Lake District
Pennine Hills
Scafell Pike
Isle of Man
Irish Sea
The Republic of Ireland
Dublin
Anglesey
Liverpool
Manchester
Leeds
York
Humber River
Sheffield
England
Snowdon
Shannon River
Limerick
Wicklow Mountains
Cambrian Mountains
Wales
Birmingham
Severn River
Oxford
London
Cork
N
W
E
S
Cardiff
Bristol
Thames River
Isle of Wight
France
English Channel
Can you find? . . .
Stonehenge
Big Ben
Edinburgh Castle
York Minster
Land's End
Can you find? . . .
a coal mine
an airplane factory
a soccer game
a car factory
an oil rig

Western Europe

The countries of Western Europe, which includes the British Isles, are some of the richest and most powerful in the world. In the north, most people live in towns and work in factories and offices. All sorts of crops and fruits are grown by farmers in the warm south. Many famous painters and composers were born in these countries. Tourists have plenty to look at in the big cities.

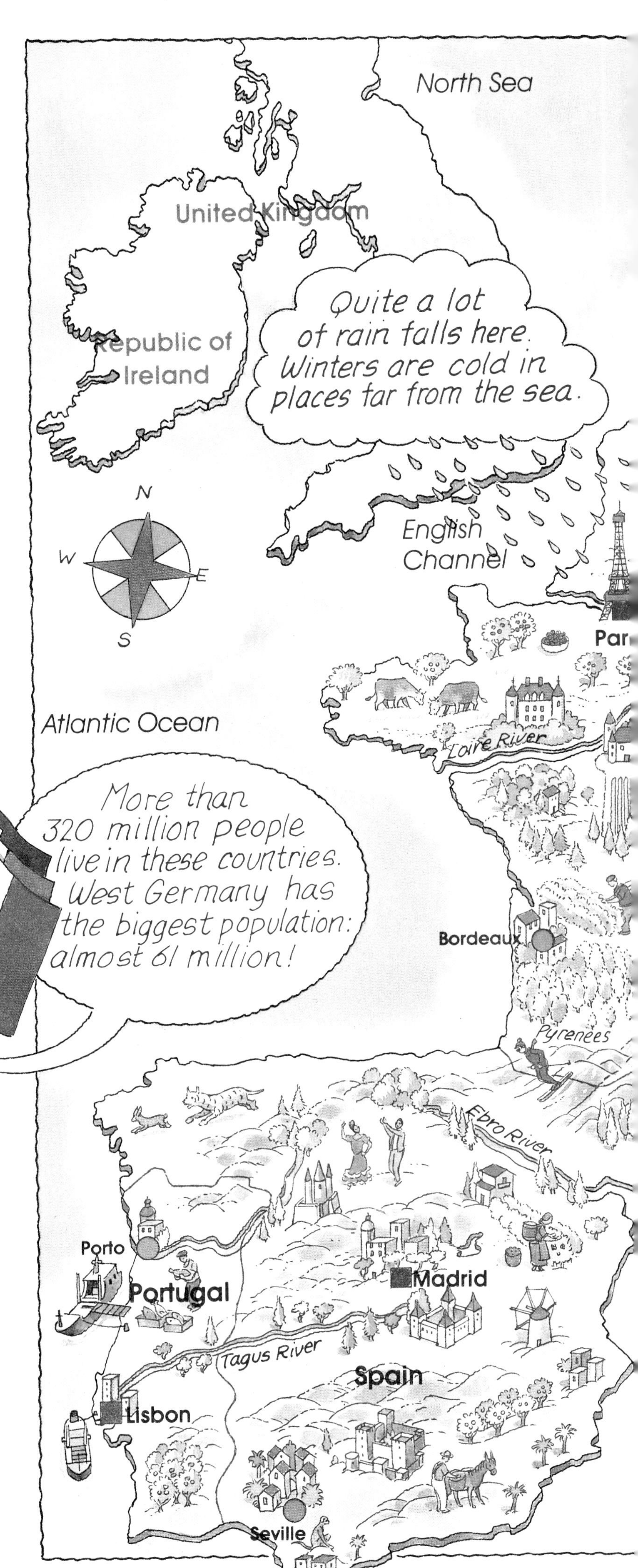

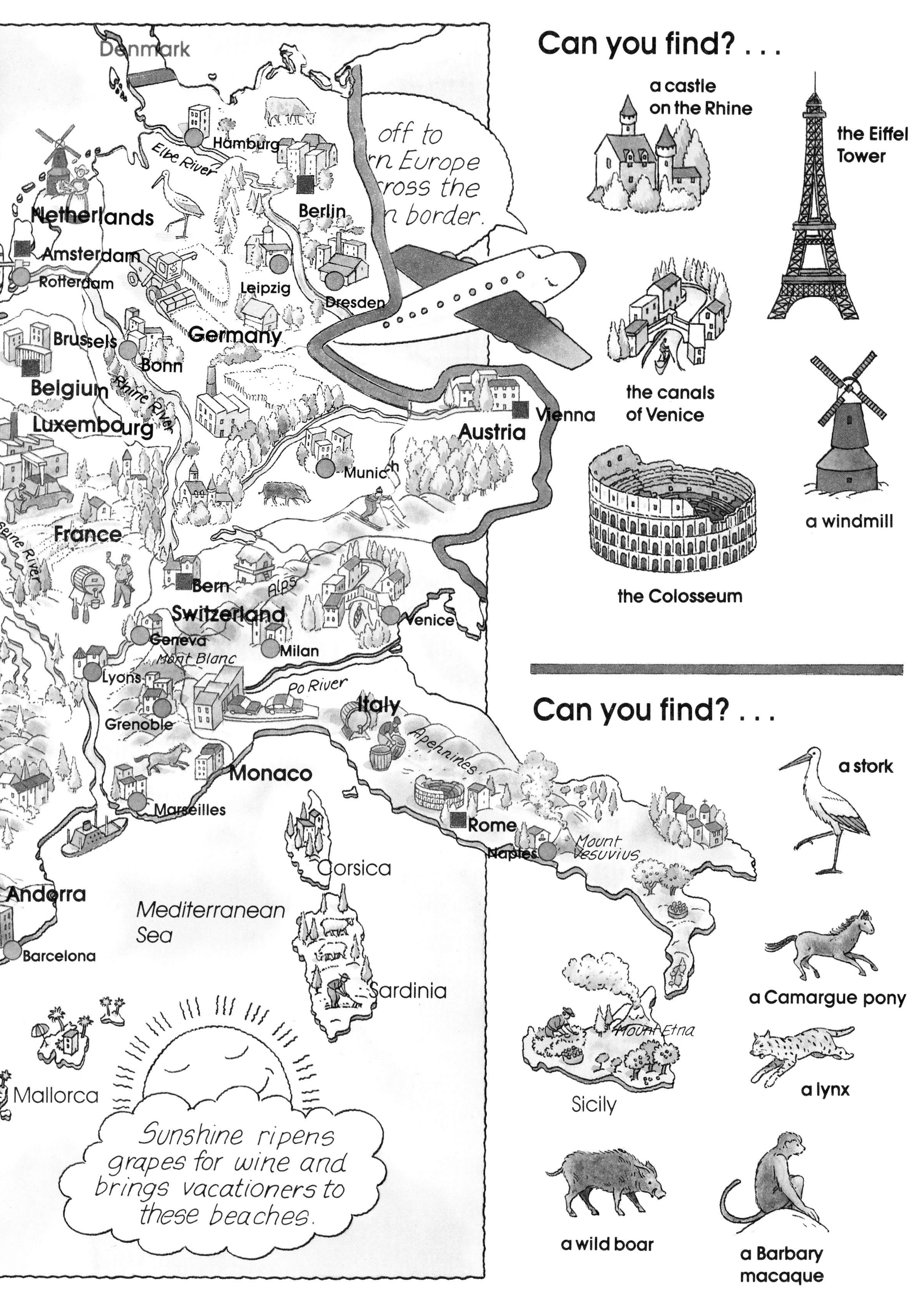
Denmark
Hamburg
Elbe River
Berlin
off to
rn Europe
cross the
n border.
Netherlands
Amsterdam
Rotterdam
Leipzig
Dresden
Brussels
Germany
Bonn
Belgium
Rhine River
Luxembourg
Vienna
Austria
Munich
France
Seine River
Bern
Alps
Switzerland
Venice
Geneva
Mont Blanc
Milan
Lyons
Po River
Grenoble
Italy
Apennines
Monaco
Marseilles
Rome
Naples
Mount Vesuvius
Corsica
Andorra
Mediterranean Sea
Barcelona
Sardinia
Mount Etna
Mallorca
Sicily
Sunshine ripens grapes for wine and brings vacationers to these beaches.
Can you find? . . .
a castle on the Rhine
the Eiffel Tower
the canals of Venice
the Colosseum
a windmill
Can you find? . . .
a stork
a Camargue pony
a lynx
a wild boar
a Barbary macaque

Eastern Europe

Many Eastern European cities have beautiful old buildings. Outside the towns, the lives of some country people have changed little for hundreds of years. The people in many Eastern European countries are less well off than in Western Europe. You might go to Greece or Yugoslavia for a sunny vacation. Greece is a hot Mediterranean country that has a long history.

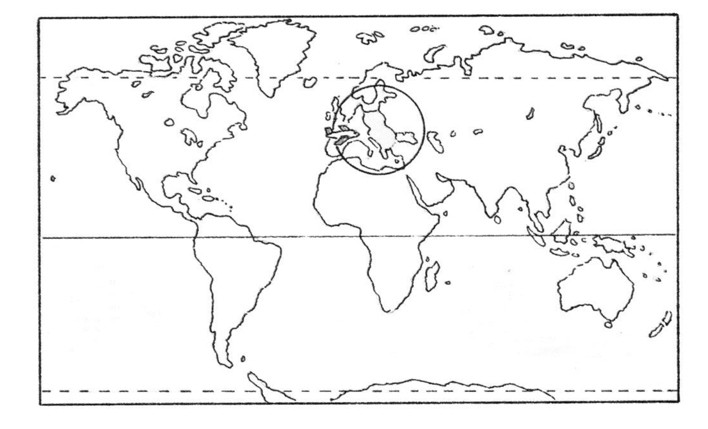

Can you find? . . .

a glass blower

a coal mine

a brewery

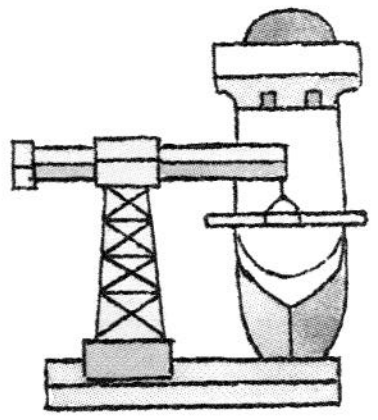

a shipyard

a goat herd

the Danube River

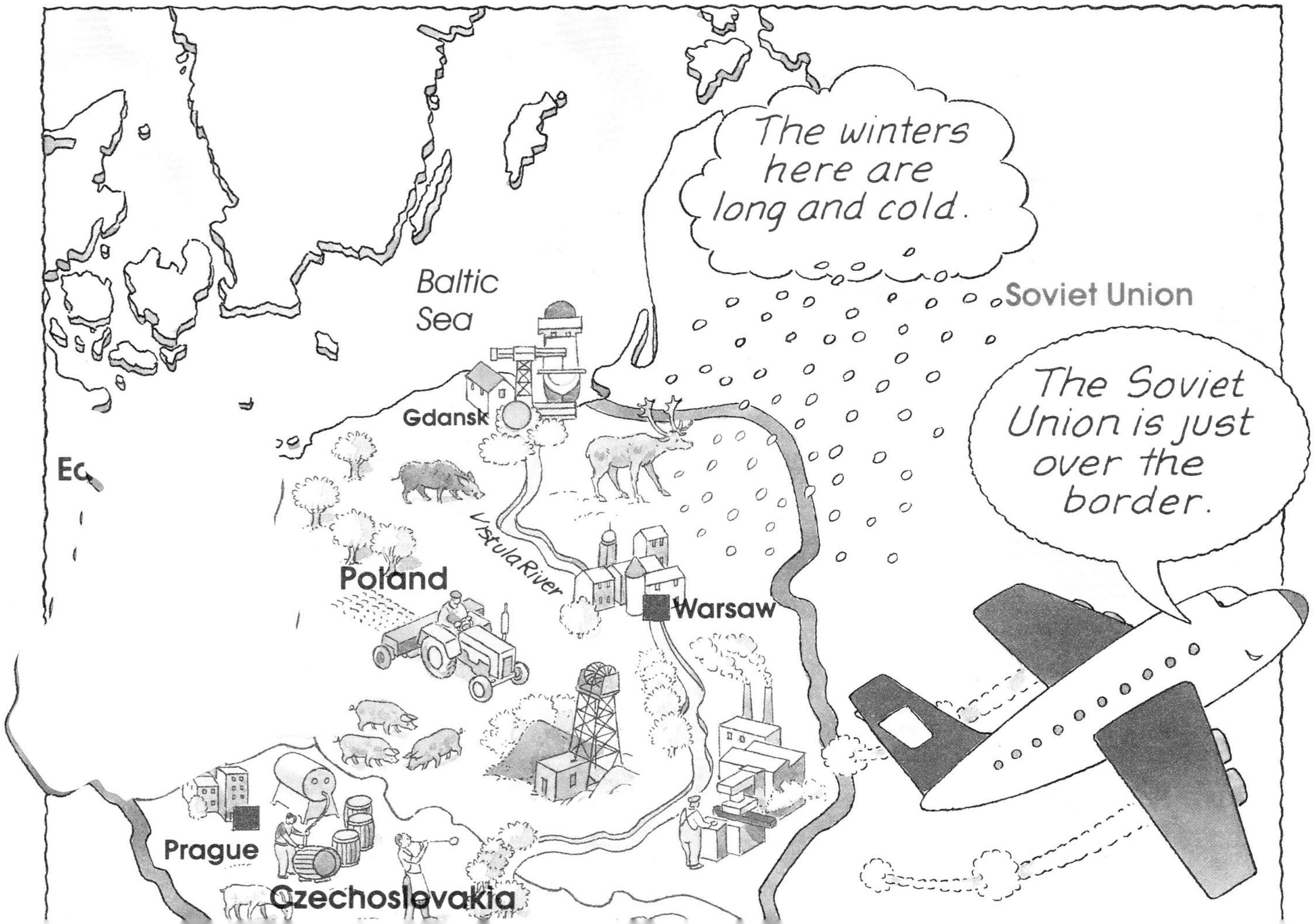

The winters here are long and cold.

Baltic Sea

Soviet Union

The Soviet Union is just over the border.

Gdansk

Vistula River

Poland

Warsaw

Prague

Czechoslovakia

About
150 million people
live in these countries.
Poland is the biggest
with nearly
40 million.

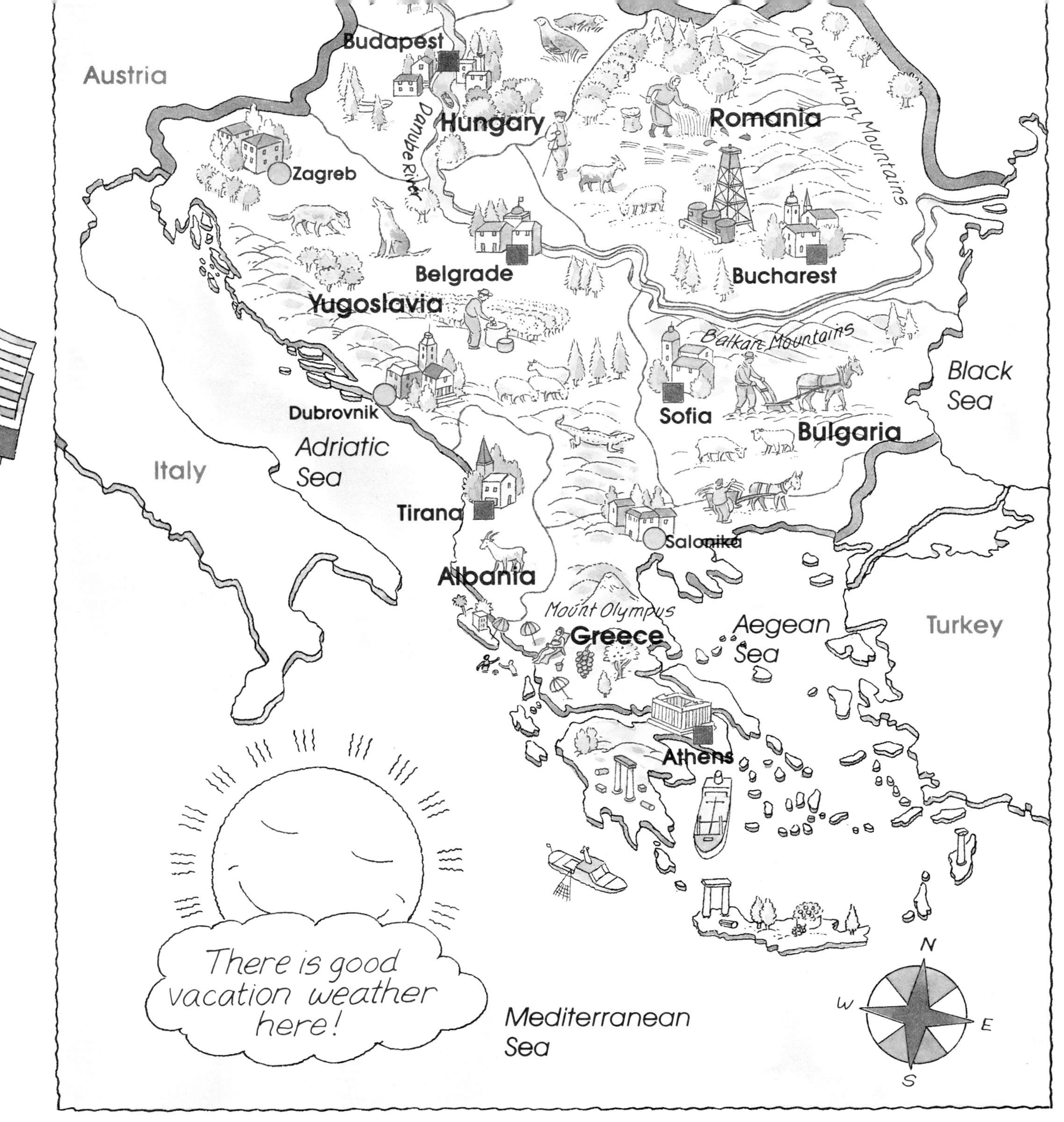

Did you know? . . .

They have a different alphabet in Greece and Yugoslavia.

Ε Λ Λ Α Δ Α

This is GREECE in Greek

ЈУГОСЛАВИЈА

This is YUGOSLAVIA in Serbo-Croat

Union
Fifteen Socialist republics make up the Soviet Union. It is the biggest country in the world. It stretches across Europe and Asia. Everybody learns to speak Russian, but many of the republics have their own languages.
Russia written in Russian looks like this
РОССИЯ
The imaginary line that follows the Ural Mountains down to the Caspian Sea shows where Europe and Asia meet.
Arctic Ocean
N
W
E
S
Leningrad
Minsk
Dnieper River
Kiev
Moscow
Odessa
Gorki
Kharkov
Ural Mountains
Don River
Volga River
Ob River
Caucasus Mountains
Caspian Sea
Tbilisi
Baku
Omsk
Trans-Siberian Railway
Novosibirsk
Yenisey River
Aral Sea
Lake Balkhash
Tashkent
China
People vacation in the sunshine here.
I'm crossing to Iran now, in the Middle East.

Can you find? . . .
desert
mountains
forest
cold bare land
(called tundra)
icy places
grassland
(called steppes)
286 million people live in the Soviet Union. About 190 million of them live in the European part.
51° below zero is not unusual here.
Cherskiy Range
Verkhoyansk Range
Lena River
Yablonovyy Range
Lake Baykal
Vladivostok
Can you find? . . .
a reindeer
a bear
a wolf
a polar bear
a Siberian tiger

The Middle East

These are mostly desert countries where little grows and most people are poor, but oil from deep under the ground has made some countries very rich. You might recognize some of the names from Bible stories. Can you find Mount Ararat where Noah's ark was supposed to have come to rest? Jerusalem is still an important center of worship for Christians, Jews and Muslims.

Can you find? . . .

an oasis in Oman

water towers in Saudi Arabia

cave homes in Turkey

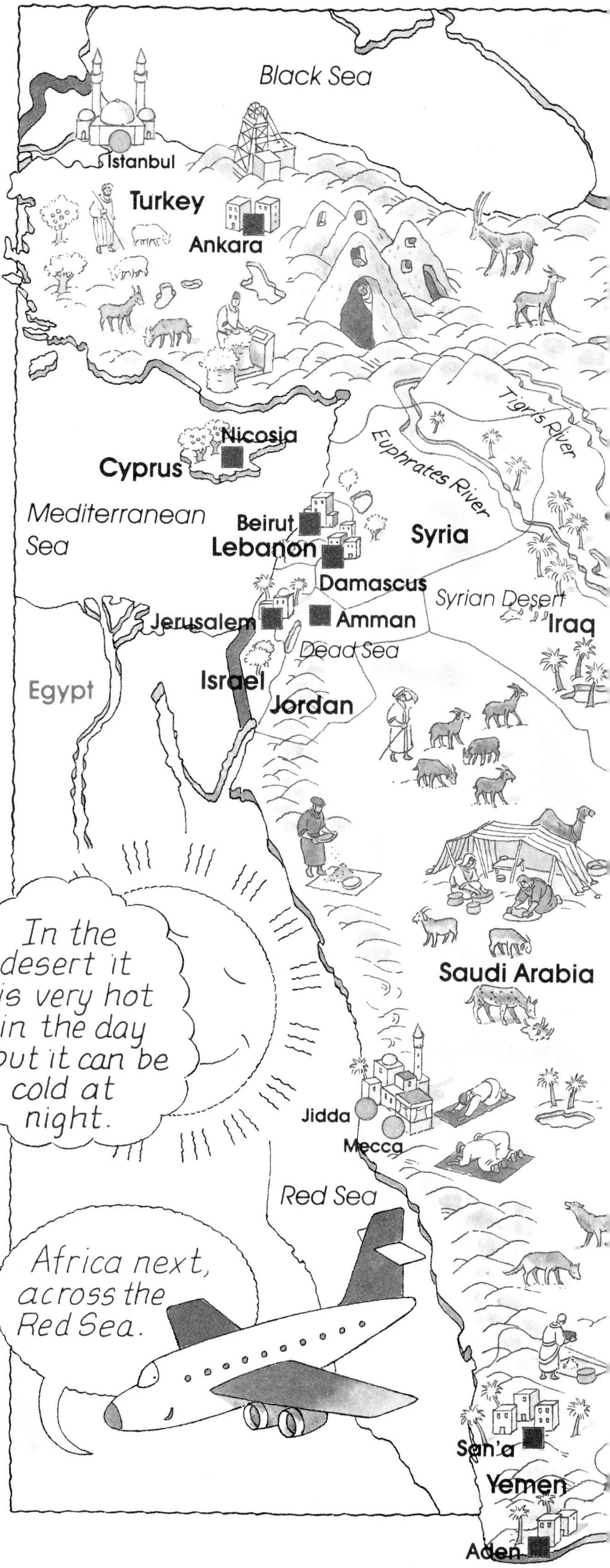

Soviet Union
Caspian Sea
N
W
E
S
Mount Ararat
Soviet Union
Tehran
Baghdad
Iran
Zagros Mountains
Basra
Kuwait
Kuwait
The Gulf
Manama
Bahrain
Qatar
Dubai
Riyadh
Abu Dhabi
United Arab Emirates
Muscat
Oman
South Yemen
Arabian Sea
About 170 million people live in these 16 countries.
a wolf
a leopard
a bear
an ibex
a hyena

Can you find? . . .

Africa

Africa is a huge, hot and beautiful continent. Most Africans live off the land, but some live in big cities. All sorts of animals, birds and plants live in the rain forests of central Africa and on the flat grassy plains on either side of it. The Sahara Desert is the hottest place in the world. It is hard to grow things there and many people living at its edge are starving.

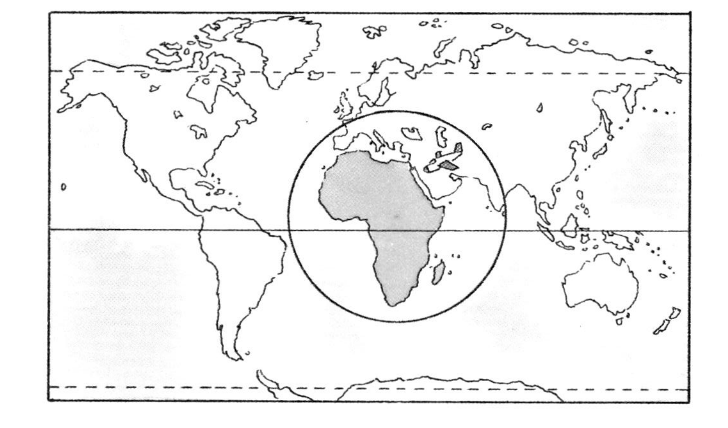

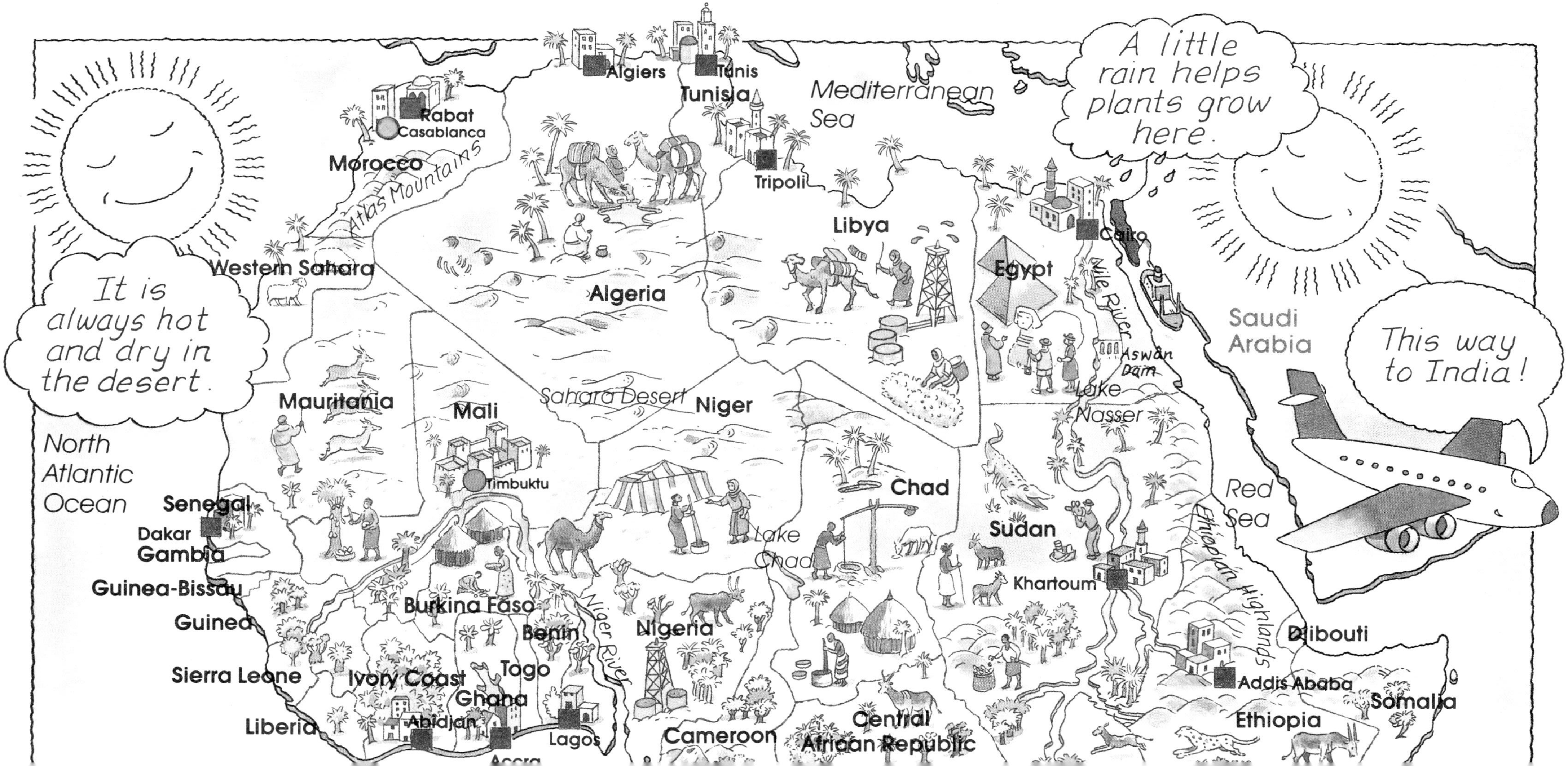

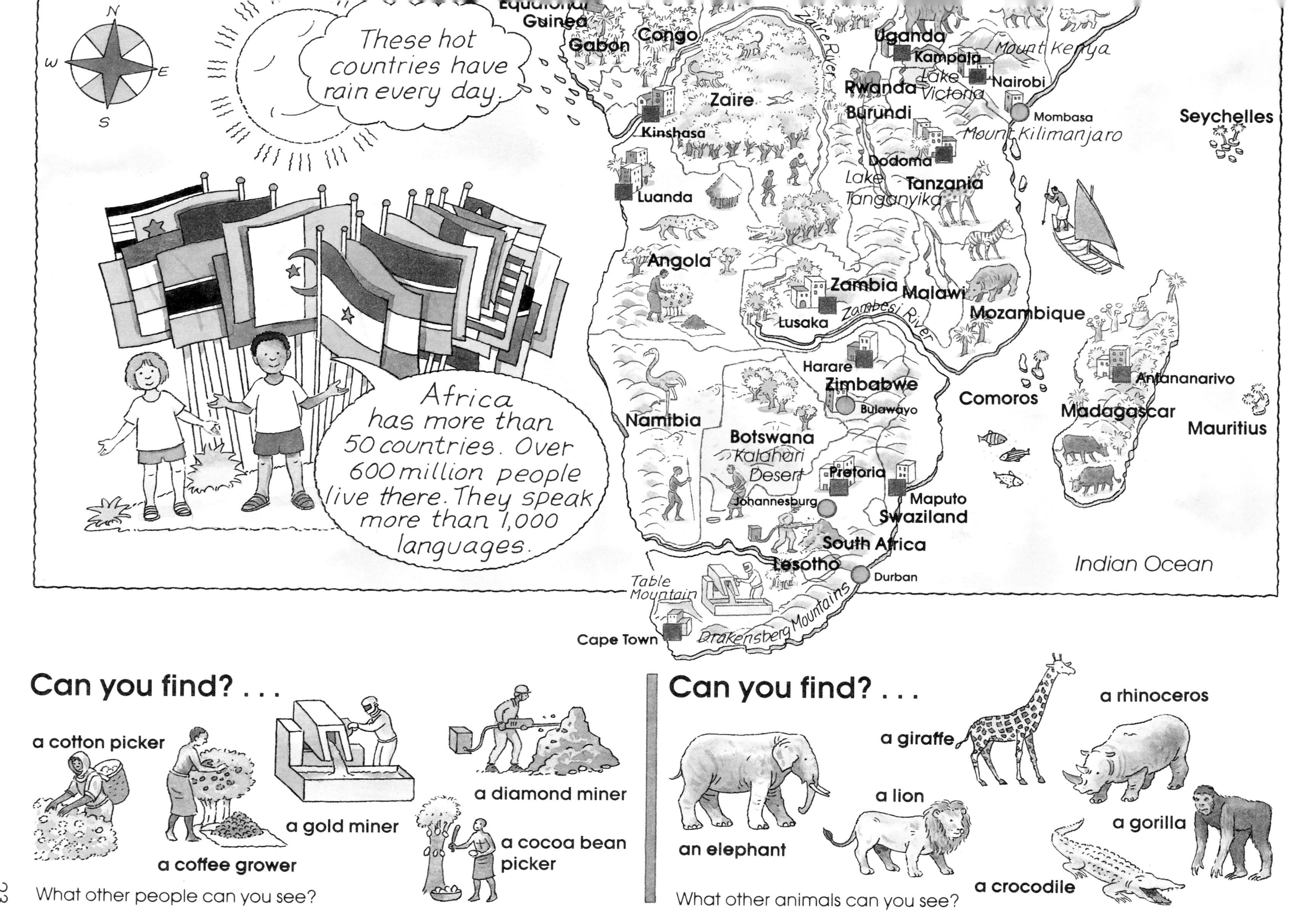
These hot countries have rain every day.
Africa has more than 50 countries. Over 600 million people live there. They speak more than 1,000 languages.
Guinea
Gabon
Congo
Zaire
Kinshasa
Luanda
Angola
Zaire River
Uganda
Kampala
Lake Victoria
Nairobi
Mount Kenya
Mombasa
Rwanda
Burundi
Mount Kilimanjaro
Seychelles
Dodoma
Lake Tanganyika
Tanzania
Zambia
Malawi
Lusaka
Zambesi River
Mozambique
Harare
Zimbabwe
Bulawayo
Comoros
Antananarivo
Madagascar
Mauritius
Namibia
Botswana
Kalahari Desert
Pretoria
Johannesburg
Maputo
Swaziland
South Africa
Lesotho
Durban
Indian Ocean
Table Mountain
Cape Town
Drakensberg Mountains
N
E
S
W
Can you find? . . .
a cotton picker
a coffee grower
a gold miner
a diamond miner
a cocoa bean picker
What other people can you see?
Can you find? . . .
an elephant
a giraffe
a lion
a rhinoceros
a gorilla
a crocodile
What other animals can you see?

Southern Asia

The Himalaya Mountains at the top of the map are the highest mountains in the world. Most families in this part of Asia live in small farming villages, but some of the poorest people live in the overcrowded cities. In India cows are sacred. They wander where they like, and no one is allowed to hurt them.

Hindu Kush
Kabul
Afghanistan
Himalaya Mountains
Islamabad
Lahore
Pakistan
Indus River
Thar Desert
Nepal
New Delhi
Karachi
India
Arabian Sea
Bombay
Madras
Colombo
Sri Lanka

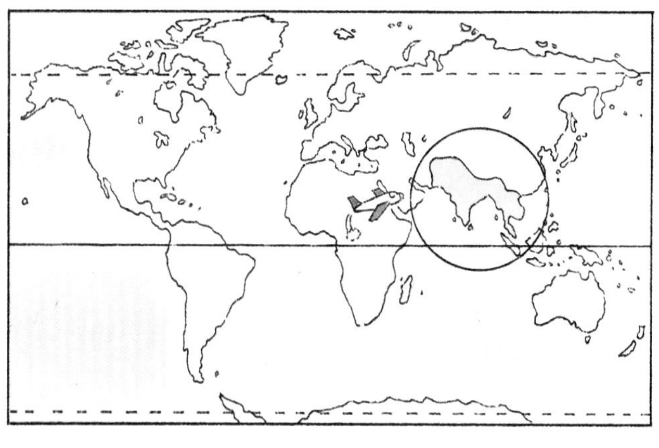

Can you find? . . .

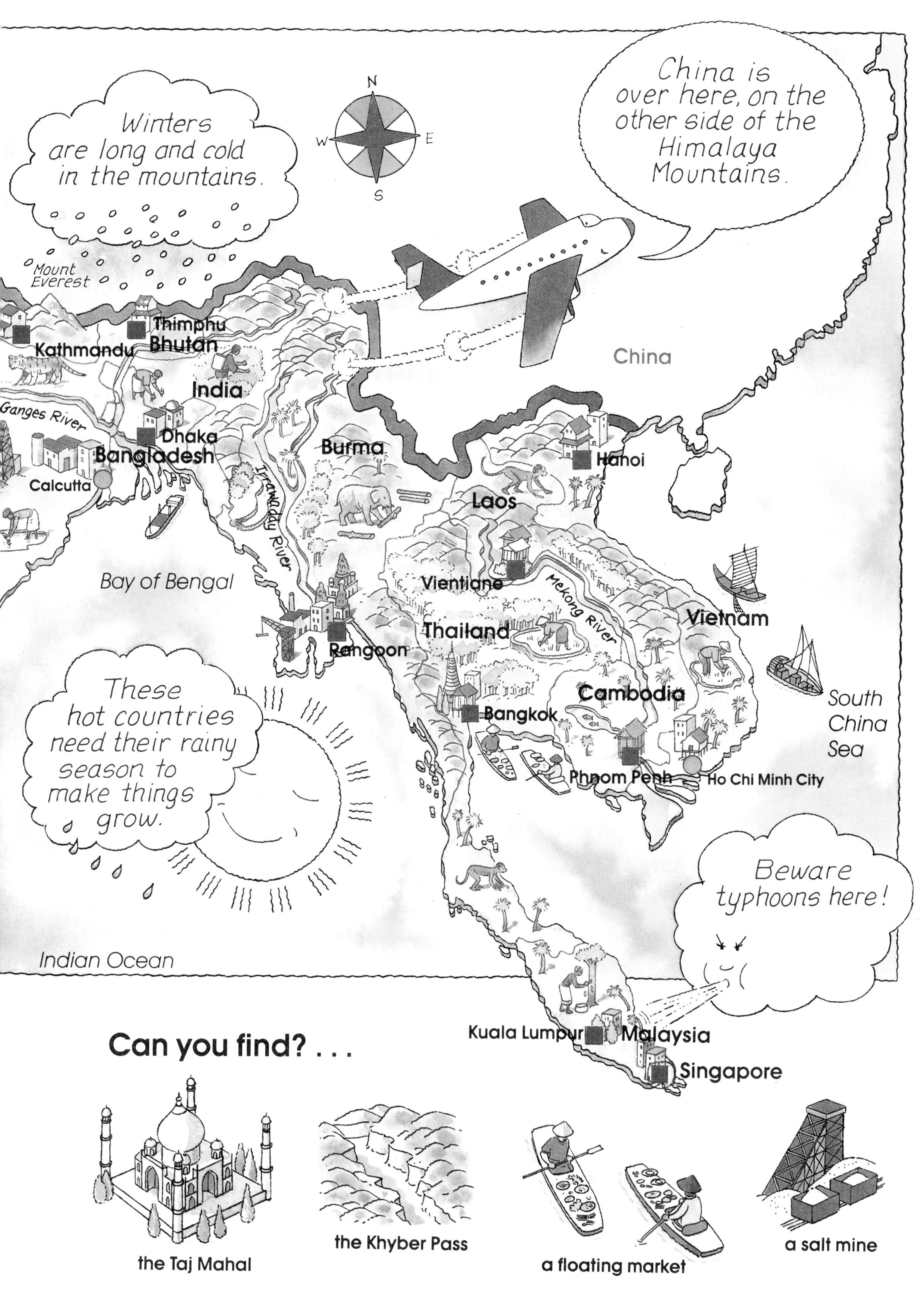
Winters are long and cold in the mountains.
China is over here, on the other side of the Himalaya Mountains.
N
W
E
S
Mount Everest
Thimphu
Bhutan
Kathmandu
China
India
Ganges River
Dhaka
Bangladesh
Calcutta
Burma
Hanoi
Laos
Irrawaddy River
Bay of Bengal
Vientiane
Mekong River
Thailand
Vietnam
Rangoon
Cambodia
South China Sea
Bangkok
Phnom Penh
Ho Chi Minh City
These hot countries need their rainy season to make things grow.
Beware typhoons here!
Indian Ocean
Kuala Lumpur
Malaysia
Singapore
Can you find? . . .
the Taj Mahal
the Khyber Pass
a floating market
a salt mine

China and Japan

One fifth of the whole world's people live in China. China's Great Wall, built more than 2,000 years ago, is so long that it can be seen from the moon!

Japan is made up of four big islands and hundreds of little ones. Your TV or radio was probably made in Japan.

Thousands of people live in the tall buildings of Hong Kong, on the southeastern coast of China.

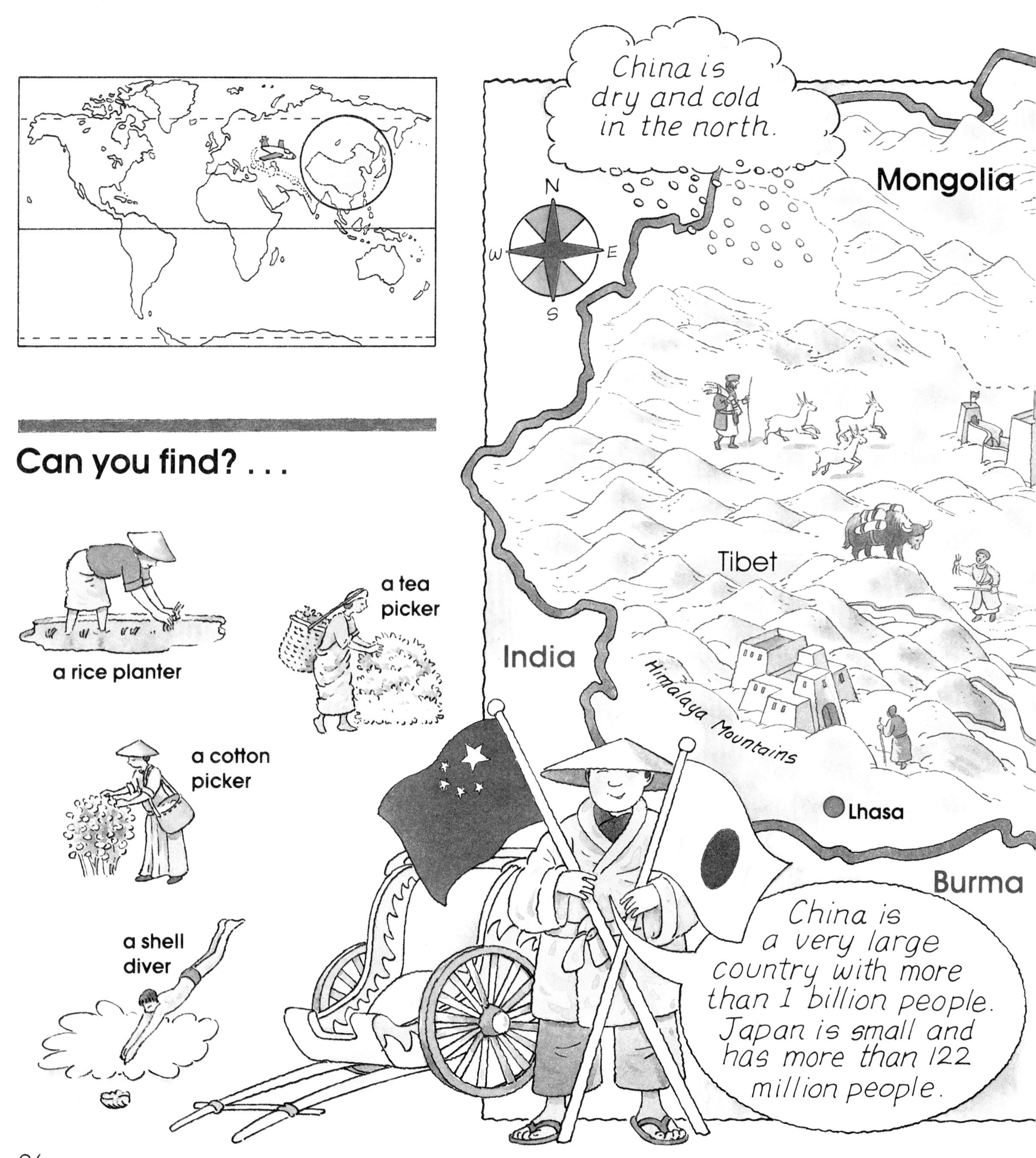

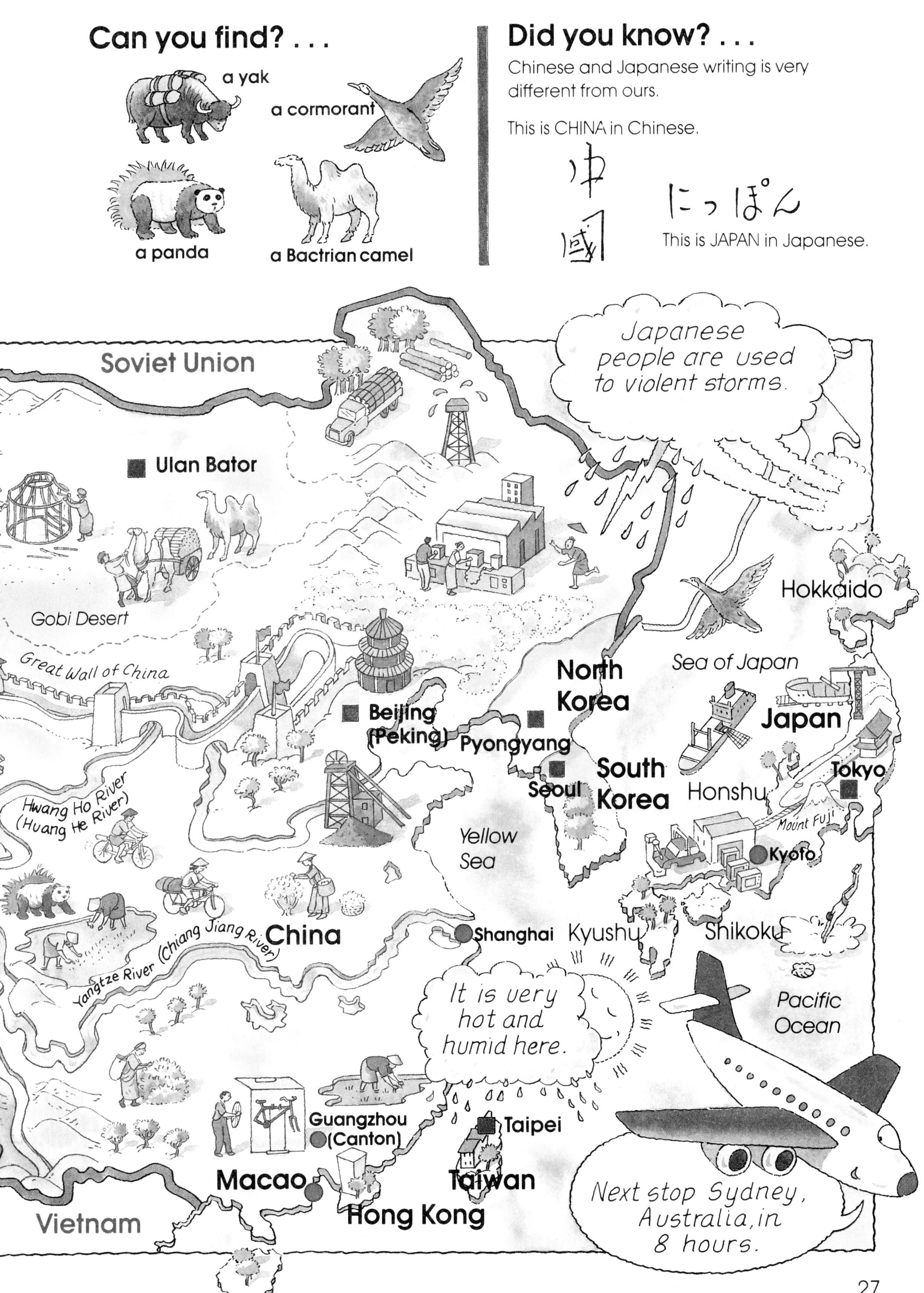
Can you find? . . .
a yak
a cormorant
a panda
a Bactrian camel
Did you know? . . .
Chinese and Japanese writing is very different from ours.
This is CHINA in Chinese.
中國
にっぽん
This is JAPAN in Japanese.
Soviet Union
Japanese people are used to violent storms.
Ulan Bator
Gobi Desert
Great Wall of China
Hokkaido
Sea of Japan
North Korea
Beijing (Peking)
Pyongyang
Japan
South Korea
Seoul
Tokyo
Honshu
Mount Fuji
Hwang Ho River (Huang He River)
Yellow Sea
Kyoto
Shanghai
Kyushu
Shikoku
China
Yangtze River (Chiang Jiang River)
It is very hot and humid here.
Pacific Ocean
Guangzhou (Canton)
Taipei
Macao
Taiwan
Hong Kong
Vietnam
Next stop Sydney, Australia, in 8 hours.

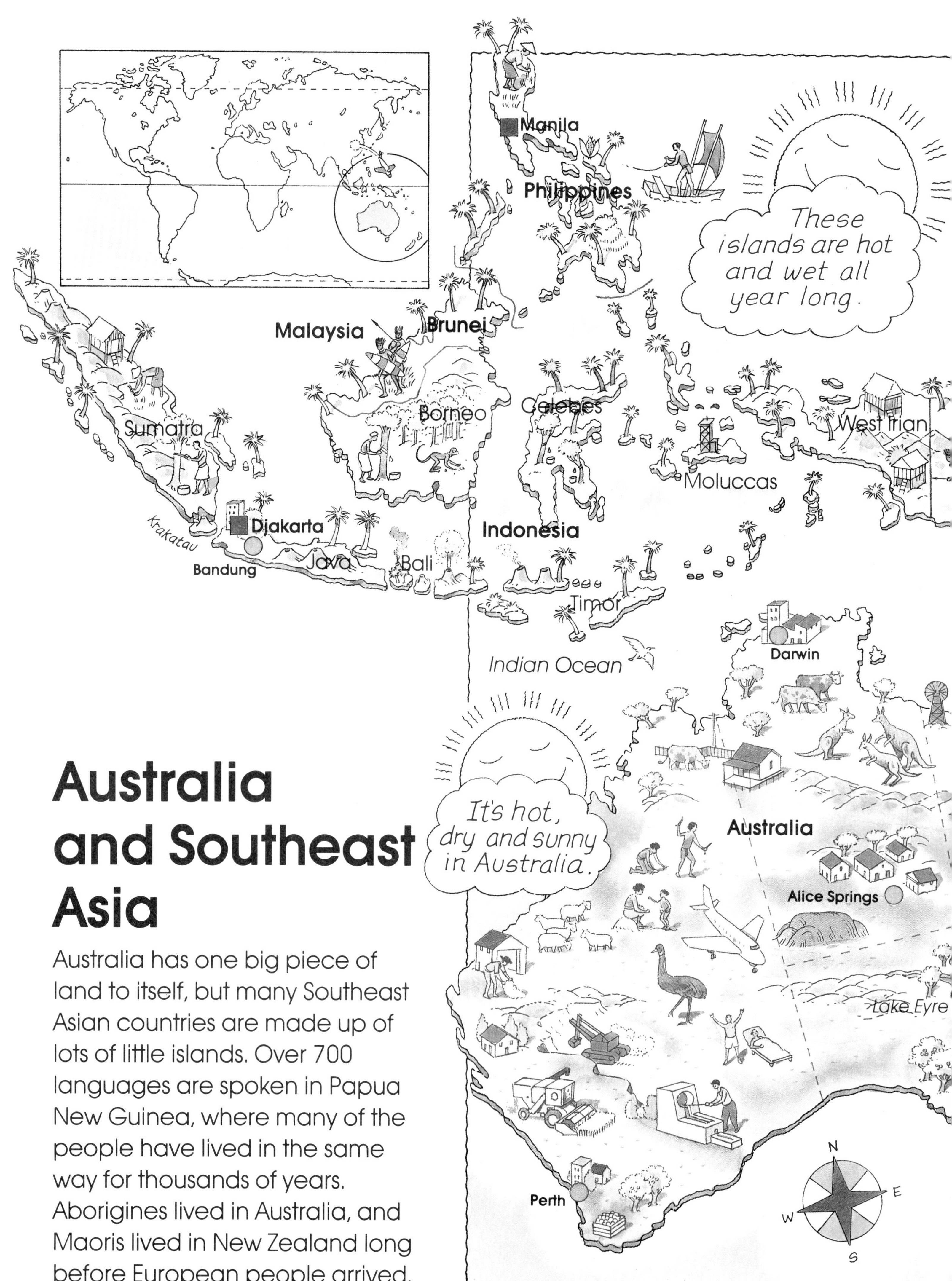

Australia and Southeast Asia

Australia has one big piece of land to itself, but many Southeast Asian countries are made up of lots of little islands. Over 700 languages are spoken in Papua New Guinea, where many of the people have lived in the same way for thousands of years. Aborigines lived in Australia, and Maoris lived in New Zealand long before European people arrived.

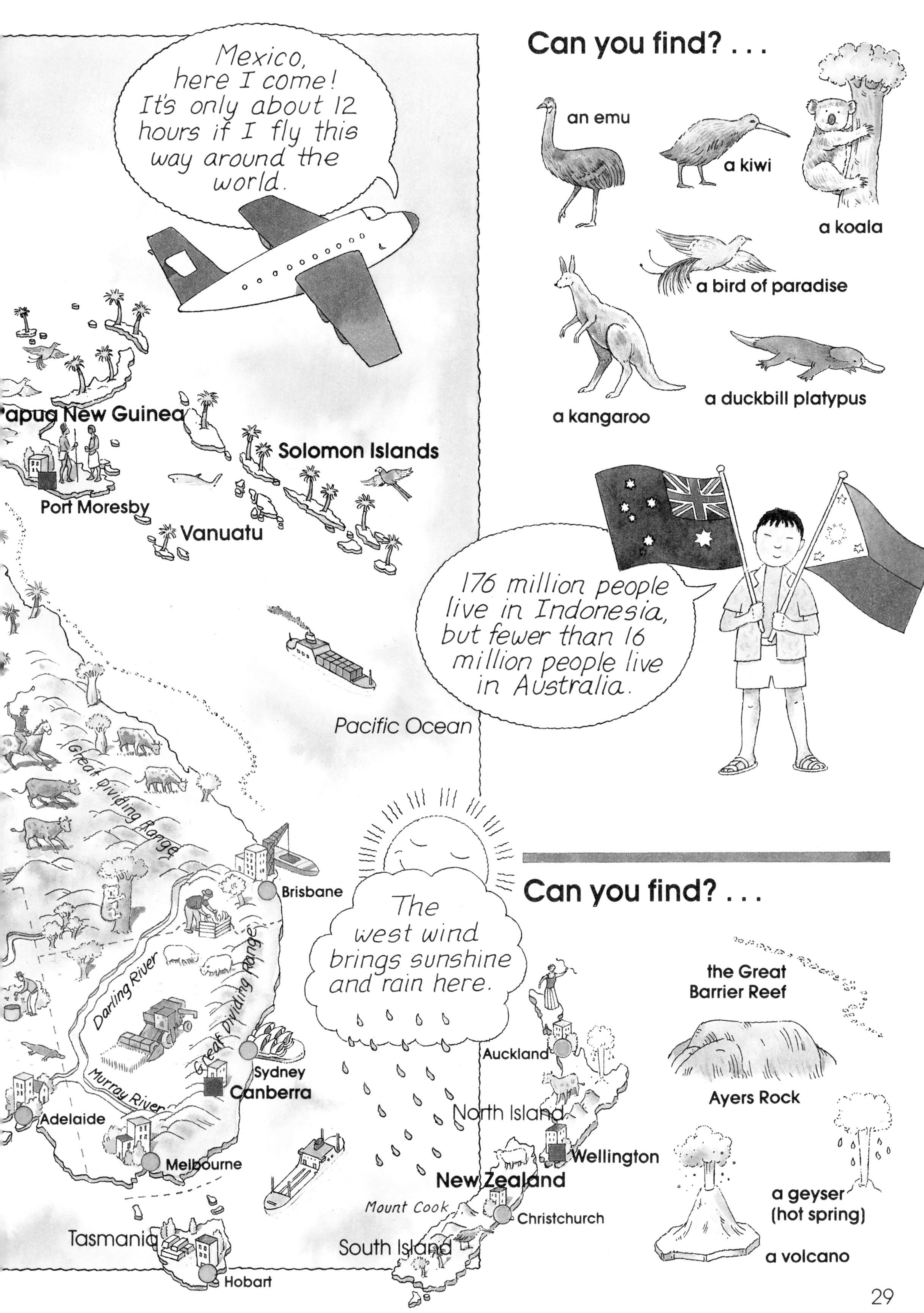

Can you find? . . .

an emu
a kiwi
a koala
a bird of paradise
a kangaroo
a duckbill platypus

Can you find? . . .

the Great Barrier Reef
Ayers Rock
a geyser (hot spring)
a volcano

Central and South America

Brazil is the biggest country in South America. This is where the rain forests of the Amazon are. Indian tribes still live deep in the forest. Many Central and South Americans are poor farmers. Most of them speak Spanish or Portuguese. Big cities are growing fast, but people who move to them often can't find work. In the islands of the Caribbean many people speak English. Vacationers enjoy the sunshine and music there.

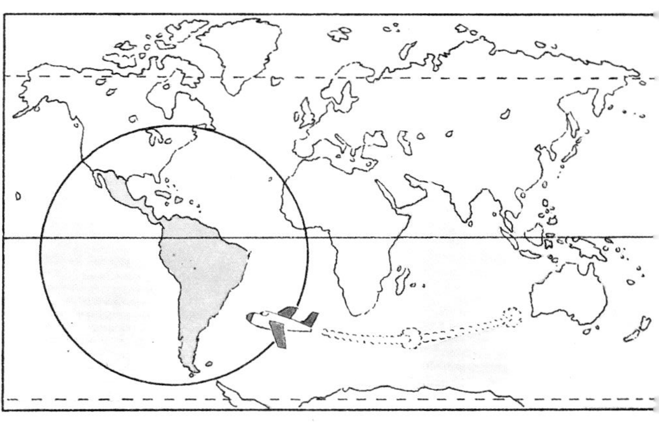

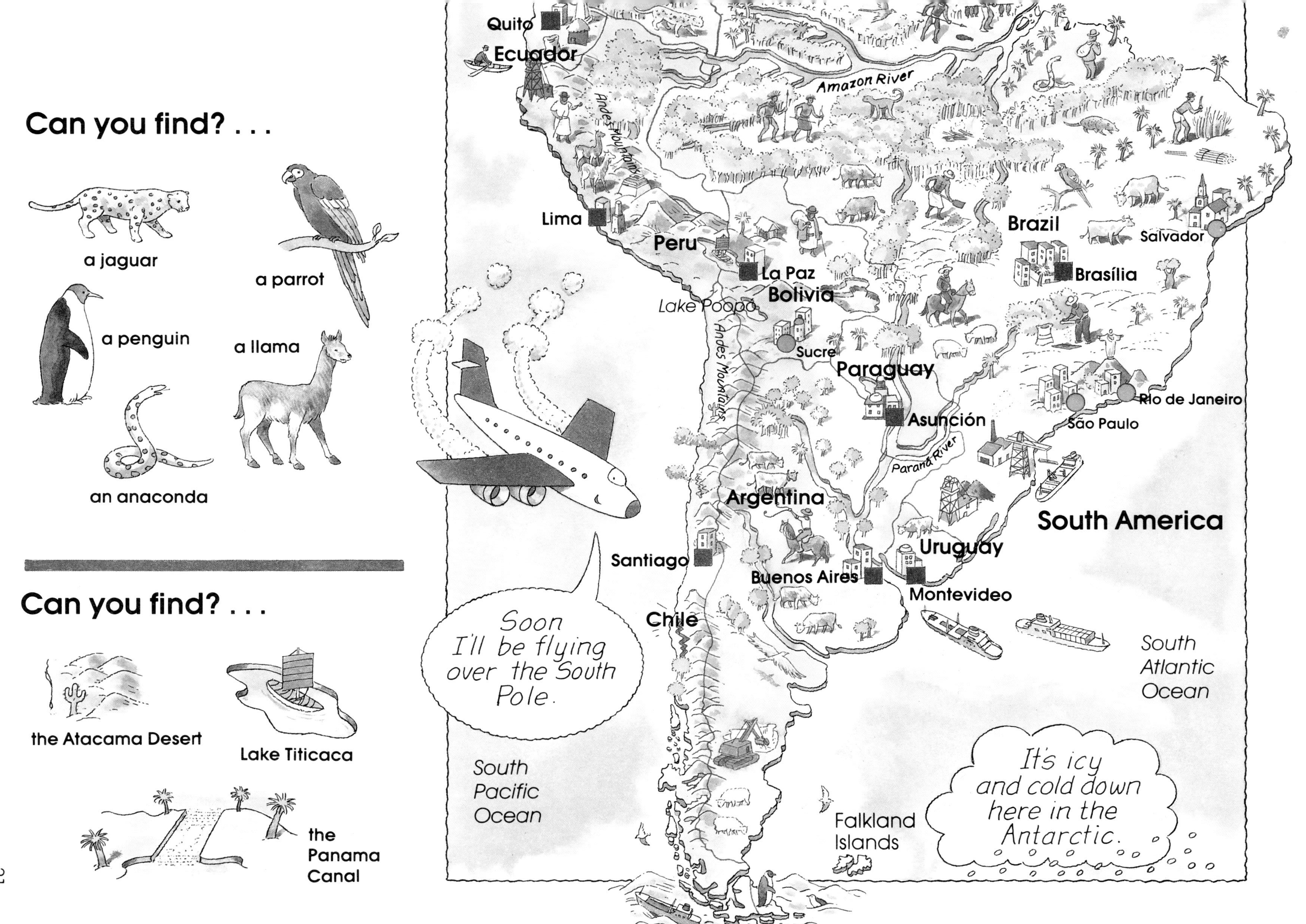
Can you find? . . .
a jaguar
a parrot
a penguin
a llama
an anaconda
Can you find? . . .
the Atacama Desert
Lake Titicaca
the Panama Canal
Quito
Ecuador
Amazon River
Andes Mountains
Lima
Peru
La Paz
Bolivia
Lake Poopo
Sucre
Paraguay
Asunción
Paraná River
Brazil
Brasília
Salvador
Rio de Janeiro
São Paulo
Argentina
Santiago
Buenos Aires
Uruguay
Montevideo
Chile
South America
South Atlantic Ocean
South Pacific Ocean
Falkland Islands
Soon I'll be flying over the South Pole.
It's icy and cold down here in the Antarctic.

The Arctic and Antarctic

These two maps show the "top" and "bottom" of the world. The North Pole is in the middle of a frozen ocean. Inuits (or Eskimos) still live in the surrounding lands. Antarctica is a whole continent covered in snow and ice that nobody owns.

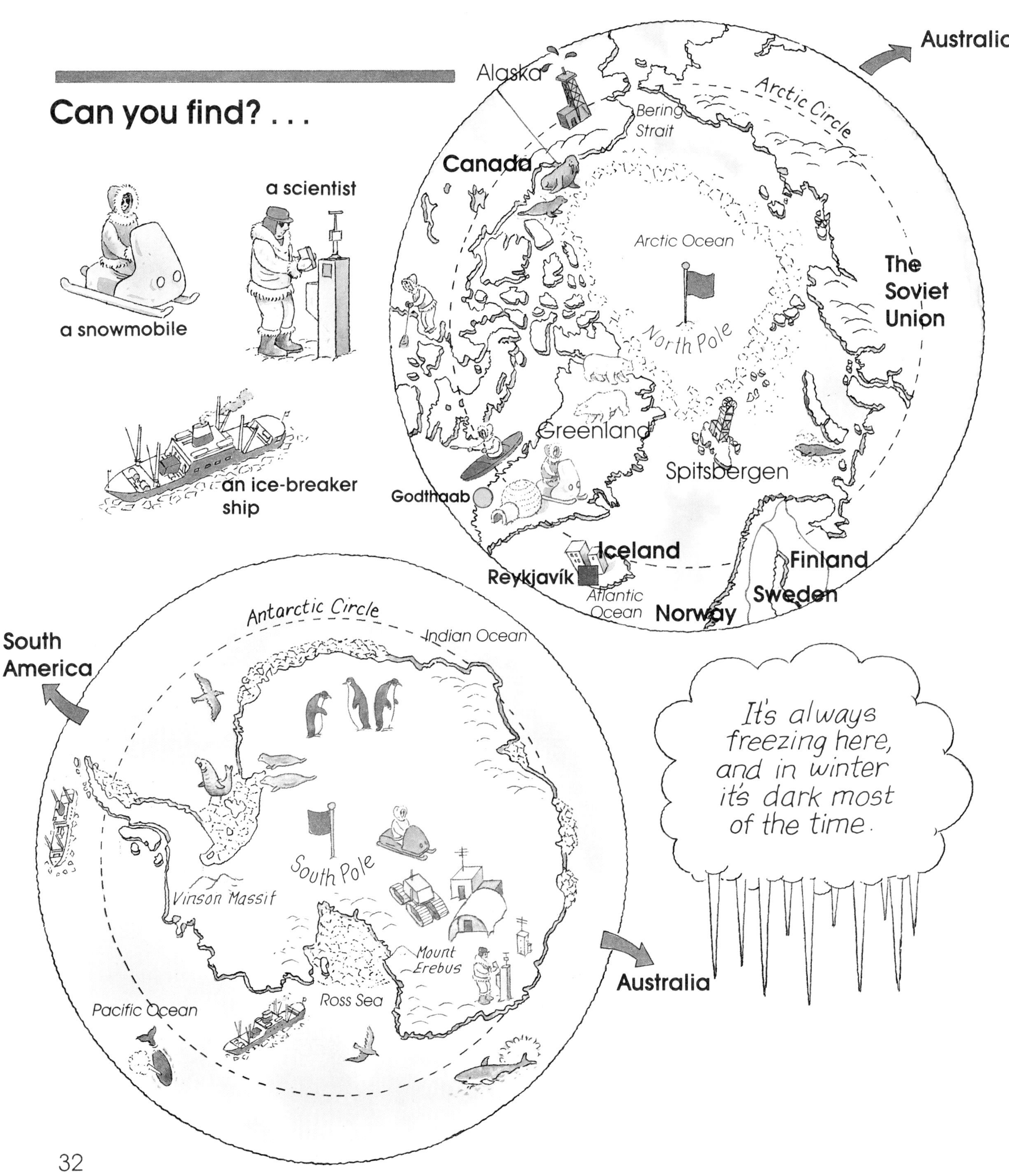

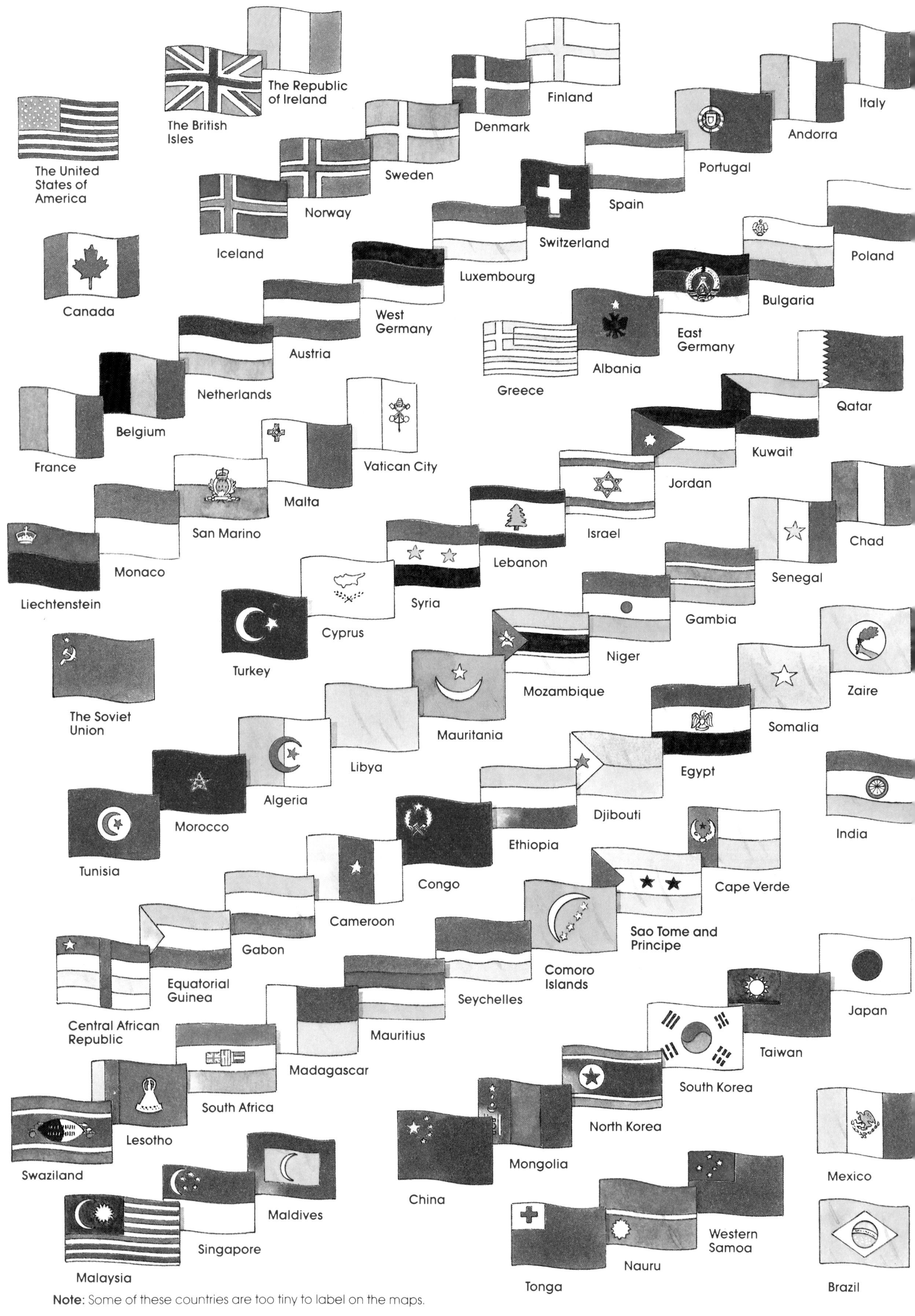

Note: Some of these countries are too tiny to label on the maps.